W9-CUY-782

Methods in Enzymology

Volume 411
DNA MICROARRAYS, PART B:
DATABASES AND STATISTICS

METHODS IN ENZYMOLOGY

EDITORS-IN-CHIEF

John N. Abelson   Melvin I. Simon

DIVISION OF BIOLOGY
CALIFORNIA INSTITUTE OF TECHNOLOGY
PASADENA, CALIFORNIA

FOUNDING EDITORS

Sidney P. Colowick and Nathan O. Kaplan

Methods in Enzymology

Volume 411

DNA Microarrays, Part B: *Databases and Statistics*

EDITED BY

Alan Kimmel & Brian Oliver

NIDDK
NATIONAL INSTITUTES OF HEALTH
BETHESDA, MARYLAND

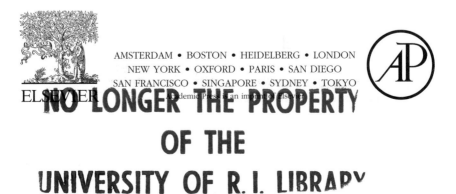

AMSTERDAM • BOSTON • HEIDELBERG • LONDON
NEW YORK • OXFORD • PARIS • SAN DIEGO
SAN FRANCISCO • SINGAPORE • SYDNEY • TOKYO
Academic Press is an imprint of Elsevier

ELSEVIER

Academic Press is an imprint of Elsevier
525 B Street, Suite 1900, San Diego, California 92101-4495, USA
84 Theobald's Road, London WC1X 8RR, UK

This book is printed on acid-free paper. ∞

For information on all Elsevier Academic Press publications
visit our Web site at www.books.elsevier.com

ISBN-13: 978-0-12-182816-5
ISBN-10: 0-12-182816-6

PRINTED IN THE UNITED STATES OF AMERICA
06 07 08 09 9 8 7 6 5 4 3 2 1

Table of Contents

CONTRIBUTORS TO VOLUME 411   ix

VOLUMES IN SERIES   xiii

1. RNA Extraction for Arrays   LAKSHMI V. MADABUSI,
                                               GARY J. LATHAM, AND
                                               BERNARD F. ANDRUSS   1

2. Analyzing Micro-RNA Expression Using   TIMOTHY S. DAVISON,
   Microarrays   CHARLES D. JOHNSON, AND
                                               BERNARD F. ANDRUSS   14

3. Troubleshooting Microarray Hybridizations   BRIAN EADS,
                                               AMY CASH,
                                               KEVIN BOGART,
                                               JAMES COSTELLO, AND
                                               JUSTEN ANDREWS   34

4. Use of External Controls in Microarray   IVANA V. YANG   50
   Experiments

5. Standards in Gene Expression Microarray   MARC SALIT   63
   Experiments

6. Scanning Microarrays: Current Methods and   JERILYN A. TIMLIN   79
   Future Directions

7. An Introduction to BioArray Software   CARL TROEIN, JOHAN
   Environment   VALLON-CHRISTERSSON,
                                               AND LAO H. SAAL   99

8. Bioconductor: An Open Source Framework   MARK REIMERS AND
   for Bioinformatics and Computational   VINCENT J. CAREY   119
   Biology

9. TM4 Microarray Software Suite   ALEXANDER I. SAEED,
                                                NIRMAL K. BHAGABATI,
                                                JOHN C. BRAISTED,
                                                WEI LIANG,
                                                VASILY SHAROV,
                                                ELEANOR A. HOWE,
                                                JIANWEI LI,
                                                MATHANGI THIAGARAJAN,
                                                JOSEPH A. WHITE, AND
                                                JOHN QUACKENBUSH   134

10. Clustering Microarray Data   JEREMY GOLLUB AND
                                                GAVIN SHERLOCK   194

11. Analysis of Variance of Microarray Data   JULIEN F. AYROLES AND
                                                GREG GIBSON   214

12. Microarray Quality Control   JAMES M. MINOR   233

13. Analysis of a Multifactor Microarray Study TOM DOWNEY   256
    Using Partek Genomics Solution

14. Statistics for ChIP-chip and DNase   PETER C. SCACHERI,
    Hypersensitivity Experiments on   GREGORY E. CRAWFORD,
    NimbleGen Arrays   AND SEAN DAVIS   270

15. Extrapolating Traditional DNA Microarray   THOMAS E. ROYCE,
    Statistics to Tiling and Protein   JOEL S. ROZOWSKY,
    Microarray Technologies   NICHOLAS M. LUSCOMBE,
                                                OLOF EMANUELSSON,
                                                HAIYUAN YU,
                                                XIAOWEI ZHU,
                                                MICHAEL SNYDER, AND
                                                MARK B. GERSTEIN   282

16. Random Data Set Generation to Support   DANIEL Q. NAIMAN   312
    Microarray Analysis

17. Using Ontologies to Annotate   PATRICIA L. WHETZEL,
    Microarray Experiments   HELEN PARKINSON, AND
                                                CHRISTIAN J. STOECKERT, JR 325

18. Interpreting Experimental Results Using   TIM BEISSBARTH   340
    Gene Ontologies

19. Gene Expression Omnibus: TANYA BARRETT AND
    Microarray Data Storage, Submission,   RON EDGAR   352
    Retrieval, and Analysis

20. Data Storage and Analysis in ArrayExpress   ALVIS BRAZMA,
                                                 MISHA KAPUSHESKY,
                                                 HELEN PARKINSON,
                                                 UGIS SARKINS, AND
                                                 MOHAMMAD SHOJATALAB   370

21. Clustering Methods for Analyzing Large   JÉRÔME HENNETIN AND
    Data Sets: Gonad Development,   MICHEL BELLIS   387
    A Study Case

22. Visualizing Networks   GEORGE W. BELL AND
                                                 FRAN LEWITTER   408

23. Random Forests for Microarrays   ADELE CUTLER AND
                                                 JOHN R. STEVENS   422

AUTHOR INDEX   433

SUBJECT INDEX   461

Contributors to Volume 411

Article numbers are in parentheses and following the name of contributors.
Affiliations listed are current.

JUSTEN ANDREWS (3), *Department of Biology, Indiana University, Bloomington, Indiana*

BERNARD F. ANDRUSS (1, 2), *Asuragen, Inc., Austin, Texas*

JULIEN F. AYROLES (11), *Department of Genetics, North Carolina State University, Raleigh, North Carolina*

TANYA BARRETT (19), *National Center for Biotechnology Information, National Library of Medicine, National Institutes of Health, Bethesda, Maryland*

TIM BEISSBARTH (18), *The Walter and Eliza Hall Institute of Medical Research, Bioinformatics Group, Victoria, Australia*

GEORGE W. BELL (22), *Bioinformatics and Research Computing, Whitehead Institute for Biomedical Research, Cambridge, Massachusetts*

MICHEL BELLIS (21), *CRBM–CNRS, Montpellier, France*

NIRMAL K. BHAGABATI (9), *The Institute for Genomic Research, Rockville, Maryland*

KEVIN BOGART (3), *Drosophila Genomics Resource Center, Indiana University, Bloomington, Indiana*

JOHN C. BRAISTED (9), *The Institute for Genomic Research, Rockville, Maryland*

ALVIS BRAZMA (20), *European Bioinformatics Institute, Wellcome Trust Genome Campus, Hinxton, Cambridge, United Kingdom*

VINCENT J. CAREY (8), *Harvard University, Boston, Massachusetts*

AMY CASH (3), *Department of Biology, Indiana University, Bloomington, Indiana*

JAMES COSTELLO (3), *Drosophila Genomics Resource Center, Indiana University, Bloomington, Indiana*

GREGORY E. CRAWFORD (14), *Department of Pediatrics, Institute for Genome Sciences and Policy, Duke University, Durham, North Carolina*

ADELE CUTLER (23), *Department of Mathematics and Statistics, Utah State University, Logan, Utah*

SEAN DAVIS (14), *Cancer Genetics Branch, National Human Genome Research Institute, National Institutes of Health, Bethesda, Maryland*

TIMOTHY S. DAVISON (2), *Asuragen Discovery Services, Asuragen, Inc., Austin, Texas*

TOM DOWNEY (13), *Partek Incorporated, Saint Louis, Missouri*

BRIAN EADS (3), *Department of Biology, Indiana University, Bloomington, Indiana*

RON EDGAR (19), *National Center for Biotechnology Information, National Library of Medicine, Bethesda, Maryland*

OLOF EMANUELSSON (15), *Department of Molecular Biophysics and Biochemistry, Yale University, New Haven, Connecticut*

MARK B. GERSTEIN (15), *Department of Molecular Biophysics and Biochemistry, Yale University, New Haven, Connecticut*

GREG GIBSON (11), *Department of Genetics, North Carolina State University, Raleigh, North Carolina*

JEREMY GOLLUB* (10), *Department of Biochemistry, Stanford University Medical School, Stanford, California*

JÉRÔME HENNETIN (21), *CRBM–CNRS, Montpellier, France*

ELEANOR A. HOWE (9), *Department of Biostatistics and Computational Biology, Dana-Farber Cancer Institute, Boston, Massachusetts*

CHARLES D. JOHNSON (2), *Asuragen Discovery Services, Asuragen, Inc., Austin, Texas*

MISHA KAPUSHESKY (20), *European Bioinformatics Institute, Wellcome Trust Genome Campus, Hinxton, Cambridge, United Kingdom*

GARY J. LATHAM (1), *Asuragen, Inc., Austin, Texas*

FRAN LEWITTER (22), *Bioinformatics and Research Computing, Whitehead Institute for Biomedical Research, Cambridge, Massachusetts*

JIANWEI LI (9), *The Institute for Genomic Research, Rockville, Maryland*

WEI LIANG (9), *The Institute for Genomic Research, Rockville, Maryland*

NICHOLAS M. LUSCOMBE (15), *European Bioinformatics Institute, Wellcome Trust Genome Campus, Hinxton, Cambridge, United Kingdom*

LAKSHMI V. MADABUSI (1), *Asuragen Discovery Services, Asuragen, Inc., Austin, Texas*

JAMES M. MINOR (12), *Agilent Technologies, Inc., Santa Clara, California*

DANIEL Q. NAIMAN (16), *Department of Applied Mathematics and Statistics, Johns Hopkins University, Baltimore, Maryland*

HELEN PARKINSON (17, 20), *European Bioinformatics Institute, Wellcome Trust Genome Campus, Hinxton, Cambridge, United Kingdom*

JOHN QUACKENBUSH (9), *Department of Biostatistics and Computational Biology, Dana-Farber Cancer Institute Boston, Massachusetts*

MARK REIMERS (8), *National Cancer Institute, Bethesda, Maryland*

THOMAS E. ROYCE (15), *Program in Computational Biology and Bioinformatics, Yale University, New Haven, Connecticut*

JOEL S. ROZOWSKY (15), *Department of Molecular Biophysics and Biochemistry, Yale University, New Haven, Connecticut*

LAO H. SAAL (7), *Institute for Cancer Genetics, College of Physicians and Surgeons, Columbia University, New York, New York*

ALEXANDER I. SAEED (9), *Department of Bioinformatics, The Institute for Genomic Research, Rockville, Maryland*

MARC SALIT (5), *Chemical Science and Technology Laboratory, National Institute of Standards and Technology, Gaithersburg, Maryland*

UGIS SARKANS (20), *European Bioinformatics Institute, Wellcome Trust Genome Campus, Hinxton, Cambridge, United Kingdom*

PETER C. SCACHERI (14), *Department of Genetics, Case Western Reserve University, Cleveland, Ohio*

VASILY SHAROV (9), *The Institute for Genomic Research, Rockville, Maryland*

GAVIN SHERLOCK (10), *Department of Genetics, Stanford University Medical School, Stanford, California*

Current affiliation: Inconix Pharmaceuticals, Inc., Mountain View, California.

MOHAMMAD SHOJATALAB (20), *European Bioinformatics Institute, Wellcome Trust Genome Campus, Hinxton, Cambridge, United Kingdom*

MICHAEL SNYDER (15), *Department of Molecular, Cellular, and Developmental Biology, Yale University, New Haven, Connecticut*

JOHN R. STEVENS (23), *Department of Mathematics and Statistics, Utah State University, Logan, Utah*

CHRISTIAN J. STOECKERT, JR. (17), *Department of Genetics, Center for Bioinformatics, University of Pennsylvania School of Medicine, Philadelphia, Pennsylvania*

MATHANGI THIAGARAJAN (9), *Department of Bioinformatics, The Institute for Genomic Research, Rockville, Maryland*

JERILYN A. TIMLIN (6), *Biomolecular Analysis and Imaging, Sandia National Laboratories, Albuquerque, New Mexico*

CARL TROEIN (7), *Computational Biology and Biological Physics, Department of Theoretical Physics, Lund University, Lund, Sweden*

JOHAN VALLON-CHRISTERSSON (7), *Department of Oncology, Lund University Hospital, Lund, Sweden*

PATRICIA L. WHETZEL (17), *Department of Genetics, Center for Bioinformatics, University of Pennsylvania School of Medicine, Philadelphia, Pennsylvania*

JOSEPH A. WHITE (9), *Department of Biostatistics and Computational Biology, Dana-Farber Cancer Institute, Boston, Massachusetts*

IVANA V. YANG (4), *Laboratory of Respiratory Biology, National Institute of Environmental Health Sciences, Research Triangle Park, North Carolina*

HAIYUAN YU (15), *Department of Molecular Biophysics and Biochemistry, Yale University, New Haven, Connecticut*

XIAOWEI ZHU (15), *Program in Computational Biology and Bioinformatics, Yale University, New Haven, Connecticut*

METHODS IN ENZYMOLOGY

VOLUME I. Preparation and Assay of Enzymes
Edited by SIDNEY P. COLOWICK AND NATHAN O. KAPLAN

VOLUME II. Preparation and Assay of Enzymes
Edited by SIDNEY P. COLOWICK AND NATHAN O. KAPLAN

VOLUME III. Preparation and Assay of Substrates
Edited by SIDNEY P. COLOWICK AND NATHAN O. KAPLAN

VOLUME IV. Special Techniques for the Enzymologist
Edited by SIDNEY P. COLOWICK AND NATHAN O. KAPLAN

VOLUME V. Preparation and Assay of Enzymes
Edited by SIDNEY P. COLOWICK AND NATHAN O. KAPLAN

VOLUME VI. Preparation and Assay of Enzymes *(Continued)*
Preparation and Assay of Substrates
Special Techniques
Edited by SIDNEY P. COLOWICK AND NATHAN O. KAPLAN

VOLUME VII. Cumulative Subject Index
Edited by SIDNEY P. COLOWICK AND NATHAN O. KAPLAN

VOLUME VIII. Complex Carbohydrates
Edited by ELIZABETH F. NEUFELD AND VICTOR GINSBURG

VOLUME IX. Carbohydrate Metabolism
Edited by WILLIS A. WOOD

VOLUME X. Oxidation and Phosphorylation
Edited by RONALD W. ESTABROOK AND MAYNARD E. PULLMAN

VOLUME XI. Enzyme Structure
Edited by C. H. W. HIRS

VOLUME XII. Nucleic Acids (Parts A and B)
Edited by LAWRENCE GROSSMAN AND KIVIE MOLDAVE

VOLUME XIII. Citric Acid Cycle
Edited by J. M. LOWENSTEIN

VOLUME XIV. Lipids
Edited by J. M. LOWENSTEIN

VOLUME XV. Steroids and Terpenoids
Edited by RAYMOND B. CLAYTON

VOLUME XVI. Fast Reactions
Edited by KENNETH KUSTIN

VOLUME XVII. Metabolism of Amino Acids and Amines
(Parts A and B)
Edited by HERBERT TABOR AND CELIA WHITE TABOR

VOLUME XVIII. Vitamins and Coenzymes (Parts A, B, and C)
Edited by DONALD B. MCCORMICK AND LEMUEL D. WRIGHT

VOLUME XIX. Proteolytic Enzymes
Edited by GERTRUDE E. PERLMANN AND LASZLO LORAND

VOLUME XX. Nucleic Acids and Protein Synthesis (Part C)
Edited by KIVIE MOLDAVE AND LAWRENCE GROSSMAN

VOLUME XXI. Nucleic Acids (Part D)
Edited by LAWRENCE GROSSMAN AND KIVIE MOLDAVE

VOLUME XXII. Enzyme Purification and Related Techniques
Edited by WILLIAM B. JAKOBY

VOLUME XXIII. Photosynthesis (Part A)
Edited by ANTHONY SAN PIETRO

VOLUME XXIV. Photosynthesis and Nitrogen Fixation (Part B)
Edited by ANTHONY SAN PIETRO

VOLUME XXV. Enzyme Structure (Part B)
Edited by C. H. W. HIRS AND SERGE N. TIMASHEFF

VOLUME XXVI. Enzyme Structure (Part C)
Edited by C. H. W. HIRS AND SERGE N. TIMASHEFF

VOLUME XXVII. Enzyme Structure (Part D)
Edited by C. H. W. HIRS AND SERGE N. TIMASHEFF

VOLUME XXVIII. Complex Carbohydrates (Part B)
Edited by VICTOR GINSBURG

VOLUME XXIX. Nucleic Acids and Protein Synthesis (Part E)
Edited by LAWRENCE GROSSMAN AND KIVIE MOLDAVE

VOLUME XXX. Nucleic Acids and Protein Synthesis (Part F)
Edited by KIVIE MOLDAVE AND LAWRENCE GROSSMAN

VOLUME XXXI. Biomembranes (Part A)
Edited by SIDNEY FLEISCHER AND LESTER PACKER

VOLUME XXXII. Biomembranes (Part B)
Edited by SIDNEY FLEISCHER AND LESTER PACKER

VOLUME XXXIII. Cumulative Subject Index Volumes I-XXX
Edited by MARTHA G. DENNIS AND EDWARD A. DENNIS

VOLUME XXXIV. Affinity Techniques (Enzyme Purification: Part B)
Edited by WILLIAM B. JAKOBY AND MEIR WILCHEK

VOLUME XXXV. Lipids (Part B)
Edited by JOHN M. LOWENSTEIN

VOLUME XXXVI. Hormone Action (Part A: Steroid Hormones)
Edited by BERT W. O'MALLEY AND JOEL G. HARDMAN

VOLUME XXXVII. Hormone Action (Part B: Peptide Hormones)
Edited by BERT W. O'MALLEY AND JOEL G. HARDMAN

VOLUME XXXVIII. Hormone Action (Part C: Cyclic Nucleotides)
Edited by JOEL G. HARDMAN AND BERT W. O'MALLEY

VOLUME XXXIX. Hormone Action (Part D: Isolated Cells, Tissues, and Organ Systems)
Edited by JOEL G. HARDMAN AND BERT W. O'MALLEY

VOLUME XL. Hormone Action (Part E: Nuclear Structure and Function)
Edited by BERT W. O'MALLEY AND JOEL G. HARDMAN

VOLUME XLI. Carbohydrate Metabolism (Part B)
Edited by W. A. WOOD

VOLUME XLII. Carbohydrate Metabolism (Part C)
Edited by W. A. WOOD

VOLUME XLIII. Antibiotics
Edited by JOHN H. HASH

VOLUME XLIV. Immobilized Enzymes
Edited by KLAUS MOSBACH

VOLUME XLV. Proteolytic Enzymes (Part B)
Edited by LASZLO LORAND

VOLUME XLVI. Affinity Labeling
Edited by WILLIAM B. JAKOBY AND MEIR WILCHEK

VOLUME XLVII. Enzyme Structure (Part E)
Edited by C. H. W. HIRS AND SERGE N. TIMASHEFF

VOLUME XLVIII. Enzyme Structure (Part F)
Edited by C. H. W. HIRS AND SERGE N. TIMASHEFF

VOLUME XLIX. Enzyme Structure (Part G)
Edited by C. H. W. HIRS AND SERGE N. TIMASHEFF

VOLUME L. Complex Carbohydrates (Part C)
Edited by VICTOR GINSBURG

VOLUME LI. Purine and Pyrimidine Nucleotide Metabolism
Edited by PATRICIA A. HOFFEE AND MARY ELLEN JONES

VOLUME LII. Biomembranes (Part C: Biological Oxidations)
Edited by SIDNEY FLEISCHER AND LESTER PACKER

VOLUME LIII. Biomembranes (Part D: Biological Oxidations)
Edited by SIDNEY FLEISCHER AND LESTER PACKER

VOLUME LIV. Biomembranes (Part E: Biological Oxidations)
Edited by SIDNEY FLEISCHER AND LESTER PACKER

VOLUME LV. Biomembranes (Part F: Bioenergetics)
Edited by SIDNEY FLEISCHER AND LESTER PACKER

VOLUME LVI. Biomembranes (Part G: Bioenergetics)
Edited by SIDNEY FLEISCHER AND LESTER PACKER

VOLUME LVII. Bioluminescence and Chemiluminescence
Edited by MARLENE A. DELUCA

VOLUME LVIII. Cell Culture
Edited by WILLIAM B. JAKOBY AND IRA PASTAN

VOLUME LIX. Nucleic Acids and Protein Synthesis (Part G)
Edited by KIVIE MOLDAVE AND LAWRENCE GROSSMAN

VOLUME LX. Nucleic Acids and Protein Synthesis (Part H)
Edited by KIVIE MOLDAVE AND LAWRENCE GROSSMAN

VOLUME 61. Enzyme Structure (Part H)
Edited by C. H. W. HIRS AND SERGE N. TIMASHEFF

VOLUME 62. Vitamins and Coenzymes (Part D)
Edited by DONALD B. MCCORMICK AND LEMUEL D. WRIGHT

VOLUME 63. Enzyme Kinetics and Mechanism (Part A: Initial Rate and
Inhibitor Methods)
Edited by DANIEL L. PURICH

VOLUME 64. Enzyme Kinetics and Mechanism
(Part B: Isotopic Probes and Complex Enzyme Systems)
Edited by DANIEL L. PURICH

VOLUME 65. Nucleic Acids (Part I)
Edited by LAWRENCE GROSSMAN AND KIVIE MOLDAVE

VOLUME 66. Vitamins and Coenzymes (Part E)
Edited by DONALD B. MCCORMICK AND LEMUEL D. WRIGHT

VOLUME 67. Vitamins and Coenzymes (Part F)
Edited by DONALD B. MCCORMICK AND LEMUEL D. WRIGHT

VOLUME 68. Recombinant DNA
Edited by RAY WU

VOLUME 69. Photosynthesis and Nitrogen Fixation (Part C)
Edited by ANTHONY SAN PIETRO

VOLUME 70. Immunochemical Techniques (Part A)
Edited by HELEN VAN VUNAKIS AND JOHN J. LANGONE

VOLUME 71. Lipids (Part C)
Edited by JOHN M. LOWENSTEIN

VOLUME 72. Lipids (Part D)
Edited by JOHN M. LOWENSTEIN

VOLUME 73. Immunochemical Techniques (Part B)
Edited by JOHN J. LANGONE AND HELEN VAN VUNAKIS

VOLUME 74. Immunochemical Techniques (Part C)
Edited by JOHN J. LANGONE AND HELEN VAN VUNAKIS

VOLUME 75. Cumulative Subject Index Volumes XXXI, XXXII, XXXIV–LX
Edited by EDWARD A. DENNIS AND MARTHA G. DENNIS

VOLUME 76. Hemoglobins
Edited by ERALDO ANTONINI, LUIGI ROSSI-BERNARDI, AND EMILIA CHIANCONE

VOLUME 77. Detoxication and Drug Metabolism
Edited by WILLIAM B. JAKOBY

VOLUME 78. Interferons (Part A)
Edited by SIDNEY PESTKA

VOLUME 79. Interferons (Part B)
Edited by SIDNEY PESTKA

VOLUME 80. Proteolytic Enzymes (Part C)
Edited by LASZLO LORAND

VOLUME 81. Biomembranes (Part H: Visual Pigments and Purple Membranes, I)
Edited by LESTER PACKER

VOLUME 82. Structural and Contractile Proteins (Part A: Extracellular Matrix)
Edited by LEON W. CUNNINGHAM AND DIXIE W. FREDERIKSEN

VOLUME 83. Complex Carbohydrates (Part D)
Edited by VICTOR GINSBURG

VOLUME 84. Immunochemical Techniques (Part D: Selected Immunoassays)
Edited by JOHN J. LANGONE AND HELEN VAN VUNAKIS

VOLUME 85. Structural and Contractile Proteins (Part B: The Contractile Apparatus and the Cytoskeleton)
Edited by DIXIE W. FREDERIKSEN AND LEON W. CUNNINGHAM

VOLUME 86. Prostaglandins and Arachidonate Metabolites
Edited by WILLIAM E. M. LANDS AND WILLIAM L. SMITH

VOLUME 87. Enzyme Kinetics and Mechanism (Part C: Intermediates, Stereo-chemistry, and Rate Studies)
Edited by DANIEL L. PURICH

VOLUME 88. Biomembranes (Part I: Visual Pigments and Purple Membranes, II)
Edited by LESTER PACKER

VOLUME 89. Carbohydrate Metabolism (Part D)
Edited by WILLIS A. WOOD

VOLUME 90. Carbohydrate Metabolism (Part E)
Edited by WILLIS A. WOOD

VOLUME 91. Enzyme Structure (Part I)
Edited by C. H. W. HIRS AND SERGE N. TIMASHEFF

VOLUME 92. Immunochemical Techniques (Part E: Monoclonal Antibodies and General Immunoassay Methods)
Edited by JOHN J. LANGONE AND HELEN VAN VUNAKIS

VOLUME 93. Immunochemical Techniques (Part F: Conventional Antibodies, Fc Receptors, and Cytotoxicity)
Edited by JOHN J. LANGONE AND HELEN VAN VUNAKIS

VOLUME 94. Polyamines
Edited by HERBERT TABOR AND CELIA WHITE TABOR

VOLUME 95. Cumulative Subject Index Volumes 61–74, 76–80
Edited by EDWARD A. DENNIS AND MARTHA G. DENNIS

VOLUME 96. Biomembranes [Part J: Membrane Biogenesis: Assembly and Targeting (General Methods; Eukaryotes)]
Edited by SIDNEY FLEISCHER AND BECCA FLEISCHER

VOLUME 97. Biomembranes [Part K: Membrane Biogenesis: Assembly and Targeting (Prokaryotes, Mitochondria, and Chloroplasts)]
Edited by SIDNEY FLEISCHER AND BECCA FLEISCHER

VOLUME 98. Biomembranes (Part L: Membrane Biogenesis: Processing and Recycling)
Edited by SIDNEY FLEISCHER AND BECCA FLEISCHER

VOLUME 99. Hormone Action (Part F: Protein Kinases)
Edited by JACKIE D. CORBIN AND JOEL G. HARDMAN

VOLUME 100. Recombinant DNA (Part B)
Edited by RAY WU, LAWRENCE GROSSMAN, AND KIVIE MOLDAVE

VOLUME 101. Recombinant DNA (Part C)
Edited by RAY WU, LAWRENCE GROSSMAN, AND KIVIE MOLDAVE

VOLUME 102. Hormone Action (Part G: Calmodulin and Calcium-Binding Proteins)
Edited by ANTHONY R. MEANS AND BERT W. O'MALLEY

VOLUME 103. Hormone Action (Part H: Neuroendocrine Peptides)
Edited by P. MICHAEL CONN

VOLUME 104. Enzyme Purification and Related Techniques (Part C)
Edited by WILLIAM B. JAKOBY

VOLUME 105. Oxygen Radicals in Biological Systems
Edited by LESTER PACKER

VOLUME 106. Posttranslational Modifications (Part A)
Edited by FINN WOLD AND KIVIE MOLDAVE

VOLUME 107. Posttranslational Modifications (Part B)
Edited by FINN WOLD AND KIVIE MOLDAVE

VOLUME 108. Immunochemical Techniques (Part G: Separation and Characterization of Lymphoid Cells)
Edited by GIOVANNI DI SABATO, JOHN J. LANGONE, AND HELEN VAN VUNAKIS

VOLUME 109. Hormone Action (Part I: Peptide Hormones)
Edited by LUTZ BIRNBAUMER AND BERT W. O'MALLEY

VOLUME 110. Steroids and Isoprenoids (Part A)
Edited by JOHN H. LAW AND HANS C. RILLING

VOLUME 111. Steroids and Isoprenoids (Part B)
Edited by JOHN H. LAW AND HANS C. RILLING

VOLUME 112. Drug and Enzyme Targeting (Part A)
Edited by KENNETH J. WIDDER AND RALPH GREEN

VOLUME 113. Glutamate, Glutamine, Glutathione, and Related Compounds
Edited by ALTON MEISTER

VOLUME 114. Diffraction Methods for Biological Macromolecules (Part A)
Edited by HAROLD W. WYCKOFF, C. H. W. HIRS, AND SERGE N. TIMASHEFF

VOLUME 115. Diffraction Methods for Biological Macromolecules (Part B)
Edited by HAROLD W. WYCKOFF, C. H. W. HIRS, AND SERGE N. TIMASHEFF

VOLUME 116. Immunochemical Techniques (Part H: Effectors and Mediators of Lymphoid Cell Functions)
Edited by GIOVANNI DI SABATO, JOHN J. LANGONE, AND HELEN VAN VUNAKIS

VOLUME 117. Enzyme Structure (Part J)
Edited by C. H. W. HIRS AND SERGE N. TIMASHEFF

VOLUME 118. Plant Molecular Biology
Edited by ARTHUR WEISSBACH AND HERBERT WEISSBACH

VOLUME 119. Interferons (Part C)
Edited by SIDNEY PESTKA

VOLUME 120. Cumulative Subject Index Volumes 81–94, 96–101

VOLUME 121. Immunochemical Techniques (Part I: Hybridoma Technology and Monoclonal Antibodies)
Edited by JOHN J. LANGONE AND HELEN VAN VUNAKIS

VOLUME 122. Vitamins and Coenzymes (Part G)
Edited by FRANK CHYTIL AND DONALD B. MCCORMICK

VOLUME 123. Vitamins and Coenzymes (Part H)
Edited by FRANK CHYTIL AND DONALD B. MCCORMICK

VOLUME 124. Hormone Action (Part J: Neuroendocrine Peptides)
Edited by P. MICHAEL CONN

VOLUME 125. Biomembranes (Part M: Transport in Bacteria, Mitochondria, and Chloroplasts: General Approaches and Transport Systems)
Edited by SIDNEY FLEISCHER AND BECCA FLEISCHER

VOLUME 126. Biomembranes (Part N: Transport in Bacteria, Mitochondria, and Chloroplasts: Protonmotive Force)
Edited by SIDNEY FLEISCHER AND BECCA FLEISCHER

VOLUME 127. Biomembranes (Part O: Protons and Water: Structure and Translocation)
Edited by LESTER PACKER

VOLUME 128. Plasma Lipoproteins (Part A: Preparation, Structure, and Molecular Biology)
Edited by JERE P. SEGREST AND JOHN J. ALBERS

VOLUME 129. Plasma Lipoproteins (Part B: Characterization, Cell Biology, and Metabolism)
Edited by JOHN J. ALBERS AND JERE P. SEGREST

VOLUME 130. Enzyme Structure (Part K)
Edited by C. H. W. HIRS AND SERGE N. TIMASHEFF

VOLUME 131. Enzyme Structure (Part L)
Edited by C. H. W. HIRS AND SERGE N. TIMASHEFF

VOLUME 132. Immunochemical Techniques (Part J: Phagocytosis and Cell-Mediated Cytotoxicity)
Edited by GIOVANNI DI SABATO AND JOHANNES EVERSE

VOLUME 133. Bioluminescence and Chemiluminescence (Part B)
Edited by MARLENE DELUCA AND WILLIAM D. MCELROY

VOLUME 134. Structural and Contractile Proteins (Part C: The Contractile Apparatus and the Cytoskeleton)
Edited by RICHARD B. VALLEE

VOLUME 135. Immobilized Enzymes and Cells (Part B)
Edited by KLAUS MOSBACH

VOLUME 136. Immobilized Enzymes and Cells (Part C)
Edited by KLAUS MOSBACH

VOLUME 137. Immobilized Enzymes and Cells (Part D)
Edited by KLAUS MOSBACH

VOLUME 138. Complex Carbohydrates (Part E)
Edited by VICTOR GINSBURG

VOLUME 139. Cellular Regulators (Part A: Calcium- and Calmodulin-Binding Proteins)
Edited by ANTHONY R. MEANS AND P. MICHAEL CONN

VOLUME 140. Cumulative Subject Index Volumes 102–119, 121–134

VOLUME 141. Cellular Regulators (Part B: Calcium and Lipids)
Edited by P. MICHAEL CONN AND ANTHONY R. MEANS

VOLUME 142. Metabolism of Aromatic Amino Acids and Amines
Edited by SEYMOUR KAUFMAN

VOLUME 143. Sulfur and Sulfur Amino Acids
Edited by WILLIAM B. JAKOBY AND OWEN GRIFFITH

VOLUME 144. Structural and Contractile Proteins (Part D: Extracellular Matrix)
Edited by LEON W. CUNNINGHAM

VOLUME 145. Structural and Contractile Proteins (Part E: Extracellular Matrix)
Edited by LEON W. CUNNINGHAM

VOLUME 146. Peptide Growth Factors (Part A)
Edited by DAVID BARNES AND DAVID A. SIRBASKU

VOLUME 147. Peptide Growth Factors (Part B)
Edited by DAVID BARNES AND DAVID A. SIRBASKU

VOLUME 148. Plant Cell Membranes
Edited by LESTER PACKER AND ROLAND DOUCE

VOLUME 149. Drug and Enzyme Targeting (Part B)
Edited by RALPH GREEN AND KENNETH J. WIDDER

VOLUME 150. Immunochemical Techniques (Part K: *In Vitro* Models of B and T Cell Functions and Lymphoid Cell Receptors)
Edited by GIOVANNI DI SABATO

VOLUME 151. Molecular Genetics of Mammalian Cells
Edited by MICHAEL M. GOTTESMAN

VOLUME 152. Guide to Molecular Cloning Techniques
Edited by SHELBY L. BERGER AND ALAN R. KIMMEL

VOLUME 153. Recombinant DNA (Part D)
Edited by RAY WU AND LAWRENCE GROSSMAN

VOLUME 154. Recombinant DNA (Part E)
Edited by RAY WU AND LAWRENCE GROSSMAN

VOLUME 155. Recombinant DNA (Part F)
Edited by RAY WU

VOLUME 156. Biomembranes (Part P: ATP-Driven Pumps and Related Transport: The Na, K-Pump)
Edited by SIDNEY FLEISCHER AND BECCA FLEISCHER

VOLUME 157. Biomembranes (Part Q: ATP-Driven Pumps and Related Transport: Calcium, Proton, and Potassium Pumps)
Edited by SIDNEY FLEISCHER AND BECCA FLEISCHER

VOLUME 158. Metalloproteins (Part A)
Edited by JAMES F. RIORDAN AND BERT L. VALLEE

VOLUME 159. Initiation and Termination of Cyclic Nucleotide Action
Edited by JACKIE D. CORBIN AND ROGER A. JOHNSON

VOLUME 160. Biomass (Part A: Cellulose and Hemicellulose)
Edited by WILLIS A. WOOD AND SCOTT T. KELLOGG

VOLUME 161. Biomass (Part B: Lignin, Pectin, and Chitin)
Edited by WILLIS A. WOOD AND SCOTT T. KELLOGG

VOLUME 162. Immunochemical Techniques (Part L: Chemotaxis
and Inflammation)
Edited by GIOVANNI DI SABATO

VOLUME 163. Immunochemical Techniques (Part M: Chemotaxis
and Inflammation)
Edited by GIOVANNI DI SABATO

VOLUME 164. Ribosomes
Edited by HARRY F. NOLLER, JR., AND KIVIE MOLDAVE

VOLUME 165. Microbial Toxins: Tools for Enzymology
Edited by SIDNEY HARSHMAN

VOLUME 166. Branched-Chain Amino Acids
Edited by ROBERT HARRIS AND JOHN R. SOKATCH

VOLUME 167. Cyanobacteria
Edited by LESTER PACKER AND ALEXANDER N. GLAZER

VOLUME 168. Hormone Action (Part K: Neuroendocrine Peptides)
Edited by P. MICHAEL CONN

VOLUME 169. Platelets: Receptors, Adhesion,
Secretion (Part A)
Edited by JACEK HAWIGER

VOLUME 170. Nucleosomes
Edited by PAUL M. WASSARMAN AND ROGER D. KORNBERG

VOLUME 171. Biomembranes (Part R: Transport Theory: Cells and Model
Membranes)
Edited by SIDNEY FLEISCHER AND BECCA FLEISCHER

VOLUME 172. Biomembranes (Part S: Transport: Membrane Isolation and
Characterization)
Edited by SIDNEY FLEISCHER AND BECCA FLEISCHER

VOLUME 173. Biomembranes [Part T: Cellular and Subcellular Transport:
Eukaryotic (Nonepithelial) Cells]
Edited by SIDNEY FLEISCHER AND BECCA FLEISCHER

VOLUME 174. Biomembranes [Part U: Cellular and Subcellular Transport:
Eukaryotic (Nonepithelial) Cells]
Edited by SIDNEY FLEISCHER AND BECCA FLEISCHER

VOLUME 175. Cumulative Subject Index Volumes 135–139, 141–167

VOLUME 176. Nuclear Magnetic Resonance (Part A: Spectral Techniques and Dynamics)
Edited by NORMAN J. OPPENHEIMER AND THOMAS L. JAMES

VOLUME 177. Nuclear Magnetic Resonance (Part B: Structure and Mechanism)
Edited by NORMAN J. OPPENHEIMER AND THOMAS L. JAMES

VOLUME 178. Antibodies, Antigens, and Molecular Mimicry
Edited by JOHN J. LANGONE

VOLUME 179. Complex Carbohydrates (Part F)
Edited by VICTOR GINSBURG

VOLUME 180. RNA Processing (Part A: General Methods)
Edited by JAMES E. DAHLBERG AND JOHN N. ABELSON

VOLUME 181. RNA Processing (Part B: Specific Methods)
Edited by JAMES E. DAHLBERG AND JOHN N. ABELSON

VOLUME 182. Guide to Protein Purification
Edited by MURRAY P. DEUTSCHER

VOLUME 183. Molecular Evolution: Computer Analysis of Protein and Nucleic Acid Sequences
Edited by RUSSELL F. DOOLITTLE

VOLUME 184. Avidin-Biotin Technology
Edited by MEIR WILCHEK AND EDWARD A. BAYER

VOLUME 185. Gene Expression Technology
Edited by DAVID V. GOEDDEL

VOLUME 186. Oxygen Radicals in Biological Systems (Part B: Oxygen Radicals and Antioxidants)
Edited by LESTER PACKER AND ALEXANDER N. GLAZER

VOLUME 187. Arachidonate Related Lipid Mediators
Edited by ROBERT C. MURPHY AND FRANK A. FITZPATRICK

VOLUME 188. Hydrocarbons and Methylotrophy
Edited by MARY E. LIDSTROM

VOLUME 189. Retinoids (Part A: Molecular and Metabolic Aspects)
Edited by LESTER PACKER

VOLUME 190. Retinoids (Part B: Cell Differentiation and Clinical Applications)
Edited by LESTER PACKER

VOLUME 191. Biomembranes (Part V: Cellular and Subcellular Transport: Epithelial Cells)
Edited by SIDNEY FLEISCHER AND BECCA FLEISCHER

VOLUME 192. Biomembranes (Part W: Cellular and Subcellular Transport: Epithelial Cells)
Edited by SIDNEY FLEISCHER AND BECCA FLEISCHER

VOLUME 193. Mass Spectrometry
Edited by JAMES A. MCCLOSKEY

VOLUME 194. Guide to Yeast Genetics and Molecular Biology
Edited by CHRISTINE GUTHRIE AND GERALD R. FINK

VOLUME 195. Adenylyl Cyclase, G Proteins, and Guanylyl Cyclase
Edited by ROGER A. JOHNSON AND JACKIE D. CORBIN

VOLUME 196. Molecular Motors and the Cytoskeleton
Edited by RICHARD B. VALLEE

VOLUME 197. Phospholipases
Edited by EDWARD A. DENNIS

VOLUME 198. Peptide Growth Factors (Part C)
Edited by DAVID BARNES, J. P. MATHER, AND GORDON H. SATO

VOLUME 199. Cumulative Subject Index Volumes 168–174, 176–194

VOLUME 200. Protein Phosphorylation (Part A: Protein Kinases: Assays, Purification, Antibodies, Functional Analysis, Cloning, and Expression)
Edited by TONY HUNTER AND BARTHOLOMEW M. SEFTON

VOLUME 201. Protein Phosphorylation (Part B: Analysis of Protein Phosphorylation, Protein Kinase Inhibitors, and Protein Phosphatases)
Edited by TONY HUNTER AND BARTHOLOMEW M. SEFTON

VOLUME 202. Molecular Design and Modeling: Concepts and Applications (Part A: Proteins, Peptides, and Enzymes)
Edited by JOHN J. LANGONE

VOLUME 203. Molecular Design and Modeling:
Concepts and Applications (Part B: Antibodies and Antigens, Nucleic Acids, Polysaccharides, and Drugs)
Edited by JOHN J. LANGONE

VOLUME 204. Bacterial Genetic Systems
Edited by JEFFREY H. MILLER

VOLUME 205. Metallobiochemistry (Part B: Metallothionein and Related Molecules)
Edited by JAMES F. RIORDAN AND BERT L. VALLEE

VOLUME 206. Cytochrome P450
Edited by MICHAEL R. WATERMAN AND ERIC F. JOHNSON

VOLUME 207. Ion Channels
Edited by BERNARDO RUDY AND LINDA E. IVERSON

VOLUME 208. Protein–DNA Interactions
Edited by ROBERT T. SAUER

VOLUME 209. Phospholipid Biosynthesis
Edited by EDWARD A. DENNIS AND DENNIS E. VANCE

VOLUME 210. Numerical Computer Methods
Edited by LUDWIG BRAND AND MICHAEL L. JOHNSON

VOLUME 211. DNA Structures (Part A: Synthesis and Physical Analysis of DNA)
Edited by DAVID M. J. LILLEY AND JAMES E. DAHLBERG

VOLUME 212. DNA Structures (Part B: Chemical and Electrophoretic Analysis of DNA)
Edited by DAVID M. J. LILLEY AND JAMES E. DAHLBERG

VOLUME 213. Carotenoids (Part A: Chemistry, Separation, Quantitation, and Antioxidation)
Edited by LESTER PACKER

VOLUME 214. Carotenoids (Part B: Metabolism, Genetics, and Biosynthesis)
Edited by LESTER PACKER

VOLUME 215. Platelets: Receptors, Adhesion, Secretion (Part B)
Edited by JACEK J. HAWIGER

VOLUME 216. Recombinant DNA (Part G)
Edited by RAY WU

VOLUME 217. Recombinant DNA (Part H)
Edited by RAY WU

VOLUME 218. Recombinant DNA (Part I)
Edited by RAY WU

VOLUME 219. Reconstitution of Intracellular Transport
Edited by JAMES E. ROTHMAN

VOLUME 220. Membrane Fusion Techniques (Part A)
Edited by NEJAT DÜZGÜNEŞ

VOLUME 221. Membrane Fusion Techniques (Part B)
Edited by NEJAT DÜZGÜNEŞ

VOLUME 222. Proteolytic Enzymes in Coagulation, Fibrinolysis, and Complement Activation (Part A: Mammalian Blood Coagulation Factors and Inhibitors)
Edited by LASZLO LORAND AND KENNETH G. MANN

VOLUME 223. Proteolytic Enzymes in Coagulation, Fibrinolysis, and Complement Activation (Part B: Complement Activation, Fibrinolysis, and Nonmammalian Blood Coagulation Factors)
Edited by LASZLO LORAND AND KENNETH G. MANN

VOLUME 224. Molecular Evolution: Producing the Biochemical Data
Edited by ELIZABETH ANNE ZIMMER, THOMAS J. WHITE, REBECCA L. CANN, AND ALLAN C. WILSON

VOLUME 225. Guide to Techniques in Mouse Development
Edited by PAUL M. WASSARMAN AND MELVIN L. DEPAMPHILIS

VOLUME 226. Metallobiochemistry (Part C: Spectroscopic and Physical Methods for Probing Metal Ion Environments in Metalloenzymes and Metalloproteins)
Edited by JAMES F. RIORDAN AND BERT L. VALLEE

VOLUME 227. Metallobiochemistry (Part D: Physical and Spectroscopic Methods for Probing Metal Ion Environments in Metalloproteins)
Edited by JAMES F. RIORDAN AND BERT L. VALLEE

VOLUME 228. Aqueous Two-Phase Systems
Edited by HARRY WALTER AND GÖTE JOHANSSON

VOLUME 229. Cumulative Subject Index Volumes 195–198, 200–227

VOLUME 230. Guide to Techniques in Glycobiology
Edited by WILLIAM J. LENNARZ AND GERALD W. HART

VOLUME 231. Hemoglobins (Part B: Biochemical and Analytical Methods)
Edited by JOHANNES EVERSE, KIM D. VANDEGRIFF, AND ROBERT M. WINSLOW

VOLUME 232. Hemoglobins (Part C: Biophysical Methods)
Edited by JOHANNES EVERSE, KIM D. VANDEGRIFF, AND ROBERT M. WINSLOW

VOLUME 233. Oxygen Radicals in Biological Systems (Part C)
Edited by LESTER PACKER

VOLUME 234. Oxygen Radicals in Biological Systems (Part D)
Edited by LESTER PACKER

VOLUME 235. Bacterial Pathogenesis (Part A: Identification and Regulation of Virulence Factors)
Edited by VIRGINIA L. CLARK AND PATRIK M. BAVOIL

VOLUME 236. Bacterial Pathogenesis (Part B: Integration of Pathogenic Bacteria with Host Cells)
Edited by VIRGINIA L. CLARK AND PATRIK M. BAVOIL

VOLUME 237. Heterotrimeric G Proteins
Edited by RAVI IYENGAR

VOLUME 238. Heterotrimeric G-Protein Effectors
Edited by RAVI IYENGAR

VOLUME 239. Nuclear Magnetic Resonance (Part C)
Edited by THOMAS L. JAMES AND NORMAN J. OPPENHEIMER

VOLUME 240. Numerical Computer Methods (Part B)
Edited by MICHAEL L. JOHNSON AND LUDWIG BRAND

VOLUME 241. Retroviral Proteases
Edited by LAWRENCE C. KUO AND JULES A. SHAFER

VOLUME 242. Neoglycoconjugates (Part A)
Edited by Y. C. LEE AND REIKO T. LEE

VOLUME 243. Inorganic Microbial Sulfur Metabolism
Edited by HARRY D. PECK, JR., AND JEAN LEGALL

VOLUME 244. Proteolytic Enzymes: Serine and Cysteine Peptidases
Edited by ALAN J. BARRETT

VOLUME 245. Extracellular Matrix Components
Edited by E. RUOSLAHTI AND E. ENGVALL

VOLUME 246. Biochemical Spectroscopy
Edited by KENNETH SAUER

VOLUME 247. Neoglycoconjugates (Part B: Biomedical Applications)
Edited by Y. C. LEE AND REIKO T. LEE

VOLUME 248. Proteolytic Enzymes: Aspartic and Metallo Peptidases
Edited by ALAN J. BARRETT

VOLUME 249. Enzyme Kinetics and Mechanism (Part D: Developments in Enzyme Dynamics)
Edited by DANIEL L. PURICH

VOLUME 250. Lipid Modifications of Proteins
Edited by PATRICK J. CASEY AND JANICE E. BUSS

VOLUME 251. Biothiols (Part A: Monothiols and Dithiols, Protein Thiols, and Thiyl Radicals)
Edited by LESTER PACKER

VOLUME 252. Biothiols (Part B: Glutathione and Thioredoxin; Thiols in Signal Transduction and Gene Regulation)
Edited by LESTER PACKER

VOLUME 253. Adhesion of Microbial Pathogens
Edited by RON J. DOYLE AND ITZHAK OFEK

VOLUME 254. Oncogene Techniques
Edited by PETER K. VOGT AND INDER M. VERMA

VOLUME 255. Small GTPases and Their Regulators (Part A: Ras Family)
Edited by W. E. BALCH, CHANNING J. DER, AND ALAN HALL

VOLUME 256. Small GTPases and Their Regulators (Part B: Rho Family)
Edited by W. E. BALCH, CHANNING J. DER, AND ALAN HALL

VOLUME 257. Small GTPases and Their Regulators (Part C: Proteins Involved in Transport)
Edited by W. E. BALCH, CHANNING J. DER, AND ALAN HALL

VOLUME 258. Redox-Active Amino Acids in Biology
Edited by JUDITH P. KLINMAN

VOLUME 259. Energetics of Biological Macromolecules
Edited by MICHAEL L. JOHNSON AND GARY K. ACKERS

VOLUME 260. Mitochondrial Biogenesis and Genetics (Part A)
Edited by GIUSEPPE M. ATTARDI AND ANNE CHOMYN

VOLUME 261. Nuclear Magnetic Resonance and Nucleic Acids
Edited by THOMAS L. JAMES

VOLUME 262. DNA Replication
Edited by JUDITH L. CAMPBELL

VOLUME 263. Plasma Lipoproteins (Part C: Quantitation)
Edited by WILLIAM A. BRADLEY, SANDRA H. GIANTURCO, AND JERE P. SEGREST

VOLUME 264. Mitochondrial Biogenesis and Genetics (Part B)
Edited by GIUSEPPE M. ATTARDI AND ANNE CHOMYN

VOLUME 265. Cumulative Subject Index Volumes 228, 230–262

VOLUME 266. Computer Methods for Macromolecular Sequence Analysis
Edited by RUSSELL F. DOOLITTLE

VOLUME 267. Combinatorial Chemistry
Edited by JOHN N. ABELSON

VOLUME 268. Nitric Oxide (Part A: Sources and Detection of NO; NO
Synthase)
Edited by LESTER PACKER

VOLUME 269. Nitric Oxide (Part B: Physiological and
Pathological Processes)
Edited by LESTER PACKER

VOLUME 270. High Resolution Separation and Analysis of Biological
Macromolecules (Part A: Fundamentals)
Edited by BARRY L. KARGER AND WILLIAM S. HANCOCK

VOLUME 271. High Resolution Separation and Analysis of Biological
Macromolecules (Part B: Applications)
Edited by BARRY L. KARGER AND WILLIAM S. HANCOCK

VOLUME 272. Cytochrome P450 (Part B)
Edited by ERIC F. JOHNSON AND MICHAEL R. WATERMAN

VOLUME 273. RNA Polymerase and Associated Factors (Part A)
Edited by SANKAR ADHYA

VOLUME 274. RNA Polymerase and Associated Factors (Part B)
Edited by SANKAR ADHYA

VOLUME 275. Viral Polymerases and Related Proteins
Edited by LAWRENCE C. KUO, DAVID B. OLSEN, AND STEVEN S. CARROLL

VOLUME 276. Macromolecular Crystallography (Part A)
Edited by CHARLES W. CARTER, JR., AND ROBERT M. SWEET

VOLUME 277. Macromolecular Crystallography (Part B)
Edited by CHARLES W. CARTER, JR., AND ROBERT M. SWEET

VOLUME 278. Fluorescence Spectroscopy
Edited by LUDWIG BRAND AND MICHAEL L. JOHNSON

VOLUME 279. Vitamins and Coenzymes (Part I)
Edited by DONALD B. MCCORMICK, JOHN W. SUTTIE, AND CONRAD WAGNER

VOLUME 280. Vitamins and Coenzymes (Part J)
Edited by DONALD B. MCCORMICK, JOHN W. SUTTIE, AND CONRAD WAGNER

VOLUME 281. Vitamins and Coenzymes (Part K)
Edited by DONALD B. MCCORMICK, JOHN W. SUTTIE, AND CONRAD WAGNER

VOLUME 282. Vitamins and Coenzymes (Part L)
Edited by DONALD B. MCCORMICK, JOHN W. SUTTIE, AND CONRAD WAGNER

VOLUME 283. Cell Cycle Control
Edited by WILLIAM G. DUNPHY

VOLUME 284. Lipases (Part A: Biotechnology)
Edited by BYRON RUBIN AND EDWARD A. DENNIS

VOLUME 285. Cumulative Subject Index Volumes 263, 264, 266–284, 286–289

VOLUME 286. Lipases (Part B: Enzyme Characterization and Utilization)
Edited by BYRON RUBIN AND EDWARD A. DENNIS

VOLUME 287. Chemokines
Edited by RICHARD HORUK

VOLUME 288. Chemokine Receptors
Edited by RICHARD HORUK

VOLUME 289. Solid Phase Peptide Synthesis
Edited by GREGG B. FIELDS

VOLUME 290. Molecular Chaperones
Edited by GEORGE H. LORIMER AND THOMAS BALDWIN

VOLUME 291. Caged Compounds
Edited by GERARD MARRIOTT

VOLUME 292. ABC Transporters: Biochemical, Cellular, and
Molecular Aspects
Edited by SURESH V. AMBUDKAR AND MICHAEL M. GOTTESMAN

VOLUME 293. Ion Channels (Part B)
Edited by P. MICHAEL CONN

VOLUME 294. Ion Channels (Part C)
Edited by P. MICHAEL CONN

VOLUME 295. Energetics of Biological Macromolecules (Part B)
Edited by GARY K. ACKERS AND MICHAEL L. JOHNSON

VOLUME 296. Neurotransmitter Transporters
Edited by SUSAN G. AMARA

VOLUME 297. Photosynthesis: Molecular Biology of Energy Capture
Edited by LEE MCINTOSH

VOLUME 298. Molecular Motors and the Cytoskeleton (Part B)
Edited by RICHARD B. VALLEE

VOLUME 299. Oxidants and Antioxidants (Part A)
Edited by LESTER PACKER

VOLUME 300. Oxidants and Antioxidants (Part B)
Edited by LESTER PACKER

VOLUME 301. Nitric Oxide: Biological and Antioxidant Activities (Part C)
Edited by LESTER PACKER

VOLUME 302. Green Fluorescent Protein
Edited by P. MICHAEL CONN

VOLUME 303. cDNA Preparation and Display
Edited by SHERMAN M. WEISSMAN

VOLUME 304. Chromatin
Edited by PAUL M. WASSARMAN AND ALAN P. WOLFFE

VOLUME 305. Bioluminescence and Chemiluminescence (Part C)
Edited by THOMAS O. BALDWIN AND MIRIAM M. ZIEGLER

VOLUME 306. Expression of Recombinant Genes in
Eukaryotic Systems
Edited by JOSEPH C. GLORIOSO AND MARTIN C. SCHMIDT

VOLUME 307. Confocal Microscopy
Edited by P. MICHAEL CONN

VOLUME 308. Enzyme Kinetics and Mechanism (Part E: Energetics of
Enzyme Catalysis)
Edited by DANIEL L. PURICH AND VERN L. SCHRAMM

VOLUME 309. Amyloid, Prions, and Other Protein Aggregates
Edited by RONALD WETZEL

VOLUME 310. Biofilms
Edited by RON J. DOYLE

VOLUME 311. Sphingolipid Metabolism and Cell Signaling (Part A)
Edited by ALFRED H. MERRILL, JR., AND YUSUF A. HANNUN

VOLUME 312. Sphingolipid Metabolism and Cell Signaling (Part B)
Edited by ALFRED H. MERRILL, JR., AND YUSUF A. HANNUN

VOLUME 313. Antisense Technology (Part A: General Methods, Methods of
Delivery, and RNA Studies)
Edited by M. IAN PHILLIPS

VOLUME 314. Antisense Technology (Part B: Applications)
Edited by M. IAN PHILLIPS

VOLUME 315. Vertebrate Phototransduction and the Visual Cycle (Part A)
Edited by KRZYSZTOF PALCZEWSKI

VOLUME 316. Vertebrate Phototransduction and the Visual Cycle (Part B)
Edited by KRZYSZTOF PALCZEWSKI

VOLUME 317. RNA–Ligand Interactions (Part A: Structural Biology Methods)
Edited by DANIEL W. CELANDER AND JOHN N. ABELSON

VOLUME 318. RNA–Ligand Interactions (Part B: Molecular Biology Methods)
Edited by DANIEL W. CELANDER AND JOHN N. ABELSON

VOLUME 319. Singlet Oxygen, UV-A, and Ozone
Edited by LESTER PACKER AND HELMUT SIES

VOLUME 320. Cumulative Subject Index Volumes 290–319

VOLUME 321. Numerical Computer Methods (Part C)
Edited by MICHAEL L. JOHNSON AND LUDWIG BRAND

VOLUME 322. Apoptosis
Edited by JOHN C. REED

VOLUME 323. Energetics of Biological Macromolecules (Part C)
Edited by MICHAEL L. JOHNSON AND GARY K. ACKERS

VOLUME 324. Branched-Chain Amino Acids (Part B)
Edited by ROBERT A. HARRIS AND JOHN R. SOKATCH

VOLUME 325. Regulators and Effectors of Small GTPases (Part D: Rho Family)
Edited by W. E. BALCH, CHANNING J. DER, AND ALAN HALL

VOLUME 326. Applications of Chimeric Genes and Hybrid Proteins (Part A:
Gene Expression and Protein Purification)
Edited by JEREMY THORNER, SCOTT D. EMR, AND JOHN N. ABELSON

VOLUME 327. Applications of Chimeric Genes and Hybrid Proteins (Part B:
Cell Biology and Physiology)
Edited by JEREMY THORNER, SCOTT D. EMR, AND JOHN N. ABELSON

VOLUME 328. Applications of Chimeric Genes and Hybrid Proteins (Part C:
Protein–Protein Interactions and Genomics)
Edited by JEREMY THORNER, SCOTT D. EMR, AND JOHN N. ABELSON

VOLUME 329. Regulators and Effectors of Small GTPases (Part E: GTPases
Involved in Vesicular Traffic)
Edited by W. E. BALCH, CHANNING J. DER, AND ALAN HALL

VOLUME 330. Hyperthermophilic Enzymes (Part A)
Edited by MICHAEL W. W. ADAMS AND ROBERT M. KELLY

VOLUME 331. Hyperthermophilic Enzymes (Part B)
Edited by MICHAEL W. W. ADAMS AND ROBERT M. KELLY

VOLUME 332. Regulators and Effectors of Small GTPases (Part F: Ras Family I)
Edited by W. E. BALCH, CHANNING J. DER, AND ALAN HALL

VOLUME 333. Regulators and Effectors of Small GTPases (Part G: Ras Family II)
Edited by W. E. BALCH, CHANNING J. DER, AND ALAN HALL

VOLUME 334. Hyperthermophilic Enzymes (Part C)
Edited by MICHAEL W. W. ADAMS AND ROBERT M. KELLY

VOLUME 335. Flavonoids and Other Polyphenols
Edited by LESTER PACKER

VOLUME 336. Microbial Growth in Biofilms (Part A: Developmental and Molecular Biological Aspects)
Edited by RON J. DOYLE

VOLUME 337. Microbial Growth in Biofilms (Part B: Special Environments and Physicochemical Aspects)
Edited by RON J. DOYLE

VOLUME 338. Nuclear Magnetic Resonance of Biological Macromolecules (Part A)
Edited by THOMAS L. JAMES, VOLKER DÖTSCH, AND ULI SCHMITZ

VOLUME 339. Nuclear Magnetic Resonance of Biological Macromolecules (Part B)
Edited by THOMAS L. JAMES, VOLKER DÖTSCH, AND ULI SCHMITZ

VOLUME 340. Drug–Nucleic Acid Interactions
Edited by JONATHAN B. CHAIRES AND MICHAEL J. WARING

VOLUME 341. Ribonucleases (Part A)
Edited by ALLEN W. NICHOLSON

VOLUME 342. Ribonucleases (Part B)
Edited by ALLEN W. NICHOLSON

VOLUME 343. G Protein Pathways (Part A: Receptors)
Edited by RAVI IYENGAR AND JOHN D. HILDEBRANDT

VOLUME 344. G Protein Pathways (Part B: G Proteins and Their Regulators)
Edited by RAVI IYENGAR AND JOHN D. HILDEBRANDT

VOLUME 345. G Protein Pathways (Part C: Effector Mechanisms)
Edited by RAVI IYENGAR AND JOHN D. HILDEBRANDT

VOLUME 346. Gene Therapy Methods
Edited by M. IAN PHILLIPS

VOLUME 347. Protein Sensors and Reactive Oxygen Species (Part A: Selenoproteins and Thioredoxin)
Edited by HELMUT SIES AND LESTER PACKER

VOLUME 348. Protein Sensors and Reactive Oxygen Species (Part B: Thiol Enzymes and Proteins)
Edited by HELMUT SIES AND LESTER PACKER

VOLUME 349. Superoxide Dismutase
Edited by LESTER PACKER

VOLUME 350. Guide to Yeast Genetics and Molecular and Cell Biology (Part B)
Edited by CHRISTINE GUTHRIE AND GERALD R. FINK

VOLUME 351. Guide to Yeast Genetics and Molecular and Cell Biology (Part C)
Edited by CHRISTINE GUTHRIE AND GERALD R. FINK

VOLUME 352. Redox Cell Biology and Genetics (Part A)
Edited by CHANDAN K. SEN AND LESTER PACKER

VOLUME 353. Redox Cell Biology and Genetics (Part B)
Edited by CHANDAN K. SEN AND LESTER PACKER

VOLUME 354. Enzyme Kinetics and Mechanisms (Part F: Detection and Characterization of Enzyme Reaction Intermediates)
Edited by DANIEL L. PURICH

VOLUME 355. Cumulative Subject Index Volumes 321–354

VOLUME 356. Laser Capture Microscopy and Microdissection
Edited by P. MICHAEL CONN

VOLUME 357. Cytochrome P450, Part C
Edited by ERIC F. JOHNSON AND MICHAEL R. WATERMAN

VOLUME 358. Bacterial Pathogenesis (Part C: Identification, Regulation, and Function of Virulence Factors)
Edited by VIRGINIA L. CLARK AND PATRIK M. BAVOIL

VOLUME 359. Nitric Oxide (Part D)
Edited by ENRIQUE CADENAS AND LESTER PACKER

VOLUME 360. Biophotonics (Part A)
Edited by GERARD MARRIOTT AND IAN PARKER

VOLUME 361. Biophotonics (Part B)
Edited by GERARD MARRIOTT AND IAN PARKER

VOLUME 362. Recognition of Carbohydrates in Biological Systems (Part A)
Edited by YUAN C. LEE AND REIKO T. LEE

VOLUME 363. Recognition of Carbohydrates in Biological Systems (Part B)
Edited by YUAN C. LEE AND REIKO T. LEE

VOLUME 364. Nuclear Receptors
Edited by DAVID W. RUSSELL AND DAVID J. MANGELSDORF

VOLUME 365. Differentiation of Embryonic Stem Cells
Edited by PAUL M. WASSAUMAN AND GORDON M. KELLER

VOLUME 366. Protein Phosphatases
Edited by SUSANNE KLUMPP AND JOSEF KRIEGLSTEIN

VOLUME 367. Liposomes (Part A)
Edited by NEJAT DÜZGÜNEŞ

VOLUME 368. Macromolecular Crystallography (Part C)
Edited by CHARLES W. CARTER, JR., AND ROBERT M. SWEET

VOLUME 369. Combinational Chemistry (Part B)
Edited by GUILLERMO A. MORALES AND BARRY A. BUNIN

VOLUME 370. RNA Polymerases and Associated Factors (Part C)
Edited by SANKAR L. ADHYA AND SUSAN GARGES

VOLUME 371. RNA Polymerases and Associated Factors (Part D)
Edited by SANKAR L. ADHYA AND SUSAN GARGES

VOLUME 372. Liposomes (Part B)
Edited by NEJAT DÜZGÜNEŞ

VOLUME 373. Liposomes (Part C)
Edited by NEJAT DÜZGÜNEŞ

VOLUME 374. Macromolecular Crystallography (Part D)
Edited by CHARLES W. CARTER, JR., AND ROBERT W. SWEET

VOLUME 375. Chromatin and Chromatin Remodeling Enzymes (Part A)
Edited by C. DAVID ALLIS AND CARL WU

VOLUME 376. Chromatin and Chromatin Remodeling Enzymes (Part B)
Edited by C. DAVID ALLIS AND CARL WU

VOLUME 377. Chromatin and Chromatin Remodeling Enzymes (Part C)
Edited by C. DAVID ALLIS AND CARL WU

VOLUME 378. Quinones and Quinone Enzymes (Part A)
Edited by HELMUT SIES AND LESTER PACKER

VOLUME 379. Energetics of Biological Macromolecules (Part D)
Edited by JO M. HOLT, MICHAEL L. JOHNSON, AND GARY K. ACKERS

VOLUME 380. Energetics of Biological Macromolecules (Part E)
Edited by JO M. HOLT, MICHAEL L. JOHNSON, AND GARY K. ACKERS

VOLUME 381. Oxygen Sensing
Edited by CHANDAN K. SEN AND GREGG L. SEMENZA

VOLUME 382. Quinones and Quinone Enzymes (Part B)
Edited by HELMUT SIES AND LESTER PACKER

VOLUME 383. Numerical Computer Methods (Part D)
Edited by LUDWIG BRAND AND MICHAEL L. JOHNSON

VOLUME 384. Numerical Computer Methods (Part E)
Edited by LUDWIG BRAND AND MICHAEL L. JOHNSON

VOLUME 385. Imaging in Biological Research (Part A)
Edited by P. MICHAEL CONN

VOLUME 386. Imaging in Biological Research (Part B)
Edited by P. MICHAEL CONN

VOLUME 387. Liposomes (Part D)
Edited by NEJAT DÜZGÜNEŞ

VOLUME 388. Protein Engineering
Edited by DAN E. ROBERTSON AND JOSEPH P. NOEL

VOLUME 389. Regulators of G-Protein Signaling (Part A)
Edited by DAVID P. SIDEROVSKI

VOLUME 390. Regulators of G-Protein Signaling (Part B)
Edited by DAVID P. SIDEROVSKI

VOLUME 391. Liposomes (Part E)
Edited by NEJAT DÜZGÜNEŞ

VOLUME 392. RNA Interference
Edited by ENGELKE ROSSI

VOLUME 393. Circadian Rhythms
Edited by MICHAEL W. YOUNG

VOLUME 394. Nuclear Magnetic Resonance of Biological Macromolecules
(Part C)
Edited by THOMAS L. JAMES

VOLUME 395. Producing the Biochemical Data (Part B)
Edited by ELIZABETH A. ZIMMER AND ERIC H. ROALSON

VOLUME 396. Nitric Oxide (Part E)
Edited by LESTER PACKER AND ENRIQUE CADENAS

VOLUME 397. Environmental Microbiology
Edited by JARED R. LEADBETTER

VOLUME 398. Ubiquitin and Protein Degradation (Part A)
Edited by RAYMOND J. DESHAIES

VOLUME 399. Ubiquitin and Protein Degradation (Part B)
Edited by RAYMOND J. DESHAIES

VOLUME 400. Phase II Conjugation Enzymes and Transport Systems
Edited by HELMUT SIES AND LESTER PACKER

VOLUME 401. Glutathione Transferases and Gamma Glutamyl Transpeptidases
Edited by HELMUT SIES AND LESTER PACKER

VOLUME 402. Biological Mass Spectrometry
Edited by A. L. BURLINGAME

VOLUME 403. GTPases Regulating Membrane Targeting and Fusion
Edited by WILLIAM E. BALCH, CHANNING J. DER, AND ALAN HALL

VOLUME 404. GTPases Regulating Membrane Dynamics
Edited by WILLIAM E. BALCH, CHANNING J. DER, AND ALAN HALL

VOLUME 405. Mass Spectrometry: Modified Proteins and Glycoconjugates
Edited by A. L. BURLINGAME

VOLUME 406. Regulators and Effectors of Small GTPases: Rho Family
Edited by WILLIAM E. BALCH, CHANNING J. DER, AND ALAN HALL

VOLUME 407. Regulators and Effectors of Small GTPases: Ras Family
Edited by WILLIAM E. BALCH, CHANNING J. DER, AND ALAN HALL

VOLUME 408. DNA Repair (Part A)
Edited by JUDITH L. CAMPBELL AND PAUL MODRICH

VOLUME 409. DNA Repair (Part B)
Edited by JUDITH L. CAMPBELL AND PAUL MODRICH

VOLUME 410. DNA Microarrays (Part A: Array Platforms and Wet-Bench Protocols)
Edited by ALAN KIMMEL AND BRIAN OLIVER

VOLUME 411. DNA Microarrays (Part B: Databases and Statistics)
Edited by ALAN KIMMEL AND BRIAN OLIVER

VOLUME 412. Amyloid, Prions, and Other Protein Aggregates (Part B)
(in preparation)
Edited by INDU KHETERPAL AND RONALD WETZEL

VOLUME 413. Amyloid, Prions, and Other Protein Aggregates (Part C)
(in preparation)
Edited by INDU KHETERPAL AND RONALD WETZEL

VOLUME 414. Measuring Biological Responses with Automated Microscopy
(in preparation)
Edited by JAMES INGLESE

VOLUME 415. Glycobiology (in preparation)
Edited by MINORU FUKUDA

VOLUME 416. Glycomics (in preparation)
Edited by MINORU FUKUDA

VOLUME 417. Functional Glycomics (in preparation)
Edited by MINORU FUKUDA

VOLUME 418. Embryonic Stem Cells (in preparation)
Edited by IRINA KLIMANSKAYA AND ROBERT LANZA

VOLUME 419. Adult Stem Cells (in preparation)
Edited by IRINA KLIMANSKAYA AND ROBERT LANZA

VOLUME 420. Stem Cell Tools and Other Experimental Protocols
(in preparation)
Edited by IRINA KLIMANSKAYA AND ROBERT LANZA

[1] RNA Extraction for Arrays

By LAKSHMI V. MADABUSI, GARY J. LATHAM, and
BERNARD F. ANDRUSS

Abstract

DNA microarrays enable insights into global gene expression by capturing a snapshot of cellular expression levels at the time of sample collection. Careful RNA handling and extraction are required to preserve this information properly, ensure sample-to-sample reproducibility, and limit unwanted technical variation in experimental data. This chapter discusses important considerations for "array-friendly" sample handling and processing from biosamples such as blood, formalin-fixed, paraffin-embedded samples, and fresh or flash-frozen tissues and cells. It also provides guidelines on RNA quality assessments, which can be used to validate sample preparation and maximize recovery of relevant biological information.

Introduction

DNA microarrays have enabled biologists to move from the realm of studying one gene at a time to understanding genome-wide changes in gene expression. The value of microarray studies has been vetted through numerous studies that have linked abnormal transcript levels with many different diseases (Archacki and Wang, 2004; Blalock et al., 2004; Borovecki et al., 2005; Dhanasekaran et al., 2001; Glatt et al., 2005; West et al., 2001). Because these types of studies will be used increasingly to create and validate diagnostic and prognostic expression signatures and to support toxicological and functional studies that underlie the regulatory filings for new drug submissions, it will become increasingly important to create standardized and robust methods for sample procurement, sample processing, and data analysis. The goal of any RNA isolation procedure is to recover an RNA population that faithfully mirrors the biology of the sample at the time of collection. Problems associated with the extraction of biologically representative RNA primarily arise from the susceptibility of RNA to degradation by ubiquitous and catalytically potent RNases. For tissues and cells, protection of RNA has traditionally been accomplished by immediate lysis using high concentrations of detergents and/or chaotropic agents and organic solvents (such as TRI reagent). These methods,

METHODS IN ENZYMOLOGY, VOL. 411
0076-6879/06 $35.00
DOI: 10.1016/S0076-6879(06)11001-0

while effective, are complex to use at point of care and suffer from low sample throughput and poor stabilization of cellular RNA for long periods. Flash freezing of the sample in liquid nitrogen and subsequent transportation on dry ice, although effective, are impractical in most clinical settings. Finally, disease specimens can present biohazard risks to the operator and constrain sample collection, thus limiting the use of best sample handling and processing practices and compromising RNA quality.

The practicality and efficacy of RNA stabilization agents such as RNA*later* to preserve the RNA in tissues, cells, and blood are gaining broad acceptance. Procedures used for collection of samples with RNA*later* are simple and can be carried out in a hospital setting with minimal training. This reagent is aqueous and nontoxic and allows convenient transportation of samples at ambient temperature. However, RNA*later* does not remove the biohazard risks associated with biosamples, and, as a result, all proper safety precautions should be observed. It is beyond the scope of this chapter to provide details on the risks associated and preventive measures to be taken when dealing with samples considered to be a biohazard. Several regulatory agencies offer guidelines on the safety issues and precautions that need to be addressed with such samples.

In addition to the handling of biological material, limitations can be imposed by the large amounts of RNA necessary for microarray experiments. As a result, samples such as tumor biopsies, formalin-fixed, paraffin-embedded (FFPE) sections, or laser microdissected samples require RNA amplification to generate adequate amounts of labeled material for microarray hybridization. The most popular and best validated approaches for amplifying RNA are based on the linear RNA amplification method developed by Eberwine (Van Gelder, 1990). This technique has been widely accepted for microarray applications and is known to preserve the original transcript ratios in the sample (Feldman *et al.*, 2002; Polacek *et al.*, 2003). In terms of RNA quality, parameters such as A260:280 measurements and Agilent RNA integrity number (RIN) are often used to gauge the quality of samples and predict their suitability for microarray studies. The minimum A260:280 or RIN number suitable for analysis varies by the array platform, number of replicates, and the experimental questions to be answered in the study.

Blood as a Biological Specimen

Blood is a highly desirable biosample for research and clinical studies for several reasons. First, blood is highly accessible and can be collected using relatively simple methods. Second, limited infrastructure is required to draw blood from a large number of patients. Finally, blood circulates throughout the entire body and thus is a vast reservoir of host biological information and an ideal specimen for experiments that aim to understand human physiology and disease. As a source of RNA, however, blood poses a number of unique challenges. The ratio of total protein to RNA in blood is roughly 100-fold greater than the ratio for most solid tissues, complicating the isolation of pure, high quality RNA. The presence of multiple cellular components in blood, each at different maturation stages in their life cycle, can lead to variation between patients. Among these various cellular components, only white blood cells (WBC) or leukocytes are nucleated and thus transcriptionally active (Fan and Hegde, 2005). However, WBC constitute only about 0.1% of the total blood cellular composition. In contrast, red blood cells (RBC) comprise ~95% of the total cell count but do not contribute to the blood gene expression program in their mature form. Immature red blood cells, known as reticulocytes, comprise only about 1% of the RBC population, yet contain significant levels of nucleic acids, particularly globin mRNA, that can contribute to the background noise in a microarray experiment. This noise can be substantial and can reduce the number of genes that are called present on microarrays.

Collection and Preservation of Blood Samples

Whole blood samples can be collected in the presence of anticoagulants such as sodium citrate, EDTA, or heparin. However, these chemicals are not effective RNA preservatives because they do not readily inhibit the RNases in blood that are the primary threat to RNA intactness and do not maintain cellular homeostasis in the sample. Indeed, the gene expression levels of many transcripts in blood stored in EDTA can change by an order of magnitude or more within a few hours (Rainen *et al.*, 2002). Rapid changes in gene expression of cytokines and transcription factors have been observed during storage for 1 to 4 h with interleukin-8 expression increasing 100-fold by 4 h (Tanner *et al.*, 2002). Additional genes such as transcription factors and pro- and anti-inflammatory genes show large changes in gene expression within a few hours to 1 day after collection (Pahl and Brune, 2002; Tanner *et al.*, 2002). The stresses caused by handling and centrifugation can alter gene expression rapidly (Haskill *et al.*, 1988). It is important to note that the purity of the RNA as measured by A260/280 is very consistent even after the extended storage of whole blood at ambient temperature, and often the intactness of ribosomal RNA bands is also well maintained although the underlying representation of many genes may have changed dramatically.

Other commonly used methods of blood collection and preservation include use of commercial products such as PAXgene tubes (PreAnalytiX GmbH, Switzerland) and the CPT tube (Becton Dickinson, NJ) for peripheral blood mononuclear cell (PBMC) collection. A brief summary of the advantages and impact of such sample collection methods on gene expression profiling has been reviewed by Fan and Hegde (2005). We find that use of RNA*later* as a preservative in conjunction with an optimized RNA isolation protocol produces excellent RNA yields and stable and reproducible expression profiles of human whole blood (Fig. 1). A concomitant increase in RNA yield and a more consistent level of percentage present calls were observed with the use of RNA*later*.

Methodologies for Globin Transcript Removal from Whole Blood

The presence of high levels of globin transcripts in RNA isolated from whole blood can affect the quality of data generated by reducing the number of present calls, decreasing call concordance, and increasing the variation in signal between samples. To circumvent the problems associated with the presence of globin transcripts, protocols have been described that reduce globin mRNA levels by either depleting these transcripts in purified RNA or fractionating blood cells to reduce the red blood

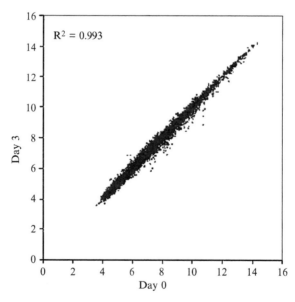

FIG. 1. RNA*later* provides room temperature stabilization of the global expression profile in human whole blood. Biological replicates of samples processed with RiboPure-Blood (Ambion) immediately after blood collection or after 3 days of storage at room temperature. The global expression level was assessed using Affymetrix human focus arrays with 10 μg aRNA input without globin reduction. Plots were constructed from signal-normalized data.

cell population, particularly reticulocytes, which are the primary reservoir of globin transcripts.

Depletion of Globin Transcripts from Whole Blood

To selectively deplete the globin transcripts from whole blood, commercial protocols were initially developed using enzymatic procedures to selectively degrade the globin transcripts. One of the first protocols was suggested by Affymetrix, Inc. This procedure described the hybridization of complementary DNA oligonucleotides to the various globin transcripts in blood followed by digestion of the RNA:DNA hybrid with RNase H. More recently, Affymetrix has launched the GeneChip Blood RNA Concentration Kit, which utilizes globin-specific peptide nucleic acid (PNA) oligomers as blocking molecules to prevent the amplification of these transcripts during T7 RNA polymerase-based linear amplification. An alternative strategy provide by Ambion is the GLOBINclear kit, which relies on the binding of biotinylated capture oligonucleotides to the RNA and uses biotin–streptavidin binding to deplete the globin transcript complex from the mixture. This method results in a dramatic reduction of the globin gene transcripts from whole blood RNA while substantially increasing the percentage present calls on Affymetrix Genechip arrays with human blood samples (Fig. 2). Thus, globin transcript reduction prevents

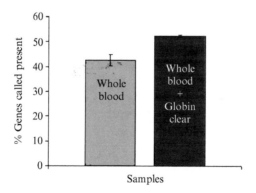

FIG. 2. GLOBINclear processing increases the sensitivity of microarrays. Quadruplicate GLOBINclear reactions were performed with pooled total RNA samples from human whole blood (from healthy donors under an IRB-approved protocol). The processed RNA was then amplified with MessageAmp II-96 to synthesize biotinylated aRNA for Affymetrix GeneChip array analysis. Quadruplicate untreated whole blood RNA samples were also amplified in parallel. Biotinylated aRNA was hybridized to Affymetrix human focus arrays. Present calls were determined using Affymetrix GCOS software with default settings. GLOBINclear processing resulted in an increase in genes called present.

the loss of significant information caused by the distorted transcript composition in human whole blood.

Fractionation of Blood

Cell fractionation methods can provide isolated total leukocyte populations or individual subsets that are substantially depleted of red cells. Historically, gradient centrifugation techniques were employed to collect subcellular fractions in blood to enrich for cells that could be used as sources of nucleic acids (Bach and Brashler, 1970; Greenberg, 1973; Phatak, 1978; Pretlow, 1971). These methods also ensured that carryover of mRNA from reticulocytes was prevented and that the associated problems with the presence of globin transcripts were avoided. Use of Ficoll–Hypaque density gradient centrifugation has been the method of choice to achieve this enrichment. Commercial products that use this technology such as the CPT vacutainer tubes reduce the processing time and difficulty of working with the gradient; however, these products do not offer RNA stabilization and thus require immediate processing to collect the RNA.

Fractionation has also been accomplished by the use of immunoselection methods that utilize antibodies to select and immobilize specific subsets of leukocytes based on unique cell surface markers. However, the cell surface antigens that distinguish and define leukocyte subsets are typically functional recognition molecules of the immune system that mediate the complex cellular interactions that make up humoral and cellular immunity. Engaging these cell surface antigens during positive immunoselection may trigger changes in mRNA levels in the selected cells. Although negative selection strategies have been used to avoid engaging leukocyte cell surface antigens in the desired population, they do not permit multiple WBC subsets to be obtained from each sample, and the negatively selected cells may be more heterogeneous compared to positively selected cells.

As an alternative to these methods, a novel filter-based leukocyte capture method marketed commercially under the name LeukoLOCK has been introduced by Ambion, Inc. This system uses leukocyte depletion filters that have been used in blood transfusion therapy to remove donor leukocytes and prevent graft-versus-host rejection in the recipient. The LeukoLOCK technology can be used in conjunction with RNA*later* to stabilize RNA in the filtered population of leukocytes for months and has proven beneficial for transportation and for recovery of RNA for use in microarray and quantitative reverse transcription polymerase chain reaction (qRT-PCR) applications. The LeukoLock procedure is outlined next; for details, please consult the manufacturer's protocol (Table I).

TABLE I
MICROARRAY DATA FOR RNA ISOLATED USING THE LEUKOLOCK PROCEDURE[a]

Sample	aRNA yield (μg)	aRNA length	GAPDH 3'/5'	β-Actin 3'/5'	% Present
LeukoLOCK	81.01	1470 bases	1.14	1.02	43.70
LeukoLOCK	73.73	1440 bases	1.26	0.99	43.90

[a] RNA was isolated from duplicates of whole blood samples fractionated using the LeukoLOCK kit. One microgram of samples was amplified using the Ebwerine technique and analyzed on Affymetrix human focus arrays. Average yields of 73 to 81 μg were obtained, and longer size amplified products around 1400 bases were observed. An 8% increase in the percentage present calls was observed using this procedure as compared to other conventional RNA isolation methods.

Overview of the LeukoLOCK Procedure for Isolation of RNA from Whole Blood Samples

Sample Collection and Capture of Leukocytes

1. Collect 9–10 ml of whole blood samples in EDTA-containing evacuated blood collection tubes.
2. Assemble the sample tube/LeukoLOCK filter apparatus.
3. Pass blood through the LeukoLOCK filter using an evacuated tube as the vacuum source. The LeukoLOCK filter captures the total leukocyte population, while plasma, platelets, and RBCs are eliminated.
4. Flush filter with phosphate-buffered saline (PBS) and RNA*later*. Flush the filter with PBS to remove residual RBCs and then with RNA*later* to stabilize leukocyte RNA.
5. Seal the LeukoLOCK filter ports with the sheath and screw cap from the transfer spike.

LeukoLOCK Filter Processing and Cell Lysis

1. Remove residual RNA*later* from the LeukoLOCK filter.
2. Flush with pH-adjusted lysis/binding solution; collect lysate in a 15-ml tube. In this step, the leukocytes that are trapped on the LeukoLOCK filter are lysed, and the lysate is flushed off the filter and collected in a 15-ml conical tube.
3. Add nuclease-free water and proteinase K, and shake for 5 min. This brief proteinase K treatment degrades cellular proteins.

RNA Isolation and Elution

1. Add RNA-binding beads and 100% isopropanol, and incubate at room temperature for 5 min.
2. Recover the RNA-binding beads by gentle centrifugation and discard the supernatant.
3. Wash with wash solution 1 and transfer the RNA-binding beads to a 1.5-ml processing tube.
4. Recover the RNA-binding beads and discard the supernatant.
5. Wash RNA-binding beads with wash solution 2/3.
6. Elute the RNA with ≤150 μl elution solution.
7. Transfer the RNA-containing supernatant to a new processing tube or other nuclease-free container appropriate for the application.
8. Store the purified RNA at $-20°$.

Use of Solid Tissues for Gene Expression Analysis

In addition to blood samples, solid tissues are used routinely for gene expression analysis. Tissues collected for clinical analyses are typically biopsy specimens, which are used for histopathological testing or molecular testing using RT-PCR. Pathological analysis of clinical tissues usually requires that the samples be fixed and embedded in paraffin to conserve the cellular architecture. As a result, specialized methods have emerged for isolating and profiling RNA from FFPE samples to more accurately illuminate the gene expression inventory of these invaluable samples.

Other options for preserving RNA profiles in freshly procured tissue include flash freezing in liquid nitrogen and RNA*later*. RNA*later* offers convenient, room temperature stabilization without compromise, even in very high resolution microarray experiments. For example, Mutter *et al.* (2004) reported that tissue could be stored in RNA*later* at room temperature for up to 3 days without introducing any systematic changes in gene expression measurements from microarray experiments. In this respect, RNA*later* was determined to be comparable to array analyses of either fresh or flash-frozen tissue.

Use of Formalin-Fixed, Paraffin-Embedded Sections for Gene Expression Analysis

A key driver for use of FFPE samples for gene expression analysis is the availability of clinical outcome data that can build retrospective relationships between gene expression patterns and disease (Lewis *et al.*, 2001). While there is a great potential for use of such samples for microarray analysis, RNA isolated from such samples is usually modified extensively by chemical adducts and degraded significantly. As a result, RNA from such fixed samples is not readily amenable to downstream enzymatic manipulations that are required for target preparation. Other problems that can introduce variability include heterogeneity in sample handling prior to formalin fixation, the age and exposure of the paraffin blocks used for sectioning, and the procedures used for RNA isolation. While it is generally believed that RNA is vulnerable to degradation during fixation, we and others find that relatively intact RNA can be isolated from samples that have been fixed in formalin for several months (Masuda *et al.*, 1999). The elevated temperatures required for paraffin embedding, however, are known to reduce the quality and yields of RNA. Other factors that can alter the integrity of RNA are the age of the paraffin block and the length of time that the samples have been stored. Indeed, RNA extracted from archived FFPE blocks that are older than 10 years is typically only about 100 nucleotides in length. While such RNA targets can be assayed with some success by a real-time RT-PCR or branch DNA assay, array analyses require nonstandard amplification and/or labeling strategies, which can further distort representation in such compromised samples. Nevertheless, such protocols have been informative in identifying gene signatures. Newer microarray designs for genome-wide profiling of FFPE samples from commercial vendors such as Affymetrix allow the interrogation of smaller windows of target sequence (e.g., the X3P arrays query only the 300 most 3′ nucleotides of each transcript) compared to the standard GeneChip arrays, which can improve the array metrics for FFPE tissues.

Typical protocols for isolation of RNA from FFPE samples consist of three main steps: deparaffinization using an organic compound such as xylene, proteinase K digestion to remove the protein–RNA cross-links and release of the RNA, phenol extraction, or a column-based purification to recover the nucleic acids. Variations to such protocols involve navigating various proteinase K digestion time points and optimization of the column-based purification of the RNA. An overview of one such protocol from the RecoverAll Total Nucleic Acid Isolation Kit is presented. This product enables the recovery of total RNA, including micro-RNA and genomic DNA, providing microarray analysis options for several classes of nucleic acids.

RecoverAll Total Nucleic Acid Isolation Procedure

DEPARAFFINIZATION

1. Assemble FFPE sections equivalent to a 80-μm or 35-mg unsectioned core.
2. Add 1 ml 100% xylene, mix, and incubate for 3 min at 50°.

3. Centrifuge for 2 min at maximum speed, and discard the xylene.
4. Wash the pellet twice with 1 ml 100% ethanol and air dry.

PROTEASE DIGESTION STEP

1. Add digestion buffer and protease.
2. Incubate at 50° for 3 h for RNA isolation and 48 h for DNA isolation.

FILTER CARTRIDGE BINDING STEP

1. Add 480 μl isolation additive and vortex.
2. Add 1.1 ml 100% ethanol and mix.
3. Pass the mixture through a filter cartridge.
4. Wash with 700 μl of wash 1.
5. Wash with 500 μl of wash 2/3, and then centrifuge to remove residual fluid.
6. Add DNase I mix to each filter cartridge and incubate for 30 min.
7. Wash with 700 μl of wash 1.
8. Wash twice with 500 μl of wash 2/3, and then centrifuge to remove residual fluid.
9. Elute nucleic acid with 2 × 30 μl elution solution or nuclease-free water.

Use of Solid Tissue Clinical Specimens for Gene Expression Analysis

Expression profiling of biopsied tissues offers a direct and privileged view of how gene expression changes are correlated with disease. Intact RNA, however, is often difficult to recover from fresh or flash-frozen biopsy samples. RNase activities are particularly acute in many human tissues, especially pancreas and other secretory organs. These RNases must be inactivated quantitatively and rapidly during the sometimes laborious process of releasing and deproteinizing RNA from the complex cellular architecture. The most common approach is to disrupt tissue samples using a polytron, mill, or Dounce in a chaotropic solution that breaches cellular substructures, rapidly inactivates RNases, and strips proteins from nucleic acid. A popular product is TRI reagent (MRC, Inc.), also known as TRIzol, a monophasic solution of phenol and guanidine thiocyanate. Tissue samples are disrupted in the TRI reagent, followed by recovery of the aqueous phase and alcohol precipitation of total RNA. Another common method is to grind the tissue in molar concentrations of guanidine isothiocyanate, followed by RNA purification using a glass filter column (e.g., RNeasy, Qiagen Inc.). Perhaps the most stringent method is to combine both approaches; indeed, the Affymetrix GeneChip expression analysis technical manual recommends TRI reagent RNA extraction, followed by a glass filter column cleanup step. No one method is universally preferred, and sample preparation studies with clinical tissues have concluded that both TRI reagent (Roos-van Groningen *et al.*, 2004) and RNeasy (Egyhazi *et al.*, 2004) can provide suitable RNA yields and/or quality.

A novel and user-friendly alternative method that can be used with many animal tissues is the multienzymatic liquefaction of tissue (MELT) RNA isolation system (Ambion). This technology is a hands-free methodology for the rapid digestion of fresh or frozen tissue in a closed tube. Up to 10 mg of tissue can be digested at room temperature using a unique formulation that includes a potent RNase inhibitor and a cocktail of powerful catabolic enzymes. Following digestion, the RNA is purified using a magnetic bead procedure. Extensive array and qRT-PCR studies have been performed comparing MELT with "gold standard" methods such as TRI reagent and RNeasy with favorable results (Latham and Peltier, 2005) In addition, both animal and human tissue biopsies have been evaluated successfully in expression profiling experiments using the MELT system (data not shown).

RNA Quality Measurements for Microarray Analysis

Analysis of nucleic acid quality by absorbance spectrophotometry has been used since the inception of nucleic acid purification methodologies. Such measurements at 260 and 280 nm have been used to deduce the amount of nucleic acid and the accompanying levels of protein carryover in a sample. Absorbance ratios of 260 to 280 nm of 1.7–2.0 for RNA are often required for downstream analysis, including microarray experiments, and samples with ratios as low as 1.4 have been used successfully for gene expression analysis. The absorbance ratio measurement, however, suffers from several limitations. This ratio does not provide information about RNA intactness, which is critical to the success of any microarray experiment, nor does it definitively report the purity of the RNA and the absence of potential enzyme inhibitors.

Another measure of RNA integrity involves analysis of the ribosomal 18s and 28s species by denaturing agarose electrophoresis to determine the extent of degradation of the sample. A modification of this methodology, which involves capillary electrophoresis of the RNA on an Agilent 2100 bioanalyzer and the concurrent measurement of the RNA concentrations along various positions (RNA sizes) of the electropherogram to obtain RIN scores, is now widely used (Schroeder *et al.*, 2006). Many times, RIN scores of 7 or higher predict satisfactory results in microarray studies, although the precise correlation between the RIN and the quality of array data is likely dependent on many factors, including the sample type and RNA isolation procedure. Fortunately, these factors can be controlled in many microarray experiments, meaning that the RIN can be an effective tool to "gate" RNA quality and suitability prior to RNA amplification and labeling. Once a satisfactory set of procedures has been identified for microarray work, we strongly recommend that standard operating procedures be written and carefully followed to minimize sources of variability that can ultimately undermine the value of the array data set. This requirement is particularly critical for interinstitutional studies that rely on multiple sites for tissue procurement and processing.

Conclusion

Gene expression profiling of clinical samples has become a paradigm for understanding disease etiology, pharmacogenomics and toxicological evaluations. Sample handling and the subsequent steps taken to preserve and isolate RNA can influence the quality and interpretation of microarray data dramatically. Because microarray analysis requires the use of multiple replicates to obtain statistically significant information, diligent assessments of sample procurement and preservation, RNA isolation, and preliminary microarray pilot studies can enable informed standard operating procedures. Vigilance with these procedures will result in more convincing expression profiling results, more enlightened conclusions, and a reduction in the resource costs associated with failed samples.

Acknowledgments

The authors thank Drs. Richard Conrad, Marianna Goldrick, and Robert Setterquist who developed Ambion's RecoverAll, LeukoLOCK, and GLOBINclear products.

References

Archacki, S., and Wang, Q. (2004). Expression profiling of cardiovascular disease. *Hum. Genom.* **1,** 355–370.

Bach, M. K., and Brashler, J. R. (1970). Subpopulations within "short-lived" and "long-lived" lymphocytes resolved by isopycnic sedimentation in Ficoll solutions coupled with a double radioisotopic labeling technique. *Exp. Cell Res.* **63,** 227–230.

Blalock, E. M., Geddes, J. W., Chen, K. C., Porter, N. M., Markesbery, W. R., and Landfield, P. W. (2004). Incipient Alzheimer's disease: Microarray correlation analyses reveal major transcriptional and tumor suppressor responses. *Proc. Natl. Acad. Sci. USA* **101,** 2173–2178.

Borovecki, F., Lovrecic, L., Zhou, J., Jeong, H., Then, F., Rosas, H. D., Hersch, S. M., Hogarth, P., Bouzou, B., Jensen, R. V., and Krainc, D. (2005). Genome-wide expression profiling of human blood reveals biomarkers for Huntington's disease. *Proc. Natl. Acad. Sci. USA* **102,** 11023–11028.

Dhanasekaran, S. M., Barrette, T. R., Ghosh, D., Shah, R., Varambally, S., Kurachi, K., Pienta, K. J., Rubin, M. A., and Chinnaiyan, A. M. (2001). Delineation of prognostic biomarkers in prostate cancer. *Nature* **412**, 822–826.

Egyhazi, S., Bjohle, J., Skoog, L., Huang, F., Borg, A. L., Frostvik Stolt, M., Hagerstrom, T., Ringborg, U., and Bergh, J. (2004). Proteinase K added to the extraction procedure markedly increases RNA yield from primary breast tumors for use in microarray studies. *Clin. Chem.* **50**, 975–976.

Fan, H., and Hedge, P. S. (2005). The transcriptome in blood: Challenges and solutions for robust expression profiling. *Curr. Mol. Med.* **5**, 3–10.

Feldman, A. L., Costouros, N. G., Wang, E., Qian, M., Marincola, F. M., Alexander, H. R., and Libutti, S. K. (2002). Advantages of mRNA amplification for microarray analysis. *Biotechniques* **33**, 906–912, 914.

Glatt, S. J., Everall, I. P., Kremen, W. S., Corbeil, J., Sasik, R., Khanlou, N., Han, M., Liew, C. C., and Tsuang, M. T. (2005). Comparative gene expression analysis of blood and brain provides concurrent validation of SELENBP1 up-regulation in schizophrenia. *Proc. Natl. Acad. Sci. USA* **102**, 15533–15538.

Greenberg, A. H. (1973). Fractionation of cytotoxic T lymphoblasts on Ficoll gradients by velocity sedimentation. *Eur. J. Immunol.* **3**, 793–797.

Haskill, S., Johnson, C., Eierman, D., Becker, S., and Warren, K. (1988). Adherence induces selective mRNA expression of monocyte mediators and proto-oncogenes. *J. Immunol.* **140**, 1690–1694.

Latham, G. J., and Peltier, H. J. (2005). MELT Total Nucleic Acid Isolation System: A new technology for hands-free tissue disruption, RNA preservation and total nucleic acid purification. *Nature Methods* **2**, i–iii.

Lewis, F., Maughan, N. J., Smith, V., Hillan, K., and Quirke, P. (2001). Unlocking the archive: Gene expression in paraffin-embedded tissue. *J. Pathol.* **195**, 66–71.

Masuda, N., Ohnishi, T., Kawamoto, S., Monden, M., and Okubo, K. (1999). Analysis of chemical modification of RNA from formalin-fixed samples and optimization of molecular biology applications for such samples. *Nucleic Acids Res.* **27**, 4436–4443.

Mutter, G. L., Zahrieh, D., Liu, C., Neuberg, D., Finkelstein, D., Baker, H. E., and Warrington, J. A. (2004). Comparison of frozen and RNALater solid tissue storage methods for use in RNA expression microarrays. *BMC Genom.* **5**, 88.

Pahl, A., and Brune, K. (2002). Gene expression changes in blood after phlebotomy: Implications for gene expression profiling. *Blood* **100**, 1094–1095.

Phatak, A. G. (1978). Various methods of lymphocyte separation and their relevance with the formation of non-immune rosettes. *J. Immunol. Methods* **20**, 109–115.

Polacek, D. C., Passerini, A. G., Shi, C., Francesco, N. M., Manduchi, E., Grant, G. R., Powell, S., Bischof, H., Winkler, H., Stoeckert, C. J., Jr., and Davies, P. F. (2003). Fidelity and enhanced sensitivity of differential transcription profiles following linear amplification of nanogram amounts of endothelial mRNA. *Physiol. Genom.* **13**, 147–156.

Pretlow, T. G. (1971). Estimation of experimental conditions that permit cell separations by velocity sedimentation on isokinetic gradients of Ficoll in tissue culture medium. *Anal. Biochem.* **41**, 248–255.

Rainen, L., Oelmueller, U., Jurgensen, S., Wyrich, R., Ballas, C., Schram, J., Herdman, C, Bankaitis-Davis, D., Nicholls, N., Trollinger, D., and Tryon, V. (2002). Stabilization of mRNA expression in whole blood samples. *Clin. Chem.* **48**, 1883–1890.

Roos-van Groningen, M. C., Eikmans, M., Baelde, H. J., de Heer, E., and Bruijn, J. A. (2004). Improvement of extraction and processing of RNA from renal biopsies. *Kidney Int.* **65**, 97–105.

Schroeder, A., Mueller, O., Stocker, S., Salowsky, R., Leiber, M., Gassmann, M., Lightfoot, S., Menzel, W., Granzow, M., and Ragg, T. (2006). The RIN: An RNA integrity number for assigning integrity values to RNA measurements. *BMC Mol. Biol.* **7,** 3.

Tanner, M. A., Berk, L. S., Felten, D. L., Blidy, A. D., Bit, S. L., and Ruff, D. W. (2002). Substantial changes in gene expression level due to the storage temperature and storage duration of human whole blood. *Clin. Lab. Haematol.* **24,** 337–341.

Van Gelder, R. N., von Zastrow, M. E., Yool, A., Dement, W. C., Barchas, J. D., and Eberwine, J. H. (1990). Amplified RNA synthesized from limited quantities of heterogeneous cDNA. *Proc. Natl. Acad. Sci. USA* **87,** 1663–1667.

West, M., Blanchette, C., Dressman, H., Huang, E., Ishida, S., Spang, R., Zuzan, H., Olson, J. A., Jr., Marks, J. R., and Nevins, J. R. (2001). Predicting the clinical status of human breast cancer by using gene expression profiles. *Proc. Natl. Acad. Sci. USA* **98,** 11462–11467.

[2] Analyzing Micro-RNA Expression Using Microarrays

By TIMOTHY S. DAVISON, CHARLES D. JOHNSON, and
BERNARD F. ANDRUSS

Abstract

The discovery of micro-RNAs (miRNAs) and the growing appreciation of the importance of micro-RNAs in the regulation of gene expression are driving increasing interest in miRNA expression profiling. Early studies have suggested prominent roles for these genetically encoded regulatory molecules in a variety of normal biological processes and diseases, particularly cancer. However, the field of miRNA expression profiling is in its infancy. Several factors, including the small size, the unknown but limited number of miRNAs, and the tissue-to-tissue and tissue-to-disease state variability in miRNA expression, make the adaptation of microarray technology to the evaluation of miRNA expression nontrivial. This chapter describes the unique features of miRNA microarray experiments and analysis and provides a case study demonstrating our approach to miRNA expression analysis.

Introduction

Micro-RNAs (miRNAs) are small (typically ~21 nucleotides), nonprotein coding RNAs transcribed from the genomes of plants and animals. These highly conserved molecules regulate gene expression by binding and modulating the translation of mRNAs that contain regions of at least

METHODS IN ENZYMOLOGY, VOL. 411
Copyright 2006, Elsevier Inc. All rights reserved.
0076-6879/06 $35.00
DOI: 10.1016/S0076-6879(06)11002-2

partial complementary to the miRNAs (for reviews, see Bartel, 2004; Pasquinelli *et al.*, 2005; Zamore and Haley, 2005). Since the discovery of the first miRNA in 1993 (Lee *et al.*, 1993; Wightman *et al.*, 1993), the list of validated miRNAs has been expanding rapidly. Over 330 human miRNA have been identified and more than 1000 sequences have been predicted to have miRNA activity (Berezikov *et al.*, 2005), with each miRNA potentially capable of regulating multiple genes (Lim *et al.*, 2003). Micro-RNAs are sequentially numbered and entered into the miRBase database (http://microrna.sanger.ac.uk; Ambros *et al.*, 2003; Griffiths-Jones *et al.*, 2006), which provides up-to-date sequence data (for both precursor and mature forms) and a variety of other information.

Although the presence and importance of miRNAs have only become apparent over the past several years, there is growing evidence that miRNAs play key roles in a variety of processes, including early development (Hornstein *et al.*, 2005; Lu *et al.*, 2005b; Reinhart *et al.*, 2000; Schulman *et al.*, 2005), cell proliferation and cell death (Brennecke *et al.*, 2003; Cheng *et al.*, 2005; Cimmino *et al.*, 2005), fat metabolism (Xu *et al.*, 2003), cell differentiation (Chang *et al.*, 2004; Dostie *et al.*, 2003; Yi *et al.*, 2006), and neuronal development (Krichevsky *et al.*, 2003; Smirnova *et al.*, 2005). A number of reviews give a comprehensive overview of the involvement of miRNAs in these processes (Alvarez-Garcia and Miska, 2005; Harfe, 2005; Klein *et al.*, 2005; Pasquinelli *et al.*, 2005; Rogaev, 2005; Wienholds and Plasterk, 2005).

Micro-RNAs Are Important Factors in Human Cancer

The apparent importance of miRNAs in regulating development and differentiation supports the idea that changes in miRNA expression and activity may contribute to oncogenesis. Indeed, several early reports correlated aberrant miRNA expression and miRNA gene sites with cancer. For example, Calin *et al.* (2002) found that more than half of the miRNAs (98 out of 186 known at the time) are in common break-point regions, fragile sites, minimal regions of loss of heterozygosity, and minimal regions of amplification known to be associated with cancer. Additionally, Michael *et al.* (2003) found that 2 of 28 miRNAs in the colorectal mucosa were downregulated significantly in 12 adenocarcinoma samples and 2 precancerous adenomatous polyps compared to matched, normal tissues. Since these early associations, a number of reports have strengthened the correlation between altered miRNA expression and various cancers (Calin *et al.*, 2004; Chan *et al.*, 2005; Ciafre *et al.*, 2005; Johnson *et al.*, 2005; Karube *et al.*, 2005; Metzler *et al.*, 2004; Takamizawa *et al.*, 2004). Further support for the importance of miRNAs in cancer comes from the identification or prediction of the genes regulated by miRNAs. For example, members of the let-7 family show reduced expression in lung cancer and are capable of repressing the translation of the oncogenic RAS protein in a cell culture model (Johnson et al., 2005). This study also reported that RAS protein is present at higher levels in lung tumors. In addition, Cimmino et al. (2005) showed that the expression of miR-15a and miR-16-1 was inversely correlated with Bcl2 (an antiapoptotic gene) expression in chronic lymphocytic leukemia (CLL). The repression of BCL2 by miR-15-a and miR16-1 was shown to induce apoptosis in a leukemia cell line model. Additionally, Volinia et al. (2006) reported that for miRNAs with altered expression in tumor samples the predicted and validated targets of these miRNAs are composed of a disproportionate number of oncogenes and tumor suppressors. These data and others suggest that miRNAs may be central factors in tumorigenesis and cancer progression.

Particularly promising for cancer diagnosis is evidence suggesting that miRNA expression profiles can be used to classify human cancers with more accuracy than mRNA expression profiles (Lu et al., 2005a). Furthermore, Calin et al. (2005) reported a panel of nine miRNAs whose high expression in CLL was strongly associated with a short interval from time of diagnosis to therapy, and low expression of those nine miRNAs was associated with a long interval from time of diagnosis to initial therapy. Thus, miRNAs may help improve the diagnosis and clinical management of cancers. A more comprehensive treatment of the roles of miRNAs in cancer and their potential as diagnostic and therapeutic markers can be found in a number of excellent review articles (e.g., Alvarez-Garcia and Miska, 2005; Croce and Calin, 2005; Esquela-Kerscher and Slack, 2006; Gregory and Shiekhattar, 2005; Hammond, 2006).

Focus of This Chapter

Reliable and accurate techniques for analyzing the global expression pattern of miRNA are critical for understanding their role in regulating gene expression and controlling both normal and disease processes. Several approaches are available for simultaneously profiling large numbers of miRNAs, including multiplex reverse transcription polymerase chain reaction (RT-PCR)-based analysis, bead-based flow cytometry assay, and microarrays.

Although microarrays have been in widespread use for some time, adapting this technology to study only a few hundred miRNA sequences is not trivial. The small size of miRNAs, averaging 21 nucleotides, provides very little sequence for labeling or designing probes. New methods have been developed for miRNA isolation, labeling, oligonucleotide probe design, data analysis, and array production. Many companies offer a wide range of miRNA services and products, including preprinted miRNA microarrays, full miRNA microarray services, miRNA probe sets, and kits for miRNA purification and labeling. Ambion, Inc. has an active program to develop innovative tools for the analysis of miRNA expression and function and offers a complete suite of miRNA-related products. To capitalize on the expertise acquired during the invention and development of these products, Ambion, Inc. began offering miRNA profiling services through its Ambion Services division (Asuragen. Inc., including Asuragen Discovery Services, was recently spun out of Ambion). This chapter provides an overview of microarray-based miRNA profiling with insights gained from our experience. This chapter focuses on expression analysis of miRNA using microarrays and describes the methods found to be optimal for performing and analyzing microarrays.

Microarray Platforms

One-Color vs Two-Color Arrays

The process of constructing an array has a major impact on the type of experimental design and analysis that can be performed. Practically, the various array formats can be classified based on whether they support one-color or two-color analysis. Most arrays for two-color analysis are produced by robotic spotting of oligonucleotide probes onto the array. Since this typically involves transferring the probes by liquid adherence to either single or arrayed pins, there is typically substantial variability between individual arrays in the amount of material spotted. As a result, individual arrays are not directly comparable without the use of a common reference sample. To perform two-color experiments with these custom-spotted microarrays, test and reference samples are each labeled with a specific fluorescent dye and cohybridized on the same array. The primary advantage of spotted arrays is that they can be custom produced in small batches for a moderate cost, facilitating changes to the probe set as new miRNAs are discovered.

As an alternative to spotted arrays, printing and fabrication technologies can be used to produce highly reproducible probe representation across many arrays and in replicate spots on each single array. These arrays permit single-color experiments and eliminate some of the analysis problems associated with the relative measurement of gene expression required by two-color cohybridization. Several companies provide custom and preprinted miRNA arrays for use with the most common microarray platforms.

miRNA Preparation for Analysis on Microarrays

Purification of miRNA

Because the quality of microarray gene expression analysis depends largely on the quality of the RNA used, robust and reproducible methods for the quantitative isolation of miRNA are essential. Many RNA isolation methods were originally designed to capture longer mRNA, whereas short RNA species were thought to contain only unimportant RNAs such as tRNA and degraded RNA fragments. However, isolation methods are now available that retain the small RNAs during total RNA isolation or enrich the small RNA population. Several companies market convenient kit formats that either enrich or retain miRNAs allowing downstream analysis.

We have found that further purification of the small RNA fraction improves the analysis of miRNA expression by reducing nonspecific hybridization to longer miRNA precursors, the homologous regions of target mRNAs, and other unrelated RNA species. A number of different methods exist for separating the small and large RNA fractions, including solid-phase extraction, microfiltration, and reverse-phase or ion-exchange chromatography. It has been our experience that these methods did not provide the levels of recovery, purity, and reproducibility necessary for high-quality analyses, so we routinely employ size fractionation using polyacrylamide gel electrophoresis (PAGE). Denaturing PAGE is being used routinely by laboratories that are isolating miRNAs from total RNA (Cummins et al., 2006; Elbashir et al., 2001; Johnson et al., 2005; Lu et al., 2005a); however, the procedure is time-consuming and the yield of miRNA is variable and rarely exceeds 50% (data not shown). Ambion, Inc. has developed a device (flashPAGE fractionator system) that can rapidly and reproducibly fractionate total RNA samples with an 80% yield of purified miRNA (Shingara et al., 2005).

Our preferred method for miRNA enrichment involves isolation of total RNA with the mirVana miRNA isolation kit (Ambion, Inc.) alone or combined with the flashPAGE fractionator system (Ambion, Inc.). Gel fractionation enriches the RNA population ranging between ~15 and 40 nucleotides in length (small RNA) approximately 10,000-fold. The relative abundance of flashPAGE-isolated miRNAs analyzed on a microarray has been shown to be representative when compared to miRNA expression in a total RNA sample analyzed by Northern blot (Shingara et al., 2005). The yield of small RNAs is typically less than 1 ng per 10 μg of total RNA from mammalian tissue, whereas immortalized cell lines typically display much lower general miRNA expression than tissues. Approximately 5 million cells can yield up to 10 μg of total RNA, a sufficient amount for miRNA array analysis.

Labeling of miRNA

After isolation of the small RNAs, they must be labeled for detection on arrays. A number of strategies have been employed for labeling miRNAs, including direct labeling, random priming, and PCR-based amplification (Babak *et al.*, 2004; Barad *et al.*, 2004; Baskerville and Bartel, 2005; Liu *et al.*, 2004; Miska *et al.*, 2004; Sun *et al.*, 2004; Tang *et al.*, 2006; Thomson *et al.*, 2004). PCR-based global amplification can increase sensitivity, but requires a number of enzymatic steps and may introduce amplification bias in the miRNA representation in a sample. Random priming is problematic because the small sequence space of miRNAs can result in nonrandom labeling using a random priming strategy. Thus, we prefer direct labeling as the simplest and most representative labeling method. A number of commercial kits are available that utilize a variety of direct labeling methods.

The most popular direct labeling methods use a tailing approach in which short, labeled sequences are enzymatically attached to the ends of the miRNAs. We use a direct labeling procedure (based on Ambion's *mir*Vana miRNA labeling kit) in which poly(A) polymerase (PAP) is used to append a mixture of unmodified and amine-modified nucleotides to the 3′ end of the miRNA (Shingara *et al.*, 2005). The tailed miRNA population is subsequently labeled with amine-reactive reagents, including fluorescent dyes, for example, Cy or Alexa dyes for direct detection, or NHS-biotin for detection with streptavidin coupled to fluorescent moieties. This method for homogeneous labeling of the miRNA fraction provides the highest specific activity without introducing bias. Labeling the mature miRNA population using this method conserves miRNA representation and allows accurate profiling at lower sample input. Experiments in which known amounts of miRNA were spiked into RNA samples indicate that this procedure permits detection of 10 pg ($\sim$3 fmol) of miRNA in 10 μg of total RNA (Shingara *et al.*, 2005). This represents as little as 0.1% of the overall miRNA population in a 10-μg total RNA sample.

Analysis of miRNA Microarray Results

Differences between miRNA and mRNA Expression Profiling

This section discusses some of the pitfalls associated with global normalization and array scaling specifically in the context of miRNA microarray studies and highlights the normalization approaches that have been prominent in the miRNA microarray literature. There are currently no published studies that address the impact of normalization methods on analysis of miRNA microarray expression profiling experiments. As a result, researchers utilizing miRNA microarray technologies have relied on normalization and scaling methods developed specifically for mRNA microarray technologies. We conclude with a case study comparing the miRNA expression profiles in human lung and placental tissues, focusing on signal quantification, normalization, power analysis, and differential expression. For convenience, we use the term "normalization" to refer to any scaling, translation, or other numerical transformation (excluding thresholding and filtering) as it pertains to expression data from a single array or set of arrays.

The primary downside to miRNA analysis is that the fundamental assumption that the same amount of miRNA is extracted from a given amount of total RNA may not be valid. There is presently no way to quantify the total amount of miRNA in a sample given that miRNA typically constitutes 0.01% of the total RNA. The development of spike-in controls to measure the capture or loss of signal during each step of the sample preparation process (including total RNA isolation, tailing, column cleaning, and fractionation of miRNA) can address process-related losses of miRNA abundance, but does not account for different proportions of miRNA within the pool of total RNA. This may have substantial consequences in the validity of the analysis if, for instance, different tissues or different disease states have variable proportions of miRNA content in a tissue- and disease-state dependent manner.

The consequences of having an unknown relative quantity of miRNA loaded onto an array are confounded by the fact that present technologies and platforms do not measure every miRNA in a sample (as only a fraction of all miRNA are known). Thus, even if it were possible to measure the overall amount of total miRNA in a sample before loading it onto the array, we know that we are not observing every miRNA, and consequently the fraction of observed miRNA relative to the total loaded miRNA is also unknown. Analytically, these issues cast doubt on forms of normalization that are based on total, mean, and median signal intensities. In addition, the use of high-density normalization approaches such as quantile normalization (Bolstad et al., 2003; Irizarry et al., 2003) are rendered ineffective when there are relatively few probes on the array.

Prior to normalization, various levels of data preprocessing are typically implemented, including averaging of replicate spots on the same array and background subtraction. These steps can have an impact on both normalization and statistical tests performed on data. Background subtraction often yields signals with negative values that are subsequently eliminated or set to a common positive threshold to allow for logarithmic transformation. Depending on the overlap between the distributions of miRNA in the sample and miRNA targets on the array platform, thresholding can result in a significant proportion (>50%) of signals being altered prior to normalization. This can have a profound effect on subsequent statistical tests, as an excess of threshold values will reduce (and potentially eliminate) variation artificially, thus potentially resulting in artificially high significance in comparative tests such as ANOVA and t tests. Furthermore, excessive thresholding signals at low values can introduce a bias that imparts an artificially inflated measure of sensitivity of the array platform.

Thresholding and data elimination impact normalizations that depend on calculating measures of central tendency such as mean, median, and mode, further confounding the lack of knowledge about relative proportions of the miRNA signal in the original sample. A hyperthresholded data set will inflate the mean and invariably have a mode (and potentially a median if >50% of the values are threshold) equal to the threshold value. Normalizing to an erroneous median jeopardizes the validity of calibrating array-to-array signal intensities. Variation in quantities of eliminated data between arrays also imparts a disparity in the definition of median with similar consequences. For example, the median of expression value derived from arrays with 20 and 40% eliminated data are equivalent to the 60 and 70th percentile of the original data set, respectively.

Finally, one of the fundamental assumptions of most normalization approaches developed for mRNA microarray technologies is that either very few genes are changing significantly between sample conditions or that there are approximately an equal number of genes with increased and decreased relative expression between sample conditions. In mRNA microarray experiments, the presence of a large number of genes with unchanging expression acts to stabilize the normalization algorithms (either explicitly or implicitly). For miRNA, this assumption breaks down on two fronts. First, we have no reason to believe that the diverse biological systems studied will have few or balanced numbers of miRNA changing in relative expression. In fact, miRNA studies of cancer (Calin et al., 2005; Lu et al., 2005a; Volinia et al., 2006; Yanaihara et al., 2006) and tissue differentiation (Babak et al., 2004; Barad et al., 2004; Garzon et al., 2006) suggest that a significant proportion of miRNA are significantly differentially expressed (up to 40%) between sample conditions. Second, with the small number of miRNA observed on present microarray platforms, there is no means for normalization techniques to be stabilized by large numbers of miRNA with unchanging expression profiles.

In order to determine the best methods for addressing these problems, we have tested over 400 permutations of background subtraction, scaling, and global normalization algorithms on designed experiments in which we know the "truth" about which miRNA are at different quantities. These experiments include a 10 × 10 Latin-square spike-in study on multiple tissue backgrounds, a dilution spike-in study, and a mixture model study (manuscript in preparation). With this combination of algorithms and experiments we have been able to quantify which method performs the best in the context of identifying differentially expressed miRNA with a lesser focus on absolute expression and fold-change differences (i.e., focusing on precision instead of accuracy). We made this decision because the confounding issues of absolute amount of initial miRNA, incomplete miRNA array coverage, and variable sample loss make accuracy a poor measure of performance for comparative analysis. To complement this approach, we are in the process of implementing alternative statistical tests to identify differentially expressed miRNA that are based on robust nonparametric approaches. These could be used alone or in addition to present methods to validate analytical processes further.

Methods of Normalization for miRNA Microarray Experiments

It should be emphasized that normalization is the process of removing nonbiological sources of variation between array experiments. In general, the choice of normalization methodologies for microarray experiments is known to have a profound impact on accuracy, precision, and overfitting (Argyropoulos et al., 2006). Consequently, downstream tests for differential expression, the development of classifiers, and data mining are highly dependent on the choice of data processing. Even in the well-developed field of mRNA microarray analysis, the appropriate choice of normalization methods is still a matter of debate (for reviews, see Allison et al., 2006; Quackenbush, 2002). This issue is further compounded in the field of miRNA microarray analysis by the fact that the amount of miRNA derived from biological samples is neither quantitative nor known relative to the original abundance of total RNA, as discussed in the previous section. To circumvent these issues, researchers have adopted "tried-and-true" methodologies from mRNA microarray expression analysis in hopes that the assumptions and approximations are not violated. These approaches included median scaling (both global and/or chip specific), scaling to spiked-in controls, and logarithmic or other variance-stabilizing normalizations.

The most popular miRNA microarray normalization approach involves the preprocessing of data by averaging of technical replicates and background subtraction followed by median normalization and logarithmic transformation. This is the approach taken by Garzon et al. (2006), Volinia et al. (2006), Yanaihara et al. (2006), and Calin et al. (2005). In their study profiling miRNA expression in human tissues, Barad et al. (2004) took a similar approach except that postnormalization thresholding based on negative controls was imposed.

A different approach was taken by Lu *et al.* (2005a) in their comprehensive assessment of miRNA expression profiles in human cancers across both array- and bead-based platforms. They incorporated a two-stage scaling strategy based on spiked-in controls that was intended to normalize between platforms and then account for differences in labeling efficiency. Data were subsequently thresholded and $\log_2$ transformed in order to both stabilize variance and place expression data in the familiar log space where absolute differences in expression correspond to fold changes.

Babak *et al.* (2004) took a novel approach to normalization of miRNA microarray data by first spatially detrending correlations between spot intensity and position via high-pass filtering (Shai *et al.*, 2003). This was followed by global transformation using variance stabilizing normalization (VSN) (Huber *et al.*, 2002). The VSN transformation is derived from the appropriate assumption of a quadratic relationship between mean expression and variance of microarray data. The resulting intensity transformation is a hyperbolic-arcsine (arcsinh) (denoted by h) that replaces the more traditional logarithmic transformation. This has a number of advantages: because the arsinh function is continuous across zero, there is no need to either establish (often arbitrary) thresholds or instigate data elimination for low-intensity signals. Furthermore, the VSN algorithm includes a calibration step to bring multiple array intensities into register for comparative analysis. h-transformed data have the property that at large intensity, differences in h (Δh) coincide with log-ratio values. However, at lower intensity, Δh values are contracted toward zero, departing from the log-ratio value. This imparts a minor shrinkage of values in comparison to $\log_2$-transformed values, thus effectively sacrificing accuracy for precision, which may be considered beneficial given the lack of knowledge in quantification of the total miRNA content in a sample. This trade-off is used commonly in mRNA array normalization approaches (Bolstad *et al.*, 2003). When using the VSN transformation on any microarray data set, one must relinquish the natural definition of fold change across the entire range of expression values and focus on statistical significance for determining differential expression.

Finally, in light of the variable effects of normalization on miRNA microarray data, we would like to emphasize the need to corroborate conclusions derived from miRNA microarray experiments with follow-up quantitative assays such as qRT-PCR and/or Northern blot experiments.

Case Study: miRNA Microarray Expression Analysis of Human Lung and Placental Tissues

A typical miRNA microarray experiment performed at Asuragen Discovery Services is described. Five samples from human lung and placenta were selected for comparison. Both lung and placenta have been shown to be highly enriched for miRNA (Barad *et al.*, 2004; Slack and Weidhaas, 2006). Ambion's *mir*Vana miRNA bioarrays (Ambion, Inc., Austin, TX) were used exclusively throughout this study. These bioarrays contain probes targeting a comprehensive selection of human, mouse, and rat miRNA from miRBase (Griffiths-Jones *et al.*, 2006) and a set of novel, proprietary putative miRNAs.

Sample preparation included quality assessment of the tissue-specific total RNA with subsequent miRNA enrichment using the Ambion, Inc. flashPAGE Fractionator and Reaction Clean-up Kit (Ambion, Inc). The miRNA targets were labeled using the *mir*Vana miRNA Labeling Kit (Ambion, Inc.). Bioarrays were processed using the hybridization and wash components supplied in the *mir*Vana miRNA Bioarray Essentials Kit.

Scanning and Data Extraction

All bioarrays were scanned with a Molecular Devices GenePix auto-loader 4200AL microarray scanner. All scans were performed at 5 μm resolution in order to achieve high-quality data. Scanner settings are optimized to achieve an appropriate balance of sensitivity and dynamic range for each array.

Sample Size Calculation

miRNA array analysis involves the generation and recovery of data from hundreds of features (miRNA probes) from multiple samples. Gleaning meaningful and robust statistical information from such experiments requires the use of multiple biological replicates. Calculation of the correct number of biological replicates for miRNA array studies is very complicated; each gene has its own associated measurement, noise, and subsequent standard deviation. Thus sample size decision is a balancing act between the power of the statistical test, differences you are looking for, and economics of experimental size. Using a custom-developed MATLAB script (The Mathworks, Inc.) based on the R library-ssize (http://www. bioconductor.org/repository/devel/vignette/ssize.pdf), we computed the required number of samples and estimated the power of the experiment given a fixed sample size and expected fold change. The key component of this method is the generation of cumulative plots of the percentage of miRNA achieving a desired power as a function of sample size or fold change. Figure 1A and B show the predicted levels of detection for these tissues and the number of biological replicates that are necessary.

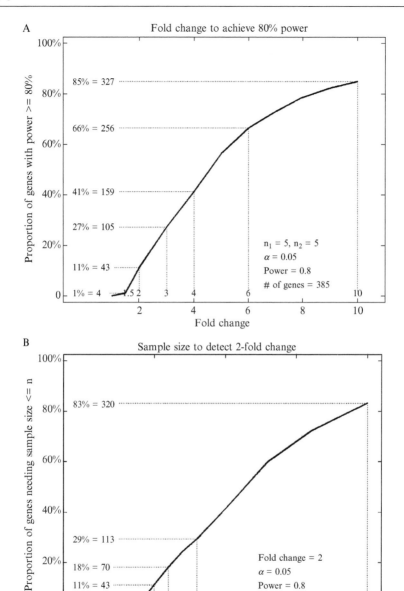

FIG. 1. Power analysis. Two types of plots are generated for power analysis. (A) The fold difference can be determined statistically at a given power and replication level. (B) The effect of varying the number of replicates and the resulting number of miRNA that can be expected to find dissimilar by twofold and at a power of 0.8. (See color insert.)

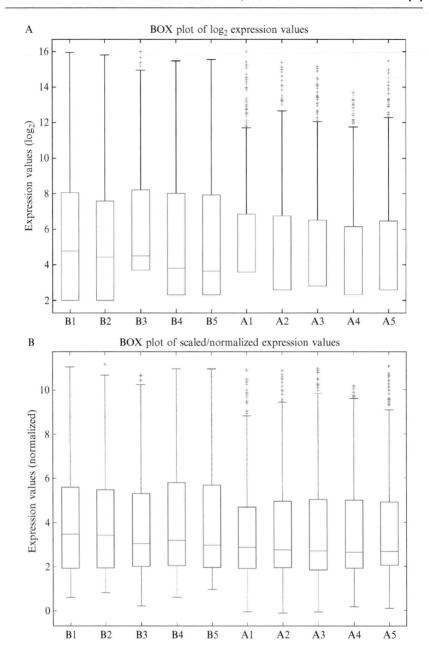

A — BOX plot of log$_2$ expression values

B — BOX plot of scaled/normalized expression values

Calculation of Array-Specific Thresholds

For each array, the minimum observable threshold is determined by examining the foreground minus background median intensities for "EMPTY" spots. We define this minimum threshold as the 5% symmetric trimmed mean plus 2 standard deviations across all "EMPTY" spots on the array. Spots with an averaged foreground minus median background intensity less than this threshold are considered "absent."

Global Normalization

Asuragen Discovery Services has validated that a global VSN procedure is best for most experimental designs. Figure 2A and B illustrate the before and after consequences of VSN normalization.

Statistical Differential Analysis

Once the miRNA are quantified and the signal values are normalized, a hypothesis test was applied to the tissue groups. Using a two-sample t test it was determined whether there are differentially expressed miRNA across groups. A two-sample t test is carried out for every gene and multiplicity correction is followed to control the false discovery rate (FDR) at 0.05. Given the nature of data and the statistical test selected, it is important to adjust for multiple testing errors. Reporting uncorrected p values for each gene over a certain threshold can be deceptive. These problems arise, for example, when 500 t tests are performed during the course of a typical two-condition array study. Using the traditional $p = 0.05$ without multiple testing correction could potentially lead to 25 miRNA being classified as significantly different when, in fact, they are not. To account for this, a FDR p value adjustment is used (Benjamini and Hochberg, 1995). The FDR is defined to be the expected value of the ratio of the number of erroneously rejected true hypotheses over the number of rejected hypotheses. The Benjamini and Hochberg step-up procedure rejects H (1)... H(k) with k being the largest i for which $P(i) <= q*i/m$, and this procedure controls the FDR at level q when $P(i)$ are independent. FDR corrections

FIG. 2. Box-whisker plots: The distribution of expression values for lung (replicates A1–A5) and placenta (replicates B1–B5) samples. The ends of the box represent the 25 and 75th percentiles, and the red line bisecting each box is the median signal for each array. (A) After thresholding and $\log_2$ transformation. (B) After VSN transformation. (See color insert.)

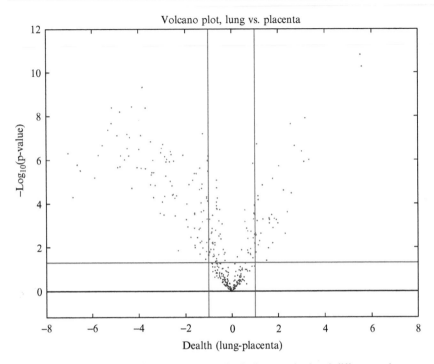

Fig. 3. Volcano plots. This plot represents both the magnitude of differences between groups and the statistical significant of those differences. To delineate the degree of change between groups there are two vertical red lines at Δh (A-B) = ±1. There is a single horizontal red line at a negative $\log_{10}$ (p value) = 1.3, which corresponds to an unadjusted p value of 0.05. miRNA above this line can be considered statistically significant at a p value of 0.05 in the absence of correction for multiple testing. miRNA found statistically significant after 5% FDR correction are color coded red instead of blue. (See color insert.)

allow for the creation of statistically reliable gene lists with a reliable (and controllable) estimate of the number of false positives. After FDR correction, 192 miRNA were shown to be expressed differentially between human lung and placenta tissues. The differences between tissues can best be represented using a volcano plot (Fig. 3).

Hierarchical Clustering

In addition to statistical tests, unsupervised clustering is often used to discern patterns within the results. This analysis calculates a measure of similarity between each point in a cluster and all the points in a neighboring

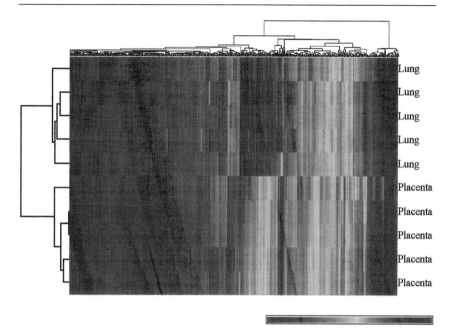

FIG. 4. Hierarchical clustering. Data are represented as a dendrogram or tree graph with the closest branches of the tree representing genes with similar gene expression patterns. The most common implementation is the agglomerative hierarchical clustering, which starts with a family of clusters with one sample each and merges the clusters iteratively, based on some distance measure, until there is only one cluster left. In this example, Euclidean distance is used as the distance metric. Array and/or sample qualities can be approximated using hierarchical clustering. Ideally, common samples should cluster into similar classes. (See color insert.)

cluster. The two clusters that are closest to each other (i.e., are most alike) are connected to form the higher order cluster. Data are represented as a dendrogram or tree graph with the closest branches of the tree representing genes with similar gene expression patterns. Ideally, common samples should cluster into similar classes. Figure 4 shows the cluster gram for lung vs placenta tissues; both tissues clustered together as expected, with red and green representing high and low signal intensities, respectively.

Conclusion

The field of miRNA analysis in many ways is still in its infancy as are the types of biological questions that are being addressed with this technology. Nevertheless, there is a great deal of interest in conducting miRNA tissue profiling. The levels of miRNAs have been shown to be altered significantly in various cancers and disease states, necessitating the use of miRNA profiling to discover signatures or biomarkers. Such discovery has great potential to identify candidates that can be used in developing therapeutics and diagnostics. There is also considerable interest from a mechanistic perspective in gaining a greater understanding of how miRNAs impact diverse biological processes in both normal and disease conditions. Heretofore, systems biology only focused on DNA, large RNA, proteins, and metabolites, but appreciation of the impact of miRNAs is adding a new dimension to the field. At the forefront of such discovery is the need for new statistical tools and methodologies to identify differentially expressed miRNAs and to identify miRNA signatures, which, together with the appropriate algorithms, will facilitate the diagnosis of disease, patient stratification, and the development of improved cancer therapies.

Acknowledgments

The authors thank Drs. Michael Wilson and Edward Sekinger at Asuragen Discovery Services for stimulating discussion and constructive comments on this manuscript.

References

Allison, D. B., Cui, X., Page, G. P., and Sabripour, M. (2006). Microarray data analysis: From disarray to consolidation and consensus. *Nature Rev. Genet.* **7**(1), 55–65.

Alvarez-Garcia, I., and Miska, E. A. (2005). MicroRNA functions in animal development and human disease. *Development* **132**, 4653–4662.

Ambros, V., Bartel, B., Bartel, D. P., Burge, C. B., Carrington, J. C., Chen, X., Dreyfuss, G., Eddy, S. R., Griffiths-Jones, S., Marshall, M., Matzke, M., Ruvkun, G., and Tuschl, T. (2003). A uniform system for microRNA annotation. *RNA* **9**, 277–279.

Argyropoulos, C., Chatziioannou, A. A., Nikiforidis, G., Moustakas, A., Kollias, G., and Aidinis, V. (2006). Operational criteria for selecting a cDNA microarray data normalization algorithm. *Oncol. Rep.* **15**(4), 983–996.

Babak, T., Zhang, W., Morris, Q., Blencowe, B. J., and Hughes, T. R. (2004). Probing microRNAs with microarrays: Tissue specificity and functional inference. *RNA* **10**, 1813–1819.

Barad, O., Meiri, E., Avniel, A., Aharonov, R., Barzilai, A., Bentwich, I., Einav, U., Gilad, S., Hurban, P., Karov, Y., Lobenhofer, E. K., Sharon, E., Shiboleth, Y. M., Shtutman, M., Bentwich, Z., and Einat, P. (2004). MicroRNA expression detected by oligonucleotide microarrays: System establishment and expression profiling in human tissues. *Genome Res.* **14**, 2486–2494.

Bartel, D. P. (2004). MicroRNAs: Genomics, biogenesis, mechanism, and function. *Cell* **116**, 281.

Baskerville, S., and Bartel, D. P. (2005). Microarray profiling of microRNAs reveals frequent coexpression with neighboring miRNAs and host genes. *RNA* **11**, 241–247.

Benjamini, Y., and Hochberg, Y. (1995). Controlling the false discovery rate: A practical and powerful approach to multiple testing. *J. R. Stat. Soc. Ser. B* **57**, 289–300.

Berezikov, E., Guryev, V., van de Belt, J., Wienholds, E., Plasterk, R. H. A., and Cuppen, E. (2005). Phylogenetic shadowing and computational identification of human microRNA genes. *Cell* **120**, 21.

Bolstad, B. M., Irizarry, R. A., Astrand, M., and Speed, T. P. (2003). A comparison of normalization methods for high density oligonucleotide array data based on bias and variance. *Bioinformatics* **19**(2), 185–193.

Brennecke, J., Hipfner, D. R., Stark, A., Russell, R. B., and Cohen, S. M. (2003). Bantam encodes a developmentally regulated microRNA that controls cell proliferation and regulates the proapoptotic gene hid in Drosophila. *Cell* **113**, 25–36.

Calin, G. A., Dumitru, C. D., Shimizu, M., Bichi, R., Zupo, S., Noch, E., Aldler, H., Rattan, S., Keating, M., Rai, K., Rassenti, L., Kipps, T., Negrini, M., Bullrich, F., and Croce, C. M. (2002). Frequent deletions and down-regulation of micro-RNA genes miR15 and miR16 at 13q14 in chronic lymphocytic leukemia. *Proc. Natl. Acad. Sci. USA* **99**, 15524–15529.

Calin, G. A., Ferracin, M., Cimmino, A., Di Leva, G., Shimizu, M., Wojcik, S. E., Iorio, M. V., Visone, R., Sever, N. I., Fabbri, M., Iuliano, R., Palumbo, T., Pichiorri, F., Roldo, C., Garzon, R., Sevignani, C., Rassenti, L., Alder, H., Volinia, S., Liu, C. G., Kipps, T. J., Negrini, M., and Croce, C. M. (2005). A MicroRNA signature associated with prognosis and progression in chronic lymphocytic leukemia. *N. Engl. J. Med.* **353**, 1793–1801.

Calin, G. A., Liu, C. G., Sevignani, C., Ferracin, M., Felli, N., Dumitru, C. D., Shimizu, M., Cimmino, A., Zupo, S., Dono, M., Dell'Aquila, M. L., Alder, H., Rassenti, L., Kipps, T. J., Bullrich, F., Negrini, M., and Croce, C. M. (2004). MicroRNA profiling reveals distinct signatures in B cell chronic lymphocytic leukemias. *Proc. Natl. Acad. Sci. USA* **101**, 11755–11760.

Chan, J. A., Krichevsky, A. M., and Kosik, K. S. (2005). MicroRNA-21 is an antiapoptotic factor in human glioblastoma cells. *Cancer Res.* **65**, 6029–6033.

Chang, S., Johnston, R. J., Jr., Frokjaer-Jensen, C., Lockery, S., and Hobert, O. (2004). MicroRNAs act sequentially and asymmetrically to control chemosensory laterality in the nematode. *Nature* **430**, 785–789.

Cheng, A. M., Byrom, M. W., Shelton, J., and Ford, L. P. (2005). Antisense inhibition of human miRNAs and indications for an involvement of miRNA in cell growth and apoptosis. *Nucleic Acids Res.* **33**, 1290–1297.

Ciafre, S. A., Galardi, S., Mangiola, A., Ferracin, M., Liu, C. G., Sabatino, G., Negrini, M., Maira, G., Croce, C. M., and Farace, M. G. (2005). Extensive modulation of a set of microRNAs in primary glioblastoma. *Biochem. Biophys. Res. Commun.* **334**, 1351–1358.

Cimmino, A., Calin, G. A., Fabbri, M., Iorio, M. V., Ferracin, M., Shimizu, M., Wojcik, S. E., Aqeilan, R. I., Zupo, S., Dono, M., Rassenti, L., Alder, H., Volinia, S., Liu, C. G., Kipps, T. J., Negrini, M., and Croce, C. M. (2005). miR-15 and miR-16 induce apoptosis by targeting BCL2. *Proc. Natl. Acad. Sci. USA* **102**, 13944–13949.

Croce, C. M., and Calin, G. A. (2005). miRNAs, cancer, and stem cell division. *Cell* **122**, 6–7.

Cummins, J. M., He, Y., Leary, R. J., Pagliarini, R., Diaz, L. A., Jr., Sjoblom, T., Barad, O., Bentwich, Z., Szafranska, A. E., Labourier, E., Raymond, C. K., Roberts, B. S., Juhl, H., Kinzler, K. W., Vogelstein, B., and Velculescu, V. E. (2006). The colorectal microRNAome. *Proc. Natl. Acad. Sci. USA* **103**(10), 3687–3692.

Dostie, J., Mourelatos, Z., Yang, M., Sharma, A., and Dreyfuss, G. (2003). Numerous microRNPs in neuronal cells containing novel microRNAs. *RNA* **9**, 180–186.

Elbashir, S. M., Lendeckel, W., and Tuschl, T. (2001). RNA interference is mediated by 21- and 22-nucleotide RNAs. *Genes Dev.* **15,** 188–200.

Esquela-Kerscher, A., and Slack, F. J. (2006). Oncomirs: MicroRNAs with a role in cancer. *Nature Rev. Cancer* **6,** 259–269.

Garzon, R., Pichiorri, F., Palumbo, T., Iuliano, R., Cimmino, A., Aqeilan, R., Volinia, S., Bhatt, D., Alder, H., Marcucci, G., Calin, G. A., Liu, C. G., Bloomfield, C. D., Andreeff, M., and Croce, C. M. (2006). MiRNA fingerprints during human megakaryocytopoiesis. *Proc. Natl. Acad. Sci. USA* **101,** 5078–5083.

Gregory, R. I., and Shiekhattar, R. (2005). MicroRNA biogenesis and cancer. *Cancer Res.* **65,** 3509–3512.

Griffiths-Jones, S., Grocock, R. J., van Dongen, S., Bateman, A., and Enright, A. J. (2006). miRBase: MicroRNA sequences, targets and gene nomenclature. *Nucleic Acids Res.* **34,** D140–D144.

Hammond, S. M. (2006). MicroRNAs as oncogenes. *Curr. Opin. Genet. Dev.* **16,** 4–9.

Harfe, B. D. (2005). MicroRNAs in vertebrate development. *Curr. Opin. Genet. Dev.* **15,** 410–415.

Hornstein, E., Mansfield, J. H., Yekta, S., Hu, J. K., Harfe, B. D., McManus, M. T., Baskerville, S., Bartel, D. P., and Tabin, C. J. (2005). The microRNA miR-196 acts upstream of Hoxb8 and Shh in limb development. *Nature* **438,** 671–674.

Huber, W., Von Heydebreck, A., Sultmann, H., Poustka, A., and Vingron, M. (2002). Variance stabilization applied to microarray data calibration and to the quantification of differential expression. *Bioinformatics* **18**(Suppl. 1), S96–S104.

Irizarry, R. A., Bolstad, B. M., Collin, F., Cope, L. M., Hobbs, B., and Speed, T. P. (2003). Summaries of Affymetrix GeneChip probe level data. *Nucleic Acids Res.* **31**(4), e15.

Johnson, S. M., Grosshans, H., Shingara, J., Byrom, M., Jarvis, R., Cheng, A., Labourier, E., Reinert, K. L., Brown, D., and Slack, F. J. (2005). RAS is Regulated by the let-7 microRNA Family. *Cell* **120,** 635.

Karube, Y., Tanaka, H., Osada, H., Tomida, S., Tatematsu, Y., Yanagisawa, K., Yatabe, Y., Takamizawa, J., Miyoshi, S., Mitsudomi, T., and Takahashi, T. (2005). Reduced expression of Dicer associated with poor prognosis in lung cancer patients. *Cancer Sci.* **96,** 111–115.

Klein, M. E., Impey, S., and Goodman, R. H. (2005). Role reversal: The regulation of neuronal gene expression by microRNAs. *Curr. Opin. Neurobiol.* **15,** 507–513.

Krichevsky, A. M., King, K. S., Donahue, C. P., Khrapko, K., and Kosik, K. S. (2003). A microRNA array reveals extensive regulation of microRNAs during brain development. *RNA* **9,** 1274–1281.

Lee, R. C., Feinbaum, R. L., and Ambros, V. (1993). The *C. elegans* heterochronic gene lin-4 encodes small RNAs with antisense complementarity to lin-14. *Cell* **75,** 843–854.

Lim, L. P., Glasner, M. E., Yekta, S., Burge, C. B., and Bartel, D. P. (2003). Vertebrate microRNA genes. *Science* **299,** 1540.

Liu, C. G., Calin, G. A., Meloon, B., Gamliel, N., Sevignani, C., Ferracin, M., Dumitru, C. D., Shimizu, M., Zupo, S., Dono, M., Alder, H., Bullrich, F., Negrini, M., and Croce, C. M. (2004). An oligonucleotide microchip for genome-wide microRNA profiling in human and mouse tissues. *Proc. Natl. Acad. Sci. USA* **101,** 9740–9744.

Lu, J., Getz, G., Miska, E. A., Alvarez-Saavedra, E., Lamb, J., Peck, D., Sweet-Cordero, A., Ebert, B. L., Mak, R. H., Ferrando, A. A., Downing, J. R., Jacks, T., Horvitz, H. R., and Golub, T. R. (2005a). MicroRNA expression profiles classify human cancers. *Nature* **435,** 834.

Lu, J., Qian, J., Chen, F., Tang, X., Li, C., and Cardoso, W. V. (2005b). Differential expression of components of the microRNA machinery during mouse organogenesis. *Biochem. Biophys. Res. Commun.* **334,** 319–323.

Metzler, M., Wilda, M., Busch, K., Viehmann, S., and Borkhardt, A. (2004). High expression of precursor microRNA-155/BIC RNA in children with Burkitt lymphoma. *Genes Chromosomes Cancer* **39,** 167–169.

Michael, M. Z., O'Connor, S. M., van Holst Pellekaan, N. G., Young, G. P., and James, R. J. (2003). Reduced accumulation of specific microRNAs in colorectal neoplasia. *Mol. Cancer Res.* **1,** 882–891.

Miska, E. A., Alvarez-Saavedra, E., Townsend, M., Yoshii, A., Sestan, N., Rakic, P., Constantine-Paton, M., and Horvitz, H. R. (2004). Microarray analysis of microRNA expression in the developing mammalian brain. *Genome Biol.* **5,** R68.

Pasquinelli, A. E., Hunter, S., and Bracht, J. (2005). MicroRNAs: A developing story. *Curr. Opin. Genet. Dev.* **15,** 200–205.

Quackenbush, J. (2002). Microarray data normalization and transformation. *Nature Genet.* **32**(Suppl.), 496–501.

Reinhart, B. J., Slack, F. J., Basson, M., Pasquinelli, A. E., Bettinger, J. C., Rougvie, A. E., Horvitz, H. R., and Ruvkun, G. (2000). The 21-nucleotide let-7 RNA regulates developmental timing in *Caenorhabditis elegans*. *Nature* **403,** 901–906.

Rogaev, E. I. (2005). Small RNAs in human brain development and disorders. *Biochemistry* **70,** 1404–1407.

Schulman, B. R., Esquela-Kerscher, A., and Slack, F. J. (2005). Reciprocal expression of lin-41 and the microRNAs let-7 and mir-125 during mouse embryogenesis. *Dev. Dyn.* **234,** 1046–1054.

Shai, O., Morris, Q., and Frey, B. J. (2003). Spatial bias removal in microarray images. University of Toronto Technical Report PSI-2003–21.

Shingara, J., Keiger, K., Shelton, J., Laosinchai-Wolf, W., Powers, P., Conrad, R., Brown, D., and Labourier, E. (2005). An optimized isolation and labeling platform for accurate microRNA expression profiling. *RNA* **11,** 1461–1470.

Slack, F. J., and Weidhaas, J. B. (2006). MicroRNAs as a potential magic bullet in cancer. *Fut. Oncol.* **2,** 73–82.

Smirnova, L., Grafe, A., Seiler, A., Schumacher, S., Nitsch, R., and Wulczyn, F. G. (2005). Regulation of miRNA expression during neural cell specification. *Eur. J. Neurosci.* **21,** 1469–1477.

Sun, Y., Koo, S., White, N., Peralta, E., Esau, C., Dean, N. M., and Perera, R. J. (2004). Development of a micro-array to detect human and mouse microRNAs and characterization of expression in human organs. *Nucleic Acids Res.* **32,** e188.

Takamizawa, J., Konishi, H., Yanagisawa, K., Tomida, S., Osada, H., Endoh, H., Harano, T., Yatabe, Y., Nagino, M., Nimura, Y., Mitsudomi, T., and Takahashi, T. (2004). Reduced expression of the let-7 microRNAs in human lung cancers in association with shortened postoperative survival. *Cancer Res.* **64,** 3753–3756.

Tang, F., Hajkova, P., Barton, S. C., Lao, K., and Surani, M. A. (2006). MicroRNA expression profiling of single whole embryonic stem cells. *Nucleic Acids Res.* **34,** e9.

Thomson, J. M., Parker, J., Perou, C. M., and Hammond, S. M. (2004). A custom microarray platform for analysis of microRNA gene expression. *Nature Methods* **1,** 47–53.

Volinia, S., Calin, G. A., Liu, C.-G., Ambs, S., Cimmino, A., Petrocca, F., Visone, R., Iorio, M., Roldo, C., Ferracin, M., Prueitt, R. L., Yanaihara, N., Lanza, G., Scarpa, A., Vecchione, A., Negrini, M., Harris, C. C., and Croce, C. M. (2006). A microRNA expression signature of human solid tumors defines cancer gene targets. *Proc. Natl. Acad. Sci. USA* **103**(7), 2257–2261.

Wienholds, E., and Plasterk, R. H. (2005). MicroRNA function in animal development. *FEBS Lett.* **579,** 5911–5922.

Wightman, B., Ha, I., and Ruvkun, G. (1993). Posttranscriptional regulation of the heterochronic gene lin-14 by lin-4 mediates temporal pattern formation in *C. elegans*. *Cell* **75,** 855–862.

Xu, P., Vernooy, S. Y., Guo, M., and Hay, B. A. (2003). The Drosophila microRNA Mir-14 suppresses cell death and is required for normal fat metabolism. *Curr. Biol.* **13,** 790–795.

Yanaihara, N., Caplen, N., Bowman, E., Seike, M., Kumamoto, K., Yi, M., Stephens, R. M., Okamoto, A., Yokota, J., Tanaka, T., Calin, G. A., Liu, C. G., Croce, C. M., and Harris, C. C. (2006). Unique miRNA molecular profiles in lung cancer diagnosis and prognosis. *Cancer Cell.* **9**(3), 189–198.

Yi, R., O'Carroll, D., Pasolli, H. A., Zhang, Z., Dietrich, F. S., Tarakhovsky, A., and Fuchs, E. (2006). Morphogenesis in skin is governed by discrete sets of differentially expressed microRNAs. *Nature Genet.* **38,** 356–362.

Zamore, P. D., and Haley, B. (2005). Ribo-gnome: The big world of small RNAs. *Science* **309,** 1519–1524.

[3] Troubleshooting Microarray Hybridizations

By BRIAN EADS, AMY CASH, KEVIN BOGART,
JAMES COSTELLO, and JUSTEN ANDREWS

Abstract

Microarray experiments are being performed more widely than ever before, but even seasoned investigators can experience technical problems with hybridizations. This chapter provides guidelines for recognizing, rectifying, and avoiding common trouble areas. Specifically, it addresses frequent complications related to artifacts of printing, RNA sample preparation and quality, fluorophore labeling, hybridization conditions, and posthybridization washes. Emphasis is placed on investigating problems though a combination of appropriate controls and image analysis, where diagnostic plots of data quality are used to illustrate characteristics of acceptable and unsatisfactory hybridizations. This chapter also discusses resources available to microarray users hoping to improve the sensitivity and specificity of their experiments.

Introduction

The experimental power of a microarray assay stems from a miniaturized format that allows for massively parallel analyses on a transcriptome or genome-wide scale (Schena *et al.*, 1995). This miniaturization, however, is a double-edged sword, as it also gives rise to a unique set of technical challenges. Hybridization to nucleic acids immobilized on a solid substrate has been used widely for several decades (Grunstein and Hogness, 1975;

METHODS IN ENZYMOLOGY, VOL. 411
0076-6879/06 $35.00
DOI: 10.1016/S0076-6879(06)11003-4

Southern, 1975), and membrane hybridization is robust and routine in most molecular biology laboratories (Sambrook and Russell, 2001). Regardless of the particulars of scale or substrate, the principles governing hybridization between labeled nucleic acid in solution and immobilized nucleic acid remain the same (Anderson and Young, 1985). The chief operational differences between standard membrane hybridizations and microarray hybridizations are (i) the nucleic acid is immobilized on glass rather than a membrane (termed the probe in the case of microarrays); (ii) the labeled nucleic acid is generated from RNA via reverse transcription rather than a DNA template (termed the target in the case of microarrays); (iii) the label is a fluorescent dye rather than a radioisotope; and (iv) the scale is reduced by two to three orders of magnitude. The cumulative effect of these differences renders microarray hybridizations significantly less robust than membrane hybridizations.

Novice and seasoned investigators alike encounter episodic technical problems during microarray hybridization. Because microarray experiments are commonly performed using limited samples and involve numerous procedural steps, it is often impractical or more likely impossible to monitor quality control at each component step. Consequently, when problems arise, investigators are faced with the challenge of inferring cause from limited data; in the worst case, working off the resultant image from a completed experiment. There are numerous modes of failure, ranging from catastrophic failure (e.g., giving no signal) to subtle defects that may go undetected (Fig. 1). This chapter provides a guide to common problems that may be encountered with dual-channel cDNA platforms resulting from these complications inherent to microarray experimentation.

This chapter is arranged into three general themes: printing, sample labeling, and hybridization. We touch on microarray fabrication and sample preparation only where they impinge upon hybridization quality control and analysis. Instead of providing an exhaustive or inclusive set of guidelines for all possible problems that investigators may encounter, we outline common problems in routine microarray hybridization, how to diagnose them, and how to mitigate or avoid them. By necessity, this guide is a biased and incomplete subset of possible pitfalls; however, it is hoped that including a general discussion of the most common problems, and how to prevent them, will result in investigators avoiding them altogether.

General Considerations

There are many factors requiring careful attention during the course of an array experiment in order to avoid common problems and ensure high-quality data. Protocols for the variety of available platforms are readily

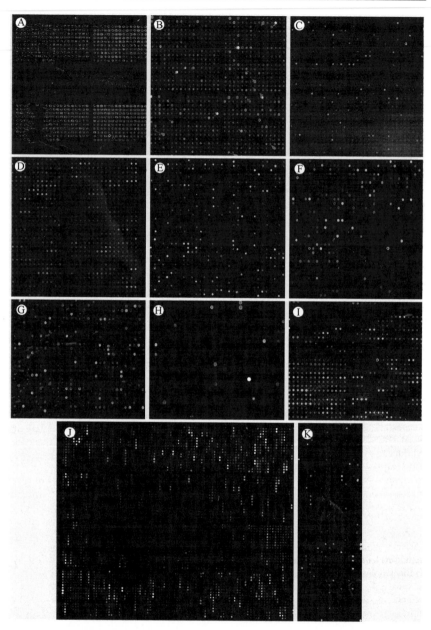

FIG. 1. (A) Slide substrate peeling can be caused by tip contact during printing. High hybridization temperatures used with SDS buffers contribute to the peeling; using a 50% formamide buffer and lowering the temperature to 42° may help with this problem. Uneven

accessible in the literature and on the internet (see later). It is advisable that investigators be familiar with standard molecular biology practices and techniques before undertaking microarray experiments (Sambrook and Russell, 2001). As with any molecular technique, good results begin with scrupulous cleanliness and attention to detail. Particular consideration should be given to ensuring the integrity of critical reagents, such as reverse transcriptase, nucleotides, dyes, and arrays. Keeping track of reagent batch numbers will help in tracing problems back to specific reagents. Special care must be taken to avoid RNase contamination when handling RNA. In the event that problems do arise, carefully controlled tests of each step should pinpoint the problem. In the case of subtle or persistent problems, advice may be sought from helpful online discussion groups (see later).

---

slide thickness or improper pin alignments can cause this problem by physically contacting the surface of the slide during deposition. Note the obvious scratch marks as well, caused by physical damage posthybridization. (B) Comet tailing of particular spots (especially highly expressed spots) can be caused by evaporation from print plates or by excessively stringent posthybridization washes. In addition, the even and low-intensity fluorescence of the majority of spots may be a cause for concern regarding background or contamination issues. Printing in a humidity-controlled environment and monitoring wash stringency carefully can help prevent comet tails. (C) Nonspecific, high background may have a number of causes. In this case, sample labeling or cleanup is a possible culprit. If this is demonstrably not the case, wash steps should be considered a possible source. (D) Background most likely caused during the final washing steps. Some types of background fluorescence are unlikely to cause serious problems for data quality, such as that seen here. (E) A high-quality microarray hybridization demonstrating low background, good dynamic range, excellent spot morphology, and strong signal. (F) Edge effects are typically seen when the target at array margins is not available for hybridization; make sure sufficient buffer is present to cover the entire array surface. This effect can also occur if the array is allowed to dry even partially before being fully washed. (G) Speckling is a nuisance because background variation in this region will be high; the pattern appears to be a splash or spill on the array, most likely at or after the final wash. (H) A low overall signal can be caused by poor RNA quality, poor labeling, or hybridization and washing stringency. The bright spot on this array represents an external control and can be used to verify that externally produced RNAs were reverse transcribed and labeled with high efficiency in the same reaction with the experimental sample. In this case, suspect areas include RNA/cDNA quality and dye incorporation (amino-allyl dye coupling) of the sample. (I) Black holes, or spots where intensity of the spot is lower than surrounding background, can be caused by problems with incomplete blocking during prehybridization, followed by incomplete washes, or by poor sample quality. (J) Bubbles or restricted flow of hybridization buffer over the array can cause areas of very low fluorescence. Increasing gain to the laser or photomultiplier tubes does not solve this problem, as it is related to a paucity of sample binding on the spots. Careful measurement of hybridization volumes and placement of coverslip are important; be sure array is kept level as it is loaded and hybridized and that temperature is constant during hybridization. (K) A trail of fluorescence indicates that the final wash was incomplete and dried on the slide before it could be removed.

Printing

The quality of microarray data is directly dependent on the quality of the microarrays used. Because the subject of microarray production has been reviewed extensively in this volume and elsewhere (Bowtell and Sambrook, 2002; Holloway *et al.*, 2002), comments are restricted to problems during array production, which become apparent after an experimental hybridization.

Spot morphology, the availability of the spotted probe to hybridize, and background fluorescence are largely determined by a matrix of interdependent printing parameters, including slide surface chemistry, spotting buffer composition, DNA concentration, and postprocessing protocols (Hedge *et al.*, 2000; Wrobel *et al.*, 2003). Slides prepared in-house with a poly-L-lysine coating are subject to between- and within-batch variation and, in some cases, peeling of the surface substrate at higher hybridization temperatures (Fig. 1A). To an extent, this may be overcome by using a formamide-based hybridization buffer, which permits lower temperature incubations at equivalent hybridization stringency. Commercial sources of slides typically have less variability, although occasionally problems attributable to poor slide coating are seen. A more common issue is that slides with different surface chemistries, or even slides from different manufacturers with the same surface chemistry, display varying amounts of autofluorescence. Scanning a new slide batch prior to and after printing can be useful if high background is observed (Hardiman, 2003).

The concentration of DNA in a spot is an important determinant of the amount subsequently available for hybridization. More DNA is not necessarily better, as both the amount of DNA and the extent of cross-linking determine how much is available for hybridization. Thus DNA concentration and cross-linking amounts need to be determined empirically for a given slide chemistry. Excess DNA at a spot often results in "comet tailing" (Fig. 1B), a problem that can also occur for other reasons, including poor slide chemistry, inadequate cross-linking, or inadequate postprocessing washes. Spot morphology is also strongly influenced by the combination of spotting buffer and slide surface chemistry (Diehl *et al.*, 2001).

Postprocessing steps after array fabrication can also directly affect array hybridization quality (Bowtell and Sambrook, 2002). Performing washing steps quickly and with sufficient agitation is critical for efficient removal of excess printed DNA. If not removed, this excess DNA can subsequently be washed off during stringent posthybridization washes, thereby contributing to background problems. Also, appropriate postprocessing conditions are important in maintaining spot morphology by preventing spot smearing or comet tails. It is crucial for all washing agents, especially sodium dodecyl sulfate (SDS), to be removed from the arrays at the end of postprocessing, as any residual wash buffer may contribute to background.

Carryover of DNA from one spot to another during printing is a problem that also needs to be addressed at the fabrication step (Bowtell and Sambrook, 2002). A simple way to detect carryover is to print spotting buffer, a positive control, and then spotting buffer sequentially side by side. Carryover would be detected as a higher fluorescence of the spotting buffer following the positive control as compared to the spotting buffer printed before the positive control. Pin cleaning methods, number of slides and spots printed, and buffer used in washing can all be varied to decrease or remove carryover between printed samples. Print runs showing evidence of carryover must be discarded or interpreted with great care.

Sample Preparation and Labeling

In our experience, the single most important factor in the success of an array experiment is the quality of input material. RNA integrity proves the adage "garbage in, garbage out" as far as data quality is concerned. Partial RNA degradation can be a pernicious problem if it goes undetected, as differences in transcript stability result in false positives when partially degraded and undegraded samples are compared on a microarray (Auer et al., 2003). The main issue in isolating intact RNA is to keep nuclease contamination to a minimum and to purify RNA away from chaotropic salts and organics used during extraction, which can interfere with subsequent reverse transcription (Sambrook and Russel, 2001). The Trizol reagent (Invitrogen) is widely used and very effective in maintaining the integrity of RNA. It has been reported that RNA isolated from snap-frozen tissue using Trizol may be refractory to reverse transcription, and subsequent purification of the same RNA samples using glass fiber filter columns (RNeasy, Qiagen) restores its ability to be reverse transcribed (Dumur et al., 2004). We therefore prefer initial extraction with Trizol and subsequent purification on RNeasy columns. Treatment with DNase and subsequent removal of the activity of the enzyme are also essential for quantitative gene expression analysis.

The quality and yield of RNA preparations should be assayed routinely. Spectrophotometry can be used to assess the purity of samples, with a 260/280 measurement indicating protein contamination if less than 1.8 and a 260/230 measurement indicating phenolic compound contamination if less than 0.6 (Zhang et al., 2004). However, spectrophotometry is not an indicator of whether the RNA is intact (full length) and should be used in concert with electrophoresis. The size distribution of RNA samples is an indicator of integrity and should be assayed routinely by electrophoresis using either denaturing agarose gels or a specialized capillary electrophoresis apparatus (Bioanalyzer, Agilent). Denaturing agarose gels can be run without formaldehyde by simply using 20 mM guanidinium isothiocyanate and $1\times$ TBE in the gel and denaturing the RNA in 3 volumes of deionized formamide and $1\times$ TBE (Goda and Minton, 1995). In either case, the relative quantities of the large and small ribosomal bands (28S/18S ratio, calculated as the relative fluorescence of each band) gives an indication of RNA integrity. In undegraded samples the larger 28S band should be about twice as fluorescent as the smaller 18S band. For samples from invertebrate tissues, we typically see two closely migrating bands. Significantly degraded samples display ratios of 1.2–1.4 or lower and noticeable smears at lower molecular weight. It must be noted, however, that RNA samples degraded to the point where they will give misleading microarray results can still have a passing 28S/18S ratio of 1.8 (Auer et al., 2003). Two methods for a more discriminating interpretation of electropherograms are to (i) calculate the ratio of the average degradation peak and the 18S peak (Auer et al., 2003), or (ii) calculate the ribosomal peaks as a percentage of total RNA (should be >30%) (Dumur et al., 2004). Should RNA degradation be suspected, a more time-consuming but definitive test is Northern blotting. For storage up to several months, samples suspended in RNase-free water at $-80°$ are stable, provided they are not subjected to freeze-thaw cycles, whereas ethanol precipitation is preferable for long-term storage.

Once the integrity and concentration of RNA have been confirmed, it is useful to add "spike-in" controls. These are exogenous RNAs with corresponding probes on the array, which will not cross-hybridize to the sample of interest. Several commercial kits of this sort are available (e.g., Stratagene, Ambion). An alternative for some investigators may be to obtain cloned DNA inserts of known sequence from an unrelated organism and use these vectors to produce synthetic RNA by in vitro transcription and DNA for spotting by PCR. Spiking controls are a useful way to assay parameters of an array experiment that are otherwise difficult to determine. For example, RNAs spiked at a 1:1 ratio in each channel should show this pattern after array normalization. Spikes added at varying concentrations and alternative ratios (e.g., 1:5, 5:1) may also be used to assess detection limits and the degree of ratio compression after normalization in an experiment.

Because reverse transcription of RNA can be an important source of variability from assay to assay, it is important to establish consistent conditions. To assess the efficiency and yield of the reaction, electrophoresis of labeled, purified cDNA is a highly useful control. This step requires access to a fluorescence imager, such as a Typhoon (Molecular Dynamics) or an AlphaImager (Alpha Innotech). This serves several purposes: first, it provides a visual assessment of how much unincorporated dye (which can be correlated with higher background) remains after cleanup. Second, it gives a clear estimate of the length distribution of each sample, as well as a quantification of fluorescence intensity in each sample. These numbers should correlate well with spectrophotometric numbers for dye incorporation. For example, failure of the reverse transcription reaction would show a high-intensity, low molecular weight product with little or no cDNA longer than a few hundred base pairs in length. Other labeling techniques, such as dendrimers (Genisphere), are less amenable to gel electrophoresis. Two successive hybridizations are used in this technique to first hybridize the reverse-transcribed cDNA and then label these samples on the array. Although this increases the number of steps involved, in our experience this labeling technology is robust. Should a quality control check be desired, unlabeled cDNA may be run on a Bioanalyzer RNA chip to verify a successful reaction. An alternative indirect control for both RNA integrity and reverse transcription is to include probes designed to mach the 5' and 3' ends of transcripts for a handful of genes that are both ubiquitously expressed and uniformly spliced. When primed with oligo(dT), full-length cDNAs will hybridize to both the 5' and the 3' probes, whereas incomplete cDNA will largely hybridize only to the 3' probe. The ratios of the fluorescence intensities of the 5' and 3' probes therefore provide an indirect measure of the size distribution of the cDNA, and also the integrity of the RNA (Dumur *et al.*, 2004).

For two-channel hybridizations, balancing the amount of DNA or dye added in each channel is important. Opinions differ about whether adding an equal amount of DNA or dye is preferable. Whichever method is chosen, it is important to maintain consistency throughout an experiment so that arrays are more comparable to one another. Downstream normalization is important to remove systematic dye bias prior to data analysis (Holloway *et al.*, 2002). In addition, it is advisable to monitor how much is added to the arrays, especially in terms of picomoles of dye, as high background can result from adding excessive amounts of dye, and low signal from a paucity of sample. A safe range in our system is 40–200 pmol of dye per channel. Another useful measure to track in labeled samples is frequency of incorporation (FOI), which is defined as the number of labeled nucleotides incorporated per 1000 nucleotides of cDNA. The equation for calculating FOI is as follows: (pmol of dye incorporated

×324.5)/ng of cDNA.[1] Typically, FOI measurements of 20–50 indicate samples suitable for array hybridization. We routinely use the microarray assay on the NanoDrop spectrophotometer (NanoDrop), which requires only 1 μl of sample and provides information on DNA concentration and dye incorporation.

Background Fluorescence

A common problem is that off-spot background limits the ability to reliably detect and quantify low abundance transcripts. Increasing the gain to the photomultiplier tube (PMT) does not help, as background increases in parallel with signal. Only enhancing specificity and sensitivity or reducing background will improve data quality. High background can have many possible sources. For example, the problem could be due to a probe that is not sufficiently purified away from unincorporated dye, or simply too much probe; close monitoring of input sample should prevent these problems. With poor sample quality the detection of low-intensity transcripts will often be truncated; when spiking controls report the expected intensity values, thus ruling out dye coupling as a problem, reverse transcription or RNA quality remain as suspect areas (Fig. 1H). Another fairly common problem is a "black hole" of fluorescence (Fig. 1I). In this case, off-spot pixels are much brighter than on-spot pixels, most likely due to improper blocking of the slide during prehybridization (see later) or to poor target quality as opposed to slide quality.

Other possible causes for high and uniform background include improper hybridization buffer composition, or low stringency of hybridization or washes (Kamberova and Shah, 2003). During prehybridization, slides should be treated with blocking agents such as bovine serum albumin to prevent nonspecific binding (Bowtell and Sambrook, 2002). Hybridization buffers should also include blocking agents such as denatured nonspecific DNA, for example, from human Cot-1, salmon sperm, or calf testes (100 μg/ml), as well as a protein component, such as I-Block (Tropix) at 1 mg/ml. Choice of blocking agents should be based on the experimental system. If spots contain extensive oligo(A/T) tracts (as is common with many cDNA-derived libraries), it is advisable to block with oligo(T/A) as well. All buffers should be freshly made using molecular biology-grade water, filtered with 0.22-μm filters, and should not be reused. Any particulate matter coming into contact with the array between hybridization and scanning will likely be visible in the scan.

---

[1] http://www.corning.com/Lifesciences/technical_information/techDocs/troubleshootingUltra GAPS_ProntoReagents.asp.

At lower hybridization temperatures on cDNA arrays, such as below 42°, the sample may become irreversibly bound to the array. To prevent this, increase hybridization temperatures to 55° with nonformamide buffers, and 45–50° or below using a 50% formamide buffer. Figure 1C shows a nonspecific, high background, which could be caused by poor sample quality, inadequate blocking or washing, or poor labeling. Optimal wash parameters are best determined empirically, as stringency seems to be somewhat platform and sample dependent. Washing slides individually in 50-ml plastic screw-cap conical tubes is convenient and has given good results. If following published protocols closely does not provide satisfactory results, extend time, increase temperature (although not dramatically, perhaps 5°), or decrease salt concentration of the washes. A standard protocol is two washes at 50° for 15 min each in 2× SSC/0.1% SDS, two washes in 0.2× SSC/0.1% SDS at room temperature for 10 min, and two washes in 0.1× SSC at room temperature for 5 min. Below 0.1× SSC the target can easily be stripped from the slide if washed for over 3 min. This can result in a background smear similar to comet tailing. For this reason, rather than lowering the salt concentration in the final wash, it is preferable to increase temperature or time. Final wash and dry procedures can have dramatic effects on background levels; some prefer to isopropanol dip their slides immediately before centrifuging. It is fairly common to see patterns of fluorescent background that appear to be wash related (Fig. 1D, G, K). In any case it is important to dry the slide completely before the final wash buffer has a chance to collect on part of the slide; once drying is complete, any fluorescence on the slide is essentially irreversibly bound. Repeated washing of the slide once it has dried will reduce signal with the same efficacy that background is reduced. It should also be noted that exposing the fluorescent dyes to light will result in photobleaching, so keeping the samples in the dark while performing washes is preferable. This can be done by simply wrapping the 50-ml conical tubes in aluminum foil.

It has also been noted in our laboratory and by others (Wit and McClure, 2004) that subtracting local background from the intensity of a spot (when data quality is good and nonuniform background is negligible) is often not advisable, for both theoretical and empirical reasons. In part this is because better correlations are often seen between microarray ratio data and vetted external data (such as RT-PCR) and replicate correlations tend to increase as well. While the possibility remains of a confounding spatial effect inflating correlations artificially, this seems unlikely, and the association between foreground and background intensity may well be evidence that some fluorescence called as background by the software is actually signal.

Often background is not uniform, but appears instead only in a few subarrays. This is due to problems either during hybridization or during washing. Uneven mixing of sample across the surface of the slide, stationary bubbles forming under the coverslip (Fig. 1J), partial drying of sample, and condensation around coverslip edges (Fig. 1F) can all cause nonuniform background. To avoid these problems, it is important to mix samples well prior to pipetting them onto the microarray and to ensure sufficient buffer to fill the area between the coverslip and the array. Keeping the slide and buffer warmed to hybridization temperature as sample is applied also helps reduce bubble formation. Oil or residue on coverslips can be washed off with 100% ethanol, followed by rinsing well in nuclease-free water. For hybridization, the atmosphere in the chamber should be kept humid by applying dilute SSC buffer or water to a reservoir in the chamber (be careful that liquid does not wick onto the slide surface). Hybridization chambers should be fully submerged in a water bath during incubation; incubation in an air oven can lead to drying of the sample. It is also good practice to gently agitate solutions during washing to keep liquid moving over the surface of the array. Other types of background can be caused by particulate contaminants that strongly fluoresce; these should be removed by making sure all solutions are filtered and kept scrupulously clean and that the arrays are well washed during prehybridization. Important precautions include wearing only powder-free nitrile gloves and touching arrays and coverslips only by the edges; any contact on the surface of the slide by forceps or fingers is typically apparent upon scanning. Standard coverslips are not used because they inhibit movement of fluid across the slide surface; LifterSlips (Erie Scientific) are preferred because they have small ridges on one face to provide separation between the coverslip and the microarray surface.

Hybridization Quality Assessment

Most users confronted with their first microarray image have a common question: is this hybridization of good quality? There is no simple answer because multiple criteria are typically important in order to extract reliable data. However, there are several important guidelines for assessing hybridization quality. Many scanner softwares have built-in quality control tests that allow the user to flexibly set criteria or use preset filters (Kamberova and Shah, 2003). For example, the GenePix software of Axon scanners can perform a number of quality-control tests and then filter out spots not meeting certain criteria, such as saturation, percent above background, or standard deviation of background. These "flagged" spots can be omitted from further analysis or normalization. GenePix also has an array quality control test that measures a number of array-wide parameters, such as signal-to-noise ratio, percent present spots, median background, and so on. In our laboratory, we use a custom script written in R (downloadable at: http://dgrc.cgb.indiana.edu/microarrays/analysis.html), which performs a normalization and reports a variety of statistics on both pre- and postnormalized data. It should be noted that this script is designed for DGRC microarrays and will not be directly applicable to other platforms; however, the source code is made available at the above URL and can be modified to suit any two-channel microarray platform data.

Regardless of particulars used to filter spots of low quality, an assessment of array quality should include several types of data. First, what are the average foreground and background of all spots, and of positive and negative controls? Rather than using false-color images from scanning software to judge quality by eye, it is preferable to use graphical summaries of data distributions, such as foreground and background in each channel. We find the Bioconductor packages Marray and Olin (available at http://www.bioconductor.org) and other functions in the basic R package to be highly useful for this purpose. A good hybridization will have a narrow, high peak of background at low-intensity values, well separated from a broader peak of signal at higher intensity (Fig. 2A and B). By comparison, low-quality hybridizations show poor differentiation of signal and background peaks (Fig. 2C and D). Spiking control spots should have ratios in the appropriate ranges and should demonstrate good reproducibility ($r^2 > 0.8$). Scatterplots and boxplots (Fig. 2I) are useful to show print-tip and subarray effects. Classical "MA" plots of log ratio vs intensity (Fig. 2J) can reveal dye, print-tip, or subarray effects. Data with more pronounced bias will generally require more aggressive normalization due to dye effects that do not reflect underlying transcript levels.

Graphical representations of array foreground and background levels, often called heat maps, are another useful way to visualize the common effect of location-dependent background. First, spatial (x–y) plots of foreground and background across the slide may highlight spatial trends. Spatial plots of ratios (Fig. 2E–G) depict values of all pixels on the array, and in Fig. 2E, a clear bias to the red channel is apparent. This difference is due to a fairly low (twofold) difference in overall background levels. A good scan (Fig. 2F) does not show such a pattern. After normalization, the mix of ratios across the slide no longer shows a location-dependent effect (Fig. 2G). These often appear to be artifacts of an incomplete wash, as shown in Fig. 2H. With both location- and intensity-dependent dye biases, a normalization approach that can deal with these problems is required; we find the R package Olin (Futschik and Crompton, 2004) to be highly useful in this respect. Generally, arrays with background intensities of 200 or lower (on a scale from zero to 65,535 units) and exhibiting a "long tail" distribution of

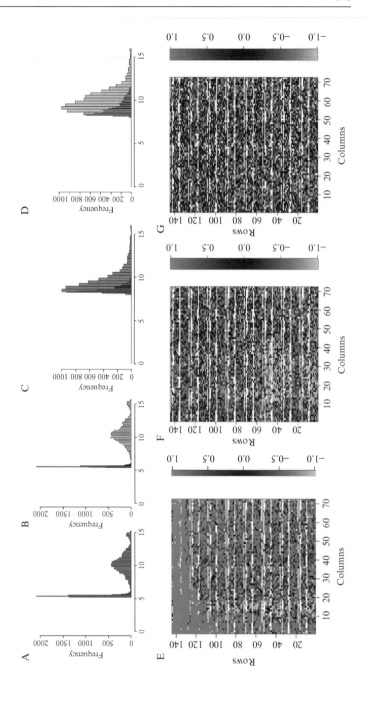

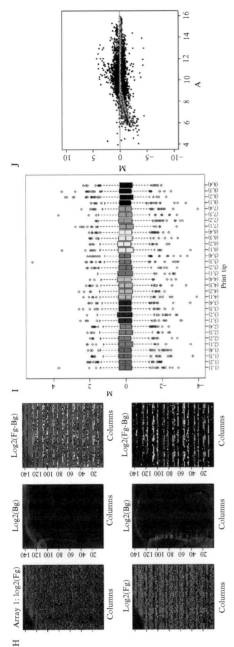

Fig. 2. Histogram of frequency of $\log_2$ (intensity) values in a hybridization of higher quality (A and B) and of lower quality (C and D). Note the distinct separation of the low-intensity and higher intensity counts for the higher quality array, and the normal distribution of signal. In the lower quality array, signal (lighter color) and background (darker color) are spread across much of the same range, which makes spot finding difficult or impossible. For each channel, noise should be near a $\log_2$ (intensity) of 6 or 7, corresponding to about 130 counts of background in a 2^{16}-bit scanning system. (E, F, and G) Spatial plots of $\log_2$ ratio values for all pixels on the array. Ratios for each pixel are plotted according to x–y coordinates of pixel location (note scale bar). (E) The array displays high red background to the left and top of the array and is red on the right side as well. This array will require aggressive normalization to remove the spatial artifact. (F) An array with lower overall background and with much less spatial bias. (G) The same array as E after normalization using Olin (Futschik and Crompton, 2004). (H) False-color image of foreground and background in red and green channels plotted across the x–y dimensions of the array. This displays $\log_2$ intensity values in foreground, background, and $\log_2$(foreground−background) and can reveal otherwise invisible trends. (I) Box plot of spot ratios within each print-tip group. Such plots are helpful for revealing artifacts related to printing and how well normalization copes with their removal. (J) M-A (ratio-intensity) plot of all spots on the array, with regression lines colored according to the subarray. In this case, intensity is centered around a mean of zero, which is reasonable, and there is relatively little intensity-dependent bias in average ratio values. (See color insert.)

spots with high intensity are of good quality. Coefficients of variation in background around spots should generally be 10% or less at intensities above 8 (log scale), and signal-to-noise ratios should be greater than 10. A good array (Fig. 1E) looks by eye to have low, even background; good spot morphology and intensity range; and a strong signal.

Concluding Remarks

Problems with microarray hybridizations range from catastrophic failure (no signal) to data of apparently high quality that nevertheless exhibits unacceptable technical variation. The ability to detect meaningful biological variation is limited by the extent to which this technical variation can be controlled and minimized. Such control can only be achieved by vigilant attention to detail and consistent monitoring of quality metrics.

Acknowledgments

We are indebted to many staff and users of the Drosophila Genomics Resource Center and the Center for Genomics and Bioinformatics who have contributed ideas, experiences, and images. This could not have been written without the invaluable assistance of Elizabeth Bohuski, Claire Burns, Angela Burr, John Colbourne, Zhao Lai, Monica Sentmenant, and Yi Zhou. We gratefully acknowledge financial assistance from the Indiana Genomics Initiative and by NIH Grant 1 P40 RR017093 to J.A.

Internet Resources

Protocols

DGRC: http://dgrc.cgb.indiana.edu/microarrays/protocols.html
UMinn: http://www.agac.umn.edu/microarray/protocols/protocols.htm
TIGR: http://www.tigr.org/tdb/microarray/protocolsTIGR.shtml
NHGRI: http://research.nhgri.nih.gov/microarray/protocols.shtml
Pat Brown: http://cmgm.stanford.edu/pbrown/protocols/
Rockefeller: http://www.rockefeller.edu/genearray/protocols.php

Discussion Groups

Yahoo: http://groups.yahoo.com/group/microarray
Stanford Microarray forum: http://cmgm.stanford.edu/cgi-bin/cgiwrap/
taebshin/dcforum/dcboard.cgi
Email Listserve: send message to GENE-ARRAYS@ITSSRV1.UCSF.
EDU

References

Anderson, M. L. M., and Young, B. D. (1985). Quantitative filter hybridization. *In* "Nucleic Acid Hybridization A Practical Approach" (B. D. Hames and S. J. Higgins, eds.), pp. 73–111. IRL Press, Oxford.

Auer, H., Lyianarachchi, S., Newsom, D., Klisovic, M., Marcucci, G., and Kornacker, K. (2003). Chipping away at the chip bias: RNA degradation in microarray analysis. *Nature Genet.* **35,** 292–293.

Bowtell, D., and Sambrook, J. (2002). "DNA Microarrays: A Molecular Cloning Manual." Cold Spring Harbor Laboratory Press, Cold Spring Harbor, NY.

Diehl, F., Grahlmann, S., Beier, M., and Hoheisel, J. (2001). Manufacturing DNA microarrays of high spot homogeneity and reduced background signal. *Nucleic Acids Res.* **29,** E38.

Dumur, C., Nasim, S., Best, A., Archer, K., Ladd, A., Mas, V., Wilkinson, D., Garrett, C., and Ferreira-Gonzalez, A. (2004). Evaluation of quality-control criteria for microarray gene expression analysis. *Clin. Chem.* **50,** 1994–2002.

Futschick, M., and Crompton, T. (2004). Model selection and efficiency testing for normalization of cDNA microarray data. *Genome Biol.* **5**(8), R60.

Goda, S., and Minton, N. (1995). A simple procedure for gel electrophoresis and Northern blotting of RNA. *Nucleic Acids Res.* **23**(16), 3357–3358.

Grunstein, M., and Hogness, D. S. (1975). Colony hybridization: A method for the isolation of cloned DNAs that contain a specific gene. *Proc. Natl. Acad. Sci. USA* **68,** 2559–2563.

Hardiman, G. (2003). "Microarrays Methods and Applications: Nuts and Bolts." DNA Press, Eagleville PA.

Hegde, P., Qi, R., Abernathy, K., Gay, C., Dharap, S., Gaspard, R., Hughes, J., Snesrud, E., Lee, N., and Quackenbush, J. (2000). A concise guide to cDNA microarray analysis. *Biotechniques* **9,** 548–556.

Holloway, A., van Laar, R., Tothill, R., and Bowtell, D. (2002). Options available, from start to finish, for obtaining data from DNA microarrays II. *Nature Genet.* **32**(Suppl.), 481–489.

Kamberova, G., and Shah, S. (2003). "DNA Array Image Analysis: Nuts and Bolts." DNA Press, Eagleville, PA.

Sambrook, J., and Russell, D. (2001). "Molecular Cloning: A Laboratory Manual." Cold Spring Harbor Laboratory Press, Cold Spring Harbor, NY.

Schena, M., Shalon, D., Davis, R., and Brown, P. (1995). Quantitative monitoring of gene expression patterns with a complementary DNA microarray. *Science* **270,** 467–470.

Southern, E. M. (1975). Detection of specific sequences among DNA fragments separated by gel electrophoresis. *J. Mol. Biol.* **98,** 503–517.

Wit, E., and McClure, J. (2004). "Statistics for Microarrays." Wiley, New York.

Wrobel, G., Schlingemann, J., Hummerich, L., Kramer, H., Lichter, P., and Hahn, M. (2003). Optimization of high-density cDNA-microarray protocols by 'design of experiments'. *Nucleic Acids Res.* **31,** e67.

Zhang, W., Shumlevich, I., and Astola, J. (2004). "Microarray Quality Control." Wiley, New York.

[4] Use of External Controls in Microarray Experiments

By IVANA V. YANG

Abstract

DNA microarray analysis has become the most widely used technique for the study of gene expression patterns on a genomic scale. Microarray analysis is a complex technique involving many steps, and a number of commercial and in-house developed arrays and protocols for data collection and analysis are used in different laboratories. Inclusion of external or spike-in RNA controls allows one to evaluate the variability in gene expression measurements and to facilitate the comparison of data collected using different platforms and protocols. This chapter describes what external controls are, which collections of spike-in controls are available to researchers, and how they are implemented in the laboratory. Applications of external controls in the assessment of microarray performance, normalization strategies, the evaluation of algorithms for gene expression analysis, and the potential to quantify absolute mRNA levels are discussed.

Introduction

The ability to study genome-wide transcription using microarrays has become the most widely used technique for the study of gene expression patterns on a genomic scale and has revolutionized both basic and clinical science fields (Ramaswamy and Golub, 2002; Staudt and Brown, 2000). Microarray analysis is a complex, multistep technique involving not only array manufacture, but also sample isolation, labeling, and hybridization as well as subsequent data analysis. Systematic variation can occur at any step of the process and affect gene expression measurements. Furthermore, different investigators use different commercial or in-house developed arrays and protocols for data collection and analysis. One way to evaluate the variability in gene expression measurements and to facilitate the comparison of data collected in laboratories throughout the world is to utilize external or spike-in controls in each experiment. Such controls can be used to assess sensitivity and accuracy of relative gene expression changes, linear range of the particular microarray platform, hybridization specificity, and consistency within and across arrays. External controls were introduced in

METHODS IN ENZYMOLOGY, VOL. 411
Copyright 2006, Elsevier Inc. All rights reserved.

early genome-wide gene expression profiling studies (Lockhart *et al.*, 1996) but their full potential in evaluating and standardizing microarray experiments is just starting to be recognized. At present, microarrays are used to monitor relative gene expression changes but inclusion of spike-in controls could potentially allow one to measure absolute mRNA levels. Microarray data sets containing sufficient numbers of measurements on spike-in controls can additionally be used for comparison of existing methods and development of new algorithms for image processing, data normalization, and detection of differential expression. This chapter describes what external controls are, how they are implemented in the laboratory, and how they are utilized for the assessment of microarray performance and evaluation of data analysis methodologies.

Description and Availability of External Controls

External controls are *in vitro*-synthesized RNA molecules that are added in defined amounts and ratios to the RNA samples to be assayed. Each control is designed to hybridize with high specificity to the control probes included on the array but not to probes complementary to the genome being studied. For mammalian arrays, these control genes are typically selected from bacterial or plant biochemical pathways that are not present in metazoan organisms. Alternatively, they are artificial genes (or "alien" sequences) based on sequences from nontranscribed genomic regions (intronic or intergenic). In addition to avoiding cross-hybridization with endogenous probes, the basic requirement for spike-in cRNA is to resemble the sample RNA in sequence properties, such as length, GC content, and the presence of the poly(A) tail (van Bakel and Holstege, 2004). Similarly, DNA probes for exogenous controls must be comparable to endogenous probes regarding design, length, method of fabrication, and cross-hybridization potential (see Chapter 4 in volume 410 by Kriel *et al.*, 2006).

Control genes are typically cloned into a vector containing an RNA polymerase promoter (e.g., SP6) and a stretch of d(A):d(T) in the multiple cloning site. This allows the cRNA to be prepared by *in vitro* transcription from the SP6 promoter and contain a synthetic poly(A) tail at the 3' end of the inserted sequence. The poly(A) tail is necessary for oligo d(T) priming of the reverse transcription (RT) of RNA into complementary DNA (cDNA), the first step in target labeling. These vectors can also be used to prepare DNA for spotting on cDNA arrays. For oligonucleotide arrays, probes are designed and synthesized in the same manner as all of the other probes present on the array.

Several different collections of external controls have been developed by commercial and nonprofit entities (Table I). Affymetrix control reagents for eukaryotic GeneChip arrays consist of poly(A)-tailed RNA encoding five *Baeillus subtilis* genes (*dap, lys, phe, thr,* and *trp*) to be spiked in the sample RNA prior to reverse transcription (see Chapter 1 in volume 410 by Dalma-Weiszhausz *et al.*, 2006). Bacteria containing recombinant plasmids with the *B. subtilis* genes are available from the American Type Culture Collection (ATCC) to researchers who would like to include this control set on their own spotted microarrays. GE Healthcare/Amersham Biosciences also offers a set of poly(A)-tailed bacterial RNA samples to be used for monitoring the performance of their CodeLink whole genome bioarrays (http://www4.amershambiosciences.com/APTRIX/upp00919.nsf/ Content/WD%3AExternal+RNA+co%28274354027-B500%29?Open Document&hometitle=WebDocs). This set includes synthetic mRNAs encoding six bacterial genes: *araB, entF, fixB, gnd, hisB,* and *leuB.* Agilent Technologies control set for their inkjet-synthesized 60-mer oligonucleotide arrays consists of 10 semisynthetic mRNAs that are made up of a portion of the adenovirus E1a gene and tagged at the end with a unique 60-mer, followed by a poly(A) tail (Hughes *et al.*, 2001; see Chapter 2 in volume 410 by Wolber *et al.*, 2006).

Researchers who fabricate their own microarrays have access to several other collections of external controls. TIGR Arabidopsis cRNA Spiking Control Resource (The Institute for Genomic Research) consists of 10 *Arabidopsis thaliana* genes (*CAB, LTP4, LTP6, NAC, PRKase, rbcL, RCA, RCP, TIM,* and *XCP2*) that have been subcloned into the pSP64poly(A) vector. The vector set is available at no charge to researchers at nonprofit institutions. cRNA and oligonucleotides for spotting representing the same 10 genes are also available commercially as the SpotReport Oligo Array Validation System (Stratagene). National Institutes of Aging (NIA) and Agilent Technologies have collaboratively assembled a set of controls based on intergenic and intronic *Saccharomyces cerevisiae* sequences (Carter *et al.*, 2005). The 60-mer oligonucleotide probes for these controls have been added to Agilent's mouse whole genome microarrays, and vectors for generating spike-in cRNAs have been made available through the ATCC. This collection can be used with endogenous RNA from almost any organism because the controls are artificial genes composed of DNA sequences from yeast intergenic and intronic regions. Lucidea Universal ScoreCard (GE Healthcare/Amersham Biosciences) is a set of 23 controls also compatible with samples from most organisms because they are, similarly to the NIA/Agilent controls, based on yeast intergenic regions. They have been shown by the manufacturer to perform independently, i.e., with no cross-reactivity, on human, mouse, rat, yeast, *A. thaliana,* and

TABLE I
SPIKE-IN RNA CONTROL COLLECTIONS FOR EUKARYOTIC MICROARRAYS

Spike-in collection	Control type	Provider	Microarray type	Web site
GeneChip eukaryotic poly(A) RNA control kit	Bacterial gene sequences	Affymetrix	Affymetrix GeneChip arrays	http://www.affymetrix.com/ support/technical/manual/ expression_manual.affx
External RNA controls for monitoring performance of CodeLink Whole Genome bioarrays	Bacterial gene sequences	GE Healthcare/ Amersham Bioscience	CodeLink Bioarrays	http://www4. amershambiosciences.com/ APTRIX/upp00919.nsf/ Content/WD%3AExternal+ RNA+co%2827434027-B500%29?OpenDocument& hometitle=WebDocs
Agilent Technologies spike-in set (Hughes et al., 2001)	Adenovirus E1 gene sequence tagged with unique 60-mers	Agilent Technologies	Inkjet synthesized 60-mer microarrays	http://www.chem. agilent.com/ scripts/pds.asp?/ page=38553
TIGR Arabidopsis cRNA spiking control resource (Wang et al., 2003)	Plant gene sequences	The Institute for Genomic Research	Spotted oligo and cDNA arrays	http://pga.tigr.org/ Arab_ctrl.shtml
SpotReport Oligo array validation system	Plant gene sequences	Stratagene	Spotted oligo arrays	http://www.stratagene.com/ products/showProduct.aspx? pid=528

(continued)

TABLE I (continued)

Spike-in collection	Control type	Provider	Microarray type	Web site
NIA/Agilent yeast spike-in controls[a] (Carter et al., 2005)	Artificial genes (yeast intergenic and intronic sequences)	National Institute of Aging/Agilent Technologies	Inkjet synthesized oligonucleotide and spotted cDNA/ oligonucleotide arrays	
Lucidea Universal ScoreCard[a]	Artificial genes (yeast intergenic sequences)	GE Healthcare/ Amersham Bioscience	Spotted oligonucleotide arrays	http://www1.amersham biosciences.com/APTRIX/ upp00919.nsf/Content/ WD:Lucidea%20Univers (202385379-B500)?Open Document&hometitle= DrugScr
SpotReport Alien Oligo array validation system	Artificial genes	Stratagene	Spotted oligonucleotide arrays	http://www.stratagene.com/ products/display Product.aspx?pid=527

[a] Species-independent set that can be used for prokaryotic and eukaryotic arrays.

Escherichia coli arrays. Finally, the SpotReport Alien Oligo Array Validation System (Stratagene) is a comparable group of 10 artificial genes that is also useful in assessing the performance of oligonucleotide-based plant, animal, or microbial microarrays.

However, the full potential of external RNA controls in the evaluation of microarray platforms and standardization of array data can only be achieved if the same controls are present on all commercial and custom microarrays. Efforts to create a common collection of spike-in controls are underway; a good example is the activity of the External RNA Controls Consortium (ERCC) (http://www.affymetrix.com/community/standards/ercc.affx). ERCC is composed of scientists from private, academic, and government sectors whose goal is the creation of "well-characterized and tested RNA spike-in controls useful for evaluating sample and system performance, to facilitate standardized data comparisons among commercial and custom microarray platforms." ERCC is planning on putting together a set of 100 platform-independent controls that will be useful for the evaluation of reproducibility, sensitivity, and robustness in gene expression analysis (Salit, 2006).

Assesment of Array Performance Using External RNA Controls

Although external controls can be spiked in at different stages of the experiment, adding them directly to the RNA sample at the beginning of the experiment is most useful because it allows one to monitor all of the downstream steps. From a quality control standpoint, it is beneficial to have additional controls added at a later stage, such as hybridization, to be able to pinpoint where problems might have occurred in unsuccessful experiments.

External cRNA controls are added to endogenous RNA at predetermined amounts and ratios. The amount of exogenous RNA is generally expressed as picograms of the control per microgram of total endogenous RNA, which is often derived from the desired number of copies of the transcript per cell. For example, assuming that there are 360,000 mRNA transcripts per cell, 20 pg of total RNA per cell, and that 1 pg of total RNA is mRNA, Kane *et al.* (2000) derived that 0.15 pg of the external control per microgram of total RNA should be added to achieve one copy per cell spiking. Several assumptions are made in this calculation and one must keep in mind that these numbers are only estimates and different cell types under different conditions will express lower or higher amounts of RNA. Furthermore, intensities of signals obtained on microarrays do not correlate directly with transcript abundances due to differences in hybridization efficiencies.

The amounts of different control targets in the spike-in set are usually chosen to cover the entire range of endogenous mRNA abundance. Several genes are added in low copy numbers (5–50) to mimic rare transcripts (0.001–0.01% mRNA), others are added at higher copy per cell numbers (100–400) to represent moderately abundant mRNAs (0.025–0.1% mRNA), and a few may be added at very high (>500) copy numbers for highly expressed transcripts (>0.15%), although hybridization of a large amount of RNA may result in signal saturation. Similarly, different control cRNAs can be added in varying Cy5/Cy3 ratios in two-color array experiments. Figure 1 shows an example in which four controls are added in equal amounts in the two channels (yellow), three are spiked in at 3:1 Cy5:Cy3 ratios (red), and three are added in at 1:3 Cy5:C3 ratios (green). The controls added in at 1:1 ratios are designed for normalization purposes in instances where global normalization approaches may not be appropriate, that is, microarrays containing small numbers of genes or comparisons in which most genes on the array are expressed differentially (Hill *et al.*, 2001; Wang *et al.*, 2003). Alternatively, these spike-ins may be used to ensure that global normalization approaches have been successful in removing dye bias (Qin and Kerr, 2004).

Controls that are added in different amounts in Cy3 and Cy5 channels are typically used to assess sensitivity of the microarray to detect differential expression of endogenous transcripts expressed at a range of abundances. For example, nearly all endogenous transcripts in the self vs self-hybridization of a mouse lung sample shown in Fig. 1A have Cy5:Cy3 ratios between 3:1 and 1:3, suggesting that transcripts that are over- or under-expressed in the mouse lung vs the Universal Mouse Reference sample (Fig. 1B) at greater than threefold (blue) can be detected as expressed differentially on this particular array platform. Including additional controls spiked in at different ratios (e.g., 2.5:1, 2:1, 1.5:1) would be useful in further defining sensitivity limits for detecting differential expression of transcripts present at different abundance levels. Analogous to using external controls to assess sensitivity for detecting differential expression in two-color hybridization assays, *B. subtilis* transcripts have been used to assess the linear range of the Affymetrix GeneChip platform in the absence and presence of endogenous eukaryotic RNA (Chudin *et al.*, 2002).

External RNA controls can also be used for quality control/assurance purposes. In the case of in-house spotted oligonucleotide or cDNA arrays, printing controls at multiple positions on each slide allows for detection of any spatial gradients and artifacts. In general, any deviations from expected input ratios are indicative of problems in array printing, target labeling, or hybridization conditions. For example, a quality control metric could be designed to capture information such as the mean Cy5/Cy3 ratio and the

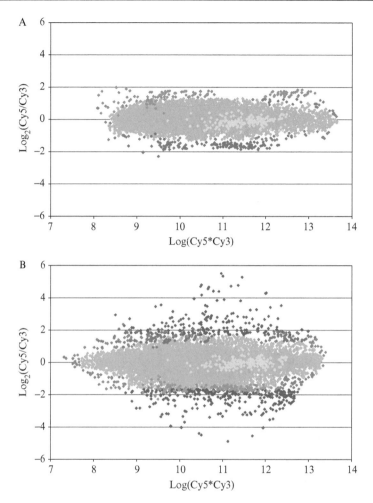

FIG. 1. Ratio–intensity (RI) plots of hybridizations of a mouse lung RNA sample to itself (A) or to the Universal Mouse Reference RNA (Stratagene) (B). Ten *A. thaliana* cRNA controls were spiked into mouse lung and reference RNA samples at 1:1 (yellow; 50:50, 100:100, 200:200, and 300:300 in terms of copies per cell), 3:1 (red; 30:10, 120:40, and 450:150), and 1:3 (green; 40:120, 80:240, and 100:300) Cy5:Cy3 ratios. Genes that are identified as differentially expressed in B are shown in blue. RNA was labeled by incorporation of the aminoallyl linker during reverse transcription followed by a coupling of Cy3- or Cy5-NHS esters, and labeled targets were hybridized to a spotted oligonucleotide microarray containing the mouse Operon set (probes for ~17,000 mouse genes) printed once and probes for the 10 *A. thaliana* genes printed multiple times on the array. (See color insert.)

coefficient of variation (CV) for repeated measurements of each control. Mean ratios differing from input ratios or exhibiting large CV would suggest that problems may have occurred in the experiment. External controls are also valuable in the evaluation of existing and validation of new microarray laboratory protocols. For instance, two published studies validated printing of oligonucleotides instead of polymerase chain reaction (PCR) products using this strategy (Kane *et al.*, 2000; Wang *et al.*, 2003), and another study used spike-ins to compare available cDNA labeling methods (Badiee *et al.*, 2003). Finally, studies attempting to estimate absolute transcript copy numbers on microarrays using external RNA are beginning to appear in the literature (Carter *et al.*, 2005). In this study, investigators estimated absolute mRNA levels in endogenous mouse RNA based on the standard curve they constructed using the yeast exogenous RNA. They compared copy numbers estimated on microarrays to those obtained by quantitative RT-PCR and found a good correlation for a large number of transcripts but not all of them. Further improvements in probe designs and other parameters associated with microarray hybridization assays, particularly to account for alternative transcript splicing, will be needed to obtain accurate mRNA levels for all genes.

Methods for Synthesis and Utilization of External RNA Controls

Laboratories that wish to make their own external control sets will need to subclone sequences of interest into a vector containing a promoter for an RNA polymerase (SP6, T3, or T7) and a poly(A) tail, such as pSP64poly(A) (Promega, Madison, WI). Methodology for subcloning is described in some detail in Wang *et al.* (2003), but can also be found in any standard molecular biology manual. Alternatively, investigators can obtain control sets that have already been assembled and are available for public use (described earlier). To generate cRNAs containing a 3' poly(A) tail, constructs must be linearized with a restriction enzyme and *in vitro* transcribed from the RNA polymerase promoter. MEGAscript kits for SP6, T3, and T7 RNA polymerases (Ambion, Austin, TX) yield large amounts of pure cRNA if the protocols included in the kit are followed. *In vitro*-synthesized cRNA should be quantified using a spectrophotometer and run on a gel or the Bioanalyzer (Agilent Technologies) to ensure proper size distribution.

To make pools for spiking into endogenous RNA samples, one needs to calculate how much of each cRNA transcript is needed to achieve desired copy numbers and fold changes between channels or arrays, add the appropriate amount of each cRNA, and bring them up to the final volume in diethylpyrocarbonate-treated water. It is crucial to have very accurate dispensing in this step, so serial dilutions should be used if quantities of stock solutions that need to be added are too small to dispense accurately. The simplest way to avoid pipetting small volumes is to make a large pool of cRNAs, divide the solution into small aliquots, and store the aliquots at −80°. Two separate pools of cRNAs are typically made for Cy3 and Cy5 channels in two-color hybridization assays. Control cRNA pools are added to total RNA prior to mRNA extraction in protocols that use mRNA for reverse transcription; if total RNA is used for labeling, one can add control cRNA directly to the RT reaction. Commercial control sets generally contain spike-in RNA samples in solution and DNA for spotting in lyophilized form. cRNA needs to be diluted and pools made as outlined earlier and following instructions provided by the manufacturer.

Evaluation of Data Analysis Methodology Using Spike-In Data Sets

For a control data set to be useful in evaluating data analysis methods, the data set must contain a sufficient number of external controls and experiments. Four such data sets (Table II) have been used in published studies focusing on either comparisons of existing or the development of new algorithms for gene expression analysis. Two data sets that have been used extensively in the development of methodology for Affymetrix GeneChips are the cRNA spike-in sets from Affymetrix and GeneLogic. The Affymetrix Latin Square data set consists of three technical replicates of 14 different complex human background RNAs with 42 controls added at a range of concentrations and at twofold changes between arrays. The GeneLogic data set is similar but it has 11 cRNAs added at varying fold changes between arrays in a total of 98 hybridizations. Another more recently collected data set on Affymetrix arrays contains 3860 synthetic RNA molecules; 1309 are spiked in at different concentrations and fold changes between the two samples and the rest are present at identical relative concentrations. Finally, a data set of 10 *A. thaliana* controls spiked into mouse liver RNA and hybridized to spotted oligonucleotide arrays has been collected by six different laboratories within the Toxicogenomics Research Consortium (www.niehs.nih.gov/dert/trc) (Qin and Kerr, 2004).

Analogous to the comparison of the existing and validation of new laboratory protocols using data collected on external controls, these four data sets have been used to evaluate both available algorithms and new methods for the analysis of microarray data. Data sets collected on Affymetrix GeneChips have been particularly valuable in developing tools to summarize expression levels from multiple probes for each gene, scale and normalize summarized expression measurements, and identify differentially

TABLE II
SPIKE-IN DATA SETS USED IN EVALUATING MICROARRAY DATA ANALYSIS METHODS

Data set	Platform	Web site
Affymetrix Latin square experiment	HG-U133A Affymetrix GeneChip	http://www.affymetrix.com/ support/technical/ sample_data/datasets.affx
GeneLogic spike-in experiment	HG-U95Av1 Affymetrix GeneChip	http://www.genelogic.com/ media/studies/index.cfm
The golden spike-in experiment (Choe et al., 2005)	DrosGenome1 Affymetrix GeneChip	http://www.elwood9.net/spike
Toxicogenomics Consortium (TRC) (Qin and Kerr, 2004)	Spotted oligonucleotide arrays	http://dir.niehs.nih.gov/ microarray/trc/

expressed genes on this platform (Bolstad et al., 2003; Choe et al., 2005; Irizarry et al., 2003a,b). For example, Irizarry and co-workers (2003a) compared Microarray Analysis Suite (MAS) 5.0, dChip, and Robust Multi-array Analysis (RMA) algorithms for summarizing probe level data based on three criteria: the precision of the measures of expression, the consistency of fold change estimates, and the specificity and sensitivity of the ability to detect differential expression and concluded that the RMA algorithm outperformed the other two methods. More recently, Choe et al. (2005) conducted an extensive investigation of different options for background correction, probe level summaries, normalization, and statistical tests for differential expression and identified a combination of analysis methods that allows detection of 70% of true positives at <10% false discovery rate. Similarly, methods for background correction, normalization, and identification of differential expression on two-color microarray data have been evaluated using the A. thaliana spike-in data set (Qin and Kerr, 2004). The findings of this study support the use of intensity-based normalization methods such as Lowess but question the practice of local background subtraction. These investigators also compared a number of statistical tests for differential expression and concluded that algorithms such as significance analysis of microarrays outperform traditional Student's t test when there are few replicates. Finally, these data sets have been valuable in implementing novel approaches for the assessment of differential expression, including a method that uses information from within-slide replicates to improve the precision of estimating genewise variances (Smyth *et al.*, 2005), a Bayesian hierarchical mixture model (Newton *et al.*, 2004), a multivariate analysis based on a T2 statistic (Lu *et al.*, 2005), and a distance synthesis method that integrates different statistics for gene ranking and selection (Yang *et al.*, 2005).

Concluding Remarks

Microarrays have evolved from a specialized and expensive tool used in small numbers in a few laboratories to a standard technique available and affordable to many researchers. Although external controls were introduced in early microarray experiments and are now present on most commercial and custom microarrays, their potential has not been fully realized. Further advances in standardization of microarray data (Bammler *et al.*, 2005) will require inclusion of common external standards and controls in every microarray experiment. Integration of common RNA standards and external controls into all gene expression profiling studies will facilitate analysis of data sets collected on different platforms using different laboratory protocols and analytical approaches. In addition, the use of external controls may enable reporting of absolute mRNA levels rather than relative gene expression changes in microarray experiments.

Acknowledgments

The author thanks Ed Lobenhofer for critical reading of the article and Paul Wolber for providing information on Agilent's control set.

References

Badiee, A., Eiken, H. G., Steen, V. M., and Lovlie, R. (2003). Evaluation of five different cDNA labeling methods for microarrays using spike controls. *BMC Biotechnol* **3**, 23.
Bammler, T., Beyer, R. P., Bhattacharya, S., Boorman, G. A., Boyles, A., Bradford, B. U., Bumgarner, R. E., Bushel, P. R., Chaturvedi, K., Choi, D., Cunningham, M. L., Deng, S., Dressman, H. K., Fannin, R. D., Farin, F. M., Freedman, J. H., Fry, R. C., Harper, A., Humble, M. C., Hurban, P., Kavanagh, T. J., Kaufmann, W. K., Kerr, K. F., Jing, L., Lapidus, J. A., Lasarev, M. R., Li, J., Li, Y. J., Lobenhofer, E. K., Lu, X., Malek, R. L., Milton, S., Nagalla, S. R., O'Malley, J. P., Palmer, V. S., Pattee, P., Paules, R. S., Perou, C. M., Phillips, K., Qin, L. X., Qiu, Y., Quigley, S. D., Rodland, M., Rusyn, I., Samson, L. D., Schwartz, D. A., Shi, Y., Shin, J. L., Sieber, S. O., Slifer, S., Speer, M. C., Spencer, P. S., Sproles, D. I., Swenberg, J. A., Suk, W. A., Sullivan, R. C., Tian, R., Tennant, R. W., Todd, S. A., Tucker, C. J., Van Houten, B., Weis, B. K., Xuan, S., and Zarbl, H. (2005). Standardizing global gene expression analysis between laboratories and across platforms. *Nature Methods* **2**, 351–356.
Bolstad, B. M., Irizarry, R. A., Astrand, M., and Speed, T. P. (2003). A comparison of normalization methods for high density oligonucleotide array data based on variance and bias. *Bioinformatics* **19**, 185–193.

Carter, M. G., Sharov, A. A., VanBuren, V., Dudekula, D. B., Carmack, C. E., Nelson, C., and Ko, M. S. (2005). Transcript copy number estimation using a mouse whole-genome oligonucleotide microarray. *Genome Biol.* **6,** R61.

Choe, S. E., Boutros, M., Michelson, A. M., Church, G. M., and Halfon, M. S. (2005). Preferred analysis methods for Affymetrix GeneChips revealed by a wholly defined control dataset. *Genome Biol.* **6,** R16.

Chudin, E., Walker, R., Kosaka, A., Wu, S. X., Rabert, D., Chang, T. K., and Kreder, D. E. (2002). Assessment of the relationship between signal intensities and transcript concentration for Affymetrix GeneChip arrays. *Genome Biol.* **3,** RESEARCH0005.

Dalma-Weiszhausz, D. D., Warrington, J., Tanimoto, E. Y., and Miyada, C. G. (2006). The Affymetrix GeneChip® paltform: An overview. *Methods Enzymol.* **410,** 3–28.

Hill, A. A., Brown, E. L., Whitley, M. Z., Tucker-Kellogg, G., Hunter, C. P., and Slonim, D. K. (2001). Evaluation of normalization procedures for oligonucleotide array data based on spiked cRNA controls. *Genome Biol.* **2,** RESEARCH0055.

Hughes, T. R., Mao, M., Jones, A. R., Burchard, J., Marton, M. J., Shannon, K. W., Lefkowitz, S. M., Ziman, M., Schelter, J. M., Meyer, M. R., Kobayashi, S., Davis, C., Dai, H., He, Y. D., Stephaniants, S. B., Cavet, G., Walker, W. L., West, A., Coffey, E., Shoemaker, D. D., Stoughton, R., Blanchard, A. P., Friend, S. H., and Linsley, P. S. (2001). Expression profiling using microarrays fabricated by an ink-jet oligonucleotide synthesizer. *Nature Biotechnol.* **19,** 342–347.

Irizarry, R. A., Bolstad, B. M., Collin, F., Cope, L. M., Hobbs, B., and Speed, T. P. (2003a). Summaries of Affymetrix GeneChip probe level data. *Nucleic Acids Res.* **31,** e15.

Irizarry, R. A., Hobbs, B., Collin, F., Beazer-Barclay, Y. D., Antonellis, K. J., Scherf, U., and Speed, T. P. (2003b). Exploration, normalization, and summaries of high density oligonucleotide array probe level data. *Biostatistics* **4,** 249–264.

Kane, M. D., Jatkoe, T. A., Stumpf, C. R., Lu, J., Thomas, J. D., and Madore, S. J. (2000). Assessment of the sensitivity and specificity of oligonucleotide (50mer) microarrays. *Nucleic Acids Res.* **28,** 4552–4557.

Kreil, D. P., Russell, R. R., and Russell, S. (2006). Microarray oligonucleotide probes. *Methods Enzymol.* **410,** 73–99.

Lockhart, D. J., Dong, H., Byrne, M. C., Follettie, M. T., Gallo, M. V., Chee, M. S., Mittmann, M., Wang, C., Kobayashi, M., Horton, H., and Brown, E. L. (1996). Expression monitoring by hybridization to high-density oligonucleotide arrays. *Nature Biotechnol.* **14,** 1675–1680.

Lu, Y., Liu, P. Y., Xiao, P., and Deng, H. W. (2005). Hotelling's T2 multivariate profiling for detecting differential expression in microarrays. *Bioinformatics* **21,** 3105–3113.

Newton, M. A., Noueiry, A., Sarkar, D., and Ahlquist, P. (2004). Detecting differential gene expression with a semiparametric hierarchical mixture method. *Biostatistics* **5,** 155–176.

Qin, L. X., and Kerr, K. F. (2004). Empirical evaluation of data transformations and ranking statistics for microarray analysis. *Nucleic Acids Res.* **32,** 5471–5479.

Ramaswamy, S., and Golub, T. R. (2002). DNA microarrays in clinical oncology. *J. Clin. Oncol.* **20,** 1932–1941.

Salit, M. (2006). Standards in gene expression microarray experiments. *Methods Enzymol.* **411,** 63–78.

Smyth, G. K., Michaud, J., and Scott, H. S. (2005). Use of within-array replicate spots for assessing differential expression in microarray experiments. *Bioinformatics* **21,** 2067–2075.

Staudt, L. M., and Brown, P. O. (2000). Genomic views of the immune system. *Annu. Rev. Immunol.* **18,** 829–859.

van Bakel, H., and Holstege, F. C. (2004). In control: Systematic assessment of microarray performance. *EMBO Rep.* **5,** 964–969.

Wang, H. Y., Malek, R. L., Kwitek, A. E., Greene, A. S., Luu, T. V., Behbahani, B., Frank, B., Quackenbush, J., and Lee, N. H. (2003). Assessing unmodified 70-mer oligonucleotide probe performance on glass-slide microarrays. *Genome Biol.* **4,** R5.

Wolber, P. K., Collins, P. J., Lucas, A. B., De Witte, A., and Shannon, K. W. (2006). The Aligent *in-situ*-synthesized microarray platform. *Methods Enzymol.* **410,** 28–57.

Yang, Y. H., Xiao, Y., and Segal, M. R. (2005). Identifying differentially expressed genes from microarray experiments via statistic synthesis. *Bioinformatics* **21,** 1084–1093.

[5] Standards in Gene Expression Microarray Experiments

By MARC SALIT

Abstract

The use of standards in gene expression measurements with DNA microarrays is ubiquitous—they just are not yet the kind of standards that have yielded microarray gene expression profiles that can be readily compared across different studies and different laboratories. They also are not yet enabling microarray measurements of the known, verifiable quality needed so they can be used with confidence in genomic medicine in regulated environments.

Introduction

This chapter highlights current applications and roles of standards in enabling successful microarray studies and describes emerging standards activities intended to bridge the gap to permit microarrays to realize their potential as a key technology in genomic medicine.

As described throughout this volume and others, DNA microarrays are powerful tools to perform genome-wide gene expression screening and contrast experiments. DNA microarray experiments are expensive investments in terms of supplies (arrays and associated reagents) and time in the laboratory and in subsequent analysis of the copious data measured.

Also as described in numerous publications, comparing results from different DNA microarray gene expression profiling experiments is challenging and inconsistent. Even interpreting the variety of comparisons is challenging and currently fraught with subjective assertion (Bammler *et al.*, 2005; Barczak *et al.*, 2003; Hughes *et al.*, 2001; Irizarry *et al.*, 2005; Kothapalli *et al.*, 2002; Kuo

METHODS IN ENZYMOLOGY, VOL. 411
0076-6879/06 $35.00
DOI: 10.1016/S0076-6879(06)11005-8

et al., 2002; Larkin *et al.*, 2005; Mah *et al.*, 2004; Mecham *et al.*, 2004; Petersen *et al.*, 2005; Rogojina *et al.*, 2003; Shi *et al.*, 2005a; Shippy *et al.*, 2004; Tan *et al.*, 2003; Yauk *et al.*, 2004; Yuen *et al.*, 2002). If it is difficult to compare meaningfully results from different experiments, what can be asserted about the ability to reproduce results of a study from laboratory to laboratory? As the field is maturing, performance and understanding are improving, but expectations are also rising.

It has been speculated that "standards" for microarray experiments will solve these challenges.

While it is clear that standards have a role to play in the gene expression profiling with DNA microarray enterprise, they are unlikely to be a complete anodyne. This chapter describes how standards might be used to provide an improved understanding of DNA microarray results and their quality and it also discusses requirements for being able to compare microarray results across and between studies.

Variability

Aside from the biological complexity of gene expression, some of the challenges in interpreting and using DNA microarray gene expression profiles arise from the measurement process itself. Biological questions can best be addressed when the measurement process is well understood and controlled and results can be compared and understood more meaningfully. Figure 1 contains a model for the measurement process, useful to identify sources of variability (each process step is subject to variability), and a useful framework for the discussion of standards. The major segments of the process are annotated with a description of standards approaches that are in place or in development, the details of which are discussed in the body of this chapter.

A generic microarray gene expression measurement process model includes RNA isolation, an optional mRNA amplification step, microarray target preparation, selection of microarray probe content, hybridization to a microarray, detection of hybridized targets, calculation of an expression measure from the detection data, and biostatistical analysis of the expression measures. Each process element introduces measurement variability. Some elements are as fundamental as the bioinformatics of identifying probe sequences that bind to an mRNA corresponding to a gene, whereas others are as seemingly trivial as the time-dependent oxidative degradation of the fluorescent label reagent attached to the sample RNA being characterized in the experiment.

Standards are often used to manage variability. An appropriate standard inserted into a process element can help bring that element into

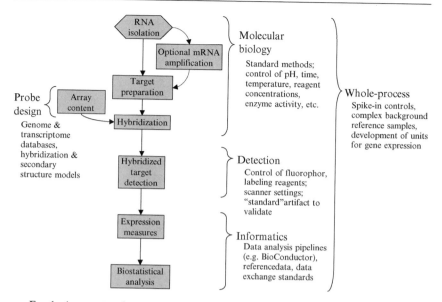

FIG. 1. An annotated process model for microarray gene expression measurements.

control, assuring consistent performance from study to study, and even from laboratory to laboratory. Quantitative assessment of variability becomes more tractable once a standard is used in a process element.

In experienced microarray laboratories, variability is managed with the adoption of carefully designed protocols and methods. In practice, these protocols and methods act as standards—*standard methods*. The chapters of Section II in the companion volume (vol. 410) present a variety of approaches that can be considered as practical standard methods and protocols. Successful microarray experiments demonstrate that these standards can be sufficient to manage parameters within the course of a single study. While standard methods can be effective, it can be difficult to demonstrate objectively that they have been without having a standard material to compare against.

Control of parameters solely with standard methods can be difficult across different laboratories and across different reagent lots, especially when some of the parameters/settings relate to the activity of enzymes. It is also nearly impossible to establish post hoc the parameter settings of an experiment. Even if the settings could be determined post hoc, knowing how the results depend on the settings, and adjusting the data accurately, would be impractical. Thus arise some of the challenge of using results across studies.

Some approaches taken to address similar problems in other measurement areas are measurement traceability, method validation, and measurement uncertainty.

A Digression: Traceability, Validation, and Uncertainty

Advances in the practice of chemical measurement science over the last several decades have established a system by which measurements can be brought into control, results can be compared, and the quality of results understood. Lessons learned from this system can be applied directly to microarray measurements.

There are three technical elements of the system: *measurement traceability* (Eurchem/CITAC, 2003; Salit, 2005), *method validation* (Green, 1996; Holcombe, 1998), and *uncertainty quantification* (Ellison, 2000). The widespread implementation of quality systems based in principle on ISO 9000, and now in compliance with ISO 17025, is a fourth, *procedural* element contributing to contemporary practice in chemical measurement.

Traceability is the property of a measurement result, whereby that result can be compared with some reference, usually a recognized standard of some sort. Method validation is the provision of objective evidence that a measurement is fit for its intended purpose—for a microarray measurement, that would be demonstration that an expression measure reflects a relative measure of the transcript concentration. Quantitative uncertainty estimates are statistical measures that characterize the range of values within which a true value is asserted to lie; these estimates are usually associated with a confidence interval, often 95%, by convention. Standards play a variety of roles in traceability, validation, and uncertainty estimation.

In practice, traceability is often how a measurement result "gets its units." Results that are traceable to a common reference can be compared. For example, the mass of a sample determined on a calibrated laboratory balance will be traceable, usually through a chain of measurements, all the way to the artifact kilogram in a vault at the International Bureau of Weights and Measures (BIPM) in Sèvres, France. The mass of that sample can be expressed in kilograms and can be compared to the mass of other samples, also expressed in kilograms. Results measured in different places, on other balances, at other times, can be compared meaningfully and quantitatively.

Method validation in chemistry typically establishes the specificity, linearity, accuracy, precision, dynamic range, detection limit, quantitation limit, and robustness of a measurement. In addition to these measurements of technical performance, good validation also provides evidence that the measurement process is stable over some period of time. This information, taken together, provides the scientist with reasonable expectations about the quality and reliability of the result. Armed with reasonable expectations, a result can be interpreted meaningfully and reasonable judgment can be applied regarding the suitability for intended use.

Quantitative uncertainty estimates are an essential element of a result, allowing meaningful comparison with other results and providing a quantitative measure of quality. Even when two results are different numerically, are they truly different? With a numerical estimate of the most likely value of a result (often the mean of several measurements), accompanied by an estimate of the range of likely values (often estimated from the standard deviation of observed values in combination with factors), meaningful assertion can be made regarding agreement or disagreement of results.

The role that standards play in traceability is obvious: they are used as the common references to link the quantitative value of measurement results. A reference might be of grand scope, such as the artifact kilogram at the BIPM, affectionately called "*Le Gran K*," or might be of minor scope, such as a particular monoclonal antibody used in a home-brew assay for a clinically relevant protein in a clinical laboratory. The scope of the reference dictates the scope of comparison supported by the traceability. Results can be compared to other results that are traceable to a common reference. In our example of the monoclonal antibody assay, results can only be compared to other results determined with that particular antibody.

If method validation is intended to provide the scientist with reasonable expectations about the result quality to be expected from the method, then it is essential that the samples introduced to probe the method be well trusted, well characterized, and reliable. Standards are usually ideal to use as the well-trusted samples, with known values, in method validation. These standards are often "Certified Reference Materials," whose properties of interest have been "certified"—having at least one relevant property whose value in the material has been established as homogeneous, which has been determined by a procedure that establishes traceability to some stated reference, with a reported uncertainty at a stated level of confidence.

The uncertainty of a result is typically quantified in pieces. The list of pieces, usually called an "uncertainty budget," would include components arising from the assortment of elements that give rise to variability in the measurement. There might be elements for uncertainty in volume from pipetting, uncertainty in quantity of nucleic acid from spectrophotometry, from variability in RNA integrity, from efficiency of labeling and amplification, from performance of the microarray, and from performance of the microarray scanner. Many of these elements would be quantified during method validation, where procedure performance was characterized with standard samples.

Standards in Traceability, Validation, and Uncertainty for DNA Microarray Gene Expression Profiles

There were hints in our digression as to how traceability, validation, and uncertainty might be relevant in DNA microarray measurements. Certainly, these concepts are already in place, even if not self-consciously.

Standards and Traceability

For example, traceability is the foundation of a common approach of microarray analysis, the "reference design" two-color microarray experiment. In this experiment, the reference RNA used in all arrays is the "standard" to which all results are traceable. By design, the sample RNA is compared to the reference, which is used in common to permit sample-to-sample comparison. Some laboratories have developed large, multistudy batches of reference RNA for the organisms/tissues under study, increasing the scope of traceability beyond a single study to permit comparisons over time and studies. Similar considerations have led to commercially prepared reference RNA, which has been developed and widely adopted for commonly studied species (human, rat, mouse) (Novoradovskaya et al., 2004). At the time of this writing, there is as yet no effort to establish such a multigene reference material as a certified reference material.

Another aspect of traceability (as noted earlier) relates to the practice of using common units to express quantities. There is a conversation about establishing a unit for gene expression, hosted by the "Gene Expression Units Working Group" of the UK Measurements for Biotechnology (MfB) program, a program of the UK Department of Trade and Industry, hosted by LGC, Ltd. This working group is exploring the requirements and opportunities to establish a unit to express magnitude of gene expression, tied to the International System of Units (SI). Expressing quantities in common units enables comparisons of absolute (as opposed to relative) magnitudes, and tying the units to the SI relates the quantities to a stable, coherent, and maintained system. This effort would enable measurement comparability by providing a common means of reporting gene expression magnitude.

Standards and Validation

A series of workshops hosted by the U.S. National Institute of Standards and Technology (NIST) spawned an ad hoc consortium, the External RNA Control Consortium (ERCC). This industry-led,

NIST-hosted consortium is dedicated to supporting method validation by making available a reference set of RNA controls, protocols for their use, and a statistical analysis tool for data interpretation that can be used across microarray platforms and quantitative reverse transcriptase polymerase chain reaction (RT-PCR) (Baker et al., 2005; Cronin et al., 2004). Chapter 4 in this volume discusses the application of external RNA controls in more depth. These materials can be used to establish the technical performance of a gene expression experiment, in particular to help respond to a call by FDA scientists for the demonstration of "...sufficient sensitivity, specificity, reproducibility, robustness, reliability, accuracy, precision and clinical relevance of the chosen microarray platform application" to support regulation (Petricoin et al., 2002).

The ERCC standards are intended to be added, or "spiked-in" to a sample, and tracked through the measurement process. To distinguish these RNA molecules from endogenous mRNA from mammalian species and common model organisms, they are derived from plant, bacterial, and random antigenomic sequences. The reference set is intended to contain approximately 100 sequences, to be selected from a larger library of candidate sequences, based on their performance in a thorough cross-platform testing experiment (ERCC, 2005).

At the time of this writing, these spike-in control standards will be unique in that the microarray probe content for the reference sequences will be widely commercially available upon their introduction. NIST intends to make clones of the sequences for the standards available as a Standard Reference Material (a certified reference material from NIST).

The stated intent of the ERCC is for these spike-in controls to be added to samples immediately after RNA isolation, at the stage of total RNA. For an array application, the controls would typically be added in a mixed pool, with different RNAs at different concentrations, permitting a "dose-response" curve to be measured from the spike-ins, in each sample. This application provides method validation and quality measures, focused on technical performance, by tracking the spike-ins through the process and assuring that the measured signals for the spike-ins are consistent with expectations. This information does not validate the biological or clinical inference of the measurement, but can give a "red light" or a "yellow light" (but not a "green light"). The protocols and statistical analysis tool will be limited to supporting this use.

However, even before their availability, there are a variety of proposals to use these reference sequences (and standards created from them) to do finer-grained validation of the process. Such approaches might use subsets of the reference set to monitor RNA isolation, mRNA amplification, monitor microarray target preparation distinctly from target labeling, and for use in "normalization" of array signals. This finer-grained validation would likely employ reagents specific to particular microarray platforms, for example, relying on spike-in transcripts prelabeled with a particular labeling moiety to validate separately the labeling/target preparation step from the hybridization step. The scope of the direct ERCC work has been limited to areas of common interests among the ~60 member organizations, focused on three gene expression platforms: quantitative RT-PCR and both one- and two-color microarrays. Thus, "off-label" validation applications will likely await release of the reference sequences and innovative development in the marketplace.

Method validation is at the core of a project being led by the U.S. Food and Drug Administration, the "microarray quality control" (MAQC) project (Shi, 2005). This public–private–academic project is conducting a large-scale array experiment on multiple microarray platforms, at multiple sites, with two complex RNA samples. The study will be composed of approximately 1000 microarray hybridizations, as well as several other gene expression measures, including quantitative RT-PCR of ~1000 genes. This study may serve to describe a variety of measures of the mRNA populations of these samples, to present a "snapshot" of gene expression measurement technology, and to establish a pair of materials useful for method validation.

One criterion for the selection of the complex RNAs in the MAQC study was availability of sufficient supply material to establish them as references over a period of several years. Additional selection criteria included the goal of having a pair of materials that showed a range of differential gene expression at a range of concentration. Samples selected from results of a pilot experiment are both commercially available materials: one manufactured from a composite of human cell lines and the other extracted from a human brain composite.

These materials are being measured in a titration of ratios of 1:0, 0.75:0.25, 0.25:0.75, and 0:1. A common experiment design is being used at every participating measurement site, with five technical replicates of each of the four samples. Measuring mixtures of materials offer an opportunity to assess linearity of expression measures, one important component of accuracy.

More important perhaps than the characterization of the materials with the diverse set of measurement tools in the study may be the availability of large quantities (anticipated to last for several years) of homogeneous preparations of the complex RNAs. With sufficient quantities available, these may be useful reference materials for performance demonstration and method validation. An example use of such materials might be to demonstrate performance equivalence of alternate protocols for microarray target preparation, permitting a laboratory to modify, refine, or optimize their procedures with confidence. This is exactly the sort of circumstance where method validation can play a role in assuring measurement quality and integrity.

Standards and Uncertainty

In the analysis of microarray results, a precision-only estimate of uncertainty is typically used to select differentially expressed genes—a t test is used to identify genes that differ between control and test groups with statistical significance (of course the t test compares the precision with the difference between the means, but the effect of measurement uncertainty is based only on the observed precision). This precision-only estimate is reasonably appropriate for identifying differential expression between groups in a study, as it can be reasonably assumed that other elements of uncertainty are the same among samples and between classes.

In fact, using the observed precision as a measure of uncertainty may sometimes be a reasonable approximation. An important rule of thumb when quantifying uncertainty is that components with magnitude smaller than about one-third of the largest can safely be ignored. If variability is dominated strongly by biological variation of the mRNA concentration for the gene under study, other elements of uncertainty can be neglected.

As microarray gene expression measurement matures, a more complete uncertainty budget will emerge, with elements assessed quantitatively in the method validation process. The role of standards in quantifying measurement uncertainty will be determined by their role in method validation.

Data Exchange Standards

The initial impacts of standards in microarray measurements arose through the pioneering efforts of the Microarray Gene Expression Data (MGED) society (Ball *et al.*, 2002, 2004a,b,c). This group has developed a standard to describe a microarray experiment, the "minimum amount of information about a microarray experiment" (MIAME). MGED describes MIAME as the information "...needed to enable the interpretation of the results of the experiment unambiguously and potentially to reproduce the experiment." MGED has successfully encouraged adoption of the MIAME standard through its outreach to the community of scientific journals, many of which require publication of microarray data along with meeting MIAME requirements.

The stated focus of MGED is "... on establishing standards for microarray data annotation and exchange, facilitating the creation of microarray databases and related software implementing these standards, and promoting the sharing of high quality, well annotated data within the life sciences community." In addition to the development and maintenance of the MIAME standard, there is significant work ongoing on the "microarray and gene expression" (MAGE) standard for representation of microarray expression data itself. Taken together, these standards will broadly facilitate the exchange of microarray information. Examples of public gene expression data repositories that exploit the MIAME and MAGE standards are the Gene Expression Omnibus and ArrayExpress, described in Chapters 19 and 20 of this volume, respectively.

Data exchange is the initial predicate to the ability to mine data from different microarray studies in meta studies. Ultimately, common units for gene expression, standards to act as common references for mRNA concentration, standards and protocols for method validation, and quantitative uncertainty estimates will all be needed as infrastructure, in addition to data exchange, to enable the meta-study enterprise.

Standards in the Gene Expression Process Model

Examining the generic process model in Fig. 1, one observes that, to a lesser or greater degree, traceability, method validation, and quantification of uncertainty are deployed in each process segment.

The Molecular Biology Segment: RNA Isolation, mRNA Amplification, Target Preparation, and Hybridization

Standards to establish traceability of RNA concentration are not used in contemporary microarray gene expression measurements. However, other standards used to establish traceability are "hidden in plain sight" throughout the experiment—used to control the various parameters that influence results, such as temperature, time, pH, and concentration of reagents throughout the molecular biology elements of the experiment, from RNA isolation through hybridization. In particular, for sensitive gene expression experiments it is vital that conditions be reproduced throughout a microarray study. This is accomplished by setting parameters on equipment calibrated using traceable standards, whether that calibration is obvious or not. Calibration of a pH meter used to adjust buffer pH is an obvious step, whereas calibration of a spectrophotometer, laboratory timer, or wall clock is often transparent, although essential. Traceability of the values of all quantities that influence the measurement assures that those values and settings can be reported and reproduced.

These molecular biology steps are routinely validated in microarray experiments, either directly in a study, when being commissioned in a laboratory, or through the use of "kits" or procedures validated elsewhere. The degree of rigor applied to validation varies with the practices and experience of the laboratory and with the requirements of the study. The reader is referred to the Eurachem/CITAC guide on validation for a thorough discussion of a variety of concepts and practices used in validation (Holcombe, 1998).

Only "intuitive" uncertainty estimation is typically practiced for these molecular biology elements. The development and validation of the process have established that, when performed under controlled conditions (as described earlier), the uncertainty of the overall process is not dominated by these operations.

Array Content

No standards exist yet for describing microarray probe content. Public and private genome sequences are well established for model organisms, and these genomes are the *de facto* standards against which the oligonucleotide probes in common commercial application are designed. These genomes are also used to assemble a stable reference for gene identification and characterization, "RefSeq," which is the current definitive reference for gene sequences (Pruitt *et al.*, 2005). When designed in this manner, one can assert that the probe content is "traceable" to the genome database.

The quality of the public human genome database meets the target uncertainty of $< 0.01\%$, fewer than one single base error per 10,000 bases (Schmutz *et al.*, 2004). This work may serve as validation evidence for the carefully assembled human genome, but does not serve to validate all genomes; significant errors can remain in genome assembly and in less well-characterized genomes (Salzberg and Yorke, 2005). Sequence polymorphism in the organisms under study may vary considerably, although laboratory studies of model organisms often employ carefully bred strains with controlled diversity, and human polymorphism is estimated to be about 1 base in 1000 (Kwok *et al.*, 1996). Additionally, open questions exist about mapping observed mRNAs to the genome and should be considered as an as-yet uncharacterized source of potential confusion (Furey *et al.*, 2004).

Along with the sequence content, predictive models for the thermodynamics of hybridization and secondary structure can be considered as standards for array content. The "nearest-neighbor" predictive models act as *de facto* standards for estimating microarray probe-target affinity and selection of hybridization conditions (Markham and Zuker, 2005; SantaLucia, 1998). Model-based predictors of secondary structure are used to avoid selection of probe sequences that are subject to nonlinear conformations (Zuker, 2003). More comprehensive approaches more specific to microarray probe design have been described that embody elements of prediction for both hybridization thermodynamics and secondary structure, as well as estimates of selectivity (Markham and Zuker, 2005; Rouillard *et al.*, 2003). To the degree that models are consistent and accurate, array design standardized against them should exhibit smaller variability and better comparability.

Detection

Detection of hybridized targets is typically done with an optical fluorescence measurement. The "standards" used in this process element are typically of very limited scope—the fluorophor and labeling reagent batches and the microarray scanner used in a study. Variability in reagents and fluorophors, and scanner-to-scanner variability, pose limitations on the ability to compare results across studies through traceability. These factors need not be barriers to validation and uncertainty estimation within a study, but they present a challenge to more global validation and uncertainty estimates.

Variability in microarray scanner performance has been reported, most notably with the setting of detection parameters (photomultiplier tube and laser excitation power) (Lyng *et al.*, 2004; Shi *et al.*, 2005b). A commercially available artifact with typically used microarray fluorophors spotted in a concentration gradient has been used to characterize scanners (Zong *et al.*, 2003). While this artifact has been used in the aforementioned references, there has been no report of using this artifact for traceable calibration, for formal validation, or for uncertainty estimation.

Expression Measure Estimation and Biostatistical Analysis

Standards have played an important role in validation of these ultimate gene expression profile analysis stages. These standards have been of two sorts: common analysis software/analysis pipelines and reference data sets.

A prominent software/analysis pipeline for microarray data biostatistical analysis is BioConductor (Gentleman *et al.*, 2004), described in depth in Chapter 8 of this volume. Bioconductor is an open source and open development software project for the analysis and comprehension of genomic data based on the open source statistical analysis package, *R*. The Bioconductor analysis pipeline has become a *de facto* standard for microarray data analysis and has an active community that provides a lively and collaborative peer-review environment for biostatistical algorithm development and application. A lively email discussion group provides expert technical and scientific support.

Reference data sets have had a particularly strong impact for the Affymetrix GeneChip platform, where numerous approaches have been developed to summarize results from multiple short (25 base oligonucleotide) probes into expression measures for a gene. Affymetrix has made publicly available several multiarray data sets containing sets of genes externally added, or "spiked-in," to a complex background in a Latin square experimental design (Liu *et al.*, 2002). These "Latin square" experiments permit performance evaluation and validation through the assessment of observed versus expected gene spike-in levels. For example, these data, used as a reference, have spawned an effort to objectively compare summary expression measure algorithm performance (Cope *et al.*, 2004).

The Future

As microarray technology matures, the role of standards will become better defined, and standard methods, reference materials, reference data, and models will take on more explicit roles in traceability, method validation, and uncertainty estimation. With this maturity and standards infrastructure, microarray result quality will be more confidently quantitatively assessed, and it will be possible to compare meaningfully results measured in different laboratories, at different times, with different equipment.

References

Baker, S. C., Bauer, S. R., Beyer, R. P., Brenton, J. D., Bromley, B., Burrill, J., Causton, H., Conley, M. P., Elespuru, R., Fero, M., Foy, C., Fuscoe, J., Gao, X., Gerhold, D. L., Gilles, P., Goodsaid, F., Guo, X., Hackett, J., Hockett, R. D., Ikonomi, P., Irizarry, R. A., Kawasaki, E. S., Kaysser-Kranich, T., Kerr, K., Kiser, G., Koch, W. H., Lee, K. Y., Liu, C., Liu, Z. L., Lucas, A., Manohar, C. F., Miyada, G., Modrusan, Z., Parkes, H., Puri, R. K., Reid, L., Ryder, T. B., Salit, M., Samaha, R. R., Scherf, U., Sendera, T. J., Setterquist, R. A., Shi, L., Shippy, R., Soriano, J. V., Wagar, E. A., Warrington, J. A., Williams, M., Wilmer, F., Wilson, M., Wolber, P. K., Wu, X., and Zadro, R. (2005). The External RNA Controls Consortium: A progress report. *Nature Methods* **2**(10), 731–734.
Ball, C., Brazma, A., Causton, H., Chervitz, S., Edgar, R., Hingamp, P., Matese, J. C., Icahn, C., Parkinson, H., Quackenbush, J., Ringwald, M., Sansone, S. A., Sherlock, G., Spellman, P., Stoeckert, C., Tateno, Y., Taylor, R., White, J., and Winegarden, N. (2004a). An open letter on microarray data from the MGED society. *Microbiology* **150**(Pt 11), 3522–3524.
Ball, C., Brazma, A., Causton, H., Chervitz, S., Edgar, R., Hingamp, P., Matese, J. C., Parkinson, H., Quackenbush, J., Ringwald, M., Sansone, S. A., Sherlock, G., Spellman, P.,

Stoeckert, N., Tateno, Y., Taylor, R., White, J., and Winegarden, N. (2004b). Standards for microarray data: An open letter. *Environ. Health Perspect* **112**(12), A666–A667.

Ball, C. A., Brazma, A., Causton, H., Chervitz, S., Edgar, R., Hingamp, P., Matese, J. C., Parkinson, H., Quackenbush, J., Ringwald, M., Sansone, S. A., Sherlock, G., Spellman, P., Stoeckert, C., Tateno, Y, Taylor, R., White, J., and Winegarden, N. (2004c). Submission of microarray data to public repositories. *PLoS Biol.* **2**(9), E317.

Ball, C. A., Sherlock, G., Parkinson, H., Rocca-Sera, P., Brooksbank, C., Causton, H. C., Cavalieri, D., Gaasterland, T., Hingamp, P., Holstege, F., Ringwald, M., Spellman, P., Stoeckert, C. J., Jr., Stewart, J. E., Taylor, R., Brazma, A., and Quackenbush, J. (2002). Standards for microarray data. *Science* **298**(5593), 539.

Bammler, T., Beyer, R., Bhattacharya, S., Boorman, G., Boyles, A., Bradford, B., Bumgarner, R., Bushel, P., ChaturvediK Choi, D., Cunningham, M., Deng, S., Dressman, H., Fannin, R., Farin, F., Freedman, J., Fry, R., Harper, A., Humble, M., Hurban, P., Kavanagh, T., Kaufmann, W., Kerr, K., Jing, L., Lapidus, J., Lasarev, M., Li, J., Li, Y., Lobenhofer, E., Lu, X., Malek, R., Milton, S., Nagalla, S., O'malley, J., Palmer, V., Pattee, P., Paules, R., Perou, C., Phillips, K., Qin, L., Qiu, Y., Quigley, S., Rodland, M., Rusyn, I., Samson, L., Schwartz, D., Shi, Y., Shin, J., Sieber, S., Slifer, S., Speer, M., Spencer, P., Sproles, D., Swenberg, J., Suk, W., Sullivan, R., Tian, R., Tennant, R., Todd, S., Tucker, C., Van Houten, B., Weis, B., Xuan, S., and Zarbl, H. (2005). Standardizing global gene expression analysis between laboratories and across platforms. *Nature Methods* **2**(5), 351–356.

Barczak, A., Rodriguez, M., Hanspers, K., Koth, L., Tai, Y., Bolstad, B., Speed, T., and Erle, D. (2003). Spotted long oligonucleotide arrays for human gene expression analysis. *Genome Res.* **13**(7), 1775–1785.

Cope, L. M., Irizarry, R. A., Jaffee, H. A., Wu, Z., and Speed, T. P. (2004). A benchmark for Affymetrix GeneChip expression measures. *Bioinformatics* **20**(3), 323–331.

Cronin, M., Ghosh, K., Sistare, F., Quackenbush, J., Vilker, V., and O'Connell, C. (2004). Universal RNA reference materials for gene expression. *Clin. Chem.* **50**(8), 1464–1471.

Ellison, S. L. R., Rosslein, M., and Williams, A. (eds.) (2000). "Quantifying Uncertainty in Analytical Measurement."

Eurchem/CITAC. (2003). "Traceability in Chemical Measurement: A Guide to Achieving Comparable Results in Chemical Measurement."

External RNA Control Consortium (ERCC). (2005). Proposed methods for testing and selecting the ERCC external RNA controls. *BMC Genom.* **6**(1), 150.

Furey, T. S., Diekhans, M., Lu, Y., Graves, T. A., Oddy, L., Randall-Maher, J., Hillier, L. W., Wilson, R. K., and Haussler, D. (2004). Analysis of human mRNAs with the reference genome sequence reveals potential errors, polymorphisms, and RNA editing. *Genome Res.* **14**(10B), 2034–2040.

Gentleman, R. C., Carey, V. J., Bates, D. M., Bolstad, B., Dettling, M., Dudoit, S., Ellis, B., Gautier, L., Ge, Y., Gentry, J., Hornik, K., Hothorn, T., Huber, W, Iacus, S., Irizarry, R., Leisch, F., Li, C., Maechler, M., Rossini, A. J., Sawitzki, G., Smith, C., Smyth, G., Tierney, L., Yang, J. Y., and Zhang, J. (2004). Bioconductor: Open software development for computational biology and bioinformatics. *Genome Biol.* **5**(10), R80.

Green, J. M. (1996). A practical guide to analytical method validation. *Anal. Chem.* **68**, 305A–309A.

Holcombe, D. (1998). "The Fitness for Purpose of Analytical Methods: A Laboratory Guide to Method Validation and Related Topics." Teddington, Middlesex, UK.

Hughes, T., Mao, M., Jones, A., Burchard, J., Marton, M., Shannon, K., Lefkowitz, S., Ziman, M., Schelter, J., Meyer, M., Kobayashi, S., Davis, C., Dai, H., He, Y., Stephaniants, S., Cavet, G., Walker, W., West, A., Coffey, E., Shoemaker, D., Stoughton, R., Blanchard, A., Friend, S., and Linsley, P. (2001). Expression profiling using microarrays fabricated by an ink-jet oligonucleotide synthesizer. *Nat. Biotechnol.* **19**(4), 342–347.

Irizarry, R., Warren, D., Spencer, F., Kim, I., Biswal, S., Frank, B., Gabrielson, E., Garcia, J., Geoghegan, J., Germino, G., Griffin, C., Hilmer, S., Hoffman, E., Jedlicka, A., Kawasaki, E., Martinez-Murillo, F., Morsberger, L., Lee, H., Petersen, D., Quackenbush, J., Scott, A., Wilson, M., Yang, Y., Ye, S., and Yu, W. (2005). Multiple-laboratory comparison of microarray platforms. *Nature Methods* **2**(5), 345–350.

Kothapalli, R., Yoder, S., Mane, S., and Loughran, T. J. (2002). Microarray results: How accurate are they? *BMC Bioinform.* **3**, 22.

Kuo, W., Jenssen, T., Butte, A., Ohno-Machado, L., and Kohane, I. (2002). Analysis of matched mRNA measurements from two different microarray technologies. *Bioinformatics* **18**(3), 405–412.

Kwok, P. Y., Deng, Q., Zakeri, H., Taylor, S. L., and Nickerson, D. A. (1996). Increasing the information content of STS-based genome maps: Identifying polymorphisms in mapped STSs. *Genomics* **31**(1), 123–126.

Larkin, J., Frank, B., Gavras, H., Sultana, R., and Quackenbush, J. (2005). Independence and reproducibility across microarray platforms. *Nature Methods* **2**(5), 337–344.

Liu, W. M., Mei, R., Di, X., Ryder, T. B., Hubbell, E., Dee, S., Webster, T. A., Harrington, C. A., Ho, M. H., Baid, J., and Smeekens, S. P. (2002). Analysis of high density expression microarrays with signed-rank call algorithms. *Bioinformatics* **18**(12), 1593–1599.

Lyng, H., Badiee, A., Svendsrud, D. H., Hovig, E., Myklebost, O., and Stokke, T. (2004). Profound influence of microarray scanner characteristics on gene expression ratios: Analysis and procedure for correction. *BMC Genom.* **5**(1), 10.

Mah, N., Thelin, A., Lu, T., Nikolaus, S., Kuhbacher, T., Gurbuz, Y., Eickhoff, H., Kloppel, G., Lehrach, H., Mellgard, B., Costello, C., and Schreiber, S. (2004). A comparison of oligonucleotide and cDNA-based microarray systems. *Physiol. Genom.* **16**(3), 361–370.

Markham, N. R., and Zuker, M. (2005). DINAMelt web server for nucleic acid melting prediction. *Nucleic Acids Res.* **33**(Web Server issue), W577–W581.

Mecham, B., Klus, G., Strovel, J., Augustus, M., Byrne, D., Bozso, P., Wetmore, D., Mariani, T., Kohane, I., and Szallasi, Z. (2004). Sequence-matched probes produce increased cross-platform consistency and more reproducible biological results in microarray-based gene expression measurements. *Nucleic Acids Res.* **32**(9), e74.

Novoradovskaya, N., Whitfield, M. L., Basehore, L. S., Novoradovsky, A., Pesich, R., Usary, J., Karaca, M., Wong, W. K., Aprelikova, O., Fero, M., Perou, C. M., Botstein, D., and Braman, J. (2004). Universal reference RNA as a standard for microarray experiments. *BMC Genom.* **5**(1), 20.

Petersen, D., Chandramouli, G., Geoghegan, J., Hilburn, J., Paarlberg, J., Kim, C., Munroe, D., Gangi, L., Han, J., Puri, R., Staudt, L., Weinstein, J., Barrett, J., Green, J., and Kawasaki, E. (2005). Three microarray platforms: An analysis of their concordance in profiling gene expression. *BMC Genom.* **6**(1), 63.

Petricoin, E. F., 3rd, Hackett, J. L., Lesko, L. J., Puri, R. K., Gutman, S. I., Chumakov, K., Woodcock, J., Feigal, D. W., Jr., Zoon, K. C., and Sistare, F. D. (2002). Medical applications of microarray technologies: A regulatory science perspective. *Nature Genet.* **32**(Suppl.), 474–479.

Pruitt, K. D., Tatusova, T., and Maglott, D. R. (2005). NCBI Reference Sequence (RefSeq): A curated non-redundant sequence database of genomes, transcripts and proteins. *Nucleic Acids Res.* **33**, D501–D504.

Rogojina, A., Orr, W., Song, B., and Geisert, E. J. (2003). Comparing the use of Affymetrix to spotted oligonucleotide microarrays using two retinal pigment epithelium cell lines. *Mol. Vis.* **9**, 482–496.

Rouillard, J. M., Zuker, M., and Gulari, E. (2003). OligoArray 2.0: Design of oligonucleotide probes for DNA microarrays using a thermodynamic approach. *Nucleic Acids Res.* **31**(12), 3057–3062.

Salit, M. L., and Turk, G. C. T. (2005). Raceability of single element calibration solutions. *Anal. Chem.* **77,** 136A–141A.

Salzberg, S. L., and Yorke, J. A. (2005). Beware of mis-assembled genomes. *Bioinformatics* **21**(24), 4320–4321.

SantaLucia, J., Jr. (1998). A unified view of polymer, dumbbell, and oligonucleotide DNA nearest-neighbor thermodynamics. *Proc. Natl. Acad. Sci. USA* **95**(4), 1460–1465.

Schmutz, J., Wheeler, J., Grimwood, J., Dickson, M., Yang, J., Caoile, C., Bajorek, E., Black, S., Chan, Y. M., Denys, M., Escobar, J., Flowers, D., Fotopulos, D., Garcia, C., Gomez, M., Gonzales, E., Haydu, L., Lopez, F., Ramirez, L., Retterer, J., Rodriguez, A., Rogers, S., Salazar, A., Tsai, M., and Myers, R. M. (2004). Quality assessment of the human genome sequence. *Nature* **429**(6990), 365–368.

Shi, L. (2005). MicroArray Quality Control (MAQC) Project. U.S. Food and Drug Administration.

Shi, L., Tong, W., Fang, H., Scherf, U., Han, J., Puri, R., Frueh, F., Goodsaid, F., Guo, L., Su, Z., Han, T., Fuscoe, J., Xu, Z., Patterson, T., Hong, H., Xie, Q., Perkins, R., Chen, J., and Casciano, D. (2005a). Cross-platform comparability of microarray technology: Intra-platform consistency and appropriate data analysis procedures are essential. *BMC Bioinform.* **6**(Suppl. 2), S12.

Shi, L., Tong, W., Su, Z., Han, T., Han, J., Puri, R. K., Fang, H., Frueh, F. W., Goodsaid, F. M., Guo, L., Branham, W. S., Chen, J. J., Xu, Z. A., Harris, S. C., Hong, H., Xie, Q., Perkins, R. G., and Fuscoe, J. C. (2005b). Microarray scanner calibration curves: Characteristics and implications. *BMC Bioinform.* **6**(Suppl 2), S11.

Shippy, R., Sendera, T., Lockner, R., Palaniappan, C., Kaysser-Kranich, T., Watts, G., and Alsobrook, J. (2004). Performance evaluation of commercial short-oligonucleotide microarrays and the impact of noise in making cross-platform correlations. *BMC Genom.* **5**(1), 61.

Tan, P., Downey, T., Spitznagel, E. J., Xu, P., Fu, D., Dimitrov, D., Lempicki, R., Raaka, B., and Cam, M. (2003). Evaluation of gene expression measurements from commercial microarray platforms. *Nucleic Acids Res.* **31**(19), 5676–5684.

Yauk, C., Berndt, M., Williams, A., and Douglas, G. (2004). Comprehensive comparison of six microarray technologies. *Nucleic Acids Res.* **32**(15), e124.

Yuen, T., Wurmbach, E., Pfeffer, R., Ebersole, B., and Sealfon, S. (2002). Accuracy and calibration of commercial oligonucleotide and custom cDNA microarrays. *Nucleic Acids Res.* **30**(10), e48.

Zong, Y. W. Y., Zhang, S., and Shi, Y. (2003). How to evaluate a microarray scanner. *In* "Microarray Methods and Applications: Nuts and Bolts" (H. G. Eagleville, ed.), pp. 99–114. DNA Press.

Zuker, M. (2003). Mfold web server for nucleic acid folding and hybridization prediction. *Nucleic Acids Res.* **31**(13), 3406–3415.

[6] Scanning Microarrays: Current Methods and Future Directions

By Jerilyn A. Timlin

Abstract

The microarray platform is a powerful tool for conducting large-scale, high-throughput gene expression experiments. However, careful attention to detail throughout the five major steps in the microarray process—design, printing, hybridization, scanning, and analysis—must be used to ensure that reliable and accurate conclusions are obtained from data. The act of scanning the array has received the least attention of all parts of the microarray process, despite it being a critical quality-limiting component. This chapter specifically addresses the effects of scan parameters and limitations of the scanning technology divided into two categories: instrumentation effects (those that arise from the scanning instrumentation itself) and user-controller parameters (those that an operator chooses) for the most common microarray platform—the two-color cDNA microarray printed on a glass substrate. Significant research efforts have gone into developing microarray analysis techniques, but the field is ripe for research to characterize the variability and errors introduced by the scanning process itself, the scanner instrumentation, and the user. Implications of these errors for large-scale, multiple slide and multiple laboratory experiments are discussed. Wise choices for scanning parameters and consideration of instrument specifics will ultimately increase data reliability and reduce the need for complex preprocessing mechanisms prior to the extraction of expression information. In addition, emerging technologies such as surface plasmon imaging, resonance light scattering, and hyperspectral imaging are presented briefly as promising, complementary techniques to traditional scanning methods.

Introduction

The microarray platform is a powerful tool for conducting large-scale, high-throughput gene expression experiments. However, careful attention to detail throughout the five major steps in the microarray process—design, printing, hybridization, scanning, and analysis—must be used to ensure that reliable and accurate conclusions are obtained from data. Several excellent reference books are available that provide thorough coverage of the entire

METHODS IN ENZYMOLOGY, VOL. 411
0076-6879/06 $35.00
DOI: 10.1016/S0076-6879(06)11006-X

microarray process (Schena, 1999, 2000, 2003). Ideally, the design phase should begin with a robust statistical design of the experiment and print layout including repeat genes and replicate arrays with and without dye swapping to permit assessment of the degree of confidence in the results (Hegde *et al.*, 2000; Kerr and Churchill, 2001). The printing process has been shown to introduce significant variability in data such as pin-dependent artifacts and correlations, spatial location effects, and nonuniform substrate background, as well as contaminating fluorescence (Balazsi *et al.*, 2003; Brown *et al.*, 2001; Martinez *et al.*, 2003; Tseng *et al.*, 2001). High-density oligomer arrays (such as Affymetrix GeneChips) are generally free from printing artifacts due to their *in situ* synthesis with photolithographic methods, but do have unique considerations surrounding cross-hybridization and probe selection. With all microarrays, the laboratory protocols for RNA extraction and DNA hybridization (Neal and Westwood, 2006; Browstein, 2006) should be scrutinized thoroughly to understand and minimize sources of variation (such as operator, time, and probe availability) and thus produce the highest quality microarrays. The act of scanning the array has received the least attention of all parts of the microarray process, despite it being a critical quality-limiting component. In contrast, the data analysis step of the microarray process has been the subject of much research. Many normalization and analysis methods have been developed, and a variety of reviews and articles are available to assist the microarray user (Hegde *et al.*, 2000; Quackenbush, 2001, 2002; Schena, 2003; Schuchhardt *et al.*, 2000; Wu *et al.*, 2001; Yang *et al.*, 2000). Often the advanced analysis methods reported in the literature seek to improve array results by compensating for problems that arise as a result of the scanning process or as a limitation of the scanner being used. A better understanding of the effects of the sources of variation within a microarray experiment will ultimately lead to improvements in the quality of array data and subsequent biological inference. This chapter specifically addresses the effects of scan parameters and limitations of the scanning technology. Figure 1 presents the steps in the microarray process and accents the focus of this chapter.

Scanning a microarray is a fairly simple task to execute, but it involves selection of a variety of parameters that can have profound effects on the resulting data. These effects can be divided into two categories: instrumentation effects (those that arise from the scanning instrumentation itself) and user-controller parameters (those that an operator chooses). This chapter presents scanning methods for the most common microarray platform: the two-color cDNA microarray printed on a glass substrate, although many of the considerations discussed within are generally applicable to other varieties of microarrays, including high-density oligomer arrays and protein arrays.

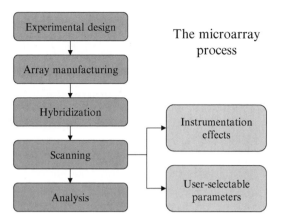

FIG. 1. Illustration of the microarray process. Experimental design is meant to include both design of the array and the hybridization experiments.

Overview of the Scanning Process

For the purposes of this chapter the process of scanning a microarray begins with a glass substrate that has been prepared with printed or synthesized DNA probes in known spatial locations. These probes have previously undergone a hybridization process whereby two different varieties of target DNA, typically labeled with a red and a green fluorescent dye, have had an opportunity to bind with their complementary probe. The overall goal of microarray scanning is to produce images that accurately locate and quantitate the amount of red and green fluorescent molecules on the microarray as these should correspond closely to the relative amounts of gene expression in the test and control samples. These images will serve as inputs into analysis methods for extracting the differences in the patterns of gene expression between the two varieties of target DNA.

Scanner instrumentation is detailed elsewhere (Schena, 1999, 2003). There are many commercially available microarray scanners, each differing slightly in configuration and capabilities, and although only the most common—the two-color microarray scanner utilizing photomultiplier tube (PMT) detection technology—is considered here, many of the points discussed within this chapter are applicable to other scanner varieties. In the two-color microarray scanner with PMTs, the red and green fluorescent molecules at each spatial pixel of the microarray are excited with independent lasers (simultaneously on some instruments and in subsequent scans on others), and discrete bands of the emitted red and green photons are passed through optical filters to independent PMTs for detection.

The detectors and computer transform the photons received into a digital value and typically a 16-bit tiff image corresponding to the intensity and location of each color of fluorescent molecule is created. Table I lists some of the commercially available microarray scanners and outlines their major features. It is important to consider that even if all the steps that create a microarray work without flaw (a rare occurrence, indeed), the process of scanning an array introduces its own uncertainty onto resulting data. Most microarray scanner manufacturers and array facilities have scanning protocols available on the web or in print. While these discuss the details and nuances of the scanner they are written for and are therefore very useful, they should not be considered as a substitute for careful considerations of your particular experiment and desired data. It is critical to make appropriate choices based on the experiment at hand to ensure that a minimal amount of variability is imparted to microarray data from scanning as well as throughout the entire multistep microarray process.

User-Controlled Parameters

Array Handling

Microarrays should be handling with gloves at all times throughout the manufacture and scanning process to minimize the potential for fingerprint contamination. The oils deposited by fingerprints fluoresce very strongly, especially in the red channel, and the resulting image will contain small bright flecks from the tiny pools of oil that severely complicate analysis and can compromise the resulting spot intensity values. Arrays should be stored in a sealed, dark environment to minimize collection of dust and potential dye degradation. It has been shown that ozone may be detrimental to the Cy5 label used in many microarray hybridizations (AgilentTechnologies, 2004; Fare et al., 2003). In general, microarray slides are not considered robust over the long term and should be scanned as soon as possible after hybridization to ensure minimal degradation from ozone, photobleaching, and chemical interactions. A more detailed discussion of photobleaching follows in the next section.

When scanning a microarray be sure to consult the microarray scanner's documentation regarding positioning of the slide and focus control. Some commercial microarray scanners place the slides in the slide holding tray hybridization side up, some hybridization side down. Confocal scanners are useful because they reject undesired emissions originating from above and below the printed spots, creating an image with a less out-of-focus signal (10-μm depth of field) (Schermer, 1999). However, the confocal nature makes these instruments extremely sensitive to focus position,

TABLE I

COMMERCIALLY AVAILABLE MICROARRAY SCANNERS FOR SCANNING GLASS SLIDE MICROARRAYS[a]

Company/scanner	Lasers	Detector	Confocal?	Pixel size (μm)	Scan type: Time	Focus control?	Autoloader available? (# of slides)	Additional filters available?
Agilent Technologies DNA Microarray Scanner G2565B	Two-color format; 532 nm, 633 nm	Two 16-bit PMTs (adjustable 1–100%)	Yes	5 or 10	Two-color simultaneous data acquisition: 8 min/slide at 10 μm resolution	Yes (dynamic adjustment)	Yes (48)	No
Applied Precision, LLC ArrayWoRx	White light, four-color UV to NIR (filter selectable)	One 14-bit cooled CCD camera	No	3.25–26	One-color sequential data acquisition: 5–7 min for a two-color slide at 10 μm resolution	Yes	Yes (25)	Yes (up to 8)
Biomedical Photometrics, Inc. DNAscope LM	Two-color format; 532 and 635 nm (488 nm also available)	One 16-bit PMT (adjustable)	Yes	2, 5, 10, 20, 30	One-color sequential data acquisition: ~7 min/slide/channel at 10 μm resolution	Yes	No	Yes (up to 10)
Biomedical Photometrics, Inc. DNAscope V	Two-color format; 532 and 635 nm (three additional available)	Two 16-bit PMTs (adjustable)	Yes	2, 5, 10, 20, 30	Two-color simultaneous data acquisition: 4 min/slide/channel at 10 μm resolution	Yes (dynamic adjustment)	No	Yes (up to 10)
Genomic Solutions Gene TAC UC4	Two-color format; 532 and 635 nm	One 16-bit PMT (adjustable)	Partial (dark field)	1–100 (5 for entire slide)	One-color sequential data acquisition: ~5 min/slide/channel at 10 μm resolution	No	Yes (4)	Yes (up to 3)
Genetix aQuire	Two-color format; 532 and 639 nm (488 nm also available)	Two 16-bit PMTs (adjustable)	Yes	5, 10, 20, 30	Two-color simultaneous data acquisition: 6.5 min/slide at 10 μm resolution	Yes (1-μm steps)	No	Yes (up to 6)

(continued)

TABLE I (*continued*)

Company/scanner	Lasers	Detector	Confocal?	Pixel size (μm)	Scan type: Time	Focus control?	Autoloader available? (# of slides)	Additional filters available?
Molecular Devices GenePix 4000B	Two-color format; 532 and 635 nm	Two 16-bit PMTs (adjustable)	No	5, 10, 20, 40, 60, 80, 100	Two-color simultaneous data acquisition: 6.5 min/slide at 10 μm resolution	Yes (1-μm steps)	No	No
Molecular Devices GenePix Professional 4200 and 4200 A	Up to four; 488, 532, 594, and 635 nm	One 16-bit PMT (adjustable)	No	5, 10, 20, 40, 60, 80, 100	One-color sequential data acquisition: 4 min/slide/ channel at 10 μm resolution	Yes (1-μm steps)	Yes (36)	Yes (up to 16)
Perkin Elmer ProScan Array and ProScan Array HT	Up to four; 488, 543, 594, and 633 nm	One 16-bit PMT (adjustable)	No	5, 10, 20, 30, 50	One-color sequential data acquisition: 5 min/slide/ channel at 10 μm resolution	Yes	Yes (20)	Yes (up to 11)
Perkin Elmer Scan Array Gx	Two-color format; 543 and 633 nm	One 16-bit PMT (adjustable)	No	5, 10, 20, 30, 50	One-color sequential data acquisition: 5 min/slide/ channel at 10 μm resolution	Yes	No	No
Tecan Group, Ltd. LSReloaded	Up to four; 488, 532, 594, and 633 nm	One or two 16-bit PMT (adjustable)	No (depth of focus is selectable)	4, 6, 10, 20, 40	Two-color simultaneous data acquisition: 4 min/ slide at 10 μm resolution	Yes	Yes (4)	Yes (up to 28)

[a] All the scanners included report sensitivities for Cy3 and Cy5 at or better than 0.1 chormophores/μm^2. This table is believed to be an accurate representation of the most common microarray scanners at the time of this writing and is not necessarily all inclusive. Because scanner instrumentation and features are changed frequently to keep up with progressions in the field, the reader is urged to contact the companies directly for up-to-date information and available models. Many of the manufacturers are willing to provide custom solutions, such as filters or scanning resolutions, to meet specific application needs.

and if a slide is not placed in the appropriate focal plane of a confocal scanner or if the software is not configured correctly to find the best focus, most of the true signal can be rejected, leading to very weak signal and poor scanner performance. Variations in slide thickness can significantly affect results in experiments involving multiple slides, but even slight variations in substrate thickness over the length of a single slide can have a large effect on the resulting image intensities with a confocal microarray scanner. Nonconfocal scanners, also known as wide-field microarray scanners, do not have the same sensitivity to focus position ($\sim$60-μm depth of field), but often have higher degrees of signal contamination from glass substrate fluorescence and dust.

Spatial Resolution, Signal Averaging, and Pixel Dwell Time

The spatial resolution or pixel size of a scanner, the degree of signal averaging via multiple scans, and pixel dwell time all work in concert with the laser power and PMT voltage setting to determine the amplitude of the signal collected in each pixel. The term "signal averaging" has been used in the microarray literature to refer to more than one type of signal averaging, such as results from averaging replicate spots printed multiple times within an array and/or over multiple arrays and from using bootstrapping methods to increase the reliability of expression values. The signal averaging of interest to the scanning of microarrays arises from performing multiple scans of an area on the array and averaging the results (also referred to as line averaging). Increasing the pixel size or dwelling longer on each pixel can increase the number of number of photons collected and averaging over multiple scans can decrease the uncertainty in the measurement, but these come at the expense of increased time to scan a slide and potential for damage via photobleaching. Dye photobleaching is reported in the literature from many laboratories, has been observed to range from 1 to 20%, and varies depending on the incident laser flux and exposure levels with the Cy5 molecules photobleaching easier than the Cy3 molecules (Malicka *et al.*, 2002; Nguyen *et al.*, 2002). A complete discussion of the photobleaching process can be found in Schena (2003). The photobleaching effect appears to be extremely variable on microarrays; while many laboratories report high levels of photobleaching, in some of our own studies we have scanned slides multiple times with only small amounts of sample damage. It is advisable for any microarray user to minimize the number of scans performed on a microarray; if an experimental study will require multiple scans the array stability issues should be assessed before the experiment is underway.

Spatial resolution is a critical setting for many microarray experiments. Microarray scanners for glass substrate microarrays are typically available with spatial resolutions ranging from 5 to 50 μm, with most scanners offering selectable resolutions within that range. Some instruments even offer 1 or 2 μm, such as the DNAScope scanner (GeneFocus/Biomedical Photometrics, Inc.). In principle, the 10-μm spatial resolution of most scanners should be more than sufficient to resolve the typical microarray features ($\sim$80–150 μm in diameter), but in practice higher spatial resolutions are often deemed necessary in order to facilitate feature location, background subtraction and normalization, and remove effects of dust, spot abnormalities, and array imperfections. In addition, as array printing and synthesis technology advances, microarray features sizes are getting smaller and smaller, approaching the limits of scanner resolutions. Highly developed imaging technology offering spatial resolutions at the diffraction limit already exists for optical microscopy and related fields, but to accommodate higher spatial resolution configurations in microarray scanners, additional developments would be necessary to avoid sacrificing speed of acquisition, handle the much larger file sizes, and analyze the higher fidelity images.

PMT and Laser Settings

In modern scanners, often both the PMT voltage and the laser power are adjustable via software menus. Each manufacturer differs in implementation and adjustments may scale linearly or nonlinearly with the software controls. Thus, instrument documentation should be read thoroughly before adjusting the settings. Both of these settings have the same effect—an increase or decrease in the fluorescence emission intensity observed—but the trade-offs are very different. A laser power that is too high can affect sample integrity (see earlier discussion of photobleaching), whereas a PMT setting that is too high can introduce nonlinearities and additional noise. Ultimately the real information provided by a scanner is limited by the bit depth of the scanner's detector. Saturated image pixels occur if settings for laser power or PMT voltage are set such that the real signal is greater than the bit depth (4096 counts for a 12-bit scanner; 65,535 counts for a 16-bit scanner). These saturated pixels are unusable and should be excluded from any subsequent analysis. While an occasional saturated pixel may occur from the presence of dust or impurity and most likely does not skew the analysis, routine inclusion of data that are not within the dynamic range of the scanner will lead to erroneous data and should be avoided. The dynamic range of the scanner should be considered thoroughly when choosing an instrument.

There are many schools of thought when it comes to PMT settings and the right answer is often experiment dependent. The most common scanning practice is to perform a preview scan (using lower spatial resolution and laser power) and use that image to select PMT voltages for the red and green channels that balance the intensities in the two channels while minimizing the number of saturated pixels. This practice developed from the assumption that the preponderance of the genes on a typical microarray is not expressed differentially and therefore the majority of the pixel signals should be about equal in the two channels. While this can be appropriate in some instances, there are many microarray experiments that will not follow this assumption, such as microarrays that are measuring yeast genes exiting the quiescent state. Arrays from an experiment of this type often have high numbers of genes that are expressed differentially.

In addition, one often performs large microarray studies involving more than one microarray slide and each slide can have a dynamic range of six orders of magnitude or more. Therefore, it can be helpful to scan each array twice to capture both the high- and the low-intensity information. Dudley and co-workers (2002) have reported success using this strategy to extend the dynamic range of the array experiment. Lyng and co-workers (2004) have also proposed a dual-scan strategy as a way to improve data reliability by ensuring that the spot values acquired are contained within the usable range of the scanning system (the range in which the expression ratios were independent of PMT voltages). In their work this range was between 200 and 50,000 counts for two different scanner manufacturers. Arrays are scanned once, ensuring that no pixels are saturated, and then again at a higher PMT voltage to permit the lower intensity values to be recorded accurately. A calibration factor is calculated from a subset of the middle intensity spot values on the two images and the saturated pixel values are corrected.

We have noted that it is typical protocol (suggested in most scanner manufacturer documentation) to adjust the PMT voltage settings for each channel for every slide in a study involving multiple slides for comparison. This serves to maximize the usable dynamic range of each slide, but it also introduces the requirement for normalization techniques that operate without flaw to "rescale" data on each slide to a common basis set. In an example study from our group, yeast microarrays were prepared to assess the effect within and across cultures. Results of this study determined that by using the same PMT voltage settings to scan every slide, the need for normalization of data was virtually eliminated as shown by the much lower standard deviations of the parameter estimates from our measurement model shown in Fig. 2. Given the complexities and problems often associated with data normalization, it is almost always beneficial to maintain PMT settings throughout the course of a multiple slide experiment, unless extremely large variations in dynamic range are anticipated, forcing one to adjust PMT settings to minimize the degree of pixel saturation.

Measurement model

$$Y_{ijk} = \mu_i + \alpha_j + \beta_{k(j)} + \varepsilon_{ijk}$$

Y_{ijk} = Log2(R/G) measurement of i^{th}gene (k^{th} rep within j^{th} culture)

i = Gene label

j = Culture label

k = Rep (withinculture) label

μ_i = Overall average of ith gene

α_j = Global effect of jth culture (across all genes)

$\beta_{k(j)}$ = Global effect of kth rep within j^{th} culture (across all genes)

ε_{ijk} = Specific effect of ith gene (k^{th} rep within j^{th} culture)

PMT settings adjusted	PMT settings constant
$\hat{\sigma}_\alpha = 0.575$ (2 dof)	$\hat{\sigma}_\alpha = 0.081$ (2 dof)
$\hat{\sigma}_\beta = 0.179$ (3 dof)	$\hat{\sigma}_\beta = 0.090$ (3 dof)

FIG. 2. Description of microarray experiment measurement model and results for array scanned with PMT settings optimized for each array and with the PMT setting held constant for all arrays.

Avoiding on-the-fly balancing of PMT channels may very well reduce the need for many of the complex and risky normalization procedures commonly applied to microarray data, providing more accurate and reliable results.

Image Display Characteristics

It is important to understand the particulars of your scanner software display and how they affect what a user sees on the computer monitor. Typical microarray images are 16-bit data (see discussion in previous subsection regarding data depth and dynamic range limitations), but they are represented in the software as 8-bit images through the use of either compression or reduction techniques. The process whereby these 8-bit images are generated can alter the image dramatically and a user should be aware of this effect, especially because the displayed image is often used to select subsequent scan parameters, such as PMT or laser settings. In some scanner software this setting is user configurable. Sixteen-bit images can be compressed to 8-bit images, which retain the range of pixel values, but all of the data are scaled accordingly. This compression can produce poor results with the high dynamic range of microarray data, depending on the nature of the scaling. Image reduction techniques offer an alternative to compression, but do not display the full range of intensities. The three common practices of image reduction employed in microarray scanning software are to preserve the brightest intensities, dropping the lowest 8-bits of data for display purposes; preserve the weakest intensities, dropping the highest 8 bits of data; and preserve the middle intensities by dropping the top 4 and bottom 4 bits of data. Real 16-bit data are used in microarray data analysis calculations so image compression or reduction does not affect the final resulting expression values. They only affect the image as viewed on the computer monitor, which can lead to erroneous conclusions by the user if the user is unaware of the effect.

Instrumentation/Hardware Effects

Several key components of microarray scanning instrumentation and software have the potential to introduce additional variability to microarray data. Two most notable errors are signal contamination and scanner bias. Additional errors can be imparted to data via image misalignment and scanner geometry effects, but these are generally small with modern instruments and are not presented in detail here. Most commercial software can maintain adequate alignment tolerances of ± 1 pixel between images, and scanning and random geometry errors are observed at level far below $\pm 2\%$. Several scanners have calibration slides for calibrating spatial resolution and red/green channel output characteristics. In addition, third party slides are available commercially for calibrating sensitivity to red and green dyes. Calibration slides should be run at regular intervals as per manufacturers' instructions to calibrate the instrument and monitor instrument functionality as alignment and geometry errors can compromise data significantly if the tolerances stated earlier are not met. In addition, calibration slides can be used to PMT settings for experimental runs if the calibration slide covers a dynamic range similar to the experiment planned. This is particularly attractive because it eliminates the need for risky adjustment of PMT voltages during the experiment (see earlier discussion).

Signal Contamination

Signal contamination can be broadly defined as the inclusion of intensity in an image pixel that does not arise from fluorescence of the analyte of interest. The signal recorded in a microarray image pixel using commercially available filter-based scanning technology is generally superposition of the signal from the fluorescent dye of interest, signal from other fluorescent species emitting in the same spectral band pass, and noise (electronic and statistical). A thorough discussion of noise can be found in many analytical instrumentation textbooks, as well as in Schena (2003), and is not repeated here. With microarrays the extraneous signal due to electronic or statistical noise is often small compared to the signal from other fluorescent species. Common sources of spectral signal contamination are the glass substrate, dust particles, and spectral cross talk from the other fluorescent labels in multicolor experiments. The glass substrate contributes a uniform spatial signal across the surface of the slide and therefore can be compensated for by local or global background subtraction methods. Typically, dust particles affect a small area (one to two pixels), but their effects on the resulting microarray spot ratios cannot be ignored (Minor, 2006). Options for removing the effect of dust from data include thresholding methods and outlier detection methods or the use of median values rather than mean values. (Median values are generally more robust to multiple effects seen in microarrays, including errors in representing the spot shape or diameter.) Dust on the backside of the slide can be burdensome when using a nonconfocal scanner and it is recommended that the slide be handled appropriately to minimize these artifacts. Gentle cleaning of the backside of the slide can be done if needed.

Spectral cross talk can also occur from unknown interferents or from additional labels used in multicolor experiments. Unlike the glass background and dust, spectral cross talk is often a spot-specific effect and thus is more difficult to isolate. In most experiments, dyes with well-separated excitation and emission spectra are used, and microarray scanners utilize filters that ensure that the spectral cross talk from other dyes used in the microarray experiment is small, but never absent. Figure 3 shows emission spectra of commonly used microarray dyes Cy3 and Cy5 with typical filter transmission spectra overlaid. Although these filters are representative of the filters in common microarray instruments and can give an idea of the degree of cross talk expected between these two dyes, it should be noted that some popular scanners employ long-pass filters instead of band-pass filters. While a long-pass filter potentially collects more photons because it passes all the photons of a longer wavelength than the cut-on wavelength, this type of filter inherently increases the degree of spectral cross talk and may not be worth the trade-off. Even with the highest quality filter, spectral cross talk will be observed at some level in multicolor experiments from the dyes used, and the degree of cross talk inherently increases with the number of colors used. An excellent (and relevant) description of spectral cross talk in confocal microscopy can be found at http://www.olympusfluoview.com/theory/bleedthrough.html. Correction for spectral cross talk is not performed regularly; however, the effect could be modeled easily at a basic level, as spectra of the dyes used are known and the instrument response per channel can be determined. Wang and co-workers (2004) have described the effect in detail for two-color microarrays and presented

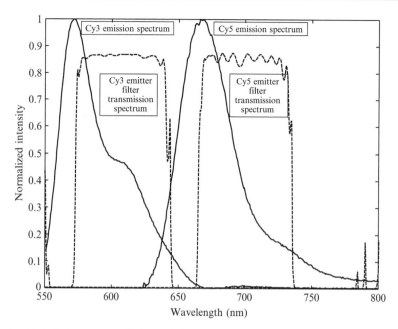

FIG. 3. Emission spectra of Cy3 and Cy5 labels overlaid with typical filter transmission spectra for these dyes to illustrate spectral overlap and cross talk. Filter spectra shown are for the Cy3 emission filter HQ610_75 (Chroma Technology Corp) and for the Cy5 emission filter HQ700_75 (Chroma Technology Corp.) and are representative of the filters employed commonly in some commercial microarray scanners.

a correction strategy. Their microarray scanner exhibited cross talk ratios of 1.25 and 1.51 % for the green and red channels, respectfully, when Cy3 and Cy5 were used. The consequence of cross talk at these levels is a limited range of ratios (100 to 0.01) if no correction is performed.

In recent literature spot-specific signal contamination has also been reported in the green channel that was neither the dye of interest nor an additional dye (Martinez et al., 2003; Timlin et al., 2005). The presence of this contaminant, most likely from the buffer solutions used during printing, was particularly detrimental because it was spot specific, variable, and persisted to varying degrees posthybridization. Prescanning the array will identify, but not correct, the problem. Resulting green intensity values were extremely skewed, and there is evidence in that report as well as in archived data sets that this is a widespread problem with microarrays. The presence of this or any other spot-specific, variable contaminant may be responsible for the lack of reproducibility of dye-flip experiments at low intensities and for the need for intensity-dependent normalization procedures. Hyperspectral scanning is the only way currently to identify and correct for contaminants of this kind. Hyperspectral scanning technology is discussed in more detail later.

Additionally, signal contamination can occur from the analyte of interest when it binds nonspecifically to either a probe that is not a perfect match or outside the printed DNA spot. The use of negative controls can sometimes indicate the relative levels of nonspecific binding of the target sequences to printed probe, but this does not provide quantitative information. High or variable background levels are generally good indicators of a serious problem with nonspecific binding to the glass substrate. If this is present it is important (1) that the resulting microarray spot data not be background subtracted, as the background levels are not necessarily a measure of the background signal under the printed spot and will only serve to add additional variance to data (Scharpf *et al.*, 2004), and (2) that the hybridization protocol be optimized to minimize the nonspecific binding.

Scanner Bias

It has been proposed that many parts of the entire microarray process could introduce bias in one channel or the other of a two-color scanner. The relative rates of dye incorporation are known to be different (t Hoen *et al.*, 2003), and the red label typically degrades faster than the green label (Malicka *et al.*, 2002). The presence of signal contamination in one channel, such as the spot-specific green contaminant noted earlier, is another example leading to bias in the resulting values. Bias in one of the channels (from any source contaminant, instrumentation, etc.) could be responsible for the nonlinearities or intensity-dependent effect in the log ratios of gene expression data. It has been shown that the scanning instrumentation itself and image analysis software could contribute bias to a microarray experiment (Bengtsson *et al.*, 2004). Bengsston *et al.* (2004) used multiple PMT gain settings to confirm the presence of a small but noteworthy bias that varies slightly between arrays and between scanners that appears to be originating from the detector hardware. They proposed a scanning protocol consisting of multiple scans at varying PMT settings and subsequent data calibration with a method based on a constrained affine model to minimize the effect on the expression ratios. At first glance, bias may appear to affect the weakest intensity spots most severely and thus one may be tempted to simply set a higher minimum threshold for the analysis. However, bias can propagate much further, as many normalization protocols rely on ratios skewed by the bias to compute normalization factors.

Alternative Scanning Technologies Provide Advantages

Research has presented alternative technologies that possess unique advantages to the traditional laser excitation, filter-based microarray scanner platform. Two of these, resonance light scattering (RLS) and surface plasmon resonance (SPR), are nonfluorescence-based, optical imaging techniques. Because they do not monitor fluorescence, the disadvantages of organic fluorophores, such as lack of photostability, spectral cross talk, and differential incorporation, are eliminated. However, each of these techniques does require its own microarray manufacturing and processing technologies and its own dedicated imaging system. SPR is an optical imaging technique in which the signal derived from monitoring changes in the local refractive index when a biomolecule is adsorbed onto a metallic surface, typically gold. Research has shown SPR to be sensitive and applicable to probing *in situ* and *ex situ* interactions between biomolecules (DNA—DNA, DNA—RNA, DNA—protein, and protein—protein) in real time. In addition, the surface analytical technique SPR has the potential for revealing the kinetics of biomolecular interactions, although nonspecific binding may be problematic. SPR has been covered in detail (Nelson *et al.*, 2001), and the applications of SPR to protein arrays have been reviewed elsewhere (Cutler, 2003; Gloekler and Angenendt, 2003).

RLS is also an optical imaging technique capitalizing on the ability of small metallic particles such as gold and silver to generate intense monochromatic light when illuminated with white light. The wavelength of the resulting monochromatic light can be controlled by controlling the diameter and composition of the metallic particles, thus permitting the manufacture of a series of metallic tags that can be used to identify different targets similar to their organic fluorophores counterparts. Commercial RLS-based scanning systems are available from Genicon Sciences (San Diego, CA) and it has been shown to be a very sensitive detection platform for bacterial pathogen detection (Francois *et al.*, 2003). It is clear that RLS can provide ultrasensitive detection for a variety of analytical-sensing applications.

Unlike RLS and SPR, hyperspectral scanning (HSS) is based on fluorescence emission. HSS has several advantages over traditional microarray scanning methods, which are univariate, filter-based techniques employing optical filters that pass a discrete band of photons to a single-point detector. A hyperspectral scanner collects an entire fluorescence emission spectrum for each image pixel. When coupled with multivariate data extraction techniques (Kotula *et al.*, 2003), HSS allows fluorescent species to be identified and quantified on the basis of spectral shape. This feature permits isolation of the fluorescent signal from the analyte of interest and fluorescent contaminants and background emissions and can potentially increase the number of microarray experimental conditions that can be compared on a single substrate with the use of multiple, spectrally overlapped fluorophores. A complete description of hyperspectral imaging and its benefits can be found elsewhere (Kotula *et al.*, 2003; Michalet *et al.*, 2003; Schultz *et al.*, 2001; Zimmerman *et al.*, 2003). HSS for microarrays was first published in 2001 (Schultz *et al.*, 2001), and recent literature has presented the characterization of a HSS system optimized for scanning glass substrate microarrays (Sinclair *et al.*, 2004) and the application of that system to extract the underlying emission spectra of extraneous fluorescent species from glass microarrays (Timlin *et al.*, 2005). The HSS method leads to more accurate data because the signal monitored is solely from the analyte of interest and does not include contributions from extraneous fluorescent species, making HSS an indispensable tool for diagnosing problems with microarrays and improving microarray quality control. The multivariate analysis employed

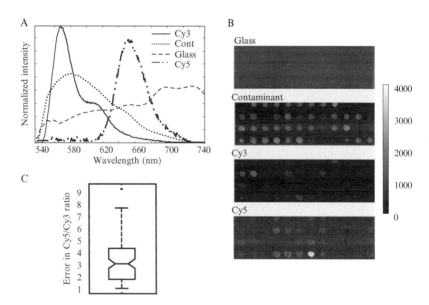

FIG. 4. Results of hyperspectral scanning and multivariate analysis of a microarray slide with a spot-specific green channel contaminant. (A) Four pure component spectra (Cy3, Cy5, glass, green channel contaminant) extracted from the multivariate analysis. (B) Independent component concentration maps (images) corresponding to each extracted spectral species shown in A. (C) Calculated percentage error in Cy5/Cy3 ratio of commercial scanner data resulting from the presence of the green channel contaminant that is confounded in the green channel of the commercial scanner. (Percentage error is calculated by dividing the R/G ratio from the commercial scan by the Cy5/Cy3 ratio calculated from hyperspectral imaging data.) Spots are ∼200 μm in diameter.

can operate with little or no a priori knowledge, a necessity for extracting contaminating fluorescent species and improving the accuracy and reliability of microarrays for bioresearch. As demonstrated in Fig. 4, HSS has been utilized to identify and correct for the presence of a contaminant in the green channel of commercial scanners, which skewed the microarray ratios of 75% of the spots scanned by more than a factor of 2 for this array (Martinez et al., 2003). This result was confirmed with slides from multiple manufacturers and preparation protocols, and the levels of contaminant detected varied from less than 1% of the total spot intensity to 40% or greater in some spots.

Developments of new dyes that span the visible spectrum (Shaner et al., 2004) and bio-conjugated quantum dots (Jovin, 2003) have fueled interest in hyperspectral imaging for cells and tissue and also have potential application for hyperspectral microarray scanning, as these new labels have emission and excitation properties that are extremely well suited to hyperspectral imaging. In theory the number of dyes that can be identified and imaged quantitatively with HSS is limited by spectral resolution and the spectral and spatial variation of the sample. In practice this is typically between 10 and 20 labels for the HSS system presented by Sinclair et al. (2004), depending on the sample. Ultimately for microarray experiments the upper limitation is mostly likely lower (3–6) and set by the biology and chemistry used to hybridize this many labeled strands efficiently. This area has not been explored adequately.

Specific Considerations for Multiple Slide, Multiple Scanner, and/or Multiple Laboratory Experiments

As microarray technology gains popularity there are increasing needs for large-scale projects that span multiple slides and multiple laboratories. Unfortunately, slide-to-slide variation and nonbiological variance often make even interlaboratory microarray results difficult to reproduce and assert statistical confidence. This lack of reproducibility can hamper the utility of microarrays and mask the true biological relationships sought from the experiment. In our own research we have conducted a study to understand the measurement capability of a two-color microarray scanner. This experiment allowed us to understand the levels of signals that we can measure accurately given our equipment and protocols at hand and provides information on operator effects, drift of the instrument, the stability of the hybridized chips (with respect to laser excitation), and, to some extent, reproducibility of the hybridization process. A statistical study of this type would be encouraged before any large-scale experiment was undertaken so that sources of variation can be identified and minimized. Large-scale, cross platform validation studies have begun to address these issues of laboratory-to-laboratory variation and platform-specific differences (Irizarry *et al.*, 2005; Larkin *et al.*, 2005). It is anticipated that this will be an area of active research in the coming years as data accumulate in public databases (Barrett and Edgar, 2006; Brazma *et al.*, 2006).

Conclusions

Although the act of scanning a microarray slide is often an overlooked part of the microarray process, incorrect or inappropriate scan settings can have profound effects on resulting data. Care should be taken to fully understand the effect of each of the critical parameters (both instrumental and user-selectable) on microarray data rather than simply following a protocol developed for a previous experiment. Significant research efforts have gone into developing analysis techniques, but the field is ripe for research to characterize the variability and errors introduced by the scanning process itself, the scanner instrumentation, and the user, especially for large multiple slide experiments. Improvements in this area will ultimately increase data reliability and reduce the need for complex preprocessing mechanisms prior to the extraction of expression information. Emerging technologies such as surface plasmon imaging and hyperspectral imaging offer unique capabilities and have promise as complementary techniques to traditional scanning.

Acknowledgments

The author acknowledges Edward V. Thomas and David M. Haaland for their contributions to this work in designed microarray experiments and statistical data analysis, including efforts that led to the results of the variable PMT setting experiments. In addition, thank you to Margaret Werner-Washburne for many helpful discussions about microarray technology. Sandia is a multiprogram laboratory operated by Sandia Corporation, a Lockheed Martin Company, for the United States Department of Energy under Contract DE-ACO4–94AL85000. A portion of this work was funded by the U.S. Department of Energy's Genomes to Life program (www.doegenomestolife.org) under project "Carbon Sequestration in Synechococcus Sp.: From Molecular Machines to Hierarchical Modeling" (www.genomes-to-life.org).

References

AgilentTechnologies (2004). Improving microarray results by preventing ozone-mediated fluorescent signal degradation Agilent Technologies.

Balazsi, G., Kay, K. A., Barabasi, A.-L., and Oltvai, Z. N. (2003). Spurious spatial periodicity of co-expression in microarray data due to printing design. *Nucleic Acids Res.* **31,** 4425–4433.

Barrett, T., and Edgar, R. (2006). Gene expression omnibus: Microarray data storage, submission, retrieval, and analysis. *Methods Enzymol.* **411,** 352–369.

Bengtsson, H., Jonsson, G., and Vallon-Christersson, J. (2004). Calibration and assessment of channel-specific biases in microarray data with extended dynamic range. *BMC Bioinform.* **5.**

Brazma, A., Kapushesky, M., Parkinson, H., Sarkins, U., and Shojatalab, M. (2006). Data storage and analysis in ArrayExpress. *Methods Enzymol.* **411,** 370–386.

Brown, C. S., Goodwin, P. C., and Sorger, P. K. (2001). Image metrics in the statistical analysis of DNA microarray data. *Proc. Natl. Acad. Sci. USA* **98,** 8944–8949.

Brownstein, M. (2006). Sample labeling: An overview. *Methods Enzymol.* **410,** 222–237.

Cutler, P. (2003). Protein arrays: The current state-of-the-art. *Proteomics* **3,** 3.

Dudley, A. M., Aach, J., Steffen, M. A., and Church, G. M. (2002). Measuring absolute expression with microarrays with a calibrated reference sample and an extended signal intensity range. *Proc. Natl. Acad. Sci.* **99,** 7554–7559.

Fare, T. L., Coffey, E. M., Dai, H., He, Y. D., Kessler, D. A., Kilian, K. A., Koch, J. E., Eric, L., Marton, M. J., Meyer, M. R., Stoughton, R. B., Tokiwa, G. Y., and Wang, Y. (2003). Effects of atmospheric ozone on microarray data quality. *Anal. Chem.* **75,** 4672–4675.

Francois, P., Bento, M., Vaudaux, P., and Schrenzel, J. (2003). Comparison of fluorescence and resonance light scattering for highly sensitive microarray detection of bacterial pathogens. *J. Microbiol. Methods* **55,** 755–762.

Gloekler, J., and Angenendt, P. (2003). Protein and antibody microarray technology. *J. Chromatogr. B* **797,** 229.

Hegde, P., Qi, R., Abernathy, K., Gay, C., Dharap, S., Gaspard, R., Hughes, J. E., Snesrud, E., Lee, N., and Quackenbush, J. (2000). A concise guide to cDNA microarray analysis. *Biotechniques* **29,** 548–562.

Irizarry, R. A., Warren, D., Specer, F., Kim, I. F., Biswal, S., Frank, B. C., Gabrielson, E., Garcia, J. G. N., Germino, G., Griffin, C., Hilmer, S. C., Hoffman, E., Jedicka, A. E., Kawasaki, E., Martinez-Murillo, F., Morsberger, L., KLee, H., Peterson, D., Quackenbush, J., Scott, A., Wison, M., Yang, Y., Ye, S. Q., and Yu, W. (2005). Multiple laboratory comparison of microarray platforms. *Nature Methods* **2,** 345–349.

Jovin, T. M. (2003). Quantum dots finally come of age. *Nature Biotechnol.* **21,** 32–33.

Kerr, M. K., and Churchill, G. A. (2001). Statistical design and the analysis of gene expression microarray data. *Genet. Res.* **77,** 123–128.

Kotula, P. G., Keenan, M. R., and Michael, J. R. (2003). Automated analysis of SEM X-Ray spectral images: A powerful new microanalysis tool. *Microsc. Microanal.* **9,** 1–17.

Larkin, J. E., Frank, B. C., Gavras, H., Sultano, R., and Quackenbush, J. (2005). Independence and reproducibility across microarray platforms. *Nature Methods* **2,** 337–344.

Lyng, H., Badiee, A., Svendsrud, D. H., Hovig, E., Myklebost, O., and Stokke, T. (2004). Profound influence of microarray scanner characteristics on gene expression ratios: Analysis and procedure for correction. *BMC Genom.* **5,** 1–10.

Malicka, J., Gryczynski, I., Fang, J., Kusba, J., and Lakowicz, J. R. (2002). Photostability of Cy3 and Cy5-labeled DNA in the presence of metallic silver particles. *J. Fluoresc.* **12,** 439–447.

Martinez, M. J., Aragon, A. D., Rodriguez, A. L., Weber, J. M., Timlin, J. A., Sinclair, M. B., Haaland, D. M., and Werner-Washburne, M. (2003). Identification and removal of contaminating fluorescence from commercial and in-house printed DNA microarrays. *Nucleic Acids Res.* **31,** e18.

Michalet, X., Kapanidis, A. N., Laurence, T., Pinaud, F., Doose, S., Pfughoefft, M., and Weiss, S. (2003). The power and prospects of fluorescence microscopies and spectroscopies. *Annu. Rev. Biophys. Biomol. Struct.* **32,** 161–182.

Minor, J. M. (2006). Microarray quality control. *Methods Enzymol.* **411,** 233–255.

Neal, S. J., and Westwood, T. (2006). Optimizing experiment and analysis parameters for spotted microarrays. *Methods Enzymol.* **410,** 203–221.

Nelson, B. P., Grimsrud, T. E., Liles, M. R., Goodman, R. M., and Corn, R. M. (2001). Surface plasmon resonance imaging measurements of DNA and RNA hybridization adsorption onto DNA microarrays. *Anal. Chem.* **73,** 1–7.

Nguyen, D. V., Bulak Arpat, A., Wang, N., and Carroll, R. J. (2002). DNA microarray experiments: Biological and technological aspects. *Biometrics* **58,** 701–717.

Quackenbush, J. (2001). Computational analysis of microarray data. *Nature Rev. Genet.* **2,** 418–427.

Quackenbush, J. (2002). Microarray data normalization and transformation. *Nature Genet.* **32.**

Scharpf, R. B., Iacobuzio-Donahue, C. A., and Parmigiani, G. (2004). When should one subtract background fluorescence in cDNA microarrays. *In* "Johns Hopkins University, Dept of Biostatistics Working Papers." Johns Hopkins University.

Schena, M. (1999). DNA microarrays: A practical approach. *In* "The Practical Approach Series" (B. D. Hames, ed.), p. 210. Oxford University Press, Oxford.

Schena, M. (2000). *In* "Microarray Biochip Technology," p. 298. Eaton Publishing, Natick.

Schena, M. (2003). *In* "Microarray Analysis," p. 630. Wiley, Hoboken.

Schermer, M. J. (1999). Confocal scanning microscopy in microarray detection. *In* "DNA Microarrays: A Practical Approach" (M. Shena, ed.), pp. 17–42. Oxford University Press, Oxford.

Schuchhardt, J., Beule, D., Malik, A., Wolski, E., Eickhoff, H., Lehrach, H., and Herzel, H. (2000). Normalization strategies for cDNA microarrays. *Nucleic Acids Res.* **28,** e47.

Schultz, R. A., Nielsen, T., Zavaleta, J. R., Ruch, R., Wyatt, R., and Garner, H.,R. (2001). Hyperspectral imaging: A novel approach for microscopic analysis. *Cytometry* **43,** 239–247.

Shaner, N. C., Campbell, R. E., Steinbach, P. A., Giempmans, B. N. G., Palmer, A. E., and Tsien, R. Y. (2004). Improved monomeric red, orange, and yellow fluorescent proteins derived from Discosoma sp. red fluorescent protein. *Nature Biotechnol.* **22,** 1567–1572.

Sinclair, M. B., Timlin, J. A., Haaland, D. M., and Werner-Washburne, M. (2004). Design, construction, characterization, and application of a hyperspectral microarray scanner. *Appl. Opt.* **43,** 2079–2089.

t Hoen, P. A. C., de Kort, F., van Ommen, G. J. B., and den Dunnen, J. T. (2003). Fluorescent labelling of cRNA for microarray applications. *Nucleic Acids Res.* **31.**

Timlin, J. A., Sinclair, M. B., Haaland, D. M., Aragon, A. D., Martinez, M. J., and Werner-Washburne, M. (2005). Hyperspectral microarray scanning: Impact on the accuracy and reliability of gene expression data. *BMC Genom.* **6.**

Tseng, G. C., Oh, M., Rohlin, L., Liao, J. C., and Wong, W. H. (2001). Issues in cDNA microarray analysis: Quality filtering, channel normalization, models of variations and assessment of gene effects. *Nucleic Acids Res.* **29,** 2549–2557.

Wang, L., Lu, Z., and Ni, X. (2004). Cross-talk correction in dual labeled fluorescent microarray scanning. *Chin. Opt. Lett.* **2,** 162–164.

Wu, W., Wildsmith, S. E., Winkley, A. J., Yallop, R., Elcock, F. J., and Bugelski, P. J. (2001). Chemometric strategies for normalisation of gene expression data obtained from cDNA microarrays. *Anal. Chim. Acta* **446,** 451–466.

Yang, Y. H., Buckley, M. J., Dudoit, S., and Speed, T. P. (2000). "Comparison of Methods for Image Analysis on cDNA Microarray Data," p. 40. University of California, Berkeley.

Zimmerman, T., Rietdorf, J., and Pepperkok, R. (2003). Spectral imaging and its applications in live cell microscopy. *Fed. Eur. Biochem. Soc. Lett.* **546,** 87–92.

[7] An Introduction to BioArray Software Environment

By CARL TROEIN, JOHAN VALLON-CHRISTERSSON, and LAO H. SAAL

Abstract

BioArray Software Environment (BASE) is a web-based software package for storing, searching, and analyzing locally generated microarray data and information surrounding microarray production. The workflow begins in sample management and, optionally, microtiter plate tracking and ends in visualization and analysis of entire experiments. The relative ease with which new analysis plug-ins can be added has given rise to a plethora of third-party tools, and the licensing terms (GNU GPL) encourage local modifications of the software. This introduction to BASE describes the basics of working with the software, both in general and in more detail for the various parts. It also provides some hints about more advanced usage and a section on what is needed to set up your own BASE server. The information is current as of BASE version 1.2.17b, which was released on November 6, 2005.

Introduction

With the advent of the microarray technique in the late 20th century, "high throughput" and gene expression profiling became buzzwords of the day. For microarrays to realize their potential, it is not only necessary to hybridize samples of interest to these small chips, but also to handle resulting data in a scientifically sound way. The first step in that process is tracking the actions taken in the laboratory, a mundane task for small experiments but markedly less so when hundreds of samples and tens of thousands of genes are involved. With that step taken, data derived from the actual microarray can be analyzed in a meaningful context. Dedicated software is needed for these tasks, and over the years a number of different software packages have appeared, created for tackling different aspects or flavors of microarray data management and analysis. As the motivation of the developers varies, so does the focus of the software and (maybe even more so) the licensing terms.

Some commercial, proprietary programs are produced by manufacturers of hardware, such as print robots and scanners, as their customers naturally need software to interact with these machines. Considering the cost of performing microarray experiments and the crucial role of data management,

METHODS IN ENZYMOLOGY, VOL. 411   0076-6879/06 $35.00
DOI: 10.1016/S0076-6879(06)11007-1

it is quite reasonable to spend money on getting good software so it is not surprising that other commercial software also exists. However, many of the people involved in microarray analysis come from backgrounds in statistics, computer science, or physics, where making your own computer programs is common. Because of this, a range of free/open source software is also available, and three of the more popular systems, including BioArray Software Environment (BASE), have been reviewed elsewhere (Dudoit *et al.*, 2003).

BASE (Saal *et al.*, 2002) is a system aimed at tracking and analyzing information all the way from the production of microarrays and handling of biological samples through hybridization, spot finding, and normalization, down to the visualization of analysis results. The software arose from a collaboration between experimentalists who manufacture their own spotted cDNA arrays and theoreticians who write custom analysis programs. Because of this, particular focus is given to microarray production, with tracking of microtiter plates and individual microarray slides, and to data analysis, with a plug-in interface whereby new analysis tools can be added readily. The user interface of BASE is entirely web based, not tying the users to any particular operating system. There are some requirements on the server side, as discussed later.

Since its release in 2002, BASE has attracted a large number of users, and through the project's mailing list many of these users are now providing new users with support when needed. The software is released under the GNU General Public License (FSF, 1991), which has encouraged some users to make modifications and share these with the community. In addition, there are many user-contributed analysis plug-ins, including frameworks for applying functions in the statistical language R (R Development Core Team, 2005), including the vast Bioconductor project (see Chapter 8 in volume 410 by Lausted *et al.*, 2006) to data stored in BASE. These efforts have made BASE useful to many researchers, as demonstrated by the use of the system in many published reports, such as Andersson *et al.* (2005), Björkbacka *et al.* (2004), Carmel *et al.* (2004), Eckhardt *et al.* (2005), Jönsson *et al.* (2005), Lyng *et al.* (2004), Rhee *et al.* (2005), Sollier *et al.* (2004), Sturme *et al.* (2005), and Wang *et al.* (2004).

Getting Started

Demonstration of BASE

Before committing to downloading and installing BASE, let alone actually using it, you may want to try it out. For this purpose, a demo server exists where anyone can get an account. The BASE project web site can be found at http://base.thep.lu.se/. There, in the menu to the left, you will find a link to "Demo BASE." At the main welcome page of the demo installation you should find instructions regarding whom to contact to obtain an account.

Requirements

BASE can be used from virtually any computer running a web browser of your choice, be it Mozilla Firefox, MS Internet Explorer, or something else, but on the server side it requires a Unix-like operating system. The most common choice is without doubt some flavor of Linux, but systems such as Solaris and Mac OS X are also known to work.

Effort has gone into making installation and administration of BASE as easy as possible, but both tasks can still be fairly complex. Setting up a BASE server without any "advanced" configuration changes, however, is something that most laboratories seem to accomplish with little or no help. The basic requirement is to have someone who can manage the server, that is, a person with the skills to set up a Linux box and keep it running. As a way of avoiding duplication of costs and labor and to facilitate the exchange of data, multiple groups (comprising hundreds of users) may well share a single instance of BASE.

Hardware-wise, BASE is not particularly demanding, but the amount of data associated with microarray experimentation should not be underestimated. A data file from a single array slide can weigh in at tens of megabytes, and in the course of data analysis this may be increased manifold. Having a dedicated server for BASE is convenient, although not strictly necessary. The question of what type of computer to use has no definite answer, but depends on how the system will be used. It is quite feasible for a small laboratory with a handful of users to set up BASE on a single run-of-the-mill PC. Data analysis tasks can be distributed to additional machines, should the need arise.

As an example of what a larger facility may need, the setup at Lund University currently consists of seven computers, serving some 150 users, 10–15 of whom are typically logged in at any given moment. The database contains over 6000 microarray slides, with an average of roughly 50,000 spots each. Two of the computers are dual-CPU servers with one or two gigabytes of RAM and about a terabyte of hard drive space, and the other five are inexpensive PCs used for running various analysis plug-ins. The hardware is of varying age, but a comparable new system would carry a price tag of approximately $15,000 to $20,000.

Installing BASE

The first step in installing BASE is to visit the project's web site, http://base.thep.lu.se/, where in addition to a tar/gzip archive with the current version of BASE you will find the installation documentation. This is a fairly lengthy piece of text, which provides a detailed description of what software packages are required (all are free/open source), how to configure them, how to install and configure BASE itself, and so forth. We highly recommend that before installing you read through the whole document, as it covers most sources of install-time problems as well as some issues that might not surface until later.

The single most important piece of information on the BASE web site concerns how to join the BASE mailing list. Subscribers to this mailing list include the developers and a large number of users who between them have built a substantial pool of knowledge, particularly concerning installation problems and the like. All questions related to BASE are welcome on the mailing list, and it is also where new releases and other events are usually announced first.

The Basics of BASE

Once BASE has been successfully installed on a server, pointing your web browser to http://your.base.server/ should bring up a page with a purple logo, much like the one shown in Fig. 1. In addition, there should be input boxes for login and password. If you have just installed BASE, you should log in as "root" with no password. Otherwise we assume that you have been given this information by the administrator, and that you are able to log in.

BASE has a simplistic user interface, where most things are done in a single window. This window is split into two frames, where the left frame houses a menu and some general information. Much of the web interface

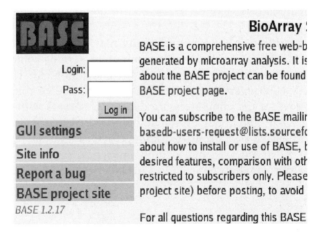

FIG. 1. The BASE login screen.

consists of *list pages* where items of some kind are presented in a searchable and sortable table. From the list pages you can reach the *view page* of individual items. Items can, in most cases, be edited, and then the view page has a link to an *edit page*, which is very similar to the corresponding *add page*, which (if it exists) can usually be reached from the list page.

For a concrete example, open the "Users" menu and click on "List users." This should bring you to a page where users are listed, although depending on your access level you may see more or less information. Clicking on a user name in the list brings up a "View user" page, and in the case that it is your own account you will be able to change your password. Most users will, of course, not have the right to create or modify user accounts, but if you head over to the "Biomaterials" menu, you can experiment with adding, removing, and listing things.

You will find that strewn about the web pages of BASE are question marks on a green background, and clicking on of these will bring up a more or less context-specific help text. Sometimes these texts are even the best available documentation for a particular function.

List pages, such as the one called "Samples," consist of a filter part at the top, a set of results from the filtering presented in a table, and some additional controls. For the most part, the way these list pages work should be intuitive, although they do have some quirks.

Filtering

On a list page, only those items that match *all* of the filtering criteria are shown, but the *In* operator allows for some OR-like filtering, in that you can give it a comma-separated list of possible values. The various fields you can filter on all have a type, such as text string, date, or integer, and as conversion from a plain string of text can be nontrivial (in particular for dates), the resulting value is listed under "Translated value." When searching on text strings (by far the most common type), you can use the wild cards * (any string of characters) and _ (any one character) together with the operators =, *In* and *Not in*. Figure 2 illustrates some of these features.

Field	Op	Value	Buttons		Translated value
Name	=	a*	Upd	Delete	a%
Used in experiment	Not in		Upd	Delete	NULL
Sample date	Between	1 may 03, last month	Upd	Delete	2003-05-01, 2006-04-16
-	=		Add/Update		
Presets Save current as new preset		Ok			

FIG. 2. Filtering samples using wildcards, dates and various operators.

For some reason, there is no *Not equal to* operator, but the *Not in* operator does the same thing, although just as with *In* you can give a whole list of values. Similarly, the *Between* and *Not between* operators take exactly two comma-separated arguments. What certainly qualifies as a quirk, but can be ignored by the novice, is that the empty string is translated to "undefined value" (*null*) when used with *In* and *Not in*. In these cases *In* and = give different results, and one has to use > instead of *Not in* to filter out strings that are defined but empty.

Filters can be saved as "presets" to avoid having to enter the same filter over and over. When you have a filter you wish to save, enter a name for the preset and click "OK." Later, you can then recall the preset, discarding the current filter, with a single click. You may want to save an empty filter as a preset, as there is no other way of removing all filtering criteria at once.

The number of matches shown per page is configurable from the "GUI settings" page that you can find in the menu. On that page you can also set various font sizes, which can be helpful if some pages get too wide or long.

As it is often possible not only to delete items, but also to undelete them again, most list pages contain the three links "Undeleted/Deleted/All" to specify what is shown. The one marked in bold indicates whether the page currently lists items that are not deleted, items that are deleted, or both. This is mostly useful if you want to locate something that you have deleted accidentally.

Ownership and Access Rights

Each user's access to different parts of BASE can be controlled by an administrator. More specifically, a number of access rights exist that often come in pairs, where one grants the right to view items of a certain kind and the other grants the right to also add or edit such items. Most items stored in BASE have an owner who controls who, if anyone, can view or edit the item. Thus, if a user is granted the rights to view and edit biomaterials, that user can still only view or edit those biomaterials whose owners have agreed to this.

The system for granting other users rights to manipulate things that you own is rather limited. Users can be arranged into groups by an administrator, with no limit on how many groups an individual user can be part of. As the owner of, say, a sample, you can grant access to the sample to the members of—at most—one group that you are also a member of. This sharing is performed from the list page for, in this case, samples, by ticking the check boxes of the samples to share and then choosing a group and either "read" or "read/write" access at the bottom of the page. It may also be possible to grant read access to all users ("world") if the administrator has given you the right to do so.

Hint: on the "My account" page in the "Users" menu it is possible to specify defaults for sharing so that all new items that you create are shared with a specific group.

User Administration

As with other functions in BASE, to administer user accounts and groups you need to be logged in as a user with the right to do so. By default, the "root" account has the *superuser* right, which implies not only a full set of access rights, but also the ability to do some things that are otherwise impossible, such as changing the owner of items or seeing who is currently logged in. When you are logged in as a superuser, a reminder of this is shown near the top of the left frame.

From the "List users" page in the "Users" menu you can create new users (see Fig. 3). The username is case insensitive, but the password is not. It is possible to create password-less accounts, but be advised that if two or more persons are logged in using the same account they can interfere with each other in harmless but annoying and confusing ways. It is therefore good practice to create one account per person.

As mentioned earlier, most of the access rights have to do with using specific parts of BASE, and many of these come in pairs. Of the ones with more special functions, "superuser" and "share with world" have already been touched upon. With "complete read access" a user is given permission to view such items that their respective owners have not shared.

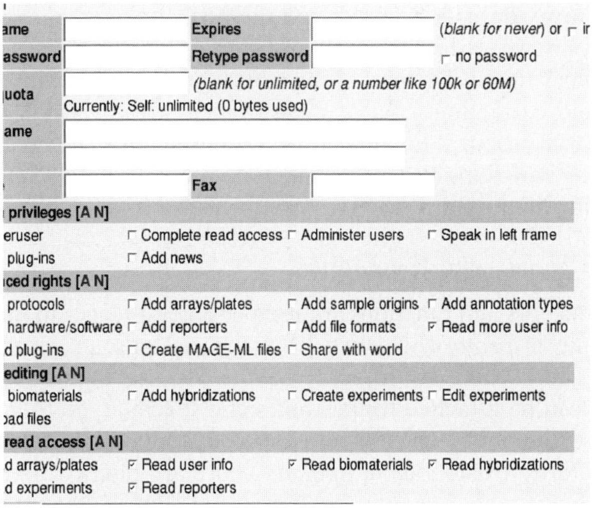

FIG. 3. The "Add user" page, displaying rights that can be granted users.

Groups of users can be created in much the same way as users, with a set of access rights that are inherited by all members. Consequently, a user has all the rights that he/she is granted personally or through any group membership. To add users to a newly created group you have to go back to the user list, mark the users to add, and choose the group in the pull-down menu at the bottom of the page. Unfortunately, because each item can only be shared with a single group, it may be necessary to create a number of groups that are simply aggregations of the users of other groups, and these groups will have to be modified separately when new users appear.

The disk quota that can be set for each user and group refers to the estimated amount of disk space used by uploaded files, raw data, and experiments. These are the things that by far tend to dominate the disk consumption. The quota system provides a way to keep track of what users or groups are hogging resources, possibly so that, for example, in an institutional setting, they can be billed for it. It can also be used to provide the strong incentive needed for users to clean up old, stale files and data once in a while.

Working with BASE

By necessity, the workflow in BASE resembles that of microarray experimentation. Production of microarrays and preparation of samples are parallel tracks that converge at the time of hybridization, wherein data can be extracted and analyzed. The production of microarrays is represented in the "Array LIMS" part of BASE, where the central players are reporters, plates, and array designs. Samples and their precursors and derivatives are collectively known as biomaterials and can, as expected, be found in the "Biomaterials" menu. The part of the workflow that deals with hybridizing, scanning, and extracting a set of measurements on the spots is available under "Hybridizations." Finally, in the "Analyze data" menu are the tools for collecting raw data into experiments and manipulating and analyzing these data in a hierarchical fashion.

Protocols, Uploads, and File Formats

Some concepts and functions are used in several places in the workflow. One example is *protocols*, which hold descriptions about experimental protocols. These come in different flavors for use in different parts of BASE and can be attached to items of some specific type. A protocol can have an associated file, which would have been uploaded by a user and at least temporarily existed as an *upload.* There are many different uses for uploaded files, usually as a medium for getting a large amount of data into BASE in one step. Files can be uploaded from those pages in the user

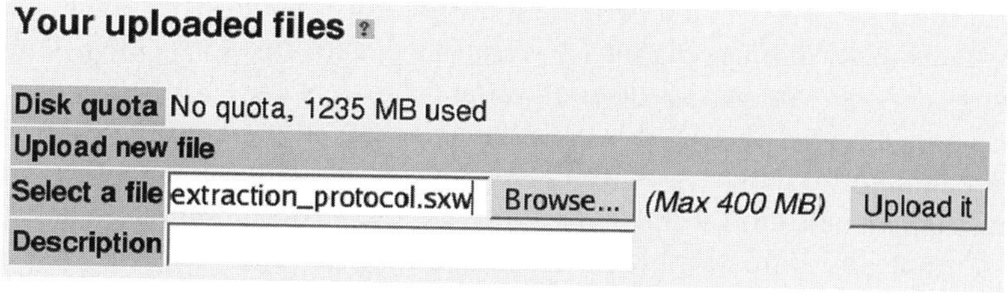

Your uploaded files ⍰

Disk quota	No quota, 1235 MB used			
Upload new file				
Select a file	extraction_protocol.sxw‖	Browse...	*(Max 400 MB)*	Upload it
Description				

FIG. 4. Files can be uploaded for later use or sharing.

interface where they are needed, but there is also a dedicated page for uploading, listing, and removing files, as shown in Fig. 4. Or rather, there are two very similar pages where one deals with uploads owned by other users. Note that the file names are required to be unique within each user's uploads, and replacing an upload with one bearing the same name requires a visit to the uploads page to remove the existing file.

Where files are uploaded to provide data in bulk, there is generally not a single universally accepted file format. Luckily, most formats used in the microarray field share enough features that it is possible to handle them with a single, configurable parser. File formats for a half-dozen different tasks can be configured from their respective file format pages. As an example, consider the "Reporter file formats" found in the "Reporters" menu (read about reporters later). As with all these file formats, each record is required to be on a single line, with the same tab- or comma-delimited format for each line of data.

On the page for adding a file format, below the usual fields for name etc., are two distinct sets of input fields. First comes the information used to identify files as being of this particular format, to find where the actual data content begins, and possibly where it ends. This is followed by a number of fields that describe what columns of the files to use for all the different pieces of information that a file of this format may contain (with some "advanced" options for concatenating columns). When defining a file format, the task is made much easier by having an uploaded file of that format to work with, especially if BASE then manages to correctly guess part of the file format. This means choosing that file in the file list near the bottom of the page or entering the new file format page from elsewhere in the user interface after already having selected an uploaded file. Doing so will bring up a table with an excerpt from the beginning of the file, shown near the end of the web page.

As a practical example, consider a file that looks like this:

```
reporter     gene
rep123       geneA
rep127       geneB
```

In this case, the first line would be useful both to identify the file format ("Line number 1 must match . . .") and as a data header ("Data header regexp . . ."). So long as the file contents are shown, it is be possible to specify these things using the drop downs in the "Use as" column of the file content table. In some special cases it is possible that users familiar with regular expressions could find it useful to specify these strings manually. With the file format and data start identified, it remains to specify what is where within each column. This can be done by mouse, picking a file column for the available fields, or manually as described in detail in the online help text.

Reporters

The term *reporter* refers to a reporter for a gene, as spotted or printed on a microarray, be it an oligonucleotide, a cDNA clone, an antibody, or something else. Operations in BASE that refer to genetic material generally do so at the level of reporters. All higher level categorizations, such as gene name, UniGene cluster, or Gene Ontology (GO) category, are simply handled as properties of the individual reporters. Each reporter has a unique, case-insensitive identifier called its reporter ID. By default it also has a fairly large number of different fields, such as gene symbol and gene name, LocusLink identifier, and accession number. This list of fields can, if need be, be customized.

Existing reporters can be viewed and manipulated via the "Reporters" menu, but new reporters can only be created by uploading something that refers to a previously unseen reporter ID. This could be a result file (raw microarray data), a microtiter plate, or a file describing the reporters on a microarray design. As this creation of new reporters happens by default, and unwanted reporters cannot be removed through the BASE web interface, you should probably not normally work as a user with the right to create new reporters.

Array LIMS

LIMS features are accessed from submenu options under main menu option "Array LIMS." Using this part of the database is entirely optional. Only laboatories that wish to track probe information, locations on microtiter plates, microtiter plate identities, their transformations from plate to plate, and well annotations (quality control information) need to use this portion of BASE. If you do not want to use the array production LIMS features, all other parts of BASE can still be used. It is also possible to forgo the plate part and only store information directly describing the arrays. However, integration of array production information with biomaterial information and analysis enables very powerful and efficient analysis of your microarray data and may, in some circumstances, be needed for minimal information about a microarray experiment (MIAME) (Brazma *et al.*, 2001) compliance.

Plates

If your laboratory wishes to track microarray production information, you must begin by first defining the *plate types* that you wish to track. For instance, a laboratory that prints IMAGE clones may have several types of plates to manage and track: bacterial stock plate, bacterial duplicate plate, bacterial growth plate, plasmid isolation plate, polymerase chain reaction (PCR) plate, PCR purification plate, and array print plate. For each plate type, events (protocols) and well annotations (such as quality control steps that are done to a plate type; i.e., gel electrophoresis of PCR products, or check for bacterial growth, or sequence verification) can be stored. Currently only plate sizes of 96 or 384 wells can be managed. If you spot proteins, antibodies, or oligonucleotides you may still want to use the Array LIMS features to store reporter annotations locally.

Once you have created plate types you can upload plates to BASE from a file. Note that doing so is one of the ways to create new reporters, which can thereafter be found in the "Reporters" menu. After having uploaded your first set of plates, you can, from the plate list page, list available plates by plate type, edit well annotations, and create daughter plates (a copy of all the probe locations into another plate of the same or another plate type). It is also possible to merge 96-well plates into 384-well plates according to a fixed pattern and to create plates with well contents hand picked from an arbitrary set of plates. Plates are used to track your probe library and ultimately to create array designs, array prints, and arrays. In this way subsequent microarray results loaded into BASE can be connected to arrays and thus to information on the probes in the LIMS.

Arrays

An *array design* in just that—a description of what goes where on a set of microarrays. From an array design it is possible to define one or more *array batches* (or prints) that describe individual production batches. These can in turn consist of many *array slides*, each carrying a unique name or bar code that can be used to track the slide when it is used.

The spots on an array design are called *features*. These link a block–column–row position on the array design with the corresponding well on a plate and the reporter found in that well. Creating the features of an array design involves uploading a print map file, which specifies what plate well is used on what position. Currently, the only supported file formats are the BioRobotics TAM/MoleculareWare MWBR format and the format used by the Molecular Dynamics Generation III array spotter. Creating features from a file in either of these formats is done by first adding the plates to the array design (from the array design's edit page) and then choosing to "Add print map." If your TAM/MWBR print map identifies the plates by name or bar code, this information is used to verify your selection of plates (and their order). Alternatively, it can be used to automatically pick the right plates, but then you lose the extra verification.

Laboratories that acquire microarrays without plate information can also add features to their array designs. This is done through what BASE calls a reporter map—a file that maps block–column–row coordinates on an array to reporter IDs. Unlike print maps, the formats of these files are user defined, created through the usual file format web interface. An example of a format that can be used is GAL (GenePix Array List), which is otherwise used as input to the GenePix spot-finding software. (Incidentally, GAL files can be generated by BASE when features have been created.) The raw data files produced by most image analysis software hold sufficient information to be used as reporter maps.

Biomaterials

Biomaterials comprise *samples, extracts,* and *labeled extracts* in BASE. The workflow allows users to enter sample information in BASE; samples can then be extracted and the extracts can be labeled and used in hybridizations. All along, information such as quantities processed and protocols used can be recorded. Sample annotations (such as clinical data, series time point, and protein value) can be added to samples and this information is connected to the data analysis part of BASE and can be retrieved by analysis and data visualization tools.

"Sample origins" are used to create definitions of sample sources. This includes specifying the organism a sample can be derived from, different tissue types, and tissue subtypes, as well as specific cell lines. The definitions created under sample origins will be available to choose from for all users when new samples are added. Unlike sample annotations (see later), the sample origins cannot, unfortunately, be used by analysis and visualization tools, but rather serve as a personal record.

Sample annotation types, simply called "Sample annotations" in the menu, are used to create properties or types that subsequently can be used by users to annotate their entered samples. Give your new annotation type a name, type, size of the input field box, and default value. The four types possible are floating point number (e.g., tumor size), integer (e.g., patient

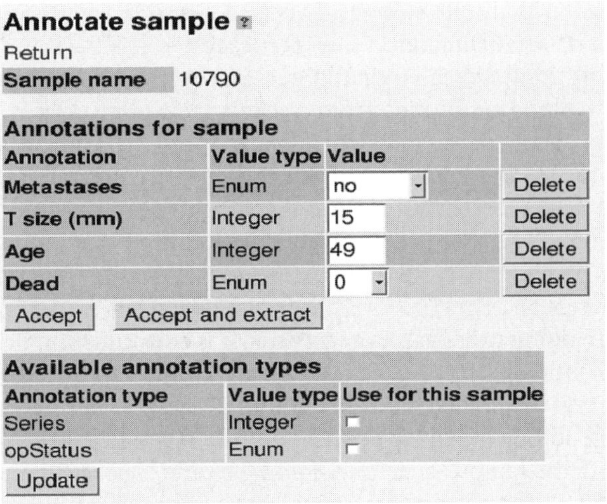

FIG. 5. Annotating a sample in terms of previously defined properties.

age), text (free text annotation), or an enumeration of possible values (e.g. protein status positive/negative, treatment a, b, or c). Sample annotations given to a sample (as in Fig. 5) will be inherited in the analysis end by the raw data sets associated with the hybridizations of that sample. In this way data analysis and visualization tools will be able to use sample annotation data to analyze and display data in an integrated and powerful fashion.

Samples are the true starting point of all data analysis in BASE. By entering a sample you can start extracting from it and label its extracts, recording quantity/quality information and methods for these steps. Then, labeled extracts can be associated together in hybridizations, to which image scans and raw data sets can be added.

Hybridizing and Scanning

The workflow from biomaterials, specifically from labeled extracts, continues with *hybridizations*. This is also the point where data from the biomaterials part meet data from the array LIMS part of BASE. A hybridization should be associated with one labeled extract for each channel (dye) of the experimental setup. When working with two channels and a sample vs reference design, the labeled extract in channel 1 represents your biological sample, whereas channel 2 holds the reference. The reason for this is that only the sample annotations for channel 1 will be available for use in the analysis end.

A single slide is sometimes scanned multiple times. To account for this, a hybridization can have one or more *scans*, each holding information about the scanner and its settings. It is also possible to attach images to a scan and mark what channel(s) they represent.

Uploading Raw Data. To finally get your microarray data into BASE, choose to "Upload result file" from a scan's view page. This requires a file format to be defined for the data file produced by your image analysis software. Most of the commonly used software creates tab-delimited files that can be handled by the generic file format creator. Only the output from ImaGene is handled as a special case, as it consists of two files that first need to be merged. Note that BASE uses the words result file, raw data set, and (internally) RawBioAssay more or less synonymously.

If you are doing dye-swap experiments, it is convenient to define one file format with, for example, Cy3 in channel 1 and Cy5 in channel 2, and another format with the opposite relation. Then, by using the right file format when adding raw data, you can get the raw data to always be stored in BASE with the sample in channel 1 and the reference in channel 2. This, of course, only applies if your experiment has a sample–reference design. As discussed later in the section on data analysis, BASE presently lacks facilities for handling more complex experimental designs, although there are possible partial workarounds.

Selecting a raw data file after you have defined a file format will bring up a page with some further options. Some of these have to do with connecting the spots to the features of an array design, whereas others affect reporter identification and creation. The purpose of knowing the array design at the time of adding raw data is threefold: the block coordinates and reporters in the result file can be verified, the raw data spots can be connected to plate wells, and the assignment of position numbers to the spots can be made independent of how the file is sorted. This requires the array design to have features, created from either a print map or a reporter map.

If an array slide has been selected for the hybridization, the corresponding array design will already be selected. If you are not tracking individual array slides it is still possible to choose an array design to use.

Each spot will be assigned a position number within the raw data set. It is highly desirable in the analysis end to have the same position number for analogous spots in different data sets, and this is ensured to be the case if you use an array design and choose to assign the position number from features. Otherwise, because the image analysis software could have missed some spots, there is no way to know for sure what block coordinates may be legal, and the position numbers are assigned from the order the spots appear in the file. Thus, you should make sure to have connected your hybridization to an array slide before you create a raw data set from it.

When a raw data set has been created, it is possible to view its spots in a table (see Fig. 6) and to plot it in some different ways. The html plot tool, which also appears in a few different places, can mainly plot one measured quantity against another, possibly coloring the points based on a third quantity. A small set of derived quantities are also available, so you could, for example, plot the channel 1 background corrected mean intensity against its channel 2 equivalent. There is also a Java plot tool, which is not supported by the core developers and thus may or may not work with various BASE updates.

If raw 16-bit TIFF images have been added to the scan the raw data set belongs to, it is possible to generate small images of the individual spots, which will be displayed in the raw data table and elsewhere. Generating these spot images requires raw data to include physical spot coordinates. Additionally, information about offset and scaling of those coordinates as compared to the TIFF images may be needed.

Data Analysis: Experiments

When data about and from your microarray experimentation have been entered into BASE, the time has come to compile, examine, analyze, and present those data. As you may have noticed, at the level of individual samples and hybridizations, it is not possible to assign objects to belong to specific "projects" (or whatever one would call such groups). Instead, BASE does this aggregation at the level of raw data set by collecting them into what we term "experiments." An experiment per se is little more than a collection of raw data sets, a hierarchy of analysis steps, and a description of the purpose of the experiment. Its power lies in the flexibility of the analysis part, the availability of analysis plug-ins for many different purposes, and the possibility of creating new plug-ins.

Everything related to experiments and analysis can be found in the "Analyze data" menu. When creating a new experiment, you need to specify

Raw data table

Return

Field	Op	Value	Buttons	Translated value
SNR ch1 median (FG-BG)/SD_BG ⋅	>	3	Upd | Delete	3
⋅	>		Add/Update	

Presets notbad ⋅ | Use | Save | Delete | Save current as new preset | Ok

LIMS info None ⋅ | **Raw data** Important ⋅ | **Reporter data** Important ⋅

<<prev next>> 1 2 3 4 5 6 10 20 30 40 50 100 200 300 400 500 1000 2000 3000 3419   (27350 hits, 8 per page)

Pos	Reporter	Cluster	Gene	Spot size	Flags	Sp	Reporter name ⌄
15090	H200011098	Hs.277624	ZZEF1	100	0	▣	ZZEF1 : I(E) Zinc finger, ZZ type with ▭
35702	H300000548	Hs.135805	ZZANK1	80	0	▣	ZZANK1 : C(E) Zinc finger, ZZ type with ▭
12439	H200006223	Hs.490415	ZYX	80	0	▣	ZYX : C(E) Zyxin
18487	H300017348	Hs.490415	ZYX	70	0	▣	ZYX : C(E) Zyxin
16499	H300003238	Hs.159249	ZXDA	80	0	▣	ZXDA : I(E) Zinc finger, X-linked, ▭
33068	H300011959	Hs.42650	ZWINT	110	-50	▣	ZWINT : ZW10 interactor
2828	H300001740	Hs.42650	ZWINT	60	0	▣	ZWINT : I(E) ZW10 interactor
1320	H300019042	Hs.42650	ZWINT	80	0	▣	ZWINT : C(E) ZW10 interactor

FIG. 6. Raw data from a single array displayed in a searchable table.

the number of channels it will contain. This is normally the number of different dyes used to label the samples. While the case of two channels is not given much special treatment by BASE, it is what most existing analysis plug-ins require.

Adding raw data sets to an experiment is done from the list of raw data sets. To see what raw data sets an experiment already has, find the "Raw data sets" tab on the experiment's page. There is nothing to prevent a raw data set from being used in several experiments. Internally, BASE stores each experiment in a separate set of database tables (using the MyISAM engine of MySQL), with the implication that activities in one experiment cannot interfere with activities in another experiment, whereas if several people work on the same experiment simultaneously, they might experience the occasional brief stall.

More often than not, the words "microarray data" refer not to the wide variety of quantities produced by the image analysis software, but to one intensity value per channel, or even to a single log ratio. Analysis within an experiment has the structure of a tree, with many different possibilities at each step, but it always starts with the extraction of intensity values from raw data. Thus, each raw data set gives rise to one *BioAssay*. BioAssays are always grouped together into *BioAssaySets*, and BioAssaySets are the basic building blocks of the analysis tree.

Creating a Root BioAssaySet. Creation of a BioAssaySet from raw data is carried out from the experiment's "Raw data sets" page. First you must select the raw data sets to use (typically all of them, by clicking the "A" in the table header). At the bottom of the page you then enter a name for the BioAssaySet and choose what kind of intensities to use. In many cases this could be background-corrected mean or median values, but other choices are also available, and it is possible for an administrator to modify the list of options (see the installation documentation). In any case, a BioAssaySet will be created and then appear on the "Analysis steps" page.

The lower part of the "Analysis steps" page displays the analysis tree (see Fig. 7), while the upper part can show information about a selected node in the tree. Initially, there will only be your newly created BioAssaySet. If you select it by clicking on its name in the tree, you will be presented with, among other things, the list of its BioAssays and several ways of browsing and visualizing the data it contains. For instance, the overview plot consists of a small plot of log ratio versus intensity for the spots of each BioAssay.

A possible desired series of analysis steps could be removal of flagged spots, normalization of the remaining values to zero median log ratio, finding a set of highly expressed genes, and doing a hierarchical clustering of those genes. These steps illustrate some of the different things that can be done in BASE, and it is illustrative to examine some of them a bit closer.

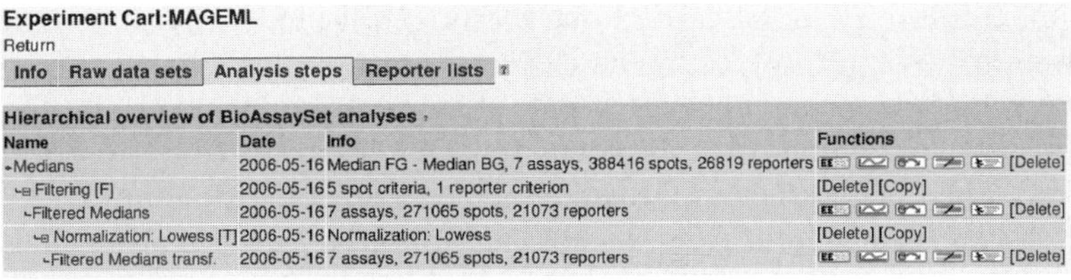

Fig. 7. A small analysis tree with filtered and normaized data.

Filtering Data. Removing flagged spots is an example of a filter applied to the spots of a BioAssaySet. Filtering, like some other functions, can be reached from one of the icons in the analysis tree. To understand how the filtering works, one needs to know how BASE stores a BioAssaySet. Every BioAssay in the set has a number of spots, and each spot in a BioAssay has a (unique) position number and an associated reporter. Most often the array slides used in an experiment have identical layouts so that all spots with a given position number also have the same reporter. With this design, a BioAssaySet can be said to have those positions and/or reporters that any one of its BioAssays has. In other words, when all spots at a certain position have been filtered away, that position is gone, and until then there could be missing values for some of the BioAssays.

Returning our attention to the filtering page, we note that the "Spot filter" part has a long list of fields that can be filtered on. Many have prefixes that indicate what aspect of the spot they refer to, such as raw data or its array LIMS annotations, and others refer to the intensity channel(s) of the spot. Under the "Gene filter" heading you have the possibility to filter on reporters. This is equivalent to including criteria for reporters in the spot filter, except that you also have access to something called "In # of Assays." This refers to the number of BioAssays that the reporter appears on after the spot filter has been applied, regardless of whether it appears in the same position in the different BioAssays (plug-ins exist for filtering on the presence of unique reporter positions).

Once you have chosen a set of filtering criteria, and possibly saved the filters as presets, you can click the accept button and watch as the spots are filtered into a new BioAssaySet. As you can then see in the analysis tree, the filter is stored as a Transformation, between the old BioAssaySet and the new one.

Note that, as elsewhere in BASE, it is only possible to combine multiple criteria with AND. All spots that pass the spot and gene filters will be used, and there is no simple way to, for example, keep *all* spots for those reporters that are differentially expressed on at least three different BioAssays. Such complex filtering can be carried out within BASE, either by some plug-in written for this purpose or by the clever use of reporter lists.

A *reporter list* can only exist as part of an experiment, but can be copied between experiments via the global reporter list page (the "Reporters" menu). Creating a reporter list can be done in several different ways: from the results of a search, from an uploaded file, or by an analysis plug-in. Regardless of how it was created, a reporter list consists of a set of reporters, each one associated with a number that may represent a score or rank. Filtering on a reporter list, where applicable, can be based on absence or presence in the list ("Reporter list") or on the score ("Reporter list score").

Normalizing and Analyzing Data. Data normalization is a tricky subject outside the scope of this text. However, what is well within that scope is how to carry out a normalization in BASE using a plug-in such as the LOWESS (LOESS) (Cleveland and Devlin, 1988) normalizer. This is one of the handful of plug-ins that are shipped with BASE. More plug-ins are available from plug-in pages of the BASE web site and from their respective developers, and the mailing list may be a good starting point if you are looking for something specific.

Just as with filtering, the first step in transforming data with a normalizer is to click the right icon in the analysis tree (the right one, with a running man) or in some other way come to the "Transformation: job" page. There you simply choose the plug-in to run—in this case the one called "Normalization: Lowess." Doing so will bring up a page with the help text of the plug-in and a set of plug-in-dependent options. Here, you can change the stiffness parameter for the LOWESS algorithm or leave it at its reasonable default value. Disregarding the other options, you can now start the normalizer and then follow the "Check the status of the job" link. When the job has finished, a new BioAssaySet will have been created as a child of the one that was to be normalized.

A similar procedure would be followed to perform, say, a hierarchical clustering. In that case a new BioAssaySet will not be created, but instead the plug-in will create some plots and other files. These can then be viewed using a special hierarchical clustering viewer that is part of BASE. In addition to producing a BioAssaySet and/or a set of files that can somehow be viewed, a plug-in may create reporter lists that can be used for filtering data.

The "Jobs" menu item leads to a page where all jobs are listed. There is a queue system that is meant to ensure that the jobs are processed in a reasonable order. As running a plug-in could potentially take a long time and use a lot of memory, there are limitations on which jobs can be executed in parallel. From the "Computation servers" page it is possible for an administrator to add and configure additional computation servers and set the maximum number of concurrent jobs and the amount of RAM allotted to them.

Analysis Plug-Ins. How to create an analysis plug-in for BASE is described in detail in a document available on the web site, but that text is somewhat minimalistic. If you are going to create a plug-in you have little choice but to read it; the design of the system is described here briefly.

An analysis plug-in consists of two parts: a plug-in definition, which is stored in the database and contains all information that BASE or the users may need, and the plug-in itself as an executable file on the BASE server. Creating your own plug-ins is meant to be easily doable in a wide range of programming languages, so the communication between BASE and the plug-in is intentionally kept very limited.

When making a plug-in, the first step is to define it through the web interface. Most importantly, BASE must know what data to export to the plug-in and what parameters (if any) it expects from the user. When a job is started, data are exported in the BASEfile format (which is described in the BASE documentation), producing a file similar to what can be created by exporting data from a BioAssaySet. The plug-in is then executed with this BASEfile as its input. Thus a parser for the BASEfile format is needed regardless of what language one uses for the plug-in. Code that deals with this problem is currently available in C++++, Java, and Perl. In addition, R/Bioconductor can be used through wrappers in Perl (see http://www.lcb.uu.se/baseplugins.php) or Java (see http://www.maths.lth.se/help/R/aroma.Base/). The output of the plug-in can be an arbitrary set of files that will be presented to the user, but to have transformed data imported back into BASE, it must again be in the BASEfile format.

MAGE-ML Export

To ensure reproducibility, many journals today require you to submit your raw microarray data upon publication of the results of your experiment. Commonly, data should be submitted to either GEO (Barrett *et al.*, 2005; see Chapter 19 in volume 410 by Lutfalla and Uze, 2006) or Array-Express (Parkinson *et al.*, 2005; see Chapter 20 in volume 410 by Hewitt, 2006). In the latter case, data must be in the MAGE-ML format (Spellman *et al.*, 2002), and in the former case MAGE-ML is one of several options. Regardless of how data are submitted, it is necessary to fulfill the requirements of MIAME (Brazma *et al.*, 2001).

Exporting data in the MAGE-ML format from BASE can be done from the "Info" tab of an experiment. The raw data sets currently associated with the experiment will be exported into the MAGE-ML file, as will everything upstream of them (such as samples and array designs), but not normalized or otherwise processed data. The exporting program is rather picky. Exporting will only work if you have used the array LIMS part of BASE; every hybridization in your experiment must be associated with an array slide whose array design has features. There may also be problems if you have customized the columns for reporters or raw data in your BASE installation. Improvements to the MAGE-ML export functionality are currently in development.

References

Andersson, A., Olofsson, T., Lindgren, D., Nilsson, B., Ritz, C., Edén, P., Lassen, C., Råde, J., Fontes, M., Mörse, H., Heldrup, J., Behrendtz, M., Mitelman, F., Höglund, M., Johansson, B., and Fioretos, T. (2005). Molecular signatures in childhood acute leukemia and their correlations to expression patterns in normal hematopoietic subpopulations. *Proc. Natl. Acad. Sci. USA* **102**, 19069–19074.

Barrett, T., Suzek, T. O., Troup, D. B., Wilhite, S. E., Ngau, W.-C., Ledoux, P., Rudnev, D., Lash, A. E., Fujibuchi, W., and Edgar, R. (2005). NCBI GEO: Mining millions of expression profiles—database and tools. *Nucleic Acids Res.* **33**, D562–D566.

Björkbacka, H., Kunjathoor, V. V., Moore, K. J., Koehn, S., Ordija, C. M., Lee, M. A., Means, T., Halmen, K., Luster, A. D., Golenbock, D. T., and Freeman, M. W. (2004). Reduced atherosclerosis in MyD88-null mice links elevated serum cholesterol levels to activation of innate immunity signaling pathways. *Nature Med.* **10**, 416–421.

Brazma, A., Hingamp, P., Quackenbush, J., Sherlock, G., Spellman, P., Stoeckert, C., Aach, J., Ansorge, W., Ball, C. A., Causton, H. C., Gaasterland, T., Glenisson, P., Holstege, F. C. P., Kim, I. F., Markowitz, V., Matese, J. C., Parkinson, H., Robinson, A., Sarkans, U., Schulze-Kremer, S, Stewart, J., Taylor, R., Vilo, J., and Vingron, M. (2001). Minimum information about a microarray experiment (MIAME): Toward standards for microarray data. *Nature Genet.* **29**, 365–371.

Carmel, J. B., Kakinohana, O., Mestril, R., Young, W., Marsala, M., and Hart, R. P. (2004). Mediators of ischemic preconditioning identified by microarray analysis of rat spinal cord. *Exp. Neurol.* **185**, 81–96.

Cleveland, W. S., and Devlin, S. J. (1988). Locally weighted regression: An approach to regression analysis by local fitting. *J. Am. Stat. Assoc.* **83**, 596–610.

Dudoit, S., Gentleman, R. C., and Quackenbush, J. (2003). Open source software for the analysis of microarray data. *BioTechniques* **34**, S45–S51.

Eckhardt, B. L., Parker, B. S., van Laar, R. K., Restall, C. M., Natoli, A. L., Tavaria, M. D., Stanley, K. L., Sloan, E. K., Moseley, J. M., and Anderson, R. L. (2005). Genomic analysis of a spontaneous model of breast cancer metastasis to bone reveals a role for the extracellular matrix. *Mol. Cancer Res.* **3**, 1–13.

Free Software Foundation (1991).GNU General Public License. http://www.gnu.org/licenses/gpl.html.

Hewitt, S. M. (2006). The application of tissue microarrays in the validation of microarray results. *Methods Enzymol.* **410**, 400–412.

Jönsson, G., Naylor, T. L., Vallon-Christersson, J., Staaf, J., Huang, J., Ward, M. R., Greshock, J. D., Luts, L., Olsson, H., Rahman, N., Stratton, M., Ringnér, M., Borg, Å., and Weber, B. L. (2005). Distinct genomic profiles in hereditary breast tumors identified by array-based comparative genomic hybridization. *Cancer Res.* **65**, 7612–7621.

Lausted, C. G., Warren, C. B., Hood, L. E., and Lasky, S. (2006). Printing your own inkjet microarrays. *Methods Enzymol.* **410**, 168–189.

Lutfalla, G., and Uze, G. (2006). Performing quantitative RT-PCR experiments. *Methods Enzymol.* **410**, 386–400.

Lyng, H., Badiee, A., Svendsrud, D. H., Hovig, E., Myklebost, O., and Stokke, T. (2004). Profound influence of microarray scanner characteristics on gene expression ratios: Analysis and procedure for correction. *BMC Genom.* **5,** 10.

Parkinson, H., Sarkans, U., Shojatalab, M., Abeygunawardena, N., Contrino, S., Coulson, R., Farne, A., Garcia Lara, G., Holloway, E., Kapushesky, M., Lilja, P., Mukherjee, G., Oezcimen, A., Rayner, T., Rocca-Serra, P., Sharma, A., Sansone, S., and Brazma, A. (2005). ArrayExpress: A public repository for microarray gene expression data at the EBI. *Nucleic Acids Res.* **33,** D553–D555.

R Development Core Team (2005). R: A language and environment for statistical computing R Foundation for Statistical Computing, Vienna, Austria.

Rhee, S. J., Walker, W. A., and Cherayil, B. J. (2005). Developmentally regulated intestinal expression of IFN-gamma and its target genes and the age-specific response to enteric Salmonella infection. *J. Immunol.* **175,** 1127–1136.

Saal, L., Troein, C., Vallon-Christersson, J., Gruvberger, S., Borg, Å., and Peterson, C. (2002). BioArray Software Environment (BASE): A platform for comprehensive management and analysis of microarray data. *Genome Biol.* **3,** software0003.1-0003.6.

Sollier, J., Lin, W., Soustelle, C., Suhre, K., Nicolas, A., Geli, V., and de La Roche Saint-Andre, V. (2004). Set1 is required for meiotic S-phase onset, double-strand break formation and middle gene expression. *EMBO J.* **23,** 1957–1967.

Spellman, P. T., Miller, M., Stewart, J., Troup, C., Sarkans, U., Chervitz, S., Bernhart, D., Sherlock, G., Ball, C., Lepage, M., Swiatek, M., Marks, W. L., Goncalves, J., Markel, S., Iordan, D., Shojatalab, M., Pizarro, A., White, J., Hubley, R., Deutsch, E., Senger, M., Aronow, B. J., Robinson, A., Bassett, D., Stoeckert, C. J., Jr., and Brazma, A. (2002). Design and implementation of microarray gene expression markup language (MAGE-ML). *Genome Biol.* **3,** research0046.1-0046.9.

Sturme, M. H., Nakayama, J., Molenaar, D., Murakami, Y., Kunugi, R., Fujii, T., Vaughan, E. E., Kleerebezem, M., and de Vos, W. M. (2005). An agr-like two-component regulatory system in *Lactobacillus plantarum* is involved in production of a novel cyclic peptide and regulation of adherence. *J. Bacteriol.* **187,** 5224–5235.

Wang, X., Wang, M., Amarzguioui, M., Liu, F., Fodstad, O., and Prydz, H. (2004). Downregulation of tissue factor by RNA interference in human melanoma LOX-L cells reduces pulmonary metastasis in nude mice. *Int. J. Cancer* **112,** 994–1002.

[8] Bioconductor: An Open Source Framework for Bioinformatics and Computational Biology

By Mark Reimers and Vincent J. Carey

Abstract

This chapter describes the Bioconductor project and details of its open source facilities for analysis of microarray and other high-throughput biological experiments. Particular attention is paid to concepts of container and workflow design, connections of biological metadata to statistical analysis products, support for statistical quality assessment, and calibration of inference uncertainty measures when tens of thousands of simultaneous statistical tests are performed.

METHODS IN ENZYMOLOGY, VOL. 411   0076-6879/06 $35.00
DOI: 10.1016/S0076-6879(06)11008-3

Introduction: Bioconductor in Brief

Bioconductor is a project devoted to the development of software and methods for statistical analysis and visualization of data from high-throughput experimental platforms in biology. The project is fully open source, with most software released under the Lesser GNU Public License (see http://www.gnu.org/licenses/licenses.html#LGPL). Most software available through the Bioconductor project is written in R, an open source data analysis environment that has become a major tool for quantitative scientists throughout the world.

Bioconductor may be viewed as a collection of specifications of containers and workflows for preprocessing and analyzing high-throughput data.

• Containers are defined for management and analysis of expression data at various levels of technical processing, for management of experiment-level metadata in the Minimum Information about a Microarray Experiment (MIAME) data model (see later; Brazma et al., 2006), for management and analysis of sample-level data and custom metadata about samples, and for the organization and manipulation of large quantities of biological annotation, such as mappings between proprietary probe identifiers and public database or ontology identifiers.

• Workflows are defined through high-level graphical user interfaces for specific forms of preprocessing and downstream analysis and through the interaction of software packages that are driven in a command-line interface. Documentation of workflows and workflow components is provided in the form of manual pages for specific modules and functions, vignettes that document multistep processes, and fully worked use case descriptions that are distributed with the software.

• Motivations and approaches of Bioconductor have been described comprehensively by Gentleman et al. (2004). This chapter provides details of philosophy, use, and future prospects of Bioconductor as a source of software and software design and distribution methods for critical methods of bioinformatics and computational biology.

Technical Details

Software Distribution

Bioconductor is rooted in the R language (Ihaka and Gentleman, 1996). Software resources are organized into packages, which are structured folders of code, documentation, and illustrative data. The distribution of packages for Bioconductor can proceed in various ways. At present, Windows and Macintosh graphical user interfaces for R include buttons for package installation over the web. These buttons allow selection of the Bioconductor software repository and selection of specific packages. Alternatively, a user may load a publicly available script (http://www.bioconductor.org/biocLite.r) into an R session, issue the command bioclite( ), and software will travel over the internet into this R distribution, where it is installed automatically and will persist until removed manually. Software modules can also be downloaded manually using a web browser pointed at http://www.bioconductor.org or can be obtained using functions provided in a Bioconductor package called *reposTools*.

Containers

There are four basic container types at present. Containers are available for preprocessed microarray data (e.g., data imported from scanner outputs or Affymetrix CEL files), for postprocessed microarray data (including detailed information on sample characteristics and treatments), for metadata about microarray experiments (principally satisfying the MIAME protocol), and for general biological metadata, such as the Gene Ontology.

The use of containers depends on the internal structures of the containers, defined in Bioconductor infrastructure packages such as Biobase, and on the accessor methods that are provided in these infrastructure packages. An accessor F for a container C is used with the syntax F(C).

Preprocessed Microarray Data. Affymetrix distributes a collection of CEL files from a Latin square design spike-in experiment. A subset of these data is distributed in the *SpikeInSubset* package of Bioconductor. The following code loads this package, loads the U133A-TAG subset provided there, and requests a report on this subset.

```
> library(SpikeInSubset)
> data(spikein133)
> spikein133

AffyBatch object
size of arrays=712x712 features (23781 kb)
cdf=HG-U133A_tag (22300 affyids)
number of samples=6
number of genes=22300
annotation=hgu133atag
```

This structure includes metadata about the samples. These metadata, often referred to as "phenotype data," even though they can involve information not typically regarded as phenotypic, can be accessed in the form of an R data.frame using the pData accessor. Here we inquire about the dimensions of

the pData component and select, using the matrix [row, column] selection
idiom, a small fraction of the metadata available for the spike-in study. This
is organized as samples in rows and attributes in columns. The attributes
are probe set identifiers, and the values of attributes are the picomolar
concentrations of the spike-in material.

```
> dim(pData(spikein133))
[1] 6 42

> pData(spikein133)[1:3, c(1, 5, 10)]
          203508_at    204959_at    207777_s_at
Expt6_R1      2            4             16
Expt6_R2      2            4             16
Expt6_R3      2            4             16
```

The preprocessed expression intensities can be accessed using the pm()
function. This submatrix is organized with *probes* in rows, and samples in
columns:

```
> pm(spikein133)[c(1, 5, 10), 1:3]
         Expt6_R1    Expt6_R2    Expt6_R3
[1,]       245.0        238        238.0
[2,]      2325.0       2238       2591.0
[3,]       541.8        445        564.8
```

For cDNA platforms, containers named marrayRaw or RGList are
used frequently. See the documentation of *marray* and *limma* packages
for details.

Processed Microarray Data. Container design for corrected and normalized expression data emphasizes tight binding of experimental data and
sample-level metadata, including probe identifiers and rich sample-level
data. These containers have been implemented through preservation of
manipulation idioms that are familiar through the use of simple data objects
in pure R.

While it is possible to work with pure R matrices to represent gene
expression experiments, Bioconductor enriches the data structure considerably. In the context of microarray data, let G denote the number of genes
measured in a microarray experiment, let N denote the number of samples
on which measurements were made, and let p denote a number of variables,
such as treatment type, sample identifier, and sample characteristics, that
identify important aspects of the experiment that should be known in any
downstream analysis. Bioconductor defines a class of objects called exprSets
that can represent all the relevant experimental data and metadata in a
unified way. Specifically, if E is an instance of the exprSet class, then exprs
(E) returns the GxN matrix of expression measures. pData(E) returns the

Nxp table of sample-level attributes, and description(E) returns a list of MIAME-defined experiment metadata attributes.

To illustrate this container concept, we interact with the golubMerge exprSet, which is supplied in the *golubEsets* package of Bioconductor. First we attach the data package and then we mention the 'golubMerge' exprSet, which combines the training and test data used in Golub *et al.* (1999).

```
> library(golubEsets)
> golubMerge
Expression Set (exprSet) with
  7129 genes
  72 samples
    phenoData object with 11 variables and 72 cases
  varLabels
    Samples: Sample index
    ALL.AML: Factor, indicating ALL or AML
    BM.PB: Factor, sample from marrow or peripheral
blood
    T.B.cell: Factor, T cell or B cell leuk.
    FAB: Factor, FAB classification
    Date: Date sample obtained
    Gender: Factor, gender of patient
    pctBlasts: pct of cells that are blasts
    Treatment: response to treatment
    PS: Prediction strength
    Source: Source of sample
```

Note that this report about golubMerge data tells the number of genes and samples and provides details on the sample-level variables that are available. The exprSet can be treated as a "two-dimensional" object:

```
> sm <- golubMerge[1:3, 1:2]
```

This computes a new exprSet with three genes and two samples. All the appropriate sample level data are carried along.

Of particular interest are the numerical values of gene expression. This is obtained using the exprs() accessor function:

```
> exprs(sm)
                 [,1]   [,2]
AFFX-BioB-5_at   -342    -87
AFFX-BioB-M_at   -200   -248
AFFX-BioB-3_at     41    262
```

Sample-level data can be accessed very conveniently using a list accessor idiom:

```
> table(pData(golubMerge)$ALL.AML
```

```
ALL    AML
47     25
```

This gives the clinical leukemia classifications of the 72 patients.

Metadata about Microarray Experiments. The MIAME data model (Brazma *et al.*, 2001) provides an informal protocol for documenting microarray data sets in a uniform manner. The "MIAME" class has slots corresponding to the MIAME fields:

```
> getClass(''MIAME'')
Slots:
```

Name:	name	lab	contact	title
Class:	character	character	character	character
Name:	abstract	url	samples	hybridizations
Class:	character	character	list	list
Name:	norm-Controls	preprocessing	other	
Class:	list	list	list	

```
Extends: ''characterORMIAME''
```

A graphical user interface (GUI) for eliciting MIAME metadata can be run from R.

The phenoData class is used to manage sample-level metadata.

```
> getClass(''phenoData'')
Slots:
```

Name:	pData	varLabels	varMetadata
Class:	data.frame	list	data.frame

General Biological Metadata. The Bioconductor approach to biological annotation is somewhat complex, reflecting a variety of objectives that are difficult to harmonize simply. Some of the most prominent aims are as follow.

- To support substantive filtering of high-throughput data structures, allowing, for example, restriction of differential expression analysis to those probes that have been associated with specific molecular functions.
- To meld statistical analysis workflows with biological interpretation so that, for example, estimated contrast coefficients can be labeled with genome or pathway annotations as desired.
- To support visualization of assay data in meaningful genomic contexts, such as chromosomal location or pathway topology.

• To support stability of underlying annotation resources for statistical analyses that may take months to complete.

The data infrastructure meeting these objectives consists of collections of R environments. An example is the *GO* (Gene Ontology) package. Upon loading this package, executing the GO( ) function produces a listing of environments and information on their contents.

```
> GO()

Quality control information for GO
Date built: Created: Tue May 17 10:04:27 2005

Mappings found for non-probe based rda files:
   GOALLLOCUSID found 9287
   GOBPANCESTOR found 9529
   GOBPCHILDREN found 4765
   GOBPOFFSPRING found 4765
   GOBPPARENTS found 9529
   GOCCANCESTOR found 1536
   GOCCCHILDREN found 561
   GOCCOFFSPRING found 561
   GOCCPARENTS found 1536
   GOLOCUSID2GO found 62424
   GOLOCUSID found 7770
   GOMFANCESTOR found 7220
   GOMFCHILDREN found 1366
   GOMFOFFSPRING found 1366
   GOMFPARENTS found 7220
   GOOBSOLETE found 1020
   GOTERM found 18285
```

Data package environments are all named according to a convention. The environment name begins with the data package name and has a suffix indicating the specific contents. For example, GOBPANCESTOR is the environment that maps from GO identifiers to ancestors (generalizations) of the associated term in the Biological Process subontology.

In conjunction with the *annotate* package, high-level reports on environment contents can be extracted. The lookUp function takes an identifier token, the name of the data package of interest, and the suffix of the name of the environment to be searched.

```
> lookUp(''GO:0000001'', ''GO'', ''TERM'')

GOID = GO:0000001
Term = mitochondrion inheritance
Definition = The distribution of mitochondria, in-
   cluding the mitochondrial genome, into daughter
   cells after mitosis or meiosis, mediated by interac-
   tions between mitochondria and the cytoskeleton.
Ontology = BP
> lookUp(''GO:0000001'', ''GO'', ''BPPARENTS'')
                  isa                      isa
''GO:0048308''          ''GO:0048311''
```

Accessing a description of the path(s) to the root of GO employs the environment GOBPANCESTOR.

The *GO* metadata package is a very general metadata resource, with information only about the gene ontology structure and content and, on some mappings, between gene catalogs and GO categories. Other metadata packages include

- *KEGG*—a series of environments providing information on the KEGG (Kyoto Encyclopedia of Genes and Genomes) pathway catalog
- *cMAP*—environments that address the NCI Cancer Molecular Analysis Project unification of KEGG and BioCarta pathway and molecule catalogs
- *humanLLMappings*—environments that encode the mapping between Entrez Gene identifiers of human genes and other systems, such as UniGene clusters and GO categories
- *YEAST*—a collection of environments that map ORF identifiers to alias gene names, enzyme codes, PubMed entries, GO, and KEGG pathway catalog entries

Workflows

A hallmark of Bioconductor's approach to software design and dissemination is the support of user-constructed custom workflows. Because the software is provided in a loosely coupled system of packages, analysts can select and sequence tasks with great freedom.

Some developers have taken advantage of the component-based design to build unified graphical user interfaces. Prominent examples are *limmaGUI* and *affylmGUI*, which allow users to step from raw scanner outputs, through quality control and gene filtering, through linear modeling for differential expression, to the development of hyperlinked lists of genes and associated annotations. Active discussion of further development occurs on the Bioconductor mailing list (https://stat.ethz.ch/mailman/listinfo/ bioconductor).

Documentation Strategies

The Bioconductor project recognized early on that broad utility would require a commitment to outstanding documentation resources. All Bioconductor software must be linked to manual pages that include executable examples. The project also developed a concept of "vignette," which is a document that combines code, narrative, and graphics to illustrate an analysis process that may involve multiple packages. Vignettes can be processed by various software components to (1) export all code illustrated in the vignette computations, (2) transform code into an interactive graphical user interface so that the user can step through sequences of computations by pushing buttons and can evaluate effects of code execution in the current R environment, and (3) transform narrative, code, and graphics into PDF format documents. Bioconductor packages *annotate, DynDoc,* and *tkWidgets* coordinate the implementation of these functionalities.

Array Preprocessing

Spotted Array Quality Control and Preprocessing

Bioconductor includes a number of packages designed primarily for spotted array data. These include *marray, limma, vsn, arrayMagic,* which builds on *vsn,* and *arrayQuality,* which builds on *marray.* The primary raw data structures are the marrayRaw class of *marray* and the *RGList* class of *limma*; the *convert* package allows users to convert between them. Several functions read in raw data from each of several types of image quantitation programs. For example, for GenePix files, `read.marrayInfo` reads in target information; `read.Galfile` reads in the GAL files, and `read.GenePix` reads in the .GPR files.

Quality control (QC) is a first step. Microarray specialists can now recognize several kinds of common defects affecting individual spots, and also problems affecting whole regions of microarrays (Minor, 2006). Spot defects are often caused by printing problems; regional variations in ratios often reflect nonuniform hybridization or washing. We think the ability to examine microarray data using statistical QC measures provides an important safety net to uncover biases or artifacts in these data. The R statistical environment provides a number of general-purpose graphical tools for statistical QC, such as box plots for visualizing distributions at various time points or for different batches of chips. The Bioconductor packages offer several special purpose QC tools adapted to dense microarray data. The *marrayRaw* class and the function image( ) conveniently allow the user to display the spatial distribution of signals on spotted microarrays. This is illustrated by the following command, which produces an image of the green spot intensities for the third array of the data stored in the *marrayRaw* instance raw.data:

```
> image(raw.data[, 3], xvar = ''maGf'', bar = TRUE)
```

The *arrayMagic* package can display many different types of quality control plots. The package *arrayQuality* displays a number of common QC graphics in one web page. One of the pages produced by the command maQualityPlots(raw.data) is shown in Fig. 1.

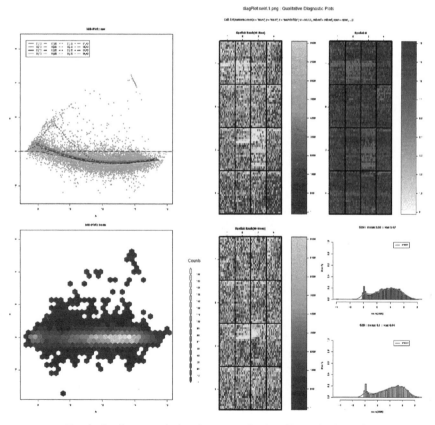

FIG. 1. Quality control plots from *arrayQuality*. (See color insert.)

The plot shown in Fig. 1 shows the ratio-intensity (M-A) plot in the upper left, with separately colored loess traces for each print-tip group. The center shows two images of the ratios across the chip: before and after normalization.

The next step in analyzing spotted arrays is normalization. The marray classes offer options for both two-channel (with each array separately), and single-channel (between array) normalization. The command

```
> norm.data <- maNorm(raw.data, norm = ''l'')
```

produces the now widely used loess normalization of red-green ratios by compensating for an estimated intensity-dependent bias. Other options include location-scale normalizations, print-tip loess, and two-dimensional loess smoothing of ratios. A further step that adjusts ratios between arrays may improve replicability between arrays.

Further analysis of spotted microarray data can be done with functions in the *limma* package (and many of the preprocessing steps can be done entirely within *limma* or using the convert utilities to transfer data between formats). The *limma* package has especially good facilities for specifying design matrices for spotted array experiments (i.e., which samples are in which dye on which slides). The function lmFit uses design matrix information together with the ratios on all slides, to give the best consensus estimate of relative mRNA abundance in each sample. The function eBayes gives estimates of probability of differential expression for each gene.

Preprocessing of Affymetrix Data

The Affymetrix GeneChip poses inviting challenges for the biostatistician, and Biconductor incorporates a wealth of statistical thinking about Affymetrix outputs. The basis for all analysis is the *affy* package, particularly the methods for reading in and organizing information from Affymetrix raw data (CEL) files.

The Affymetrix GeneChip provides a number of probes for each target mRNA; these probes are distributed over the chip surface (see Dalma-Weiszhausz *et al.*, 2006). Affymetrix provides indexing information for probes on the array via the chip definition file (CDF) for each chip type. This information is made available to the affy package by R environments, which are a way of storing key value pairs.

To read in a set of CEL files in the current directory the user types

```
> cel.data <- ReadAffy()
```

Then to obtain the default expression measures (the RMA estimates), the user types

```
> expr.set <- rma(cel.data)
```

As for spotted arrays, quality control is vital for Affymetrix chips. Thus function is served by the nuse function in the *affyPLM* package and by new packages *harshlight* and *bias.display*. These packages produce false-color images displaying the discrepancies between individual probes on a chip, and the expected values of those probes, based on averaging across chips.

The affy package divides up the process of obtaining expression estimates into four steps: background correction, normalization, adjustment for nonspecific binding, and combining ratios from different probes. The basic affy package provides a number of options for each of these steps via the `expresso` function, and users may mix and match their favorite methods. Other more specialized packages offer some different variants on the multichip method.

The `expresso` function offers two methods for estimating and compensating background. The user specifies `bg.correct.method=''mas''` to compute a regional estimate of the lowest 2% of the probe intensities and to subtract them from the original values; this procedure follows the practice of Affymetrix' own MAS5.0 software. The `''rma''` method estimates empirically the density of nonspecific hybridization over the whole chip. Then it computes a Bayesian estimate of the specific hybridization for any individual probe by averaging over the distribution of possible nonspecific signals.

The issue of normalization is contentious. The affy package offers several options: setting `normalize.method = ''constant''` invokes a scaling transformation to bring the mean of all chips into agreement, as is done by Affymetrix' MAS5.0. The `''quantiles''` method computes an estimate of the distribution of all background-corrected signals and then shoehorns all individual distributions into that shape. There is active informal discussion of this normalization; some researchers feel that it is too strong, and they note that replicates for the most abundant genes are less consistent than by a simpler normalization. However, replicates for the majority of genes are more concordant, especially for the least abundant genes.

Much probe signal comes from nonspecific hybridization. The intent of Affymetrix was that the "mismatch" probes would provide a specific estimate of hybridization of similar but not identical cRNAs to the corresponding "perfect match" probes. It has been the experience of many statisticians that computations based on PM only give more consistent results (T. Speed, personal communication). However, the user may specify either option. The current thinking is that the relevant background signal for probe intensities is the nonspecific signal, and so steps 1 and 3 should be combined. This more sophisticated approach is implemented by the separate package *gcRMA*, which estimates nonspecific signals using the model developed in Zhang *et al.* (2004).

The final step in the `expresso` paradigm is synthesis of a single gene abundance estimate from the evidence of multiple probe signals. MAS5.0 constructs a measure by computing an average of corrected PM signals independently for each chip. The user may construct the MAS5.0 measure by specifying `summary.method=''mas5''` in `expresso`. However, there is much information to be found by comparing across chips. A simple linear model for how the probe signal depends on gene abundance is that $S=af+e$, where s represents signal, a is gene abundance, and f represents the affinity of a specific probe for its target gene; e is noise. Raw data contain many outliers, and hence a robust fit is necessary. An adaptation of Tukey's median polish procedure provides a robust fit specified via the `summary.method = ''medianpolish''` option. A more sophisticated (and time-consuming) fit may be obtained by the function `fitPLM` in the `affyPLM` package.

Addressing Multiple Comparisons

When searching among thousands of genes for evidence of differential expression, there are bound to be many genes that exceed even fairly rigorous p value thresholds. Statisticians have developed several approaches to estimating and limiting the number of false positives. Two approaches that are implemented in Bioconductor are developing (1) more powerful genome-wide test statistics (*limma* and *siggenes*) and (2) methods for estimating the number of false positives (*multtest*) in a genome-wide test.

The basic idea behind moderated t test statistics is that the t statistic depends on an estimate of within-group variability. For small sample sizes (the usual case with microarray data), this variability estimate is itself highly variable; mistaken underestimates of within-group variability give rise to many false positives. However, by adjusting individual estimates of variation closer to a common value (such as their common mean), one can improve the majority of single estimates of variability, at the cost of introducing errors for a small number of genes.

The *siggenes* package implements ideas similar to those of the Statistics Applied to Microarrays program, whose approach was first described in Tusher and colleagues (2001).

The *limma* package implements an "empirical Bayes" method to estimating both variability and the probabilities that specific genes are expressed differentially. The user may choose a prior expectation of the number of changed genes (the conservative default is 1% of all genes on the array). Then the program returns a set of probabilities of differential expression for all genes.

The *multtest* package allows the user to compute several different estimates of the probability (or fraction) of false positives selected by a statistical procedure. The most useful estimates are the family-wide error rate, the false discovery rate (FDR), and the tail probability of proportion of false positives (TPPFP). The family-wide error rate is the probability of any false positive being selected by the test procedure. Researchers commonly care about the typical confidence in a large list of genes. The FDR is an estimate of the expected proportion of false positives in a list (Benjamini and Hochberg, 1995). The TPPFP provides a probability bound for the fraction of false positives.

Conclusions: Data Analysis for High-Throughput Biology and Bioconductor

In the coming decade, statistical methods will play a central role in dealing with the volume of data coming from high-throughput measurements such as microarrays. Through a collaborative process that is primarily informal, the Bioconductor project has engendered widely used software tools that address data translation and quality control, gene filtering, inference on differential expression, and phenotype prediction. The project has led to innovations in the production and dissemination of process-level documentation (vignettes and related dynamic documents) and in the important activity of reporting on statistical findings in biologically interpretable fashion, by allowing convenient binding of genomic, pathway, or functional ontology annotation to lists of features found to be statistically interesting, and by supporting straightforward creation and export of hypertext documents encoding these associations. The project has also spearheaded the use of a new object-oriented programming paradigm in R, the S4 system detailed in Chambers (1998).

Many open source software development and distribution projects have emerged in response to the challenges of the human genome project and to the excitement of postgenomic research agendas. These include projects focused on laboratory information management (BASE, Troein et al. (2006), base.thep.lu.se), scripting and programming language application (bioperl, biopython, biojava, bioruby, coordinated at open-bio.org), and data model and ontology development (open biological ontologies at obo.sourceforge.net, biopax at www.biopax.org).

The fundamental objectives of the Bioconductor project have been (a) promotion of advanced statistical technique in high-throughput biology and (b) reduction of barriers to interdisciplinary research. Through Bioconductor, statisticians have been given ready access to data examples and analysis practices in high-throughput biology. Biologists have been given access to the classic statistical analysis workflow components latent in R and also to emerging analysis strategies targeted directly at high-throughput biology. This objective has been pursued under appropriate constraints of transparency (all software and data provided by the project are "open source"). An extensive community of developers and users has emerged, united by an active mailing list and the software/documentation portal at www.bioconductor.org. We believe that the current situation of Bioconductor represents a partial achievement of the fundamental objectives. More work needs to be done to reduce the complexity of workflows, to help users match scientific needs to software capabilities, and to help developers integrate new techniques with existing structures. We are grateful to the Bioconductor Developers Core and to the many users and contributors who made this project possible.

Acknowledgment

V. Carey's contributions are supported in part by NIH Grant 1R33 HG002708, a statistical computing framework for genomic data.

References

Benjamini, Y., and Hochberg, Y. (1995). Controlling the false discovery rate: A practical and powerful approach to multiple testing. *J. R. Stat. Soc. Ser. B* **57,** 289–300.

Brazma, A., Hingamp, P., Quackenbush, J., Quackenbush, J., Sherlock, G., Spellman, P., Stoeckert, C., Aach, J., Ansorge, W., Ball, C. A., Causton, H. C., Gaasterland, T., Glenisson, P., Holstege, F. C., Kim, I. F., Markowitz, V., Matese, J. C., Parkinson, H., Robinson, A., Sarkans, U., Schulze-Kremer, S., Stewart, J., Taylor, R., Vilo, J., and Vingron, M. (2001). Minimum information about a microarray experiment (MIAME): Toward standards for microarray data. *Nature Genet.* **29,** 365–371.

Brazma, A., Kapushesky, M., Parkinson, H., Sarkins, U., and Shojatalab, M. (2006). Data storage and analysis in ArrayExpress. *Methods Enzymol.* **411,** 370–386.

Chambers, J. M. (1998). "Programming with Data: A Guide to the S Language." Springer-Verlag, New York.

Dalma-Weiszhausz, D. D., Warrington, J., Tanimoto, E. Y., and Miyada, C. G. (2006). The Affymetrix GeneChip platform: An overview. *Methods Enzymol.* **410,** 3–28.

Gentleman, R. C., Carey, V. J., Bates, D. M., Bolstad, B., Dettling, M., Dudoit, S., Ellis, B., Gautier, L., Ge, Y., Gentry, J., Hornik, K., Hothorn, T., Huber, W., Iacus, S., Irizarry, R., Leisch, F., Li, C., Maechler, M., Rossini, A. J., Sawitzki, G., Smith, C., Smyth, G., Tierney, L., Yang, J. Y., and Zhang, J. (2004). Bioconductor: Open software development for computational biology and bioinformatics. *Genome Biol.* **5,** R80URL:http://genomebiology.com/2004/5/10/R80.

Golub, T. R., Slonim, D. K., Tamayo, P., Slonim, D., Golub, T. R., and Kohane, I. S. (1999). Molecular classification of cancer: Class discovery and class prediction by gene expression monitoring. *Science* **286,** 531–537.

Ihaka, R., and Gentleman, R. (1996). A language for data analysis and graphics. *J. Comput. Graph. Statist.* **5,** 299–314.

Minor, J. M. (2006). Microarray quality control. *Methods Enzymol.* **411,** 233–255.

Troein, C., Vallon-Christersson, J., and Saal, L. H. (2006). An introduction to BioArray
  Software Environment. *Methods Enzymol.* **411**, 99–119.
Tusher, V. G., Tibshirani, R., and Chu, G. (2001). Significance analysis of microarrays applied
  to the ionizing radiation response. *Proc. Natl. Acad. Sci. USA* **98**, 5116–5121.
Zhang, L., Miles, M. F., and Aldape, K. D. (2004). A model of molecular interactions on short
  oligonucleotide microarrays: Implications for probe design and data analysis. *Nature
  Biotechnol.* **21**(7), 818–821.

[9] TM4 Microarray Software Suite

By ALEXANDER I. SAEED, NIRMAL K. BHAGABATI, JOHN C. BRAISTED,
  WEI LIANG, VASILY SHAROV, ELEANOR A. HOWE, JIANWEI LI,
  MATHANGI THIAGARAJAN, JOSEPH A. WHITE, and JOHN QUACKENBUSH

Abstract

Powerful specialized software is essential for managing, quantifying, and
ultimately deriving scientific insight from results of a microarray experiment.
We have developed a suite of software applications, known as TM4, to support
such gene expression studies. The suite consists of open-source tools for data
management and reporting, image analysis, normalization and pipeline control,
and data mining and visualization. An integrated MIAME-compliant MySQL
database is included. This chapter describes each component of the suite and
includes a sample analysis walk-through.

Introduction

The Human Genome Project was envisioned as a grand endeavor that
would change biology by providing a catalog of genes in humans and other
model organisms. Although a large number of genome sequencing projects, including that of the human genome, have been declared finished, the
collection of the sequence itself has not fundamentally altered our approach to understanding biological systems. Rather, it has been the development of techniques and technologies that allow us to analyze patterns of
expression for sets of genes, proteins, or metabolites approaching the total
number that are active in an organism at any given point in time.

Since their introduction in 1995 (Lipshutz, 1995; Schena, 1995), DNA
microarrays have matured significantly to become the most widely used
technique for the analysis of global patterns of expression and represent a
technology that is now used routinely as a means of generating testable
hypotheses prior to other studies. DNA microarrays consist of an arrayed

METHODS IN ENZYMOLOGY, VOL. 411
Copyright 2006, Elsevier Inc. All rights reserved.

0076-6879/06 $35.00
DOI: 10.1016/S0076-6879(06)11009-5

collection of probes bound to a solid substrate that are used to interrogate the levels of gene expression using hybridization to labeled nucleic acids and detection of those hybridization events. Although microarray technology is still evolving, the development of robust and reliable commercial platforms, combined with a significant decrease in the cost of an assay, has resulted in an explosion of gene expression data. The challenge of doing an expression profiling experiment is no longer in the generation of data, but rather in effectively capturing the information and using it to explore the biology of the systems under study.

In that regard, the role of software in a study involving microarrays cannot be overstated. Specialized tools are available to complement the experimental procedure and subsequent data analysis. Data management software is used to capture vital information describing the laboratory portion of a microarray experiment. Scanned microarray slides are processed and quantified using image analysis software. Normalization utilities ready data for comparisons and further analysis. Data mining and visualization tools can then help explore data from many perspectives. When used together, such software becomes a system to maximize the utility of the microarray experiment and gain better insight into the biology of interest.

We have developed a suite of software applications to support gene expression studies. This suite, called TM4, consists of a comprehensive set of tools that allow users to collect, manage, and effectively analyze data from microarray experiments. This chapter describes the TM4 suite and each of its components. The chapter concludes with an example analysis using a real data set and several analysis techniques.

The four major applications of TM4 are Madam, Spotfinder, Midas, and MeV. Each application in the suite is publicly and freely available. This includes the source code, which is OSI certified as open source under the artistic license (http://www.opensource.org/licenses/artistic-license.php).

Madam is the primary data entry, tracking, and reporting system of TM4. A series of data entry forms provide users with an organized method of recording their experimental parameters and data. Query and reporting tools present important data on a variety of entities, such as a single hybridization or an entire study. This application also serves as a repository for other tools in the data management and reporting realm. These include a polymerase chain reaction (PCR) scoring and microtiter plate loading utility, a study design tool, and a free-form SQL query window. Madam works closely in conjunction with a MIAME-compliant relational database to carry out its functions. The role of such a database is described elsewhere (Troein et al., 2006).

Spotfinder is a multichannel image analysis tool. This application provides the means to load the output of a microarray scanning operation—typically a pair of 16-bit tagged image format file (TIFF) images (Timlin, 2006). Semi-automatic grid construction and several methods to adjust the placement of each grid cell manually allow for accurate spot detection. The intensity of each spot can then be quantified and written to an output file along with related spot parameters and flags (Minor, 2006). A number of quality control displays are available, helping users detect systemic issues in slide production.

Midas is a normalization and filtering tool used to process raw data output from Spotfinder and prepare it for further analysis and data mining. Users create a project file, chaining together multiple normalization, filtering, and quality control (QC) modules, using an intuitive graphical workflow builder. The input options provide ways to consistently process single, paired, or whole studies worth of raw expression data. An intuitive graphing system illustrates the effects of normalization with a variety of detailed plots. These graphs can be embedded in a Midas summary report, a pdf-formatted file that also contains a description of the data processing procedure used.

MeV is the main data analysis and visualization tool of TM4. Users can load raw or normalized data from a variety of input file types. A broad range of algorithms is available, including those for clustering, classification, and statistical tests. The intuitive graphical interface simplifies navigation between algorithm results. An integrated scripting interface and XML-based format provides a means to analyze data sets in a regimented and reproducible fashion.

Although these applications were designed with interconnectivity in mind, each piece can be used independently of the others. Aside from the .mev format of TM4 (tab-delimited text with standardized column headers and comment rows), several other popular input and output formats are supported. While originally designed for two-dye fluorescent microarray systems, TM4 has been expanded to support other technologies, such as the Affymetrix Genechip platform (Dalma-Weiszhauz *et al.*, 2006).

A SourceForge web site (http://sourceforge.net/projects/tm4) serves as the central code repository for TM4. This site also hosts the application downloads, user mailing lists, and discussion forums. The TM4 development team actively provides technical support via email. System requirements for each application are detailed in the documentation included with the download. The entire TM4 suite, including software, documentation, and sample data, can be downloaded from http://www.tm4.org.

The TM4 suite was originally developed at The Institute for Genomic Research, under the direction of principal investigator Dr. John Quackenbush. Grants to Dr. Quackenbush for TM4 development were provided by The National Cancer Institute, The National Science Foundation, The National Heart, Lung and Blood Institute, and the NHLBI's Programs for Genomics Applications (PGA). Details regarding the ongoing development of TM4 and the teams responsible are available at the aforementioned SourceForge site. Beyond the main TM4 development team, many organizations and individuals have contributed to this open source project. Their contributions and affiliations are listed in the documentation for each application. The development of TM4 continues through collaborative efforts of groups world-wide, but with work now concentrated at three primary sites: John Quackenbush and his group at the Dana-Farber Cancer Institute and Harvard School of Public Health; members of the Pathogen Functional Genomics Resource Center's microarray software group at The Institute for Genomic Research; and Roger Bumgarner and his group at the University of Washington.

MADAM

Madam (also referred to as MADAM) is the data manager of TM4. It handles the tasks of data entry, tracking, and reporting while serving as an interface to a relational microarray database. Madam offers a series of data entry pages, which provide the user an easy method to load the database with information about their microarray experiments. Several report types display vital information about various stages of the experiment and let the user track the progress. Madam also houses several distinct tools with data management functions.

In addition to these roles, Madam is also capable of generating output in the MicroArray Gene Expression Markup Language (MAGE-ML) format. The submission of microarray data to public repositories is often required when publishing the results of a microarray study. MAGE-ML is the standard format for microarray data exchange and submission. If the user populates the database via the data entry pages correctly, Madam can generate MAGE-ML files that describe the entire microarray experiment. Two popular microarray data repositories are ArrayExpress (Brazma *et al.*, 2003, 2006) and Gene Expression Omnibus (GEO; Barrett and Edgar, 2006; Edgar *et al.*, 2002).

Madam is distributed with a MIAME-compliant relational database and the MySQL database platform. The database is a critical component for the operation of this software and nearly all of the functions of Madam involve interactions with the database in some manner. Madam cannot function without the database. Accordingly, Madam has features that assist the user with MySQL database installation and administration, including the creation of user accounts and Java Database Connectivity (JDBC) configuration.

Madam was designed with two-channel spotted arrays in mind, but efforts are currently underway to expand the interface and underlying database to accommodate other platforms as well, including Affymetrix GeneChips. Detailed operating instructions for this application can be found in the program manual included with the software distribution.

MAD: The Microarray Database

The database includes over 60 tables that store information about nearly every aspect of the microarray experiment. Some tables track the identities of the genes, clones, and oligonucleotides that compose spotted arrays and the microtiter plates in which they are located. The slide geometry is similarly important, including the number of spotted elements and printing pens, spacing between blocks, and the identities of the plates used. Each hybridization and the two probes used in each hybridization, including their origins, are also tracked. Data from outside the microarray laboratory are also stored, such as postimage analysis expression results in raw and normalized forms. The database schema (Fig. 1) illustrates each of the tables and the fields they contain, as well as the links between them.

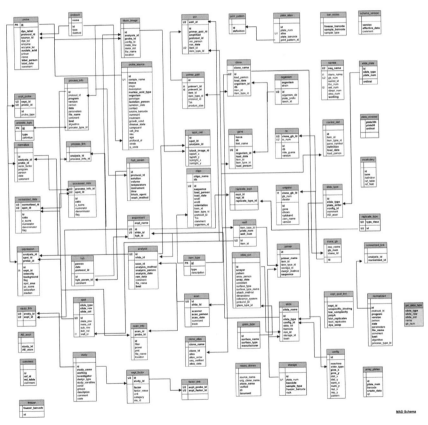

FIG. 1. Schema of the microarray database, MAD. These tables are used extensively for every function of Madam.

Many of these fields store data required by the MIAME (Minimal Information About a Microarray Experiment) specification (http://www. mged.org/Workgroups/MIAME/miame.html). The MIAME specification (Brazma *et al.*, 2001) describes data that are needed to enable the interpretation of the results of a microarray experiment in an unambiguous manner and to possibly reproduce the experiment. It follows that Madam users should try to enter information as completely and correctly as possible to ensure that the MIAME requirements are met. More information about MIAME can be found elsewhere in this volume.

Although the included database is built for the MySQL DBMS, other relational databases can be used. Madam has been connected to both Sybase and Oracle DBMS.

Madam Interface

The main Madam interface (Fig. 2) consists of four parts that are contained within an application window. A menu bar runs across the top of the window and provides access to the *File, Entry, Tools,* and *Help* menus. The *Help* menu contains the *Help Manual* menu item, a detailed and interlinked guide to the Madam interface, and all of the functions of the application. The *File, Entry,* and *Tools* menus contain items relevant to specific aspects of Madam and are described in subsequent sections.

The Navigation Panel is located on the left side of the interface. It consists of a set of tabs: *Entry, Edit, Report, Application,* and *MAGE-ML.* These correspond to each of Madam's major functions. Selecting a tab will display relevant controls for that function in the area immediately below the tabs.

The working panel is found on the right side of the interface. This is the area where most of the activity is based; the content will change depending on the task the user is working on. It can display forms for data entry and MAGE-ML writing, HTML-based reports, and some interfaces for supplementary tools.

At the bottom of the interface is the event log. This area reports important system messages, errors, and significant user activities and is persistent through all the aspects of the software.

Data Entry and Editing Pages

The task of loading data into the microarray database is facilitated through the use of data entry pages of Madam. Each page corresponds to a specific entity, such as a labeled probe or a glass slide. The data entry pages are active by default when Madam is started. To navigate here at any time, the user can click on the *Entry* tab in the *Navigation Panel.* When the

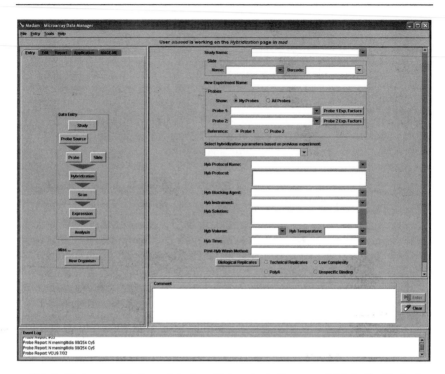

FIG. 2. Madam graphical user interface. Data entry buttons are visible in the *Navigation Panel* on the left side. The *Working Panel*, on the right, is currently displaying the *Hybridization* entry form. The *Event Log* at the bottom records recent activity and displays system messages.

Entry tab is selected, the *Navigation Panel* will display a flowchart of sorts, consisting of ordered buttons. Each of these buttons corresponds to one of the data entry pages and the layout of the buttons mimics the typical order in which each page will be used. Clicking on one of these buttons will bring up the appropriate data entry page form in the *Working Panel*. Another method to navigate to a data entry page is by selecting the desired page from the list contained in the *Entry* menu of the main menu bar.

Every data entry page shares several common features designed to simplify the process of filling out the form. All fields have a descriptive label alongside them to indicate the role of the field. Each entry field corresponds to a table and field combination in the database and this information is sometimes valuable to the user. By holding the mouse pointer over the field label for a moment and reading the tooltip, the user can learn both the table and the field name used in the database to store data entered in that part of the form.

Real-time validation is in effect and fields whose contents are not valid are noted by a red color, either within the text field itself or by a red element next to the input area of the fields. The user can view the reason by holding the mouse pointer over the field for a moment and reading the tooltip. Some fields are colored when the form is first displayed. This is because those fields are required; because they do not have values by default, they fail the validation for this reason.

Many of the drop-down lists in Madam accept input from the user either by clicking one of the entries of the list or by typing text into the field. Typically these lists are populated with data directly from the database and as such they can become quite long. The user can type a few characters into the field and hit the enter key, thus removing all items from the list that do not start with the characters already in the field.

At the bottom of every form is a text area for comments and *Clear* and *Enter* buttons. The *Enter* button indicates that the user is finished entering data in the form and wants to proceed to the next step. It is important to note that the *Enter* button will be disabled if any fields on the form are not currently valid. Clicking the *Enter* button brings up a confirmation dialog. This dialog displays a table with each field and the corresponding value noted. The user is given the opportunity to review the contents before beginning the upload of these data. If there are any errors, clicking the *Cancel* button will return the user to the entry page. If everything is as desired, clicking the *Submit* button will start the process of uploading data into the database. The progress bar and descriptive text messages in the dialog will indicate the status of this process.

The *Study* entry page captures information about a series of related microarray hybridizations (experiments) and the variables and experimental parameters involved. The *Probe Source* page is used to upload details about the biological source of a labeled probe used in hybridization. With this data, the *Probe* page can be used to enter information describing the probe itself, including the fluorescent dye used.

Describing the design of a microarray, from the geometry of the spots to the identities of each array element, is perhaps the most complex data entry task in Madam. This information is loaded using a page called *Slide*. To simplify the task, there are three different methods that can be used to create a new slide entry in the database. The user is free to choose the most appropriate method.

Two probes and one slide can be selected to form hybridization. These selections and details about the protocol and chemistry of the hybridization can be entered using the *Hybridization* page. The next page, *Scan*, is ready to record the settings used when the aforementioned slide is scanned. This page also records information about the TIFF image files that are produced by the scanner as output.

The *Expression* page is useful for uploading the raw intensities and flags calculated for each spot during the image analysis phase. This page requires a .mev format expression file as input. Information about normalization and data processing can be entered using the *Analysis* page, but the focus of this page is changing as the storage requirements for normalized and analyzed data evolve. Future releases of the software will reflect the current standards.

The *New Organism* button is set apart from the rest of the data entry buttons, as it is not used commonly. The *New Organism* entry page appears in a separate window when invoked and can be used to insert a new organism into the database. Selecting the appropriate organism is important when defining the source of a labeled probe or uploading plates using the *PCR Score* tool, described later.

Once data are uploaded, it cannot be altered using these data entry pages. Madam instead offers a set of data-editing pages. These can be accessed by clicking on the *Edit* tab of the *Navigation Panel*. Doing so will bring up a set of list boxes where the user can select the name of the entity to edit. Five options are available: *Study, Probe Source, Probe, Slide,* and *Experiment*. Selecting a name and clicking the arrow next to it will display a data-editing form in the *Working Panel*. These forms function in the same manner as the data entry forms, including real-time validation. The main difference is the replacement of the *Enter* button with the *Edit* button. Clicking *Edit* will bring up a confirmation dialog nearly identical to the data edit confirmation. Only the fields that have changed will be displayed and both the current and original values will be shown alongside the field names. Clicking the *Submit* button in this dialog will begin the process of updating the database with these edited data.

Report Generation

Through the reporting interface of Madam, a user can view and export HTML-based reports that encapsulate vital details about entities that were uploaded using the data entry pages. The user can click on the *Report* tab of the *Navigation Panel* at any time to bring up the report selection interface. The *Navigation Panel* will display five sets of controls for selecting study, experiment, slide, slide type, or probe reports. For each type there is a drop-down list and a *View* button. To generate a report of a given type, the user should select the appropriate entity identifier from the drop-down list and click the *View* button. The selected report is shown in the *Working Panel*.

Each time a report is generated its name will be added to a list near the bottom of the *Navigation Panel*. This list can be used to track the history of viewed reports and quickly recall one by clicking on the name. At the bottom of the panel there are four buttons. The *Save* button can be used to output the currently visible report as an HTML file for later viewing in a web browser. The *Print* and *Print Preview* buttons both send the visible report to a printer; the latter also shows an image approximating the report's printed appearance. Finally, the *Clear* button resets the entire report interface.

MAGE-ML Writing

MAGE-ML is a standardized XML format for microarray data that has gained wide acceptance in the community. It can be used to distribute microarray descriptions and results to colleagues or for submissions to public microarray databases. Submitting microarray data to public databases such as ArrayExpress or GEO is important when publishing the results of microarray experiments, thus giving others the ability to view your data sets and potentially reproduce the results.

Madam provides a means of writing MAGE-ML files. To do this successfully, the user must first make sure all their data have been entered accurately and as completely as possible using the data entry pages. Missing data can cause problems in the MAGE-ML files that are produced. The MIAME-compliant database that is associated with Madam is capable of storing the necessary information to produce complete MAGE-ML files.

The MAGE-ML interface is invoked by clicking on the *MAGE-ML* tab in the *Navigation Panel*. The panel will then display a tree containing the various objects that can be encoded into the MAGE-ML format. Selecting an object from the tree will bring up a form in the *Working Panel* that is specific for each object type. The user can fill in this form in the same manner as the data entry pages. Once all the fields pass the validation test, the MAGE-ML button at the bottom of the form can be clicked, thus starting the file writing process.

Each object that can be written requires the user to describe the people involved with the project, from the experiment itself to the file generation and submission. These people can be selected from a list and labeled with a role. The list is populated by using the *Preference* menu item located in the *File* menu. Names, organizations, and contact information for all the requisite people can be stored through this interface and retrieved from within the MAGE-ML writing forms.

Two chapters in this volume describe in more detail the importance of MAGE-ML and how it is used in the context of submissions to public microarray databases. Further information about the MAGE-ML object model and file format can be found at http://www.mged.org/Workgroups/MAGE/mage.html.

Related Tools

Madam serves as a home for several distinct tools that are involved for some data entry, retrieval, and management tasks. Although each part can operate outside the context of Madam, there are several advantages when they are bundled together. Perhaps the most important is the utility of having Madam regulate the database administration tasks. As such, it is not necessary for any of these smaller tools to try to establish a new database connection or to start and stop the database software. Having all the tools accessible from one location is also convenient for users and facilitates interactivity between each piece.

These six tools are accessible from the *Application* tab of the main navigation pane. The user can click the button that corresponds to the desired tool to launch it. An alternative is to use the *Tools* menu from the main menu bar and select the appropriate tool from the list. Each of these tools included with Madam is described.

ExpressConverter handles file conversion operations for the TM4 suite. It accepts a variety of common scanner input formats, including GenePix and Agilent, and converts them to the .mev format. For other file types, *ExpressConverter* offers a customizable file converter that allows the user to describe their tabular text format and then convert files of that type to .mev. Annotation files can also be created using this customized converter to complement the expression files.

ExptDesigner is a tool that can help plan a series of microarray hybridizations (Ayroles and Gibson, 2006; Neal and Westwood, 2006). Loop and reference experiment designs are supported. The user begins by selecting the probes from the database that are involved in these hybridizations. These probes will be available for selection in the design view panel, an interface that consists of two visual tools. Users can create hybridization by drawing a directional arrow from one probe to the other in the network view or by clicking a square corresponding to the two desired probes in the matrix view. Each hybridization is then added to a list that can be exported.

PCR Score is an application designed to create and manage data associated with the microtiter plates used for microarrays (Eads *et al.*, 2006). Users can upload information describing the contents of 96-well plates, including those containing oligonucleotides and PCR products. In the latter case, a scoring interface allows the user to indicate the success of the PCR reactions. The 96-well plates can then be combined into the 384-well plates often used for array printing.

Mabcos is a microarray bar-coding system. This application prints bar codes for several types of laboratory objects, including freezers and microtiter plates. Series of bar codes can be scanned, tracking the probes, plates, or slides involved in an experiment. Bar-code scanning helps ensure that the microarray printing is performed correctly. Data can be transferred from a traditional scanner or a PalmOS-based device.

Miner is a tool that writes .mev format expression files from data in the database. The user can specify an *Experiment Name* that corresponds to the hybridization for which to retrieve data. A filter based on PCR results and a buffering option for partial files are available.

The *Query Window* is an interface for submitting free-form SQL queries directly to the database. This supplements the data entry, editing, and reporting functions of Madam by providing a finer level of control over data. Users who are familiar with the schema of the database and the SQL syntax can perform a wide range of operations, including insertions and deletions. Query results and the corresponding SQL can be saved to text files.

Administration Tools

The administration aspects of Madam are handled by a related application called the *Madam Administrator*. This program can be launched from the same directory as the main Madam executable. Some common administrative tasks that can be performed include the creation or removal of local array databases, changing the JDBC connection settings used by Madam to communicate with the database, and changing the values that appear in some parts of Madam's data entry pages and MAGE-ML forms. Details of these operations are found in the program manual included with the Madam distribution.

Spotfinder

Image processing is a key component of the microarray experiment (Minor, 2006; Timlin, 2006). Each two-color spotted microarray slide will typically produce two gray-scale 16-bit images in TIFF format. Each image corresponds to a single labeling dye such that the two images complement each other spatially and need to be processed in parallel. The microarray TIFF image is the end product of the portions of the microarray experiment conducted in the laboratory.

Despite being digital media, in essence, and storing all necessary information about the conducted experiment, the image file is not a data set ready for data analysis (Minor, 2006). The goal of image analysis is to digitize microarray image files and produce output data sets for each slide. These data sets can then be normalized and used as input for clustering, visualization, and statistical analysis tools. The image processing software itself is a specialized tool that provides parallel analysis of microarray TIFF images.

Critical steps include the definition and digitization of spots, calculation of local background, and reporting intensities into output data files.

Image Analysis Goals

The challenge is spot detection and digitization or extraction of intensities. The spots on an array correspond to the genes printed on the slide during the printing step. The hybridization procedure attaches two fluorescent markers on the same target for every spot. After slide scanning at two different excitation wavelengths, two separate images are generated, one for each fluorescent marker. It is commonly accepted to refer to these two images and data extracted from them as the two channels of the microarray experiment. The fluorescent dye signal is expected to be proportional to the overall efficiency of the hybridization as well as the gene expression level. By measuring the integrated signals from both dyes on every spot, we are able to approximate the level of gene expression for both conditions of hybridization.

Two main problems must be addressed before spot intensities can be measured: spots have to be localized spatially on the image and a local background value has to be estimated for every spot. The problem of individual spot locations cannot be solved globally for the entire image. Rather, a good approach is to proceed locally by splitting the image area into subarrays, each of which consists of a group of spots or even individual spots.

Background correction of the measured microarray spot intensities is a procedure used to derive true values from raw experimental data. This correction aims to remove the additive components from multiple sources: substrate background, cross-talk from the other channel dye, nonspecific hybridization response, and so on. Because a spotted microarray image has a nonuniform background distribution over the whole image area, local background correction becomes highly desirable. It is commonly accepted that the background estimated locally for every spot is the best method, if correction is desired. Background correction normally results in expansion of the dynamic range of data for both intensities and ratios; the outcome is more prominent for low-intensity spots and almost invisible for highly expressed genes. It should be viewed as favorable data transformation, as it increases resolution on the expression ratio scale and makes it easier to distinguish the differences between genes with close expression ratios.

Background correction may result in some negative intensity values. This can occur when the local background estimate is larger than the spot intensity, as it is for some weak spots ("black holes"). These spots can be filtered out automatically by applying the signal-to-noise ratio criteria. Realistically, such filtering will not result in data loss, as the weak spots cannot be considered as reliable data points in downstream analysis.

Approaches

There are a few approaches to solving the spot location problem. All of them utilize a known geometrical pattern of printed arrays; arrays typically consist of subgrids arranged in meta-rows and meta-columns. The subgrids themselves consist of the spots that form rows and columns in a rectangular pattern, allowing simple description by a few parameters. A less common geometry of a hexagonal or "orange packing" layout of spots is also possible and can be described by a small number of parameters. Using this, one method of determining spot locations is to apply a predefined grid to the image that can be aligned manually or automatically in subsequent steps. A user-assisted semiautomatic grid alignment is also an option. Another solution is not to use a grid explicitly, but apply an automatic or semiautomatic procedure for spot positioning based on the geometric pattern of the printed spots. Two examples of this include the spot-finding procedure based on seeded region growing algorithm (Yang *et al.*, 2002b) or Fourier transform based procedure (Gaidukevitch *et al.*, 2000).

There are not many ways to estimate the local background around a spot. The most popular method is based on the assumption that the background surrounding a spot is the same as the background in the spot itself. Using the rectangular area around the spot for local background estimation by this method is possible if a grid has been defined. The only caveat of this method is the possible overestimation of the background when the pixels closest to the spot are considered part of the background. This systematic bias can be avoided by not involving pixels from the area immediately surrounding the spot in the background calculation. While this method, with some modifications, is widely used by many image analysis tools, one can argue that the real background in the spot area may stem from the nonspecific hybridization or nonlabel fluorescence signal from the target area. Measurement of nonhybridized or nondye hybridized sites has been suggested as an experimental method for the background estimate (Yang *et al.*, 2001) and, on occasion, has been used effectively in practice (Johnston *et al.*, 2004).

Spotfinder 3

Since its first release in 1999, Spotfinder has passed through many upgrades and version changes. Spotfinder 3.0, released in 2004, was a significant redesign of the traditional architecture: a multiplatform application allowing the analysis of arrays containing more than two dyes. Currently, Spotfinder executables are available for three major desktop platforms: Windows, Linux, and Mac OSX. Due to the large size of the TIFF images that are usually analyzed (often 20 MB or more), it is recommended that Spotfinder be run on computers with at least 256 MB of RAM and a CPU clock speed of at least at 800 MHz. A 16-MB video card (32+ is better) is strongly desirable for Windows desktops. The latest release, Spotfinder 3.1.0, accepts both 8- and 16-bit TIFF images, stored in separate files or one multi-TIFF image file. Data output is generated as tab-delimited text files (.mev) and platform-independent binary files for grids (.sfg) and whole raw data sets (.sfd).

Spotfinder has an intuitive graphical user interface (GUI) with a menu bar on top, dialog tool box on the lower left side, and a number of tab pages in the center of the main program window (Fig. 3). The menu bar contains *Image*, *Grid*, *Data*, and *Settings* menus. The collection of tabs gives the user access to the pages: *General*, *Overlay*, *Analysis*, *RI plot*, *Data*, and *QC view*. The *Analysis* page is the most functionally rich element of the Spotfinder GUI; all user activities in Spotfinder related to grid design and alignment are focused on the *Analysis* page. The dialog tool box is a control for alternative selection of several dialogs. *Gridding and Processing* has the controls for setting the parameters for automatic grid adjustment and choosing a segmentation method. The *Post-Processing* dialog can be used for *QC filter* settings and for turning on/off the background correction algorithm. The *Cell Editor* dialog controls can be used to activate the *Cell Editor* for interactive cell selection on

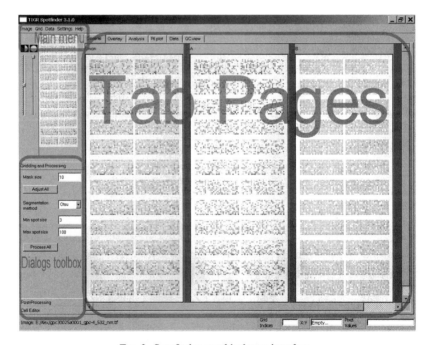

FIG. 3. Spotfinder graphical user interface.

the *Analysis* page, *RI plot* page, or *Data* page. When active, the *Cell Editor* displays the selected spot with its cell shown. The *Cell Editor* also can be used to resize and move cells, and to reprocess the selected spot.

General Steps

The following sequence is typical for a grid-based spot finding procedure using a local background calculation method over the area surrounding each spot.

Grid Composition

Designing and constructing a new grid requires the entry of several slide-specific parameters that are usually determined when the slide is printed. The user can get this information from the settings log file of the printing robot or by making measurements directly on the image file using Spotfinder. Normally the spots on the array are printed in blocks (sometimes referred to as grids or subgrids). This discussion deals with rectangular packed grids. The blocks are organized into meta-rows, which are rows of grids, and meta-columns, or columns of grids. The number of meta-rows and meta-columns on the array is defined by the number of the printing pins in the X and Y dimensions. Note that all spots in a single block are printed by an individual pin. The set of parameters needed for grid construction include the number of pins in X dimension (pinX), the number of pins in Y dimension (pinY), spot spacing (distance between spots, measured in pixels) in both X and Y dimensions, and the number of rows and columns in every block. To measure these parameters on screen, the user can switch to the *Analysis Page* and use the mouse pointer to navigate it to any spot on array, making a reading of the mouse cursor coordinates at the bottom of the program window. The initial spot spacing parameter, which defines the whole grid size, can be set with significant tolerance. As a result, blocks may be smaller or larger than the array of spots that they cover; this can be ignored for the moment.

Grid Expansion and Shrinking

The first step in grid alignment is to move the whole grid set to a desired position. The best way to place the entire grid is to keep track of the upper left-most grid at first and use it as an alignment indicator. After the upper left point of this grid is positioned in place, the spot spacing can be adjusted to fit the correct grid size. It can be done by applying *grid expand*, sequentially increasing or decreasing the spot size until each rectangle of the grid has a spot centered within. The suggested method is to watch only the upper left-most grid when applying this procedure to all subgrids on the array.

Ideally, all pins in the print head are evenly spaced; hence, all printed blocks should have a consistent offset (George, 2006). However, it is often found the pins are slightly bent; as a result, the blocks of spots are misaligned relative to each other. After positioning and sizing the upper leftmost subgrid we can proceed with adjusting the position of the other subgrids. The automatic grid adjustment procedure can be used to iteratively place the remaining subgrids. This procedure uses a mask, of a size no less than the typical spot size, to calculate the target function for every block. This mask is assigned to every spot cell and centered. The target function is defined as the integral of all the pixels in the spot cells under all the masks in the grid. Every block is moved in small steps to a maximum of one cell size up, down, left, and right. At every step the integral is calculated and stored as an element of a two-dimensional (2D) matrix. After this has been completed, the 2D matrix is searched for the maximum calculated integral value. The subgrid is then moved to the position corresponding to that maximum value. This automatic procedure requires that three conditions are met. First, every subgrid has to be set correctly in size and with the correct row and column numbers. Second, each subgrid has to be aligned roughly, within a tolerance of one cell size. Finally, at least half of the spots in the block should have significant signal (i.e., strong spots) to provide reliable landmarks to position the subgrid correctly. It is important to note that this procedure may give unsatisfactory results for subgrids with empty rows or columns on the edges, that is, the first and last columns or rows. The automatic grid adjustment can be applied repeatedly, as sometimes it is necessary to use the procedure a few times to reach sufficient grid alignment. If repeated applications do not produce satisfactory results, the only solution is to align the grid manually. The user, in this case, can use the mouse or keyboard arrow keys to position each subgrid accurately.

Spot Detection

Spot detection, also referred as spot finding or segmentation, is the next key step of image analysis. The goal of spot detection is to separate spot pixels from background ones. The image has to be segmented or divided in two subsets: one including the signal or spot pixels and the other consisting of the background pixels. A number of methods are widely used for the segmentation of microarray image spots. It has been shown (Yang *et al.*, 2001) that the choice of segmentation method applied has no significant effect on the ultimate results of image analysis. All segmentation methods used in microarray image processing can be categorized as being either histogram or shape based. The histogram-based methods do not take into account the spatial information about the analyzed spot, such as spot shape. Instead, they apply a sorting algorithm to the whole set of pixels in the cell and set a threshold that separates signal and background pixels based only on the values of the pixels. Auxiliary input parameters, such as spot size and spot pixel dynamic range, can be used to facilitate the histogram segmentation. In contrast, the shape analysis-based methods mainly rely on spatial information of the image; they look primarily on how pixels with high values are grouped together spatially. A few software tools use a method based on a seeded growing algorithm (Yang *et al.*, 2001, 2002); this is related to the shape-based methods.

The original segmentation method implemented in Spotfinder is a histogram-based algorithm that expects only a single parameter from the user: estimated spot size. This method provides good results for images with a low variation in spot size. However, images can have significant spot size variation in certain conditions. One cause could be the variation in temperature and humidity during the slide printing. Spotfinder introduced a segmentation method based on the *Otsu* algorithm (Liao *et al.*, 2001; Otsu, 1979) to address this problem. This method requires the user to input two parameters: minimum and maximum spot size on the array. The *Otsu* method runs through the original histogram of pixel values and places a threshold dividing the area into two groups—background and signal. The threshold returned by this method maximizes between-group variance. A third segmentation option in Spotfinder is a manual method in which the user interactively applies a predefined circle as mask for spot segmentation.

Spot Digitizing

Following the detection of spot boundaries by the segmentation method, the next step in image analysis is spot quantitation or digitizing. During this stage the pixels inside the spot are counted and added together to calculate the integrated intensity of each spot. The spot mean and median can also be determined.

The important issue of pixel saturation is addressed at this step. Due to the natural limitation of the dynamic range, the value of each pixel cannot exceed the 16-bit maximum, which is 65,536. If the input fluorescence signal is too high and exceeds the linear range, the corresponding pixel is assigned the maximum possible value of 65,536. In principle, the problem can be solved by rescanning the slide with different settings for sensitivity or scanner detector power. Doing so can bring all pixel values to a lower range but also may set low-intensity spots at the background level. Alternatively, if the percentage of saturated pixels is not too high, they can just be ignored and a rescan of the slide can be avoided. The check for saturated pixels is conducted when pixels inside a spot are analyzed. If any pixel in the spot is saturated at least in one channel it will be excluded, that is, removed from the spot data set in all channels. All reported values for the spots—mean, medians, integrated intensities, and their standard deviations—are computed after the saturated pixels are removed. As a measure of this correction the resulting saturation factor is reported. The saturation factor is defined as the ratio of the non-saturated pixels over the original number of pixels in the spot.

Local Background Correction

Because the direct measurement of spot background is not feasible, the local background estimate is based on some assumption. The simplest way is to assume that the background in the area surrounding a spot is the same as inside the spot itself. The local background can then be measured by analyzing the pixels just outside the spot boundary. Normally the local background is estimated by the median of these exterior pixels. To apply background correction to integrated intensities, this value must be multiplied by the spot area and subtracted from the raw integral intensity.

Reported Parameters

Spotfinder reports a number of parameters for each spot that is detected: integrated intensity, mean, median, total background (integral), background median, background standard deviation, integrated intensity standard deviation, mean standard deviation, median standard deviation, flag, QC score, and p value. These parameters are reported for each channel in the spot; if it is a two-dye array there will be two integrated intensities, etc. Some parameters are reported only once for the spot, independent of the number of channels. These include spot area, saturation factor, and total QC score.

Quality Control Parameters

The QC procedure, which analyzes and reports QC parameters, is an important part of any image processing software. Spotfinder provides a number of QC parameters: total and individual channel spot QC scores, flags assigned by the QC filter, and p values from a two sample t test. Flags assigned to each spot by the QC filter may change based on the QC filter settings. QC scores and p values, however, are independent from the QC filter settings.

The QC filter is enabled by default and performs spot shape and signal-to-noise ratio analysis based on the user-set parameters. The spot shape is analyzed under the assumption that the ideal spot has to be similar in shape to a circle. Real spots may have a less circular shape that can become even more distorted if the detected spot has originated from background fluctuations rather than a true target. The ratio of spot area to spot perimeter is calculated to check if it is significantly different from the ratio for a perfect circle of equal size. If the spot ratio deviates from that of the circle by more than 20% it can be considered a badly shaped spot. The signal-to-noise ratio check is based on the selected pixel value threshold. The threshold can be written as

$$T = \alpha M + \beta SD, \qquad (1)$$

where M is local background median, SD is background standard deviation, and parameters α and β are coefficients and can have values in the range of [0, 4]. The default settings are $\alpha = 1$, $\beta = 1$. The spot is considered strong and will pass this test if more than 50% of the pixels in the spot are higher than the selected threshold. Any spot that fails on at least one of the criteria will be flagged as a "bad" spot. Spots that pass these tests are subject to further scrutiny. The number of pixels in these spots is counted (spot area) and different warning flags are assigned to spots smaller than 50 and 30 pixels. The user may choose to disable QC filtering, ending up with an output data set that has every detected spot flagged as "good."

QC scores will be calculated and reported by Spotfinder, even if the QC filter is disabled by the user. The QC score has a value between 0 and 1; higher scores indicate better spot quality. The total QC score for each spot is the mean of QC scores for that spot from all channels. The QC score for each channel is calculated as the geometric mean of the shape and the signal-to-noise QC scores for the spot. The shape QC score is calculated in the same way as it was described earlier, but also including normalization to the spot size and scaling into the range [0, 1]. The signal-to-noise QC score is defined as the portion of pixels in spot above T, where T is defined by Eq. (1) with $M = 2$ and $SD = 0$.

Visualization of Quality Controls in Spotfinder Views

Spotfinder provides a set of visual displays for graphical presentation of QC results. The user can navigate to them using the GUI tabs *Analysis*, *RI plot*, *Data*, and *QC view*.

As a major hub of the Spotfinder interface, the *Analysis* page also graphically displays the immediate results of image processing—the contours of each detected spot are painted in one of two colors depending on the flag assigned by the QC filter: magenta for good flags (A, B, C, and S) or green for bad flags (X, Y). The shape and the location of the contour inside every cell give the user immediate visual information about the alignment of the cell and the success of spot detection (Fig. 4).

The Spotfinder *RI plot* page (Fig. 5) with "diamond plot" lines (Sharov *et al.*, 2004) can be used for quick visual examination of the slide's ratio and signal level dynamic range, correctness of background detection, and saturation correction. A ratio-intensity plot (RI plot), widely used for presentation of microarray data, is the $\log_2(MA/MB)$ plotted as function of $\log_2(MA*MB)$, where MA and MB are spot means in channel A and B, respectively. Diamond lines indicate the theoretical limits for spots with

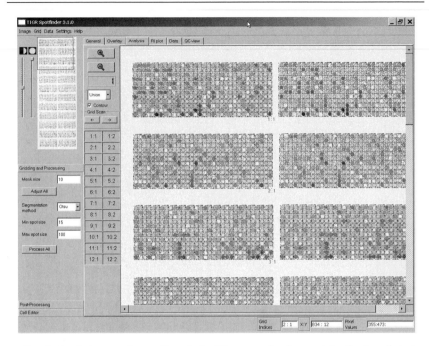

FIG. 4. Analysis page shown after whole slide processing is complete. Contour lines are colored red for good spots and green for bad ones. (See color insert.)

extreme intensities: completely saturated spots at least in one channel are limited by the right side of the RI plot diamond and spots with zero intensity in at least one channel are limited by the left side of the RI plot diamond. Correspondingly, the left-most tip of the diamond is the location of spots in which both channels produce zero intensities, and the right-most tip of the RI plot is the location of spots with complete saturation in both channels. None of the spots on the array should be expected outside of the RI plot diamond. In essence, the RI plot diamond lines are the physical limits for bit-depth limitations in one channel on the right side of diamond and zero measured signal on the left side of the diamond.

The *QC view* page allows the user to view the subgrids with individual cell rectangles colored according to a four-color scheme. Three colors—yellow, blue, and gray—are used to indicate spots with measured differential expression levels above, below, and between two chosen preset levels, respectively. These two levels of $\log_2$(ratio) are preset to 1 and −1, by default, to display those genes that are up- or downregulated by a factor of two or more in blue and yellow colors, and coloring genes that fall within that range as gray. They can be changed by the user to visualize the interesting expression ratio

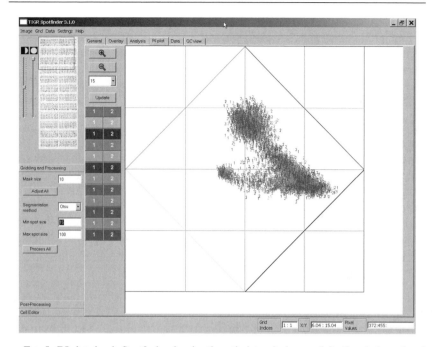

FIG. 5. RI plot view in Spotfinder showing the ratio-intensity log graph for the whole analyzed slide. The four lines forming a diamond are the limits of the log-ratio plot. Red and blue lines are the full saturation limits lines in one channel, whereas the other channel has a valid number. Yellow and green lines are the zero values limit lines at least in one channel. (See color insert.)

profile on the slide. For instance, if the user is looking for fourfold up and downregulated genes, the values of 2 and −2 should be entered on the *QC view* page. The color green is used for bad or undetected spots. Colored cells can be displayed on QC view in their true positions in the subgrid or combined together in blocks. In the latter method the area of rectangular blocks is proportional to the number of cells of a certain color on the subgrid. Relative amounts of bad/undetected color cells in subgrids are expected to be approximately the same; therefore, any noticeable increase of green color areas may indicate poor alignment of that particular subgrid.

Spotfinder Protocol Description

Program Settings

Check the program settings and change them based on the actual slide type; if the number of channels is changed, it is necessary to close and restart the program. The user also may change the visualization scale factor depending on the image size and available video card memory; it is recommended to keep default settings for the initial use. Use the menu bar to go to *Settings→General Settings*. Make sure that the *Channel Number* is set to 2 for a two-dye experiment and that the *Scale Factor* also equals 2.

Image Loading

To load two TIFF images stored in distinct files, select both TIFF files at once by holding the keyboard Ctrl key, clicking on one file and then on the other one. For loading TIFF images stored in the same file, select only that one file to load. Spotfinder automatically detects if the selected file is encoded in 8- or 16-bit format. The file names are sorted in alphabetic order for placement in channels A and B for a two-color array, and A, B, C, and D for a four-color array. The output data file columns will follow the same order. The user may swap the images in channels A and B to change the order if necessary.

Loading Existing Grid from File

A previously saved grid can be retrieved from the SFG file by clicking on the *Load grid from file* option from the *Grid* menu. The Spotfinder focus should be switched to the *Analysis* page by clicking the *Analysis* tab. This page is where the user will interact with Spotfinder for grid construction, alignment, movement, and processing tasks. If any arrays of this same type have been analyzed previously, load the grid file used previously for this slide type. The grid would likely involve only a position alignment, as it has the correct grid size but not necessarily the correct location.

Grid Construction

New array types require the construction of a new grid. The Spotfinder grid design assumes that the slide was printed by using rectangular pin (pen) settings. Every distinctive subgrid or block on the slide is printed by one dedicated pin. The pins are arranged in rectangular pattern such that the subgrids form meta-rows and meta-columns. To design a new grid, go to the menu *Grid→Compose Grid*. This will bring up the *Grid Design* dialog. In this dialog the user is asked to input eight parameters describing the geometry of the grid. These parameters are the numbers of meta-rows and meta-columns, the distance between neighboring pins in horizontal (PinX) and vertical (PinY) dimensions, the number of rows and columns in each subgrid, and the spot spacing in the horizontal and vertical dimensions. These parameters can often be retrieved from the slide print specification used by the robot that printed this slide. If this specification is not available, all parameters can be evaluated interactively on the Spotfinder *Analysis Page* by measuring relative distances with the mouse pointer. All distances are expressed in image pixels. Each of the subgrids can be moved, rotated, expanded, or shrunk interactively to fit the spots arrangements on an image. These operations can also be applied to all the subgrids simultaneously.

1. *Grid movement.* Move the whole grid set to align the upper left-most spot with the top left cell of the first (upper left) subgrid. To do this, click the right mouse button while the mouse pointer is not inside any grid. The *All Grids* menu will be activated. Choose *Move All* from the *All Grids* menu and move the mouse slowly or use keyboard arrows keys to move all subgrids simultaneously into the appropriate position. When the placement is correct, terminate the *Move* mode by pressing the *End* key on the keyboard or by clicking the left mouse button. The user can repeat this action as many times as is necessary. Undo/redo grid commands are available for convenience.

2. *Grid expansion.* If the number of rows and columns is set correctly but the subgrids do not fit the image, it may be necessary to expand or shrink the subgrids. Bring up the *All Grids* menu (as described earlier) and choose the *Expand All* command. By using the keyboard arrow keys, the user can expand or shrink all subgrids together, either horizontally or vertically. Both the expansion and the shrinking operations are performed while keeping the left and top edges of each grid fixed. Only the right and bottom edges of each subgrid are moved when these operations are performed.

3. *Changing cell size in grid.* If it is necessary, the cell size can be adjusted by selecting *Cell Size All* from the *All Grids* menu. Use the keyboard arrow keys to increase or decrease cell size in the vertical and horizontal dimensions. One arrow key press corresponds to a cell size increase or decrease by one pixel. Terminate the *Cell Size All* mode when finished by pressing the *End* key on the keyboard or by clicking the left mouse button. When increasing cell size try to avoid touching the spots by growing neighboring cells. As long as this touching can be avoided, overlapping of the adjacent cells is actually safe and desirable because it increases the area around each spot for local background calculation.

4. *Grid rotation.* Spotfinder provides the ability to rotate subgrids for better alignment of images that have an angular offset. Activate this command mode by choosing *Rotate All* from *All Grids Menu* and use the up and down arrow keys on the keyboard to rotate all grids synchronously. The rotation of each subgrid is performed around its top left corner. It is better to use only the upper left-most subgrid as an indicator of alignment. The rest of the subgrids are expected to have the same angular offset

due to the nature of the parallel arrangement of the pins during slide printing.

The user can repeat steps 1–4 in any sequence any number of times for all subgrids or any single selected subgrid to improve grid adjustment. Make sure at this point to use only the first subgrid (top left) as an indicator of proper grid size settings when the *ALL Grid* command is used.

Grid Adjustment

After setting the correct grid size the top left subgrid should be aligned and positioned correctly while the others likely have some positional offset due to the natural bending of the printing pins. These subgrids can be adjusted manually or by using the automatic procedure. The automatic procedure requires the user to provide an estimated spot size. Spot size is used to detect the location of the brightest spots that serve as landmark targets in each subgrid. The automatic grid adjustment procedure can be applied as many times as needed; in many situations it comes to satisfactory grid alignment after a few applications. However, if it fails to provide a good grid adjustment the user must adjust the grid manually by using the mouse and keyboard arrow keys. Manual subgrid position adjustment is performed by moving each grid individually with the mouse or keyboard arrow keys while in *Move* mode, which is activated from the *Move* command of the *Grid* menu.

Grid Processing

Select the segmentation method and input all required parameters. When setting minimum and maximum spot sizes for the Otsu method the rule of thumb is to set the range of spot sizes as close as possible to the visible range on the slide. However, range minimization can be potentially dangerous, as it may cause instability in the Otsu method iteration procedure. The actual spot size range on the slide can be measured interactively with the help of the mouse pointer on the *Analysis Page*. For the *Histogram* method, set the spot size slightly higher than what is expected. To start grid processing press the *Process All* button on the *Gridding and Processing* pane. Spot detection, segmentation, and local background correction steps are all performed during processing.

Grid Alignment Examination

Checking the *Contour* check box on the *Analysis Page* will show the spot boundaries. After the processing is completed, the user is able to see the contours of the spots colored in green for bad spots or magenta for good spots. If green contours appear in cells with spots that otherwise look good, this may indicate misalignment or wrong settings for the segmentation method used. Go to the *QC View Page* for subgrid alignment checking. Visually compare the relative size of the green area on different subgrids in the array. They should be approximately the same unless the array was designed intentionally with some special subgrids (e.g., all replicates are printed in one subgrid). Any subgrid with a disproportionately large green area should be checked for misalignment. If the alignment is shown to be correct, the higher number of bad spots in this subgrid can be considered indicative of the low expression of genes printed in this subgrid.

Postprocess Data Tuning

The default QC filter operation is the last step of processing. The user can change QC filter settings in the *Post-processing* dialog without having to reprocess the slide. Switch to the *Post-processing* dialog to enter new QC filter settings. Background correction can be disabled or enabled by using the check box of this dialog. The QC filter can be set more or less stringent by changing the cutoff threshold defined by the signal-to-noise ratio (see earlier discussion). By varying parameters α and β of Eq. (1) the user can set threshold T at the level where a reasonable distinction between weak spots and strong spots, produced by noise, is visible on the *Analysis page*. After making the desired changes, press the button *Update Changes*. The result can be observed in the *Analysis Page*, on the spreadsheet data table of the *Data page* and on the *RI plot page*.

Annotation Import

A variety of annotation file formats (.ann, .dat, .gal) can be loaded by Spotfinder for the purposes of displaying annotation alongside expression data and to map to an output data file. At first the user needs to construct or load the grid in Spotfinder to ensure correct mapping. To load an annotation file, go to the main menu bar and select *Data → Load Annotation File*; the selected file will be loaded and checked for the correct total number of rows (spots). Once loaded, the annotation can be viewed on the *Data* page or on the status bar at the bottom of the Spotfinder GUI for spots selected on the *Analysis* page. To map annotation to the data set in Spotfinder and in any future .mev files, go to the main menu and select *Data → Set UID from annotation* or *Data → Set DBID from annotation*. This will change default UIDs in the *MEV data* tab of the *Data* page and create an additional DBID column with DBIDs from the annotation file. The new mapping will be stored in any .mev files generated by Spotfinder in the next step.

Report Output Data

To save expression data in a .mev file, go to the menu Data → *Save Data to MEV file* and enter a file name in the *Save File* dialog. This creates a tab-delimited data file that is used by all software tools in the TM4 suite. To save grid information in a binary, platform-independent SFG file (Spotfinder Grid file), go to the menu *Grid → Save Grid in File* and input the grid file name in the dialog. The grid and all raw data can also be saved in a platform-independent, binary data SFD file (Spotfinder Data file). To create an SFD file, go to the menu *Data → Save Data to SFD file*; the file save dialog will ask for a file name to save as. The SFD file stores all processed raw data and spot contour vectors needed for graphical representation of the contours on the Spotfinder *Analysis page* in a later session. The SFD file can be used later by any user who needs to view the results with the RI plot and spot contours; the postprocessing operations can be applied to SFD data to generate a new .mev file for the same data set but with different QC filter settings.

MIDAS

Microarray analysis is a comparative analysis. In a two-color experiment, cDNA or mRNA abundances are compared between two samples. During a microarray experiment, the different samples are dyed with Cy3 (green) and Cy5 (red) fluorophores and are cohybridized to a glass slide. After scanning the slide and performing an image-processing procedure, the intensities for each spot, for both green and red channels, are recorded.

An underlying assumption in microarray analysis is that differences between the two intensities for each spot faithfully reflect the cDNA or mRNA abundance differences between the two samples. This is the basis for investigating cDNA or mRNA abundance differences in tens of thousands of spots in a microarray slide simultaneously by clustering using pattern recognition or other data mining techniques. This assumption, however, is compromised by all kinds of errors or biases introduced during the experiment and image processing. Predicting the bias, adjusting the raw intensities for each spot accordingly so that they better reflect the true picture about the cDNA/mRNA abundances is a crucial data preprocessing step before downstream analysis can be carried out. It is also important to remove those spots within an array with "unacceptable" intensities, as defined by varying criteria. These data preprocessing steps are called *Normalization and Filtering*. Other normalization and filtering methods are found elsewhere (Ayroles and Gibson, 2006; Gollub and Sherlock, 2006; Reimers and Carey, 2006).

Midas (also called MIDAS) is the data normalization and filtering tool in the TM4 microarray data analysis software suite. It contains a number of normalization and filtering modules, as well as significant gene identification modules. The software provides a user-friendly graphical scripting feature, which allows these modules to be pipelined together to form an analysis workflow. Midas has a strong graphing feature for users to investigate the analysis results. A graphical PDF analysis report can also be generated by user's request.

Building a Pipeline

A typical Midas data analysis pipeline is composed of three steps: (1) reading raw data input files, (2) defining the analysis workflow by queuing one or more analysis modules, and (3) writing processed data output files.

After an analysis pipeline is defined in the Midas Workflow window, parameters should be set in the parameter sheets associated with each module defined in the pipeline (Fig. 6). The analysis pipeline and the parameters can then be saved into a Midas project file (.prj) under a Midas project folder. The Midas project folder will be the location for all analysis results, output data files, analysis plots, reports, and error messages, if there are any. The analysis example at the end of this chapter describes

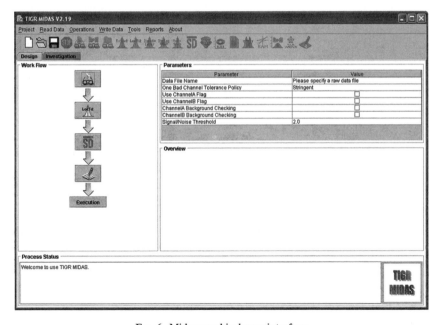

FIG. 6. Midas graphical user interface.

the steps involved in building an analysis pipeline using several popular modules.

Normalization Modules

Midas includes a number of normalization algorithms. These can be used in sequence with each other or with other modules. Choosing appropriate normalization and filtering methods can be one of the more challenging aspects of using microarrays. Applying inappropriate methods to experimental data might lead to information loss or information distortion. An example is that removing raw data by some filtering methods might have negative consequences for some downstream gradient correction methods.

A common approach to normalization is global normalization. In this approach, averages of the overall expression levels for all genes within an array across different arrays are set to be equal. This follows from the assumption that while genes can be expressed differentially, the amount of transcription is essentially similar across samples. Furthermore, it is also common to set the averaged overall expression levels for each array to be zero. This follows from the assumption that within each array, overexpressed genes and underexpressed genes are roughly balanced. Global normalization methods are mostly useful for normalizing hybridization arrays for gene profiling or similar samples comparison purposes. They might not be valid normalization approaches when the compared samples are too different across arrays or when using comparative genomic hybridization arrays.

Table I provides some general guidance for applying the right normalization and filtering methods. Keep in mind that the correction of a bias or error assumes that the experimental design and array samples do not undermine the assumptions of the applied algorithm.

Total Intensity Normalization

Total intensity normalization (Quackenbush, 2002) assumes the summed intensities for each of the two channels, channel A and channel B, for all spots within an array should be equal. If there is any observed difference, it is caused by some dye-specific systematic bias and thus should be adjusted. The algorithm calculates the factor between the two summed intensities ΣIA and ΣIB and scales intensity A (IA) or intensity B (IB) of each spot so that the goal of equal summed intensities in the two channels is achieved.

Lowess Normalization

Lowess normalization (Quackenbush, 2002; Yang et al., 2002a,c) assumes that spots having different overall intensities [measured by $\log_{10}(IA \cdot IB)$] should have different systematic bias added to their expression levels

TABLE I

GUIDELINES FOR SELECTING NORMALIZATION AND FILTERING METHODS

Issues to be addressed	Applicable methods
Averaged overall expression within an array not zero observed unexpectedly	Total intensity; iterative log-mean centering; ratio statistics; Lowess
Print tip-dependent bias observed	Standard deviation regularization
Intensity-dependent bias observed	Lowess
Nonlinear correlations observed unexpectedly between the two channel intensities (logarithm transformed)	Iterative linear regression
Inconsistent expressions between dye-swapped experiments observed	Flip-dye consistency normalization and filtering
Selecting significantly expressed genes from a single array	Slice analysis
Selecting significantly expressed genes when replicated arrays are available	Statistical methods such as "t test" and "SAM"
Noisy raw expressions	Background filtering; low-intensity filtering

(measured by $\log_2{}^{IB}/_{IA}$). Thus the goal of this normalization method is to extract the intensity-dependent systematic bias for each spot and use it to adjust the raw IA or IB for each spot. The Lowess algorithm estimates the adjustment factor for the $\log_2{}^{IB}/_{IA}$ value of a spot by finding those spots in the neighborhood of this spot, based on their intensities, and computing their commonly shared bias by a maximum likelihood technique, which applies a locally weighted model to spots' expression data in each neighborhood. A related algorithm, Loess, differs from Lowess because of the model used in the regression: Lowess uses a linear polynomial, whereas Loess uses a quadratic polynomial. The "neighborhood" is defined by a parameter called the *smoothing parameter*, which defines the percentage of all spots within a physical scope. The physical scope can be either *block*, meaning all spots printed by the same print tip, or *global*, meaning all spots on the array.

Iterative Log-Mean Centering Normalization

Iterative log-mean centering normalization (Quackenbush, 2002) assumes that the majority of the spots within an array show a balanced distribution of expression levels (measured by $\log_2{}^{IB}/_{IA}$). For these spots, their $\log_2{}^{IB}/_{IA}$ values should have a mean value of 0. Aside from this majority, a few outlier spots, those with very high or very low $\log_2{}^{IB}/_{IA}$

values, contribute significantly to the calculation of the overall $\log_2{}^{IB}/_{IA}$ mean. This algorithm uses an iterative procedure to remove the outliers and calculate the $\log_2{}^{IB}/_{IA}$ means for the outlier-removed spots until the means converge. The algorithm then scales the intensities of each spot by this converged mean value.

Iterative Linear Regression Normalization

Iterative linear regression normalization (Finkelstein *et al.*, 2000) assumes the correlation between $\log_{10}IB$ values and $\log_{10}IA$ values for all spots within a physical scope on the array displays a $y = x$ linear relationship. The physical scope can be either *block*, meaning all spots printed by the same print tip, or *global*, meaning all spots within the array. The algorithm calculates the slope and intercept between the $\log_{10}IB$ values and $\log_{10}IA$ values for spots within the specified physical scope, iteratively. During each iteration, the outlier spots, which are defined as those having $\log_{10}IB$ or $\log_{10}IA$ residuals greater than a user-defined threshold range, are removed. The final slope and intercept are achieved when the calculated correlation coefficients converge. The final slope and intercept are then used to adjust the IA and IB of each spot so that the $\log_{10}IA$ and $\log_{10}IB$ distribution displays such a linear relationship.

Standard Deviation Regularization

Standard deviation regularization (Yang *et al.*, 2002c) assumes that variances of the expression levels of the spots (measured by $\log_2{}^{IB}/_{IA}$) within different physical scopes should be the same. The physical scope can be either *block*, meaning all spots printed by the same print tip, or *global*, meaning all spots within the array. Based on this assumption, the IA and IB values of each spot are adjusted so that the same standard deviation, and thus the variance, of the $\log_2{}^{IB}/_{IA}$ values of the spots prevails among the specified physical scope. For example, the variances for all blocks on an array could be set equal to each other by this method.

Ratio Statistics Normalization

Ratio statistics normalization (Chen *et al.*, 1997) assumes that there exists a sample-independent single-mode $^{IB}/_{IA}$ distribution function with mean μ and standard deviation σ. The mean can be estimated through an iterative process described in the reference paper. The algorithm also assumes that the population of $^{IB}/_{IA}$ values for all spots in an array should approximately demonstrate a mean value of 1. The calculated mean μ can then be used to normalize the IA and IB of each spot so that the $^{IB}/_{IA}$ population mean becomes 1.

Flip-Dye Consistency Normalization and Filtering (Quackenbush, 2002)

In a pair of flip-dye arrays s1 and s2, the $\log_2{}^{IB}/_{IA}$ for any spot in s1 is expected to have an expression value of $-\log_2{}^{IB}/_{IA}$ for the corresponding spot in s2 due to the fact that the two spots are dye-swapped replicates of each other. Therefore, if the $\log_2{}^{IB}/_{IA}$ values for all spots in s1 versus $\log_2{}^{IB}/_{IA}$ values for all spots in s2 are studied for their correlation, a linear relationship is expected. The flip-dye consistency normalization algorithm checks the consistencies for each spot's expression values between s1 and s2 by calculating the $c = \log_2{}^{IB_1}/_{IA_1} - \log_2{}^{IA_2}/_{IB_2}$ histogram, where IA_1 and IB_1 denote the two-channel intensities for a spot in s1 and IA_2 and IB_2 denote the two-channel intensities of the corresponding spot in s2. By assuming this histogram follows a normal distribution with a mean of 0, those spots with c values that fall beyond a user-defined consistency range are removed. These are considered to be inconsistent data between the flip-dye replicates. The rest of the spots are output as consistent spots. For each of these consistent spots, the $\log_2{}^{IB}/_{IA}$ value is presented as the geometric mean of $\log_2{}^{IB_1}/_{IA_1}$ value and $\log_2{}^{IB_2}/_{IA_2}$ value.

Filtering Modules

Filtering modules reduce the size of the data set by removing elements that do not meet certain user-defined criteria. These modules can be added to the workflow before or after normalization modules. Filtering modules applied before normalization remove the "bad" or unreliable elements defined by certain quality control criteria to allow only "cleaner" data to be used as input for normalization procedures. Such modules include flag filtering, background filtering, and low-intensity filtering. In contrast, filtering modules applied after normalization remove elements that may not be important or interesting, given the research goals. These "postnormalization" filtering methods include in-slide replicate analysis and cross file trim, as well as the significant gene identification modules described in the following sections.

Flag Filtering

During the image processing stage, some spots might be flagged as "bad" due to a variety of reasons, such as saturation. The flag-filtering feature allows these flags be read before data are processed. Flagged spots will be excluded from any downstream processes.

Background Filtering

During the image processing stage, the user may request that background intensities be calculated along with the signal intensities IA and IB. These background intensities can be used to calculate the signal-to-noise ratios for each spot. The background filtering feature excludes those spots with signal-to-noise ratios below a user-defined threshold from the downstream processes.

Low-Intensity Filtering

The low-intensity filtering feature excludes those spots with channel A intensity IA or channel B intensity IB lower than user-defined thresholds from the downstream processes.

In-Slide Replicate Analysis

In-slide replicates are technically replicated spots printed within an array. These replicated spots are theoretically expected to demonstrate the same $\log_2{}^{IB}/_{IA}$ expression values. Observed variances among the replicates are caused by random errors. In-slide replicate analysis combines the replicated spots, which are defined as those spots in an array sharing the same annotation identifier, for example, feature name, into a single output data spot. The expression value $\log_2{}^{IB}/_{IA}$ of this combined spot is equal to the geometric mean of the $\log_2{}^{IB}/_{IA}$ values of the replicates that were combined.

Cross File Trim (Percentage Cutoff Trim)

When multiple data files with the same number of spots are analyzed together, it is often desirable to check the consistency of the expression value of a spot across all the files. This occurs after each file is normalized and filtered, but before processed data are written to output files. The consistency of a spot is calculated as a percentage of the number of files showing the spot as being "unfiltered" divided by the total number of files used.

Cross file trim allows the consistency percentage of each spot to be compared with a preset consistency threshold percentage. Spots that do not pass the threshold comparison are filtered in the output files by setting their IA and IB intensities to 0. This filtering method is also referred to as "percentage cutoff" trimming in MeV.

Significant Gene Identification Modules

Slice Analysis

It is well known that variances of expression levels of spots (measured by $\log_2{}^{IB}/_{IA}$) vary as the intensities change. This fact makes a simple "fold change" criteria for identifying differentially expressed gene within an array less than ideal.

A modified approach is to study how the expression levels of the spots are distributed across the array as the overall intensity of the spots [measured by $\log_{10}(IA \cdot IB)$] varies. In this approach, each spot is associated with a group of spots, called a "slice." A slice consists of those spots that have similar overall intensities as the query spot. The mean and standard deviation of the expression values in each slice are calculated. A differential expression z score for a spot can then be defined as the difference between the expression level of the spot and the mean expression value for the slice that the spot belongs to, divided by the standard deviation of the slice that the spot belongs to. A spot is identified as "significantly expressed" if its differential expression z score is greater than a user-defined threshold.

Slice analysis (Yang *et al.*, 2002a), a method to identify significantly expressed genes, classifies genes within a single array by their intensity-dependent differential expression z scores as described earlier.

One-Class t Test and One-Class SAM

When multiple arrays representing technical or biological replicates of the same genes are available, significantly expressed genes can be identified by applying scientific statistical analysis. Two such methods are implemented in Midas: one-class t test and one-class SAM (Chu *et al.*, 2002; Tusher *et al.*, 2001). These methods can also be found in MeV. Users who are interested in applying the one-class t test and one-class SAM are encouraged to read the corresponding sections in the MeV description and the sample analysis walk-through that follow.

Graphs and Reports

A variety of analysis graphs are plotted and saved during the execution of a Midas analysis pipeline. These graphs, such as the R-I plot (Fig. 7, left) and flip-dye diagnostic plot (Fig. 7, right), are saved within the Midas project folder and can be studied under the "Investigation" tabbed panel. Graphs of interest can be exported to the graphical PDF reports (Fig. 8) by the user's request.

MeV

After spot scanning and normalization comes the data analysis step that is usually of most interest to microarray practitioners, namely mining data to look for biologically significant patterns of gene expression (Ayroles and Gibson, 2006; Downey, 2006; Neal and Westwood, 2006; Reimers and Carey, 2006; Royce *et al.*, 2006). The MeV (MultiExperiment Viewer) software incorporates an extensive array of clustering, statistical, and visualization

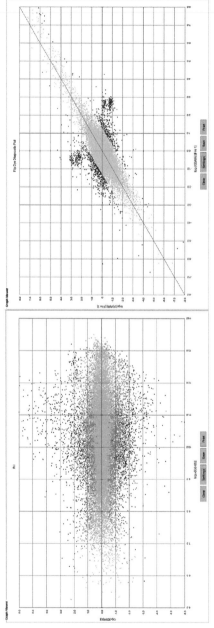

FIG. 7. (Left) An R-I plot showing significantly expressed genes classification results after slice analysis is applied. The outlier genes, which have their intensity-dependent, differential expression z score greater than twofold of standard deviation, are colored red; genes z scores below onefold of standard deviation are colored blue; the remainder are colored green. (Right) A flip-dye diagnostic plot showing consistencies about expression values between a flip-dye pair. The solid diagonal line represents the theoretical perfect consistency relationship. Genes in blue are considered to be consistent between the flip dye using twofolds of standard deviation cut as the consistency criteria. The other genes are colored red. (See color insert.)

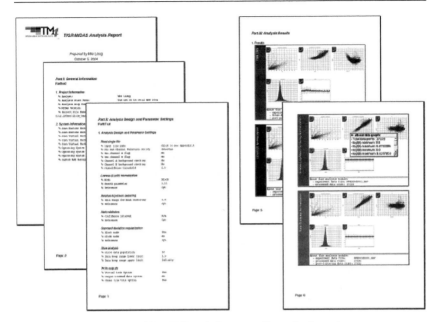

FIG. 8. Midas PDF analysis report. (See color insert.)

tools that can be used to analyze preprocessed microarray data. An intuitive and feature-rich interface makes it easy to use the software, eliminating the need for a programming or scripting language. In addition to the .mev file format used by the TM4 suite, MeV (also known as TMeV) works with file formats generated by a number of other platforms or analysis programs (Affymetrix MAS 5.0 output, RMA output, Agilent or Genepix scanner files, and a more generic tab-delimited text file format containing log ratios from multiple samples). Thus, MeV is a versatile end-stage analysis tool that can be used at the last stage of a TM4 pipeline or as a stand-alone program to analyze data that have been processed with other analysis tools.

Data Representations and Distance Metrics

In MeV, the expression level corresponding to each spot on a slide is represented as an *expression element* (Fig. 9). An expression element is typically a $\log_2$ transformation of an expression ratio in the case of two-color arrays, where a ratio is calculated by dividing the fluorescence intensity from one channel by the fluorescence intensity from the other channel for a given spot on a slide. In the case of single-channel arrays (such as Affymetrix chips), an expression element is the normalized single intensity value for a probe set. Hereafter, for convenience we use the term *gene* to refer a spot or

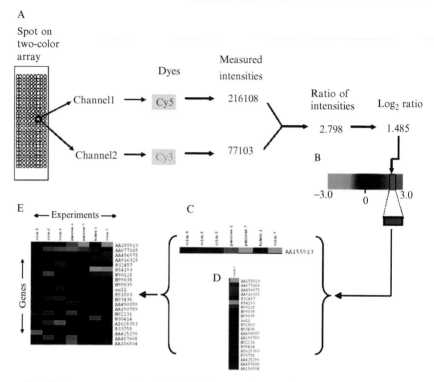

FIG. 9. Data representations in MeV. (A) Numerical and (B) false-color representations of an expression element, (C) a gene expression vector, (D) an experiment expression vector, and (E) an expression matrix. (See color insert.)

a probe set, even though the DNA sequence corresponding to that spot or probe set may not span the entire length of a gene in a biological sense. Because an *experiment* corresponds to a *slide* on which a given *hybridization* was carried out, these three terms are often used interchangeably.

An *expression vector* (Fig. 9C and D) is a set of expression elements for a given gene or experiment. For a gene expression vector, each element comes from a separate experiment in which the intensity of that spot was measured. An experiment expression vector contains the expression elements of a set of genes in a given experiment.

An *expression matrix* (Fig. 9E) in MeV is a two-dimensional array of expression elements from a set of genes over multiple experiments. By convention, each row is an expression vector from a given gene, and each column corresponds to an expression vector from a given experiment. The expression matrix in MeV (and generally in microarray data representations) is shown in a false-color view on a red–green scale by default, with green representing low expression and red representing high expression. These colors can be customized.

Another important concept is that of *distance*. A distance metric is a numerical estimate of how similar the expression patterns of two expression vectors are. The smaller the magnitude of distance, the greater the similarity of the two patterns. Many algorithms in MeV use distance metrics to put expression vectors in clusters that contain vectors of similar expression. There are many types of distance metrics, some of which use very different criteria from one another to estimate similarity. Thus, two vectors might be judged very similar by one distance metric and quite unlike one another by another metric. It is important to select a distance metric that is appropriate to the underlying question being asked. For instance, the Pearson correlation distance is appropriate when one is interested in finding genes showing similar patterns of expression over a set of experiments, such as a time course, regardless of the magnitude of expression. However, if the primary interest is in grouping together genes that have similar levels of expression (over- *vs.* underexpressed), then the Euclidean distance might be a better choice. MeV offers 11 distance metrics, any of which can be applied to the distance-based algorithms in the package.

Data Mining in MeV: A Brief Algorithm Overview

One should be aware that there is often not one "correct" analysis approach to any particular data set. What is important is to know what an algorithm is doing, how it makes decisions during cluster creation, how input parameters affect results, and what features of data may be revealed by an analysis. A powerful feature of MeV is the ability to overlay results obtained from multiple methods to reach a consensus or to reveal different aspects of data. At times finding an approach requires some level of trial and error to find methods and suitable parameters.

The mechanics of executing an analysis algorithm in MeV are quite simple. Once data are loaded, the analysis is initiated by selecting the corresponding button in the toolbar or the menu item from the "algorithms" menu. All algorithms initially open one or more dialog boxes that are used to collect input parameters. The lower left corner of each dialog contains an information button (Fig. 10), which opens a help window with information about the input parameters.

Some algorithms require rather large amounts of computer memory space and it is generally recommended to have 512 MB to 1 GB of RAM. In addition to memory requirements, several algorithms are computationally intensive and can take several minutes or, in some cases, hours to complete.

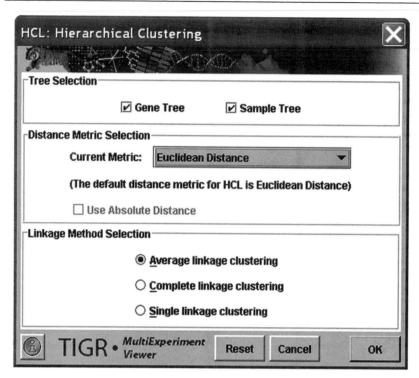

FIG. 10. HCL algorithm parameter selection dialog. The lower left of each algorithm dialog contains the parameter information button. (See color insert.)

All algorithms present progress logs or progress bars to provide a status report during algorithm execution.

MeV currently provides 24 analysis techniques. In terms of the objectives they attempt to accomplish, these algorithms can be classified into three broad categories: exploratory techniques, hypothesis testing techniques, and classification techniques. Exploratory techniques look for broad patterns in the data set; examples of algorithms in this category include hierarchical clustering (HCL) and principal components analysis (PCA). Hypothesis testing techniques use information about the experimental design to identify a subset of genes that show statistically significant differences in patterns of expression across groups of samples; examples of such techniques include TTEST, SAM, and ANOVA. Classification techniques use information about the known class membership of some genes or samples to assign the remaining genes or samples into these classes; algorithms such as SVM and KNNC fall into this category.

Alternatively, these analysis techniques can be categorized based on the nature of their underlying algorithms. These broad categories are agglomerative methods, divisive methods, methods to assess confidence in clustering results, neural network approaches, statistical tools, classification algorithms, data visualizations and component analysis, and biological theme discovery. We use these categories based on algorithm heuristic in describing some of the following algorithms. The cited references, manual, and training slides available at the TM4 web site provide greater detail about these algorithms.

Agglomerative Methods

Agglomerative methods start by considering each expression vector as a distinct and independent object. Vectors are fused into clusters based on similarity, which is determined based on the selected distance metric. A cluster so formed from two elements is then considered as a single object, a cluster of size two, rather than as two distinct elements. In subsequent rounds, objects are fused to form bigger clusters based on intercluster similarity using the same distance metric as described earlier to define intercluster distance. The method continues joining the most similar objects at each stage until all objects are assigned to one large cluster.

HIERARCHICAL CLUSTERING (HCL). Hierarchical clustering (Eisen *et al.*, 1998; Weinstein *et al.*, 1997) is likely the most widely used agglomerative method for preliminary data exploration. HCL constructs a binary tree by successively grouping the genes or samples based on similarity. A set of vectors falling under a node in the tree tend to be more similar to each other than to vectors in other sections of the tree. By observing how gene or sample expression patterns are arranged in the tree, one can select and focus on subtrees that contain consistent patterns of interest.

The HCL tree viewer in MeV permits one to dynamically select a subtree to assign to a cluster or to slice the tree into any number of distinct clusters based on a similarity value cutoff (Fig. 11). Hierarchical clustering is a popular analysis option for getting an overview of patterns in the data set.

Divisive Clustering Methods

Divisive clustering methods begin with all vectors in one cluster, which is then partitioned into distinct clusters. The objective is to create clusters such that all elements within a cluster are similar to one another, and each cluster is dissimilar to the others. No relationship is specified among the clusters.

K-MEANS/K-MEDIANS CLUSTERING (KMC). K-means clustering (Soukas *et al.*, 2000) is a divisive technique that divides the genes or samples into a set of k clusters. Initially, the vectors are assigned randomly to a predefined

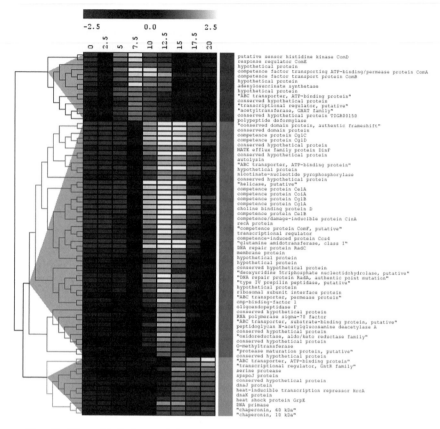

FIG. 11. Hierarchical cluster of time course data using the Pearson correlation distance metric. Prominent patterns of expression have been selected as clusters. (See color insert.)

number of clusters. The assignment is iteratively refined by shuffling vectors among clusters and updating the mean or median profile of each cluster until each vector is assigned to the cluster whose mean or median it is closest to. This method is useful when one has a reason to assume that data should partition into a specified number of clusters. During clustering analysis, vectors are sometimes divided into too many clusters, such that there are two or more clusters that have mean patterns that are similar. This suggests that those clusters should be merged. In other cases where too few clusters have been created, the clusters will tend to be large and contain quite diverse and variable patterns in each cluster. The number of clusters can be chosen by trial and error to hone in on a partitioning that appears to appropriately split data into distinct clusters.

CLUSTER AFFINITY SEARCH TECHNIQUE (CAST). The cluster affinity search technique (Ben-Dor *et al.*, 1999) partitions data into clusters that contain members guaranteed to have a minimum specified "affinity" to other members of the cluster. The affinity of a particular gene is related to the total similarity of that gene to all other genes in the cluster being created. A nice feature of this algorithm and some others like it is that the number of clusters to create is not predefined. Clusters are created until all items are assigned to clusters of the largest size possible while ensuring that all genes within a cluster have some minimal affinity for the cluster.

GENE SHAVING. Gene shaving (Hastie *et al.*, 2000) is a divisive clustering technique that partitions the genes into clusters such that genes within a cluster have low gene-to-gene variability, while having high variance across samples. Thus, a cluster of genes created by this algorithm will tend to have similar expression profiles that tend to vary substantially across samples. One important difference from many other divisive clustering techniques is that clusters from gene shaving are not always mutually exclusive so that a given gene may appear in more than one cluster. The procedure attempts to make successive clusters almost uncorrelated with previously created clusters so that if a gene appears in more than one cluster, each such cluster might highlight different aspects of the variability of that gene.

QTCLUST. QTClust (Heyer *et al.*, 1999), like CAST, is a clustering technique in which the number of clusters is not specified by the user, but is determined by two inputs: the maximum possible distance between two genes in a cluster (called the cluster diameter) and the minimum number of genes that a cluster must contain (the cluster size). For calculating cluster diameter, gene-to-gene distance is computed using a jackknifing procedure in which each sample is left out in turn. This reduces bias that might be introduced by individual outlier samples. Clusters are created in sequence, and the genes that are unassigned after the creation of a cluster are subjected to successive rounds of clustering until no more clusters can be created that satisfy both the cluster diameter and the cluster size thresholds. At the start of each round of clustering, all unassigned genes serve as potential seeds for a new cluster. The largest cluster created from all seeds in a given round is retained, and the procedure is repeated on the remaining unassigned genes. Allowing each eligible gene to serve as a potential seed for further clustering prevents the algorithm from being biased by the order in which data are presented to it.

Assessing Confidence in Clustering Results: Support Trees, Figures of Merit, and K-Means Support

Clustering algorithms are guaranteed to organize data into clusters, even when no clear patterns exist. It is therefore helpful to assign measures of confidence on the clustering results to assess whether the clustering is meaningful. This is done by repeating the clustering analysis many times with the same parameters on the same data set or a resampled data set or by gradually changing the magnitude of an input parameter and then comparing the results across all runs. MeV offers three methods to assess confidence in clustering results.

HCL SUPPORT TREES (ST). The ST module in MeV builds a hierarchical tree by the same algorithm employed by the HCL module of MeV. The difference here is that after finding the initial tree, the expression matrix is resampled with replacement many times to produce bootstrapped expression matrices. An HCL tree is built on each of these bootstrapped matrices and compared to the original tree. Each node in the original tree is assigned a value between 0 and 100, indicating the percentage of times over all resampling trials that a node containing those elements occurred in a tree obtained from a resampled matrix. These bootstrap confidence values are displayed on the tree as colors or as numerical values. Higher node values indicate that the vectors under that node clustered together frequently regardless of resampling, which indicates that the cluster represented by that node was not unduly influenced by a small subset of data.

Other algorithms in this category are figures of merit (Yeung *et al.*, 2001), which iteratively step through different values of k searching for an optimal value based on a comparison of within-cluster and between-cluster distances, and K-means support, which iterates K-means at a fixed value of k searching for stable clusters.

Machine Learning Methods

Machine learning-based clustering approaches are suitable for partitioning large data sets that contain a lot of random noise in addition to distinct expression patterns of interest. This means that these approaches are very applicable to microarray data. These approaches represent the clusters being created as a set of nodes connected as a network, where each node has a representative expression profile that is trained by data to better conform to a subset of data. As each vector is presented to the network, the node or nodes most similar to that vector adapt to become even more similar to the presented vector. By presenting the vectors to the system many times, the nodes conform to represent clusters that are inherent in data. Once the adaptation is complete, each vector is placed into a cluster related to the node with the most similar representative expression profile.

SELF-ORGANIZING MAPS. Self-organizing maps (Kohonen, 1982; Tamayo *et al.*, 1999) in MeV are a very efficient neural network implementation that permits millions of training/adaptation cycles to be run in a relatively short time. The algorithm requires an initial topology of the network, which means that an estimate of the number of expected clusters must be provided. Similar to KMC, the suitability of this estimate can be assessed based on the results of multiple runs.

SELF-ORGANIZING TREE ALGORITHM (SOTA). The self-organizing tree algorithm (Dopazo and Carazo, 1997; Herrero *et al.*, 2001) is a hybrid approach that bridges divisive and neural network approaches to produce a binary tree structure where each terminal node or leaf in the tree is a cluster. Starting with all genes in a single node or cell, the cell then divides and partitions the vectors optimally between the two offspring cells. On each division the most variable cell is split until a predetermined number of divisions or a cluster variability threshold is met. In addition to the described benefits of the machine learning methods, SOTA does not require a predetermined cluster count.

Statistical Tools for Extracting Significant Gene Lists

The basic clustering methods described previously focus on finding correlated patterns of gene expression within the data set. This is often useful for time course data or for general data mining for prevalent patterns. In the case where the experimental design contains biological or technical replicates and the samples are partitioned into discrete sets that represent experimental conditions, statistical tests can be applied to find genes that show differential expression under the conditions being studied. In addition to extracting genes of interest, each gene will have a corresponding p value describing the likelihood that the observed finding was due to chance. Microarray experiments are being designed increasingly to take advantage of statistical tools.

TTEST *(TTEST)*. MeV provides three t test (Dudoit *et al.*, 2000; Korn, *et al.*, 2001, 2004; Pan, 2002; Welch, 1947; Zar, 1999) designs: one sample, between subjects, and paired. The one-sample t test is useful for identifying genes that show consistent over- or underexpression across a series of biological or technical replicates. The between-subjects t test is useful for finding genes that are significantly differentially expressed between two independent groups of samples (e.g., two strains of mice). The paired t test can be used to find genes showing differential expression between two conditions assayed on the same samples (such as before and after administering a drug to a group of individuals).

ONE-WAY AND TWO-FACTOR ANOVA. Two types of ANOVA designs are offered: One-way (Zar, 1999), for comparison of three or more independent groups, and a completely randomized two-factor design (Keppel and Zedeck, 1989; Manly, 1997; Zar, 1999) for analyzing variation across two conditions (such as sex and strain). See Ayroles and Gibson (2006) for more about ANOVA. The t test and ANOVA modules offer error rate correction options (such as Bonferroni corrections) for multiple testing (Dudoit *et al.*, 2003), as well as false discovery rate (FDR) computations (see later).

SIGNIFICANCE ANALYSIS OF MICROARRAYS. A false discovery rate can also be computed using the popular SAM module (Tusher *et al.*, 2001; implemented as in Chu *et al.*, 2002), which includes options for five experimental designs, four of which are analogous to the *t* test and one-way ANOVA designs, while the fifth is suitable for survival data. FDR computations are often a desirable alternative to conventional statistical tests (such as *t* tests and ANOVA) in microarray data analysis. The simultaneous analysis of thousands of genes leads to highly inflated error rates for individual genes when doing conventional statistical tests. FDR analysis can help circumvent this problem by allowing the identification of a list of potentially significant genes while still keeping overall error rates low.

TEMPLATE MATCHING. Template matching (Pavlidis and Noble, 2001) is useful for finding patterns of expression that are similar to a user-specified pattern (as judged by the magnitude and sign of the correlation coefficient between the patterns of interest or the *p* value of this coefficient).

Classification Algorithms/Supervised Clustering Approaches

Supervised methods use existing biological information about specific genes or samples (the "training set") that are functionally related to "guide" the clustering algorithm. The existing information is the presumed class membership of each vector in the training set. This information is used to classify other vectors (the unknowns) based on how similar their expression patterns are to members of the training set.

SUPPORT VECTOR MACHINE (SVM). A support vector machine (Brown *et al.*, 2000) is a supervised classification method that bisects data into two distinct groups: in class and out of class. SVM uses a subset of data that is known to fall into the class of interest as examples of the class.

K-NEAREST NEIGHBOR CLASSIFIER (KNNC). KNNC (Theilhaber *et al.*, 2002) partitions data into k distinct classes, where k is a supplied number of partitions. Like SVM, KNNC uses a subset of data to use as examples of each class being partitioned.

Data Visualizations and Component Analysis

This broad category includes algorithms that attempt to simplify the interpretation of the main features of data by presenting a view of data that provides a means to consider high level structure of data.

PRINCIPAL COMPONENTS ANALYSIS. PCA (Raychaudhuri *et al.*, 2000; Downey, 2006) extracts the features in the data set that are most representative and best "explain" the nature of the variation in data. These features, known as principal components, are used to map data into 2D and

3D visualizations that can sometimes provide an intuitive view of the main aspects of variation in the data set. A related method, correspondence analysis (Fellenberg *et al.*, 2001; Culhane *et al.*, 2002), maps both genes and samples onto the same set of axes, revealing associations between genes and experiments.

GENE EXPRESSION TERRAIN MAPS (TRN). Gene expression terrain maps (Kim *et al.*, 2001) build a 3D topological terrain view where gene or sample clusters appear as peaks in the terrain. The algorithm first maps data into a two-dimensional grid such that elements that are similar are close together. The third dimension giving rise to the peaks is related to the density of the elements under the peak. This means that if many elements are similar to each other, they will appear as a tall sharp peak over a small region of the map. By using appropriate metrics, one can use TRN to get an overview of the data set and can estimate a rough idea about the number of major clusters in data.

GENE DISTANCE MATRIX. The gene distance matrix displays a 2D heat map representation of the similarity matrix. This matrix displays the distance (inverse of similarity) between any two elements (genes or samples) in the data set. When the matrix is sorted by cluster membership based on a prior clustering result, the matrix can qualitatively indicate how distinct two clusters are in terms of the expression patterns of the member. When used to assess sample distances, where the matrix is relatively small, one can interrogate the actual similarity between any pair of expression profiles.

Biological Theme Discovery

After obtaining a list of genes, an important task is to determine whether the genes have a common or connected biological role within the system being studied (Whetzel *et al.*, 2006).

EXPRESSION ANALYSIS SYSTEMATIC EXPLORER (EASE). To assist in the discovery of prevalent biological roles, MeV has an implementation of the EASE algorithm (Hosack *et al.*, 2003) for finding overrepresented biological themes in gene lists. This module compares the prevalence of a biological theme within the cluster to the prevalence of the biological theme in the population of genes from which the cluster was extracted. One must first have all of the array probes assigned various classes based on a categorical classification system, such as assignment of gene ontology terms (Ashburner, 2000). After selecting a set of genes that are "significant" in an analysis based on a statistical or other objective test, EASE compares representation of the various classes within the significant set to the representation on the entire array using Fisher's exact test to identify overrepresented categories and assign a likelihood score (p value) to each group. For example, if only 10% of the genes on the array represent energy metabolism, but 60% of the genes deemed significant are involved in energy metabolism, it is likely that this selection did not occur by chance and that energy metabolism may be mechanistically involved in the process being studied.

Interface Orientation and Selected Features

The interface of MeV is organized into four main sections (Fig. 12). The *main menu bar* (A) contains the main menus for file loading and output, data transformations and analysis, display options and utility functions, as well as other key tasks. The *algorithm tool bar* (B) organizes the algorithm module buttons into the rough algorithm categories described earlier. An abbreviated module name and graphic on each analysis button clearly indicates the analysis. The *result navigation tree* (C) is used to organize and navigate analysis results. Clicking on a node in the tree will open a viewer associated with the labeled node in the *viewer display panel* (D) to the right of the tree. The navigation tree also contains the cluster manager, script manager, and the analysis history log.

MeV has many features to help researchers extract significant information from data and clustering results. This section describes some of the most basic functions and capabilities in the order that they would be encountered during analysis.

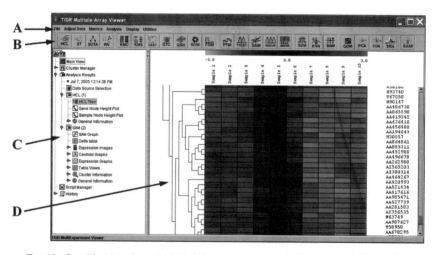

Fig. 12. Graphical interface of MeV: (A) main menu bar, (B) algorithm toolbar, (C) result navigation tree, and (D) viewer panel. A hierarchical tree viewer is shown in the viewer panel. (See color insert.)

File Loading/Data Filtering/Data Transformations

MeV supports the loading of six expression file formats, including Affymetrix, GenePix (.gpr), Agilent, and the TM4 suite's .mev format. A variety of *Data filters* can be applied to the loaded data to remove data of low quality, genes (rows) with few valid data measurements, or genes that show little change over the loaded experiments.

Data transformations can also be performed from the Adjust Data menu. These transformations include *log transformations* of expression values and *mean centering*, where each gene expression vector is shifted such that the mean of the values in each vector is zero.

Cluster Viewers

Nodes that represent clustering results are appended to the result tree as they are created. Clusters can be viewed in the viewer display panel by clicking on these nodes in the result tree. In addition to many algorithm-specific viewers, MeV provides four basic *cluster viewers* (Fig. 13) to view the expression and

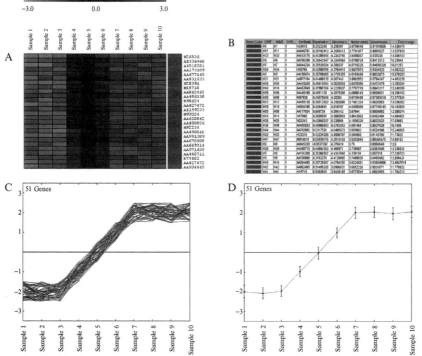

FIG. 13. Cluster viewer examples, (A) Expression image, (B) cluster table viewer, (C) expression graph, and (D) centroid graph. (See color insert.)

membership of each cluster. *Expression Images* display an expression matrix that corresponds to the subset of genes within the viewed cluster in which the expression level of each gene (row) across several experiments (columns) is displayed as a color that reflects the level of expression. The *Cluster Table Viewer* displays all gene annotation relating to the genes in the cluster and supports sorting on annotation, searches, and many other useful features. *Expression Graphs* display a graph showing the expression of each gene in the cluster over the set of loaded samples, whereas *Centroid Graphs* only show the cluster's mean expression pattern with error bars ($\pm$ 1 SD).

Cluster File Output/Cluster Storage/Cluster Operations

Once formed, clusters can be output to file in a tab-delimited text format that contains all expression and annotation data for the genes in the cluster. This format can be viewed as a spreadsheet and can be loaded easily into MeV to further visualize and mine that subset of data. Clusters can also be stored in MeV's *Cluster Manager,* which is a repository of selected clusters that can be viewed via the Cluster Manager node in the result tree. User-defined attributes such as a cluster label and description can be stored with the cluster as well as an assigned color that can be used to track the location of the cluster members during analysis. The assigned color propagates through all viewers to provide a qualitative measure of cluster overlap between analysis methods or runs. The cluster manager table provides many useful options, but the most useful are *cluster set operations,* such as cluster unions, intersections, and exclusive OR . These operations allow one to combine clusters of interest or to find genes common to two or more clusters.

Analysis Branching

A common task during analysis is to use an algorithm to reduce data to a set of interesting genes and then to extract this data subset for further analysis. We term this basic function where data are split off and analyzed as *analysis branching.* MeV provides three ways to perform analysis branching: (1) save the cluster as a file and then load it into a new MeV session as described, (2) use a feature of cluster viewers to automatically launch a new session that contains only the genes (or samples) in the cluster, and (3) right click on the cluster node in the result tree and select a check box to set that cluster as the primary data source for further analysis.

Analysis Scripting

The graphical nature of MeV lends itself to direct interaction, and it is often required that algorithms be applied several times to hone in on appropriate parameter values. An alternative to the interactive mode of MeV is the scripting mode. MeV provides graphical tools for script building, representation, and execution (Fig. 14). Once constructed, the XML

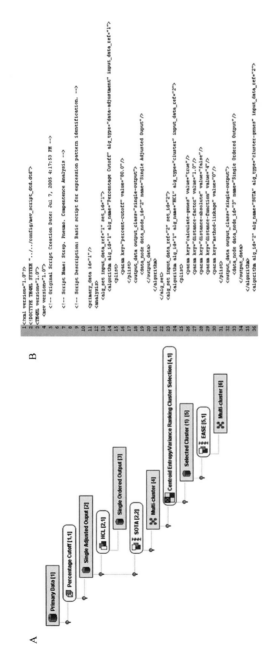

FIG. 14. Script viewers of MeV. (A) Graphical script tree viewer and (B) corresponding script XML viewer (script section).

analysis script can be saved and shared with collaborators to define analysis pipelines that reveal features of interest.

History Log

All analysis operations, from file loading, data filtering, algorithm runs, cluster storage, and file output, are recorded in a history log that describes the operation and attaches a time stamp. This serves as a detailed account of the analysis.

Sample Analysis Walk-Through

This section presents a sample Midas and MeV analysis that takes data through filtering and normalization, clustering and statistical analysis, and on to biological role analysis. To take full advantage of this walk-through it is best to download the applications and the sample data set so that one can follow along. Data for this analysis walk-through can be downloaded from this ftp site: ftp://ftp.tigr.org/pub/software/Microarray/MeV/MIE_data/. Each section indicates the proper files to use to illustrate the example. Midas and MeV can be downloaded from http://www.tm4.org/midas.html and http://www.tm4.org/mev.html.

Study Overview

This study investigates gene expression differences during ovalbumin induction of asthmatic responses. The study compared expression differences in mouse strains that are high or low responders to the stimuli in order to find genes that correlate to susceptibility or resistance. This example considers a low responder strain (CASTEi denoted as 'C' in the sample description) and a moderate responder strain (BALB\C denoted as 'B' in the sample description). For each strain there are biological duplicates for three time points: 24, 48, and 72 h. Each exposure time point had a corresponding vehicle control. The emphasis of this exercise is on the process of analysis rather than making specific claims about the nature of the findings.

Normalization Using Midas

This step filters low-quality spots using Spotfinder-generated flags, normalizes using block level Lowess and standard deviation regularization, and finally applies a flip-dye consistency filter. Because the same normalization process is repeated for each flip-dye pair (24 pairs), we will demonstrate the process on only one pair of raw files from the study as an example. The two files are contained in the sample data zip file and are labeled File_A_Sample Cy5_RefCy3.mev (file 'A') and File_B_SampleCy3_RefCy5.mev (file 'B').

These files contain the same sample and reference material but with the dye labels swapped.

Define the Analysis Pipeline

Open Midas by double clicking midas.bat in the midas directory. The analysis will proceed as follows.

    a. Read two sample .mev format files as a flip-dye pair.
    b. Execute Lowess (LocFit) normalization.
    c. Execute standard deviation regularization (SD).
    d. Perform the flip-dye consistency filter and file merge.
    e. Write result files.

Select the analysis buttons in the Midas interface to construct the pipeline by referring to Fig. 15 as a guide to help identify the buttons for each step of the process. If a button is hit in error, one can clear the graphical pipeline and start again by clicking the left-most button in the tool bar (Fig. 15).

Modify the Parameters

Once the analysis pipeline matches the one in Fig. 15, you are ready to enter and modify the analysis parameters. Click on the first icon in the

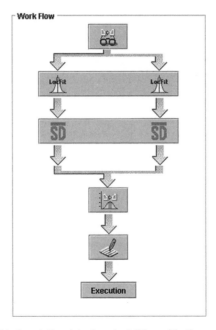

Fig. 15. The graphical scripting interface in Midas with the sample analysis pipeline indicates the order of processing operations.

pipeline that controls file loading. The parameter panel in the upper right will reflect available parameters. Select the input files by clicking on the empty field (first table cell) in the parameter panel. Use the file selection dialog to navigate to the analysis files and click on file 'A' and then ctrl-click on file 'B' to select the pair. Select the down arrow button to place the pair of files in the selection area and hit the OK button. Note that multiple file pairs can be analyzed by adding multiple pairs to the selection area. Select the check boxes to use the flags to filter low-quality data. Each of these selections will prompt a request for a flag column identifier. Just accept the default flag column header names. Review the parameters for the other parts of the pipeline by clicking on each of the remaining icons. Accept the default parameters for the other sections of the pipeline.

Select Output Reports

Select the *Reports* menu from the main menu bar and check the *Create PDF Reports* option. Now Midas is set to output a text-based result summary as well as a customizable pdf format analysis summary. The summary will contain input parameters, diagnostic plots, and numerical data related to the output such as the number of retained spots after filtering. Just before the analysis starts a dialog will be presented to customize the PDF report. For this example keep all graphs. When processing many files it is best to limit the file output to the key plots for each analysis stage, as the PDF creation requires a large amount of memory.

Execute the Analysis Pipeline

Select the *Execute* button to trigger execution. The final step is to select a project folder for output and to specify the project file name to store the pipeline and parameters. The progress of the analysis will be indicated in the analysis log at the bottom of the interface. Once the analysis is complete the diagnostic plots can be reviewed to assess the impact of the procedures.

Assessing the Results: The Investigation Panel

Open the *Investigation* panel by clicking on the tab just below the button panel to use Midas to view diagnostic plots. Use the file tree on the left side to navigate to the folders that contain the results. A right click on any plot will open a menu to allow you to view or plot the output file. The folder labeled *raw* contains the plots of data in its initial state. Plots of the same type can be overlaid to view the effect of normalization by first plotting raw data and then plotting normalized data. Some plots to try are histograms (.his) and R-I plots (.prc) in raw and post Lowess, box plots (.box) before and after standard deviation regularization, and in the flip dye folder you can view the flip dye diagnostic plots (.rrc) before and after filtering (Fig. 16).

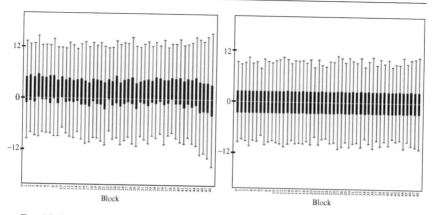

Block   Block

FIG. 16. Box plots of raw data (left) and data after Lowess normalization and standard deviation regularization (right). Note the centering effect of Lowess on block level mean log ratios. The nearly equal span of the middle quartiles of each block reflects the variance regularization following the SD regularization step.

Statistical Analysis and Clustering Using MeV

Now that data have been normalized to remove systematic bias and filtered to remove spots that are not expressed consistently we can use MeV to perform statistical analysis, clustering, and functional analysis. Data for this section started as raw mev files and were normalized and filtered as described earlier. The processed files for both strains were loaded into MeV in an order according to strain, exposure (control or experimental), and time point. The resulting expression matrix was saved to a single file to help streamline data loading for this example. The data file is in the sample data zip file and is labeled *CastEi_Balbc_combined_TDMS.txt*.

Launching MeV, File Loading, and Adjusting the Display

Double click on *tmev.bat* to launch MeV. The multiple array viewer can be resized to full screen by clicking on the maximize button in the upper right corner of the window. Select *Load Data* from the *File* menu of the multiple array viewer. The top part of the file loader interface will have a drop-down menu that is used to select the input file format. Select the second menu option labeled *Tab Delimited, Multiple Sample Files (TDMS)*. Use the file navigation tree on the left to navigate to sample data and select *CastEi_Balbc_combined_TDMS.txt* from the available files window. Selecting the file will present a portion of the file in the expression table preview panel on the right. Click on the first expression value in the upper-left position of the expression values. For this file the value happens to be NaN, as this value is missing or was filtered out. Selection of this table cell informs the loader that rows above and to the left are sample and gene annotation. Click *Load* to load the data file.

The initial main view of the expression matrix will include sample names that correspond to the original mev files. From the *Display* menu select *Sample/Column Labels* and then *Select Sample Label* to *Label by Sample Description*. The sample annotation now contains strain ID (C or B), condition (control or experimental), time, and replicate ID. To improve the appearance of the expression matrix, modify the gradient scale limits by using the *Set Color Scale Limits* option from the *Display* menu. Set the lower limit to −2.0 and the upper limit to 2.0.

Filtering out Missing Data

It is common to have genes within loaded data that have few valid intensity measurements over the loaded samples. These rows with a lot of missing data appear mostly gray in the expression matrix image. To filter these genes out, open the *Adjust Data* menu and open the *Data Filters* menu and select the *Percentage Cutoff Filter* option. Enter 85.0 in the input dialog to keep only genes for which greater than 85% of the samples have values. A data filter result node will be placed on the result tree to report the number of genes that remain and to provide a view of the conserved rows. The log of the history node will also report the filtering result. Note that 27,648 rows were loaded and after applying the filter 20,048 rows remain for further analysis.

Statistical Analysis

Because there are two strains and two conditions, a 2×2 design two-factor ANOVA can be applied if we treat all time points as being in one group. Hit the two-factor ANOVA analysis button to open the dialog. In the first dialog label factor A as strain and factor B as condition and enter two levels for each factor, as there are two strains and two conditions, control and experimental. Advance to the next dialog to make group assignments. Designate strain membership in the upper left panel by selecting group 1 for all 'C' strain samples and group 2 for all 'B' strain samples. Designate condition by placing all controls in group 1 and all experimental samples in group 2 in the group selection panel on the right. Set the critical value of p to 0.001. Near the bottom of the dialog select the check box to build HCL trees on significant genes and hit the OK button.

An HCL initialization dialog will come up to select parameters for HCL. Deselect the option to make sample trees so that samples are not reordered. Select the *Pearson Correlation* as the distance metric and hit OK.

Interaction significant genes from two-factor ANOVA in the result tree are those that show differences in response to exposure that are dependent on strain. In this example this mostly consists of genes that changed in the moderate responder strain B but not in the low responder strain C. One can view the various results by clicking on the viewer nodes in the result tree.

Dissecting Significant Genes

To further explore the interaction significant genes, click on the HCL viewer in the ANOVA result for the interaction significant cluster. Right click in the HCL cluster viewer and select *Gene Tree Properties*. Slide the slider bar to the right until the number of terminal nodes is about two or three. Select the check box labeled *Create Cluster Viewers* and hit OK. This will create clusters that correspond to subtrees with the full HCL viewer. The viewer nodes will be appended under the HCL viewer node on the result tree. Figure 17 shows the cluster centroid viewers that correspond to the two dominant patterns in the interaction significant cluster where genes were upregulated or downregulated in only the moderate responder strain B (Fig. 17).

Storing Clusters and Cluster Operations

A right click-activated menu provides a *Store Cluster* option in most cluster viewers that allows one to store clusters of interest to the cluster manager. Open a viewer other than the HCL viewer that displays all significant interaction genes, right click, and select *Store Cluster*. The cluster can be assigned attributes such as a label and a description. Selection of a cluster

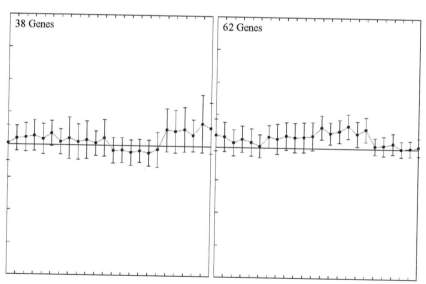

FIG. 17. Centroid graphs showing genes with a significant interaction effect. (Left) Mean profile for 38 genes that were overexpressed in the high responder strain. (Right) Mean profile for the 62 genes that were underexpressed in the responder strain. Error bars are ± 1 SD. (Two-factor ANOVA results, interaction significant genes, $p < 0.001$.)

color is required and can be used to track the genes during analysis. Stored clusters can be viewed in the cluster manager node's gene cluster table above the main analysis node in the result tree. If multiple clusters exist from various results, one can use the cluster operations in the cluster manager to perform cluster operations such as cluster unions, intersections, or exclusive OR.

Exploring Biological Themes

The EASE module can be used to investigate the biological roles within a cluster of interest. All clusters stored in the cluster manager are candidates for EASE analysis. The data directory of MeV has an EASE file system to support the analysis of this data set. Select the EASE button from the right end of the analysis tool bar or from the analysis menu. The center portion of the dialog has three tabbed panels. The first panel is used is to designate a population of genes and a cluster for analysis. Select the *Population from Current Viewer* option to define the population. Select the cluster to analyze by selecting a row in the cluster table. On the second tab check that *tc#* is selected as the gene identifier. In the bottom portion of the panel select the button to add annotation/ontology linking files and use ctrl-click to select the three GO files and the KEGG pathways file. Accept the defaults for the statistical parameters panel by hitting OK. The resulting table will list all biological roles that were identified for the cluster, and the roles will be ordered by the provided *p* value for each role. The GO hierarchical viewer will show themes in a hierarchy of specificity. MeV's manual, slide presentation in the *documentation/presentations* folder, and the EASE reference (Hosack *et al.*, 2003) will describe the parameter selections, theory basics, and the statistical details behind EASE analysis.

Further Analysis

The purpose of this section was to provide a basic sample analysis. Various other tests can be run on this data set to extract other genes of interest. The power of any analysis tool comes with the understanding of the available analysis modules and features and how they can be used to extract a variety of findings relevant to the study.

References

Ashburner, M., Ball, C. A., Blake, J. A., Botstein, D., Butler, H., Cherry, J. M., Davis, A. P., Dolinski, K., Dwight, S. S., Eppig, J. T., Harris, M. A., Hill, D. P., Issel-Tarver, L., Kasarskis, A., Lewis, S., Matese, J. C., Richardson, J. E., Ringwald, M., Rubin, G. M., and Sherlock, G. (2000). Gene ontology: Tool for the unification of biology. *Nature Genet.* **25**, 25–29.

Ayroles, J. F., and Gibson, G. (2006). Analysis of variance of microarray data. *Methods Enzymol.* **411**, 214–233.

Barrett, T., and Edgar, R. (2006). Gene Expression Omnibus (GEO): Microarray data storage, submission, retrieval and analysis. *Methods Enzymol.* **411,** 352–369.

Ben-Dor, A., Shamir, R., and Yakhini, Z. (1999). Clustering gene expression patterns. *J. Comput. Biol.* **6,** 281–297.

Brazma, A., Hingamp, P., Quackenbush, J., Sherlock, G., Spellman, P., Stoeckert, C., Aach, J., Ansorge, W., Ball, C. A., Causton, H. C., Gaasterland, T., Glenisson, P., Holstege, F. C., Kim, I. F., Markowitz, V., Matese, J. C., Parkinson, H., Robinson, A., Sarkans, U., Schulze-Kremer, S., Stewart, J., Taylor, R., Vilo, J., and Vingron, M. (2001). Minimum information about a microarray experiment (miame)-toward standards for microarray data. *Nature Genet.* **29,** 365–371.

Brazma, A., Kapushesky, M., Parkinson, H., Sarkins, U., and Shojatalab, M. (2006). Data storage and analysis in ArrayExpress. *Methods Enzymol.* **411,** 370–386.

Brazma, A., Parkinson, H., Sarkans, U., Shojatalab, M., Vilo, J., Abeygunawardena, N., Holloway, E., Kapushesky, M., Kemmeren, P., Lara, G. G., Oezcimen, A., Rocca-Serra, P., and Sansone, S. A. (2003). Arrayexpress: A public repository for microarray gene expression data at the EBI. *Nucleic Acids Res.* **31,** 68–71.

Brown, M. P., Grundy, W.N, Lin, D., Cristianini, N., Sugnet, C. W., Furey, T. S., Ares, M., Jr., and Haussler, D. (2000). Knowledge-based analysis of microarray gene expression data by using support vector machines. *Proc. Natl. Acad. Sci. USA* **97,** 262–267.

Chen, Y., Dougherty, E. R., and Bittner, M. L. (1997). Ratio-based decisions and the quantitative analysis of cDNA microarray images. *J. Biomed. Optics* **2,** 364–374.

Chu, G., Narasimhan, B., Tibshirani, R., and Tusher, V. (2002). SAM "Significance Analysis of Microarrays." Users Guide and Technical Document.http://www-stat.stanford.edu/~tibs/SAM/.

Culhane, A. C., Perriere, G., Considine, C., Cotter, T. G., and Higgins, D. G. (2002). Between-group analysis of microarray data. *Bioinformatics* **18**(12), 1600–1608.

Dalma-Weiszhauz, D. D., Warrington, J., Tanimoto, E. Y., and Miyada, C. G. (2006). The Affymetrix Gene Chip® platform: An overview. *Methods Enzymol.* **410,** 3–28.

Dopazo, J., and Carazo, J. M. (1997). Phylogenetic reconstruction using an unsupervised growing neural network that adopts the topology of a phylogenetic tree. *J. Mol. Evol.* **44,** 226–233.

Downey, T. (2006). Analysis of a multifactor microarray study using Partek Genomics Solution. *Methods Enzymol.* **411,** 256–270.

Dudoit, S., Shaffer, J. P., and Boldrick, J. C. (2003). Multiple hypothesis testing in microarray experiments. *Stat. Sci.* **18,** 71–103.

Dudoit, S., Yang, Y. H., Callow, M. J., and Speed, T. (2000). Statistical methods for identifying differentially expressed genes in replicated cDNA microarray experiments. Technical Report 2000, Statistics Department, University of California, Berkeley.

Eads, B., Cash, A., Bogart, K., Costello, J., and Andrews, J. (2006). Troubleshooting microarray hybridizations. *Methods Enzymol.* **411,** 34–49.

Edgar, R., Domrachev, M., and Lash, A. E. (2002). Gene Expression Omnibus: NCBI gene expression and hybridization array data repository. *Nucleic Acids Res.* **30**(1), 207–210.

Eisen, M. B., Spellman, P. T., Brown, P. O., and Botstein, D. (1998). Cluster analysis and display of genome-wide expression patterns. *Proc. Natl. Acad. Sci. USA* **95,** 14863–14868.

Fellenberg, K., Hauser, N. C., Brors, B., Neutzner, A., Hoheisel, J. D., and Vingron, M. (2001). Correspondence analysis applied to microarray data. *Proc. Natl. Acad. Sci. USA* **98**(19), 10781–10786.

Finkelstein, D.B, Gollub, J., Ewing, R., Sterky, F., Somerville, S., and Cherry, J. M. (2000). Iterative linear regression by sector: Renormalization of cDNA microarray data and cluster analysis weighted by cross homology. In CAMDA. http://afgc.stanford.edu/afgc_html/site2Stat.htm.

Gaidoukevitch, Y. C., Ryan, K. W., and Sequera, D. E. (2002). Method, system and product for analyzing image of an array to create an image of a grid overlay, U.S. patent 6,498,863.

George, R. A. (2006). The printing process: Tips on tips. *Methods Enzymol.* **410**, 121–135.

Jeremy Gollub, J., and Gavin Sherlock, G. (2006). Clustering microarray data. *Methods Enzymol.* **411**, 194–213.

Hastie, T., Tibshirani, R., Eisen, M. B., Alizadeh, A., Levy, R., Staudt, L., Chan, W. C., Botstein, D., and Brown, P. (2000). "Gene shaving" as a method for identifying distinct sets of genes with similar expression patterns. *Genome Biol.* **1**(2), RESEARCH0003.

Herrero, J., Valencia, A., and Dopazo, J. (2001). A hierarchical unsupervised growing neural network for clustering gene expression patterns. *Bioinformatics* **17**(2), 126–136.

Heyer, L. J., Kruglyak, S., and Yooseph, S. (1999). Exploring expression data: Identification and analysis of co-expressed genes. *Genome Res.* **9**, 1106–1115.

Hosack, D. A., Dennis, G., Jr., Sherman, B. T., Lane, H. C., and Lempicki, R. A. (2003). Identifying biological themes within lists of genes with EASE. *Genome Biol.* **4**, R70–R78.

Johnston, R., Wang, B., Nuttall, R., Doctolero, M., Edwards, P., Lu, J., Vainer, M., Yue, H., Wang, X., Minor, J., Chan, C., Lash, A., Goralski, T., Parisi, M., Oliver, B., and Eastman, S. (2004). FlyGEM, a full transcriptome array platform for the Drosophila community. *Genome Biol.* **5** research0019.1-0019.11.

Keppel, G., and Zedeck, S. (1989). "Data Analysis for Research Designs." Freeman, New York.

Kim, S. K., Lund, J., Kiraly, M., Duke, K., Jiang, M., Stuart, J. M., Eizinger, A., Wylie, B. N., and Davidson, G. S. (2001). A gene expression map for. *Caenorhabditis elegans. Science* **293**, 2087–2092.

Kohonen, T. (1982). Self-organized formation of topologically correct feature maps. *Biol. Cybernet.* **43**, 59–69.

Korn, E. L., Troendle, J. F., McShane, L. M., and Simon, R. (2001). Controlling the number of false discoveries: Application to high-dimensional genomic data. Technical report 003, Biometric Research Branch, National Cancer Institute. http://linus.nci.nih.gov/~brb/TechReport.htm.

Korn, E. L., Troendle, J. F., McShane, L. M., and Simon, R. (2004). Controlling the number of false discoveries: Application to high-dimensional genomic data. *J. Stat. Plan. Inference* **124**, 379–398.

Liao, P.-S., Chen, T.-S., and Chung, P.-C. (2001). A fast algorithm for multilevel thresholding. *J. Inform. Sci. Eng.* **17**, 713–727.

Lipshutz, R. J., Morris, D., Chee, M., Hubbell, E., Kozal, M. J., Shah, N., Shen, N., Yang, R., and Fodor, S. P. (1995). Using oligonucleotide probe arrays to access genetic diversity. *Biotechniques* **19**(3), 442–447.

Manly, B. F. J. (1997). "Randomization, Bootstrap and Monte Carlo Methods in Biology," 2nd Ed. Chapman and Hall/CRC, FL.

Minor, J. M. (2006). Microarray quality control. *Methods Enzymol.* **411**, 233–255.

Neal, S. J., and Westwood, T. (2006). Optimizing experiment and analysis parameters for spotted microarrays. *Methods Enzymol.* **410**, 203–221.

Otsu, N. (1979). A treshold selection method from gray-level histogram. IEEE Transactions on System Man Cybernetics, v. SMC-9, No. 1, 62–66.

Pan, W. (2002). A comparative review of statistical methods for discovering differentially expressed genes in replicated microarray experiments. *Bioinformatics* **18**, 546–554.

Pavlidis, P., and Noble, W. S. (2001). Analysis of strain and regional variation in gene expression in mouse brain. *Genome Biol.* **2**, research0042.1-0042.15.

Quackenbush, J. (2002). Microarray data normalization and transformation. *Nature Genet.* **32**(Suppl.), 496–501.

Raychaudhuri, S., Stuart, J. M., and Altman, R. B. (2000). Principal components analysis to summarize microarray experiments: Application to sporulation time series. Pacific

Symposium on Biocomputing 2000 Honolulu, Hawaii 452–463. Available at http://smi-web.stanford.edu/pubs/SMI_Abstracts/SMI-1999-0804.html.

Reimers, M., and Carey, V. J. (2006). Bioconductor: An open source framework for bioinformatics and computational biology. *Methods Enzymol.* **411**, 119–134.

Royce, T. E., Rozowsky, J. S., Luscombe, N. M., Emanuelsson, O., Yu, H., Zhu, X., Snyder, M., and Gerstein, M. B. (2006). Extrapolating traditional DNA microarray statistics to tiling and protein microarray technologies. *Methods Enzymol.* **411**, 282–311.

Schena, M., Shalon, D., Davis, R. W., and Brown, P. O. (1995). Quantitative monitoring of gene expression patterns with a complimentary DNA microarray. *Science* **270**(5235), 467–470.

Sharov, V., Kwong, K. Y., Frank, B., Chen, E., Hasseman, J., Gaspard, R., Yu, Y., Yang, I., and Quackenbush, J. (2004). The limits of log-ratios. *BMC Biotechnol.* **4**, 3.

Soukas, A., Cohen, P., Socci, N. D., and Friedman, J. M. (2000). Leptin-specific patterns of gene expression in white adipose tissue. *Genes Dev.* **14**, 963–980.

Tamayo, P., Slonim, D., Mesirov, J., Zhu, Q., Kitareewan, S., Dmitrovsky, E., Lander, E. S., and Golub, T. R. (1999). Interpreting patterns of gene expression with self-organizing maps: Methods and application to hematopoietic differentiation. *Proc. Natl. Acad. Sci. USA* **96**, 2907–2912.

Theilhaber, J., Connolly, T., Roman-Roman, S., Bushnell, S., Jackson, S., Call, K., Garcia, T., and Baron, R. (2002). Finding genes in the C2Cl2 osteogenic pathway by K-nearest-neighbor classification of expression data. *Genome Res.* **12**, 165–176.

Timlin, J. A. (2006). Scanning microarrays: Current methods and future directions. *Methods Enzymol.* **411**, 79–98.

Troein, C., Vallon-Christersso, J., and Saal, L. H. (2006). An introduction to BioArray Software Environment. *Methods Enzymol.* **411**, 99–119.

Tusher, V. G., Tibshirani, R., and Chu, G. (2001). Significance analysis of microarrays applied to the ionizing radiation response. *Proc. Natl. Acad. Sci. USA* **98**, 5116–5121.

Weinstein, J. N., Myers, T. G., O'Connor, P. M., Friend, S. H., Fornace, A. J., Jr., Kohn, K. W., Fojo, T., Bates, S. E., Rubinstein, L. V., Anderson, N. L., Buolamwini, J. K., van Osdol, W. W., Monks, A. P., Scudiero, D. A., Sausville, E. A., Zaharevitz, D. W., Bunow, B., Viswanadhan, V. N., Johnson, G. S., Wittes, R. E., and Paull, K. D. (1997). An information-intensive approach to the molecular pharmacology of cancer. *Science* **275**, 343–349.

Welch, B. L. (1947). The generalization of 'students' problem when several different population variances are involved. *Biometrika* **34**, 28–35.

Whetzel, P. L., Parkinson, H., and Stoeckert, C. J. (2006). Using ontologies to annotate microarray experiments. *Methods Enzymol.* **411**, 325–339.

Yang, I. V., Chen, E., Hasseman, J. P., Liang, W., Frank, B. C., Wang, S., Sharov, V., Saeed, A. I., White, J., Li, J., Lee, N. H., Yeatman, T. J., and Quackenbush, J. (2002a). Within the fold: Assessing differential expression measures and reproducibility in microarray assays. *Genome Biol.* **3**, research0062.1–0062.12.

Yang, Y.-H., Buckley, M. J., Dudoit, S., and Speed, T. P. (2002b). Comparison of methods for image analysis on cDNA microarray data. *J. Comput. Graph. Stat.* **11**(No. 1), 108–136.

Yang, Y.-H., Buckley, M. J., and Speed, T. P. (2001). Analysis of cDNA microarray images. *Brief. Bioinform.* **2**(No. 4), 341–349.

Yang, Y. H., Dudoit, S., Luu, P., Lin, D. M., Peng, V., Ngal, J., and Speed, T. P. (2002c). Normalization for cDNA microarray data: A robust composite method addressing single and multiple slide systematic variation. *Nucleic Acids Res.* **30**, e15.

Yeung, K. Y., Haynor, D. R., and Ruzzo, W. L. (2001). Validating clustering for gene expression data. *Bioinformatics* **17**, 309–318.

Zar, J. H. (1999). "Biostatistical Analysis," 4th Ed. Prentice Hall, New Jersey.

[10] Clustering Microarray Data

By JEREMY GOLLUB and GAVIN SHERLOCK

Abstract

Even a simple, small-scale, microarray experiment generates thousands to millions of data points. Clearly, spreadsheets or plotting programs do not suffice for analysis of such large volumes of data, and comprehensive analysis requires systematic methods for selection and organization of data. This chapter focuses on the concepts and algorithms of hierarchical clustering and the most commonly employed methods of partitioning or organizing microarray data, and freely available software that implements these algorithms.

Introduction

There are a multitude of different methods for analyzing microarray data, but in general, they separate into *unsupervised* and *supervised* approaches. Supervised approaches typically select genes from a data set, based on answering some directed question, using additional information such as class labels and survival times, for example, "which genes best distinguish two groups of patients?" as might be asked using the Significance Analysis of Microarrays software (SAM; Tusher *et al.*, 2001). Unsupervised approaches generally organize data based on the properties of data themselves, without reference to additional information.

Clustering algorithms, of which there are many different types, are used for unsupervised analysis. They require a method for determining the similarity of two vectors of data and rules for organizing data based on those similarities. It is useful to consider the values that make up a microarray data set as a matrix, with each row being data for a single gene and each column being data for a single array/experiment. Data for a gene in the matrix define a *gene expression vector*, which has as many dimensions as there are data points within the vector (of course, "expression" may be an inappropriate term for some experiments, but data may always be cast as a vector). Using standard mathematical metrics, the similarity (or dissimilarity) between different vectors can be then measured, and in conjunction with certain rules (an algorithm), these metrics can then be used to organize data.

METHODS IN ENZYMOLOGY, VOL. 411
Copyright 2006, Elsevier Inc. All rights reserved.
0076-6879/06 $35.00
DOI: 10.1016/S0076-6879(06)11010-1

Distance Metrics

There are many different metrics that can be used to determine the similarity between two expression vectors, and which metric is chosen will affect the outcome of a clustering analysis to a greater or lesser degree. Vendors of microarray analysis software have tended to include as many metrics as possible for the user to choose from, with perhaps inadequate guidance as to under which circumstances each is most appropriate. This trend has been followed to a lesser extent in freely available software (some of which are discussed later). This chapter describes the calculation of the more common metrics and attempts to indicate under what circumstances they might be appropriate.

Correlation Metrics

There are several different correlation metrics that can be used when clustering microarray data when a user is typically trying to determine whether genes are similarly expressed under some set of conditions. These correlation metrics can be separated into two types, parametric, and nonparametric, and their values lie between -1 and $+1$, with $+1$ indicating perfect correlation, -1 indicating perfect anticorrelation, and 0 indicating no correlation. Parametric metrics make an underlying assumption about the distribution of data (in the case of the Pearson correlation, normality is assumed), whereas nonparametric measures of correlation do not make such assumptions, typically using ranks within data. A metric that makes assumptions is more powerful than one that does not, *if those assumptions are correct* (or nearly so), whereas a nonparametric measure of similarity is likely to be more appropriate when such assumptions cannot be made safely. In the case of gene expression microarray data, the log ratio measurements do form a roughly normal distribution, and using the Pearson correlation is reasonable. However, if clustering ChIP-chip data for instance, where there may be strong signals for a small subset of elements and simply noise for the rest of the elements, the rank of a data value is most important, and thus a nonparametric measure, such as the Spearman rank correlation, or Kendall's τ (see later) is likely more appropriate. There are several statistical tests to determine the goodness of fit for data distribution to normal distribution, such as the Shapiro–Wilk test or the D'Agostino–Pearson omnibus test, which provides a p value for the hypothesis that data were drawn from a normal distribution. Such tests are implemented in various statistical packages. In addition, visual inspection of a frequency histogram of data can also be used to determine whether the distribution deviates grossly from a normal distribution.

Pearson Correlation

The Pearson correlation is probably the most frequently used metric for calculating similarity between vectors of microarray data. The Pearson correlation is insensitive to the magnitude of the compared vectors. For instance, if data for one gene describe a sine wave with an amplitude of 3 and data for another gene also describe a sine wave of identical phase, but with an amplitude of 10, they would still have a perfect correlation of 1. That is, the Pearson correlation is a measure of the similarity in shape of the two expression patterns. The Pearson correlation is calculated as

$$r = \frac{\sum_{i=1}^{n}(x_i - \bar{x})(y_i - \bar{y})}{(n-1)S_x S_y}$$

where S_x is the standard deviation of x and S_y is the standard deviation of y. Thus, if for a given i, x and y vary in the same direction, the product of those deviations will be positive, whereas when they vary from their means in the opposite direction, that product will be negative.

Variations on the Pearson Correlation

When using the standard Pearson correlation, expression vectors that are identical in shape but are offset from each other by a constant amount will have a correlation of 1. In some cases, this may be desirable (e.g., if you are only interested in the shape of the expression response to a treatment), but in cases where it is not appropriate (e.g., in a time series, derepression of transcription for some genes, compared to activation of other genes, might mean very different things and should thus be treated as different), a variation on the Pearson correlation, usually referred to as the uncentered Pearson correlation, can be used (the standard Pearson correlation is often referred to as the centered Pearson correlation in clustering software). The formula for the uncentered Pearson correlation is identical to that of the centered Pearson correlation, except that the calculation of standard deviations in the denominator assumes that the mean is zero, rather than using the real mean. In situations in which the actual ratio is as or more important than the trend, the uncentered Pearson correlation will be more informative.

Another option when employing the Pearson correlation is to use the absolute value of the correlation as the measure of similarity. This assigns vectors of opposite direction a high degree of similarity, for example, two genes describing sine waves of opposite phase will cocluster, which may be desirable if one is, for instance, interested in all genes varying in phase with the cell cycle or genes regulated positively or negatively by a transcription factor.

Spearman Rank Correlation

Spearman rank correlation is a nonparametric measure of association of two variables, based on ranks. In general terms, it answers the same question as the Pearson correlation, but is less specifically interpretable. As a nonparametric test, it makes fewer assumptions about the distribution of data in the two vectors to be compared and is thus a safer test when the Pearson correlation's assumption of normality is violated significantly. Spearman rank correlation is calculated by converting the actual values in each vector to their rank in a sorted list of those values and then comparing the ranks of the ith element of each vector for all i. Values range from -1 to 1, and in general, terms are interpreted similarly to the Pearson correlation: 1 is a perfect positive association, 0 implies no association, and -1 is a perfect negative association. The formula for calculating the Spearman rank correlation is

$$r_s = \frac{\sum_i (R_i - \bar{R})(S_i - \bar{S})}{\sqrt{\sum_i (R_i - \bar{R})^2}\sqrt{\sum_i (S_i - \bar{S})^2}}$$

where R_i is the rank of x_i among the other x values and S_i is the rank of y_i among the other y values, with ties being assigned the appropriate midrank.

Kendall's τ

Kendall's τ is an even more nonparametric measure of association than Spearman's r_s. The statistic is calculated from the relative ordering of the ranks rather than using the numerical difference between the ranks. Thus, Kendall's τ is somewhat less sensitive to actual data values than Spearman rank correlation. The statistic ranges from -1 to 1 and is interpreted similarly to Pearson correlation or Spearman rank correlation.

Distance Metrics

Euclidean Distance

The Euclidean distance between two vectors is simply the distance in space between the two end points defined by those vectors. Thus, it is sensitive to the direction of the vectors, like the Pearson correlation, and also to their magnitude. Returning to our example of two sinusoidal vectors, unlike the Pearson correlation, the Euclidean distance is sensitive to a change in magnitude, but relatively less sensitive to a small change in phase. Euclidean distance may be a more useful metric than the Pearson correlation when the magnitude of change is an important element of the analysis. The Euclidean distance is calculated as

$$\sqrt{\sum_{i=1}^{n}(x_i - y_i)^2}$$

Manhattan or City-Block Distance

The Manhattan distance between two points can be thought of in terms of the path one would have to follow between two addresses in an urban downtown, making only right-angle turns. The distance is calculated as the sum of absolute values of these orthogonal legs of the journey rather than as the familiar sum of squares of Euclidean distance. This makes the Manhattan distance less sensitive to outlier values, as each element of the vector is weighted linearly rather than quadratically. The Manhattan distance is calculated as

$$d = \sum_{i=1}^{n} |x_i - y_i|$$

If one imagines two points in space (defined by two vectors being compared), then a straight line defines their Euclidean distance, and if that line is thought of as the hypotenuse of a right-angled triangle, the Manhattan distance is the sum of the other two sides.

Agglomerative Hierarchical Clustering

Agglomerative hierarchical clustering (here just "clustering") is a simple and effective method for exploratory analysis of gene expression microarray data. An exploratory method does not specifically test any particular hypotheses, such as which genes distinguish two groups of samples and what *p* value is associated with that distinction, but instead simply allow the user to explore data in a straightforward fashion. In the case of clustering, genes with similar patterns are grouped so exploring data is much easier than if they were disorganized. Gene expression vectors are organized in a tree structure, with the goal that each vector is closest in the tree to the vectors most similar to it according to the distance metric and linkage rule (see later) chosen. Each node in the tree represents a group of similar genes, and the height of the node in the tree indicates the degree of similarity. The data matrix is then reordered according to the tree structure so that again each vector is next to similar vectors (Fig. 1). The columns of the matrix, representing individual microarrays or experimental conditions, may be clustered in the same way. For example, if the columns represent different tumor samples, clustering will tend to group tumors of the same type together, distinguish between different types by separating them, and possibly help identify subtypes of a given type (e.g., Liang *et al.*, 2005; Schaner *et al.*, 2003; Sorlie *et al.*, 2001). Clustering in both the gene and experiment dimensions may be carried out sequentially on the same matrix.

In this agglomerative (bottom up) approach, pioneered for expression data by Eisen *et al.* (1998) and Wen *et al.* (1998), the gene expression vectors are compared to each other in all pairwise combinations, generating a matrix of correlations or similarities. The largest correlation/smallest distance in the matrix defines the two most similar vectors, which are then joined to form a node. This node is then compared to each other expression vector or node (using some linkage rule; see later), and these results are added to the correlation matrix. Again, the most similar vectors/nodes are joined, and the process is repeated. Thus, single expression profiles are joined successively to form nodes, which in turn are joined further. The process continues until all individual profiles and nodes have been joined to form a single hierarchical tree. The utility of this approach is that it is simple, both to understand and to implement, and the end result can be visualized easily. Often, coordinately regulated patterns can be relatively easily discerned by eye (Fig. 1).

Rules for Comparing Nodes

As indicated earlier, a clustering algorithm needs a rule to determine how to compare a node to either a single expression vector or another node. Among the various ways in which this might be done, four are commonly implemented.

Single Linkage (aka Nearest Neighbor)

The similarity of two nodes (either of which may be a single vector) is taken as the best (highest correlation, or shortest distance) of all pairwise comparisons of the members of one node to the other. Single linkage tends to produce loose clusters with a chaining effect, as vectors may be joined to a node that are very similar to only one vector in that node. It is very computationally efficient, as after the initial similarity matrix is calculated, no further correlations or distances need be computed, but memory requirements can be large (see later).

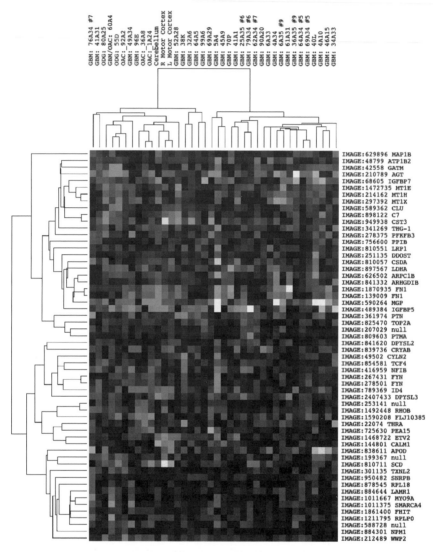

FIG. 1. Clustered gene expression data (heavily filtered data from Liang *et al.*, 2005). Blue and yellow indicate a negative and a positive log ratio, respectively. Clustering was carried out on both genes and samples (arrays) using a centered Pearson correlation metric and centroid linkage. The dendrograms or "trees" group similar vectors into nodes; the height of each node in the tree indicates the overall similarity of its members. Genes and samples both divide into two, relatively internally consistent groups. (See color insert.)

Complete Linkage (aka Furthest Neighbor)

The similarity between two nodes is recorded as the lowest similarity of all pairwise comparisons between the members of one node to the other. Complete linkage tends to produce tight clusters. It is as computationally efficient as single linkage.

Average Linkage

The similarity between two nodes is recorded as the average correlation from all pairwise comparisons between the members of one node to the other. Average linkage tends to produce clusters intermediate between single and complete linkage in terms of internal consistency. It suffers from a less efficient computation than single or complete linkage, as the average of all similarities must be calculated at each step.

Centroid Linkage

The similarity between two nodes is the similarity between the centroids of those nodes. The centroid of a node is simply calculated by averaging its constituent expression vectors. Centroid linkage can be more memory efficient than the other linkage methods (see later), but it does require additional correlations to be calculated during the tree-building stage (as many as were calculated during the creation of the initial correlation matrix). Finally, the meaning of a centroid as a summary of its node is not perfectly clear, and certainly when using the absolute correlation (either centered or uncentered), it does not make sense to use centroid linkage, as the centroid of two vectors that are anticorrelated will have the effect of creating a vector that is not like either of its constituents.

Drawbacks of Clustering

Agglomerative hierarchical clustering can lead to artifacts. For instance, when using centroid linkage, the vector that represents a node, which is calculated as the average of all vectors that belong to the node, may not reflect accurately any of the contained vectors, especially as nodes become larger. Also, irrespective of the linkage method, vectors within a node will become less similar to each other as you approach the root of the tree; that is, nodes become more heterogeneous. In addition, any suboptimal joins made early on during the running of the algorithm cannot be corrected later. A further drawback is that when clustering by experiment (columns in the matrix), the similarity between each vector is calculated over the total number of genes within the data set. Thus, even if sample A is most similar to sample B in the expression of an important set of genes, that (potentially important) fact will be obscured if sample C is most similar to sample A overall. Such a relationship could be of biological significance, and it would be preferable not to discard it. Finally, it may be the case that a hierarchical structure does not apply to data. As an alternative, there are clustering methods that partition data into more or less homogeneous groups instead of organizing data into a hierarchy. Several such partitioning methods exist; self-organizing maps (SOM) and K-means clustering are discussed next.

Data Partitioning

Self-Organizing Maps

Self-organizing maps (Kohonen, 1995) are used for partitioning data into a two-dimensional matrix of cells or partitions. Each gene and/or array is assigned to a single partition. The vectors in each partition are most similar to each other; each partition, overall, is more similar to adjacent partitions than to partitions farther away in the matrix. SOMs have been applied to gene expression data in a number of studies (e.g., Tamayo *et al.*, 1999).

To initialize a SOM the number of partitions to use must be defined, as must their geometry with respect to each other, that is, a 4 × 4 two-dimensional grid of 16 partitions. Each partition in the grid is more related to its neighbors than to distant partitions, and thus the geometry, as well as the number of partitions, will influence the outcome, that is, a 1 × 16 grid will give a somewhat different result than a 4 × 4 grid. Each partition is then assigned a seed vector by the algorithm, which has the same dimensionality as the vectors being partitioned. These seed vectors are usually initialized with random data, although as an alternative they can be seeded with the first two principal components, which capture the greatest amount of variation within the data set, and enough vectors spaced equally between them to have one seed vector per partition. Genes are then assigned to these partitions by an iterative method that manipulates these seed vectors. During each iteration, seed vectors are recalculated to represent expression data more closely, as follows. A gene is picked at random, and its vector is compared to each of the seed vectors. The seed vector to which it is most similar is then modified so that it more closely resembles the expression associated with that gene. Vectors of nearby partitions are modified similarly, but to a lesser degree. With each iteration, the amount by which the seed vectors are altered decreases, and fewer of the proximal partitions are modified each time: at the start perhaps the vectors of all partitions are affected; toward the middle of the process perhaps only the partitions immediately adjacent to the one that best matches the chosen gene vector are affected; and at the end only the best match partition vector itself is affected. Hence each iteration results in fewer vectors being modified by smaller amounts so that the map eventually converges to a solution. Finally, each gene vector is assigned to the partition with an associated vector to which it is most similar.

K-Means Clustering

K-means clustering (Everitt, 1974) partitions data in a manner similar to self-organizing maps, with the key difference being that one partition does not influence another directly. Seed vectors may be assigned in the same way as in SOMs, but then all gene/array vectors are immediately assigned to the most similar partition. The representative vector of each partition is then recalculated as the centroid of the vectors assigned to the partition, and gene/array vectors are then reassigned, possibly moving from one partition to another. This process continues until convergence is reached, with no gene or array vectors changing partitions between iterations. Because the solution may be influenced strongly by the initial seed vectors, the entire process is typically repeated many times in order to determine the stability of the solution and, ideally, find a global rather than local optimum. K-means clustering has been used successfully to analyze microarray data generated from studies of the yeast cell cycle (Tavazoie et al., 1999).

How Many Partitions to Make?

One of the main drawbacks of partitioning methods is the uncertainty in choosing an optimal number and arrangement of partitions. Several methods for determining the correct number of partitions to make have been suggested (for discussion, see Milligan and Cooper, 1985), including the Gap statistic (Tibshirani et al., 2000), which was designed with gene expression data in mind. The main goal when partitioning expression data is to reduce the within-cluster dispersion, such that each cluster is reasonably homogeneous, while at the same time the between-cluster dispersion is large (i.e., a partition is not split inappropriately into two or more similar partitions). Simply plotting the within-cluster dispersion (which for the purposes of the Gap statistic is defined as the sum of the squared Euclidean distances between all members of a cluster, divided by twice the number of members within the cluster, and then summed over all clusters) results in a line that decreases as the number of clusters increases. This makes intuitive sense—the more clusters, the less variation within each cluster, as they will have fewer and fewer members. However, looking at such a plot, there is often an elbow, or a point where the plot flattens markedly. The Gap statistic attempts to formalize detection of this point in the plot by making the plot for real data, and also for a reference distribution of data, which is created by drawing random data from the same distribution as original data. The difference between these two curves is then plotted, and the point at which their difference is maximal (the details are somewhat simplified for discussion here) is the number of clusters, k, into which data should be partitioned. The Gap statistic has not been widely implemented in common software packages, but there is a Perl library available for calculating the Gap statistic (the Statistics::Gap module), and the Gap statistic is also available in the Acuity software from Molecular Devices. In the absence of using the Gap statistic, a user can partition their data iteratively into more and more partitions to determine roughly when clusters of like patterns are being broken up and use a number of partitions just less that the number where coherent partitions are partitioned further, although of course this is a subjective process.

Computational Considerations

The run time for hierarchical clustering is proportional to the square of the number of entities being clustered (genes or experiments, depending on in which dimension the matrix is being clustered) and is, in general, related linearly to the dimensionality of the entities being clustered, for example, clustering 2000 genes will take four times as long as clustering 1000 genes, whereas doubling the number of experiments for those genes should only double the run time. In computational terms, the efficiency of hierarchical clustering is $O(n^2m)$, where n is the number of entities being clustered and m is their dimensionality. Typically, clustering a matrix of 1000 or so genes by 100 experiments should only take a few seconds, whereas clustering tens of thousands of genes by 1000 experiments can take several hours. The amount of memory required by hierarchical clustering can vary based on the linkage method. Typically, single, complete and average linkage keeps the entire matrix of correlations in memory, the size of which is proportional to the square of the number of genes. Clustering 10,000 genes results in a matrix of correlations occupying almost 400 MB of RAM, whereas 20,000 genes require 1.5 GB of RAM. Centroid linkage clustering does not need to keep all correlations in memory, and the implementation of centroid linkage clustering in XCluster (see later) only requires 2 Mb to store the correlations it calculates for our hypothetical 20,000 genes. Cluster 3.0 (de Hoon et al., 2004; see later) uses an alternative algorithm for single linkage clustering, SLINK (Sibson, 1973), whose memory requirements are related linearly to the number of entities being clustered and is significantly faster than the usual single linkage algorithm.

The run time for partitioning vectors using K-means clustering and SOM does not increase significantly based on the number of genes being partitioned, and the largest determinant is the number of iterations required. Typically, SOMs and K-means clustering only take a few minutes at most and have fairly trivial memory requirements.

Is There a Best Method and/or Best Metric?

When confronted with so many options, it would be useful if there were a simple way to determine which algorithm with which metric produces the best results. At least two approaches have been used. First, Yeung *et al.* (2001) introduced the concept of a figure of merit (FOM). This is a "leave-one-out" approach, where data from all but one array are clustered and then assessed to see how well clustered data predict data in the excluded array. This is repeated for all arrays. The more robust the clustering method is, the more predictive the clustering should be of data in the left-out experiment. Yeung *et al.* (2001) compared single, complete, and average linkage hierarchical clustering, CAST (Ben-Dor *et al.*, 1999), and K-means clustering. For real data, they found that single linkage clustering frequently performed almost identically to random assignment of genes to clusters, while the other linkage methods and CAST and K-means clustering did significantly better, performing similarly to each other. A second study (Gibbons and Roth, 2002) took a different approach and determined how coherent the biological annotation for genes within subclusters was, using different metrics and different algorithms. Again, they found that single linkage performed poorly, but they also found that average linkage performed worse than random as a cluster was cut into more subclusters. They found that as a measure of dissimilarity between gene vectors, no method outperformed Euclidean distance for ratio-based measurements or Pearson distance for nonratio-based measurements at the optimal choice of cluster number. They also showed that SOMs were the best approach for both measurement types at higher numbers of clusters. Clearly, unless there is a compelling reason to do so, single linkage should not be used for clustering microarray data, despite being an available option in many software packages.

Freely Available Clustering/Analysis and Visualization Software

A quick internet search turns up dozens of software applications for clustering. Offerings include both free and commercial software, from single-purpose applications to full-fledged statistical analysis suites, many with a range of other microarray-oriented functions from image analysis to gene annotation. To survey the territory completely would take a much larger space, and a much broader focus, than is available here.

Instead, this chapter focuses on the practical use of a family of applications narrowly tailored to hierarchical clustering of microarray data, starting with the original Cluster software from Mike Eisen at Stanford University (Eisen *et al.*, 1998), which runs on the Windows platform (Table I). One descendent, XCluster (http://genetics.stanford.edu/~sherlock/cluster.html), provides efficient computation in the context of integrated analysis pipelines; another, Cluster 3.0 (de Hoon *et al.*, 2004), provides somewhat more functionality than Cluster and can be deployed on a wide variety of computer platforms. Each of these applications relies on another program, such as TreeView (http://rana. lbl.gov/EisenSoftware.htm) or JavaTreeView (Saldanha, 2004), for graphical display of the results. After outlining the functions and options available in these programs, we will very briefly discuss two alternate approaches for microarray analysis, the MeV application of TIGR (Saeed *et al.*, 2003, 2006) and the R statistical programming language (http://www.r-project.org/; Reimers and Carey, 2006).

Cluster

Cluster (Eisen *et al.*, 1998) is a Windows application for unsupervised analysis of microarray data. It is free for academic and nonprofit use and may be downloaded from http://rana.lbl.gov/EisenSoftware.htm. The input and output of Cluster are both text files; the input file format (known as the pcl format, for preclustering) has become a *de facto* standard and is constructed easily in a spreadsheet program with genes and microarrays (individual hybridizations or conditions) represented as rows and as columns of data, respectively; the output is in the clustered data table (cdt) format, which is the pcl file with its rows and/or columns reordered, with some additional meta data used to indicated information pertaining to the hierarchical cluster. The cdt file may be accompanied by a gene tree (gtr) file and an array tree(atr) file, indicating the order and the correlation with which nodes were joined in the clustering. These files are easily parsed and displayed by TreeView. The various operations described here are organized into discrete tabs or pages in the software interface, making their use very straightforward.

Cluster provides a few simple options for filtering data prior to analysis. The object of filtering is usually to pick out genes that have responded to the experimental conditions on the basis of their pattern of expression (measurements). Genes may be filtered on the basis of the standard deviation of their measurements, on the range of measurements (greatest minus least), or on the number of microarrays in which the measurement exceeded some

TABLE I
SOME AVAILABLE CLUSTERING SOFTWARE PACKAGES

Name	Source/author	Web site	Reference
Acuity	Molecular Devices	http://www.moleculardevices.com/pages/software/gn_acuity.html	
Cluster	Michael Eisen	http://rana.lbl.gov/EisenSoftware.htm	Eisen et al. (1988)
Cluster 3.0	Michiel de Hoon	http://bonsai.ims.u-tokyo.ac.jp/~mdehoon/software/cluster/	de Hoon et al. (2004)
GeneSpring	Agilent	http://www.genespring.com	
Hierarchical Clustering Explorer	Jinwook Seo	http://www.cs.umd.edu/hcil/hce/	Seo and Shneiderman (2002)
J-Express	Molmine	http://www.molmine.com/	
MeV	TIGR	http://www.tm4.org/	Saeed et al. (2003, 2006)
XCluster	Gavin Sherlock	http://genetics.stanford.edu/~sherlock/cluster.html	

threshold. This tab also includes an option to eliminate genes that are missing too much data, which may be due to poor measurements or previous filtering of individual spots in some other software. While researchers will usually filter their data, with the goal of filtering out uninteresting genes, with a goal of clustering data for interesting ones, some caution should be exercised in interpreting the results. After "clearing out" the expression space of uninteresting genes, well-separated clusters of genes remain, precisely because the space between those clusters was emptied by filtering rather than data having clearly defined clusters (see Bryan, 2004).

Cluster also provides a handful of options for transforming and regularizing data prior to analysis, although if these steps have been carried out previously, they can be skipped. These options are intended to facilitate comparison of measurements from microarray to microarray, as between-array, systematic biases can frequently overwhelm the biological component of the measurements. The first option is to log transform data. Log transformation of ratios causes symmetric changes to be reflected in symmetric numbers. For example, a twofold increase in expression gives a ratio of 2, whereas an equivalent decrease gives a ratio of 0.5, while no change gives a ratio of 1. In these "linear" ratios, the difference between no change and a twofold increase is $2 - 1 = 1$, whereas the difference between no change and a twofold decrease is $0.5 - 1 = -0.5$. If we instead log transform the ratios, typically using base 2, no change is $\log_2(1) = 0$; a two-fold increase is $\log_2(2) = 1$; and a two-fold decrease is $\log_2(1/2) = -1$. This symmetry is required for proper functioning of most of the distance metrics, and clustering of ratiometric data should always be done after data have been log transformed. Even in the case of single-channel microarray data, in which ratios are generally not involved, the measurements are generally log normal, meaning that upon log transformation they assume a more or less symmetric distribution similar to the "normal" or "Gaussian" distribution, which causes the most common "parametric" distance metrics to function better.

The other transformation options are intended to deal with common, between-microarray, systematic biases. The most common of these is "dye bias," in which one channel of a two-color microarray is more intense than the other for technical, rather than biological, reasons (Brownstein, 2006). In addition to skewing the absolute measurement of the expression of a gene on a single microarray, if this bias differs between arrays it can distort clustering analysis greatly. If it is valid to assume that the geometric mean ratio should be 1 (or an arithmetic mean of zero, in terms of log ratios), this problem can frequently be addressed by simply adjusting the average log ratio on each array to zero, or "centering"; the software supports adjusting either the mean or the median value in this way. A similar problem afflicts the comparison of genes within each array if one sample is a biologically meaningless reference: the actual ratio measured for each gene in a given array is then meaningless or even deceptive in that only the relative change from array to array (experimental condition to experimental condition) is relevant. Because most distance metrics are nonetheless sensitive to the actual ratios, a common approach is to center the gene measurements, setting the average value for each gene, across all conditions, to zero. This operation may also be performed considering either the mean or the median of the measurements of each gene.

The final transformation option is to "normalize" genes and/or arrays. The term "normalization" is used to mean several different things in the context of microarray analysis; in this case, it means that the magnitude of the vector of measurements is set to 1 via a multiplicative adjustment of the measurements. Most of the distance metrics work with normalized vectors, behind the scenes as it were, but this option allows the normalized values to be output to the final data files.

The Cluster program supports agglomerative hierarchical clustering of genes, arrays, or both. In this tab, the user may select from a list of distance metrics: Pearson correlation (centered or uncentered), absolute value of Pearson correlation (centered or uncentered), Spearman rank correlation, or Kendall's τ. Clustering may be performed using average, complete, or single linkage. A final option is to algorithmically assign weights to genes or arrays: weighting genes unequally will affect the clustering of arrays, and vice versa. Weights are assigned by the local density of vectors, as measured by the chosen metric, so that vectors that are very similar to others are downweighted. Weights may also be assigned by the user, in the input file, for example, if there are multiple replicates of the same gene, then they should be downweighted accordingly so that they do not unduly influence the clustering of arrays. The ordering of the final tree is random in that nodes may be "flipped" arbitrarily unless the user specifies a preferred order in the input file.

K-means and K-medians clustering are available in Cluster for both genes and arrays. The user may select the number of nodes and the number of iterations to compute. No choice of distance metric is available.

Cluster supports one-dimensional SOMs along both gene and array axes. The organization of nodes is always $1 \times N$, where N is chosen by the user, and the results are used to set the preferred order of genes and/or arrays. Thus, if data are subsequently hierarchically clustered, the nodes are flipped in such a way as to approximate the order determined by the SOM as nearly as possible. In this implementation, results of the self-organizing map are very similar to those of K-means or K-medians clustering, except that the interaction between neighboring nodes produces a smoother transition from one to another.

XCluster

XCluster (http://genetics.stanford.edu/~sherlock/cluster.html) implements some of the same algorithms as the Cluster application. It is written in C and can be compiled and executed on Windows, MacOS, Linux, and Unix. XCluster has no graphical user interface (GUI); it is intended to be executed "on the command line" by other programs as part of an integrated analysis process. The advantage of XCluster, beyond its suitability for such integration, is that it is optimized for performance and a small memory footprint. Within its available options, it is generally faster than the other applications described here. The input and output file formats are similar to Cluster, except that the user may not specify a preferred order of genes or arrays in the input file.

The various command-line options passed to XCluster determine which operations will be performed. Data may be log transformed, but no other transformations and no filters are offered. Genes and arrays may be clustered hierarchically by centroid linkage using either Pearson correlation (centered or uncentered) or Euclidean distance as the distance metric. K-means clustering is offered as in Cluster, for partitioning genes only.

The self-organizing map option in XCluster differs somewhat from that of Cluster or Cluster 3.0 (described later). Two-dimensional SOMs are supported, but only genes, and not arrays, may be organized in this way. Each node will contain a subset of genes, but all arrays. Hierarchical clustering is then carried out separately within each node so the organization determined by the SOM is preserved through clustering. Each node may be clustered by both genes and arrays.

Cluster 3.0

Cluster 3.0 (de Hoon *et al.*, 2004) is a reimplementation of the original Cluster program and is very similar to it in most ways. Some additional options are available, as outlined later. The most significant advance in Cluster 3.0 is that it provides a GUI that may be run on MacOS, Linux, and Unix in addition to Windows, with the GUI part of the application being implemented natively for each different platform. Input and output file formats are identical to those for Cluster. For graphical display of results, the Java TreeView application (Saldanha, 2004) functions on all platforms on which Cluster 3.0 runs.

Cluster 3.0 has a GUI very similar to that of Cluster. In addition, it may be used as a command-line program on Unix-based operating systems (including MacOSX), which are very similar to XCluster but with some expanded options as described later. The authors also make available Perl and Python programming language interfaces to the C software library that underlies the Cluster 3.0 software for the efficient and flexible integration of the functions of Cluster 3.0 into other software.

The filtering and transformation options in Cluster 3.0 are identical to those in Cluster, with the exception that the effect of filtering may be examined before it is applied to data. This allows the user to determine whether the filters are too lax or too strict before actually altering data.

The hierarchical clustering options in Cluster 3.0 are very similar to those in Cluster, with some additions. Centroid linkage is available, in addition to average, complete, and single linkage; as mentioned earlier, the single linkage option in Cluster 3.0 is more computationally efficient than the standard algorithm. Available distance metrics are Pearson correlation and absolute Pearson correlation, centered or uncentered; Euclidean distance and harmonically summed Euclidean distance (technically not a metric, as the distance from A to C is not necessarily less than or equal to the sum of the distance from A to B and B to C); city block or "Manhattan" distance; Spearman rank correlation; and Kendall's τ.

Cluster 3.0 supports two-dimensional SOMs of both genes and arrays. Unlike the implementation in Cluster, any of the available distance metrics may be used for SOMs. This choice of metrics is also available for K-means/ K-medians clustering, which is otherwise similar to the implementation in Cluster.

Other Analysis Packages

As mentioned earlier, there is a wide variety of microarray analysis packages available, many of which implement some forms of clustering. Of particular note, as free, open-source exemplars of two different approaches to analysis software are the TIGR Multiexperiment Viewer (MeV) and the R statistical programming language.

MeV may be downloaded, as part of the TIGR TM$_4$ microarray analysis suite (Saeed *et al.*, 2003, 2006), from http://www.tm4.org/index.html. MeV is a very full-featured analysis package, providing an integrated, graphical overview of the analyses carried out and a plethora of algorithms, mostly for unsupervised analyses but with a good selection of supervised methods. For the various unsupervised analysis options, MeV provides a variety of distance metrics: Euclidean distance; Manhattan or city block distance; Pearson correlation (centered or uncentered); squared Pearson correlation (uncentered); cosine correlation; covariance; average dot product; Spearman rank correlation; and Kendall's τ. Hierarchical clustering may be carried out by single linkage, complete linkage, or weighted average linkage (weights determined by relative depth in the tree). A wide variety of other clustering methods, including SOMs and K-means/K-medians, is also offered.

The R programming language (http://www.r-project.org/) is a full-fledged programming language designed in part for, and accompanied by many functions for, statistical analysis. As such, it is a powerful tool for those with some comfort with computer programming (and a number of generally available analysis tools have been written in R as well). The BioConductor project (Gentleman *et al.*, 2004; Reimers and Carey, 2006) (http://www.bioconductor.org/) provides a large number of add-on packages for R specifically intended for microarray research. Of particular pertinence here, various functions in R support hierarchical clustering and other unsupervised analysis techniques, as well as supervised analyses. While even a brief description of R and BioConductor is beyond the scope of this chapter, they deserve mention as a powerful tool for the computational biologist.

Conclusions

There is much more to microarray data analysis than just clustering. Many experimental designs are better served by a supervised analysis or by a combination of supervised and unsupervised approaches. Nevertheless, the various forms of clustering are powerful tools for unsupervised, discovery-oriented microarray analysis. Considered, thoughtful application of these methods can lead to novel insights and important advances and has done so in many cases referenced earlier.

References

Ben-Dor, A., Shamir, R., and Yakhini, Z. (1999). Clustering gene expression patterns. *J. Comput. Biol.* **6,** 281–297.

Bryan, J. (2004). Problems in gene clustering based on gene expression data. *J. Multivar. Anal.* **90,** 44–66.

de Hoon, M. J., Imoto, S., Nolan, J., and Miyano, S. (2004). Open source clustering software. *Bioinformatics* **20,** 1453–1454.

Eisen, M. B., Spellman, P. T., Brown, P. O., and Botstein, D. (1998). Cluster analysis and display of genome-wide expression patterns. *Proc. Natl. Acad. Sci. USA* **95,** 14863–14868.

Everitt, B. (1974). "Cluster Analysis 122." Heinemann, London.

Gentleman, R. C., Carey, V. J., Bates, D. M., Bolstad, B., Dettling, M., Dudoit, S., Ellis, B., Gautier, L., Ge, Y., Gentry, J., Hornik, K., Hothorn, T., Huber, W., Iacus, S., Irizarry, R., Leisch, F., Li, C., Maechler, M., Rossini, A. J., Sawitzki, G., Smith, C., Smyth, G., Tierney, L., Yang, J. Y., and Zhang, J. (2004). Bioconductor: Open software development for computational biology and bioinformatics. *Genome Biol.* **5,** R80.

Gibbons, F. D., and Roth, F. P. (2002). Judging the quality of gene expression-based clustering methods using gene annotation. *Genome Res.* **12,** 1574–1581.

Kohonen, T. (1995). "Self Organizing Maps." Springer, Berlin.

Liang, Y., Diehn, M., Watson, N., Bollen, A. W., Aldape, K. D., Nicholas, M. K., Lamborn, K. R., Berger, M. S., Botstein, D., Brown, P. O., and Israel, M. A. (2005). Gene expression profiling reveals molecularly and clinically distinct subtypes of glioblastoma multiforme. *Proc. Natl. Acad. Sci. USA* **102,** 5814–5819.

Milligan, G., and Cooper, M. (1985). An examination of procedures for determining the number of clusters in a dataset. *Psychometrika* **50,** 159–179.

Reimers, M., and Carey, V. J. (2006). Bioconductor: An open source framework for bioinformatics and computational biology. *Methods Enzymol.* **411,** 119–134.

Saeed, A. I., Bhagabati, N. K., Braisted, J. C., Liang, W., Sharov, V., Howe, E. A., Li, J., Thiagarajan, M., White, J. A., and Quackenbush, J. (2006). TM4 microarray software suite. *Methods Enzymol.* **411,** 134–193.

Saeed, A. I., Sharov, V., White, J., Li, J., Liang, W., Bhagabati, N., Braisted, J., Klapa, M., Currier, T., Thiagarajan, M., Sturn, A., Snuffin, M., Rezantsev, A., Popov, D., Ryltsov, A., Kostukovich, E., Borisovsky, I., Liu, Z., Vinsavich, A., Trush, V., and Quackenbush, J. (2003). TM4: A free, open-source system for microarray data management and analysis. *Biotechniques* **34,** 374–378.

Saldanha, A. J. (2004). Java Treeview: Extensible visualization of microarray data. *Bioinformatics* **20,** 3246–3248.

Schaner, M. E., Ross, D. T., Ciaravino, G., Sorlie, T., Troyanskaya, O., Diehn, M., Wang, Y. C., Duran, G. E., Sikic, T. L., Caldeira, S., Skomedal, H., Tu, I. P., Hernandez-Boussard, T., Johnson, S. W., O'Dwyer, P. J., Fero, M. J., Kristensen, G. B., Borresen-Dale, A. L., Hastie, T., Tibshirani, R., van de Rijn, M., Teng, N. N., Longacre, T. A., Botstein, D., Brown, P. O., and Sikic, B. I. (2003). Gene expression patterns in ovarian carcinomas. *Mol. Biol. Cell* **14,** 4376–4386.

Seo, J., and Shneiderman, B. (2002). Interactively exploring hierarchical clustering results. *IEEE Comput.* **35,** 80–86.

Sibson, R. (1973). SLINK: An optimally efficient algorithm for the single-link cluster method. *Comput. J.* **16,** 30–34.

Sorlie, T., Perou, C. M., Tibshirani, R., Aas, T., Geisler, S., Johnsen, H., Hastie, T., Eisen, M. B., van de Rijn, M., Jeffrey, S. S., Thorsen, T., Quist, H., Matese, J. C., Brown, P. O., Botstein, D., Eystein Lonning, P., and Borresen-Dale, A. L. (2001). Gene expression patterns of breast carcinomas distinguish tumor subclasses with clinical implications. *Proc. Natl. Acad. Sci. USA* **98,** 10869–10874.

Tamayo, P., Slonim, D., Mesirov, J., Zhu, Q., Kitareewan, S., Dmitrovsky, E., Lander, E. S., and Golub, T. R. (1999). Interpreting patterns of gene expression with self-organizing maps: Methods and application to hematopoietic differentiation. *Proc. Natl. Acad. Sci. USA* **96,** 2907–2912.

Tavazoie, S., Hughes, J. D., Campbell, M. J., Cho, R. J., and Church, G. M. (1999). Systematic determination of genetic network architecture. *Nature Genet.* **22,** 281–285.

Tibshirani, R., Walther, G., and Hastie, T. (2000). Estimating the number of clusters in a dataset via the Gap statistic. *J. R. Stat. Sac. Ser. B.* **63,** 411–423.

Tusher, V. G., Tibshirani, R., and Chu, G. (2001). Significance analysis of microarrays applied to the ionizing radiation response. *Proc. Natl. Acad. Sci. USA* **98,** 5116–5121.

Wen, X., Fuhrman, S., Michaels, G. S., Carr, D. B., Smith, S., Barker, J. L., and Somogyi, R. (1998). Large-scale temporal gene expression mapping of central nervous system development. *Proc. Natl. Acad. Sci. USA* **95,** 334–339.

Yeung, K. Y., Haynor, D. R., and Ruzzo, W. L. (2001). Validating clustering for gene expression data. *Bioinformatics* **17,** 309–318.

[11] Analysis of Variance of Microarray Data

By JULIEN F. AYROLES and GREG GIBSON

Abstract

Analysis of variance (ANOVA) is an approach used to identify differentially expressed genes in complex experimental designs. It is based on testing for the significance of the magnitude of effect of two or more treatments taking into account the variance within and between treatment classes. ANOVA is a highly flexible analytical approach that allows investigators to simultaneously assess the contributions of multiple factors to gene expression variation, including technical (dye, batch) effects and biological (sex, genotype, drug, time) ones, as well as interactions between factors. This chapter provides an overview of the theory of linear mixture modeling and the sequence of steps involved in fitting gene-specific models and discusses essential features of experimental design. Commercial and open-source software for performing ANOVA is widely available.

Introduction

Since the mid-1990s, application of classical statistical methods has dramatically improved the analysis of differential gene expression across biological conditions. In the early days of microarray analysis, researchers typically adopted fold change criteria to identify genes of interest, using a convenient arbitrary cutoff value (usually a twofold difference) as a threshold. Recognizing that there is no statistical basis for such cutoffs, this approach was soon replaced by t tests for comparison of means of two samples in relation to the observed technical or biological variation. Tusher *et al.* (2001) introduced a popular method called significance analysis of microarrays that employs a modified t test associated with a permutation test to assess significance. For comparisons involving just two classes of treatment, this remains appropriate, but many experiments employ multiple levels of each factor (e.g., three drugs, four time points) and multiple factors (e.g., sex, drug, and genotype). In these cases involving complex designs, more power and flexibility are required (Cui and Churchill, 2003), and it is often desirable to contrast the contributions of each source of variation directly. Analysis of variance (ANOVA) is a suitable framework widely employed in all aspects of quantitative biology and has been quickly adapted to microarray analysis. Kerr *et al.* (2000) were the first to suggest using ANOVA to identify genes differentially expressed in the context of

METHODS IN ENZYMOLOGY, VOL. 411   0076-6879/06 $35.00

two-dye arrays, but the approach now employed most commonly uses the gene-specific modeling first proposed by Wolfinger *et al.* (2001) and, for high-density oligonucleotide arrays, Chu *et al.* (2002). This chapter presents guidelines for using ANOVAs to analyze microarray data, discussing issues associated with formulating a mixed model, the levels of replication necessary to use ANOVAs, and adjustments for multiple testing. A flow diagram in Fig. 1 illustrates the major steps involved in design and execution of a gene expression profiling experiment involving multiple classes of treatment.

Linear Modeling and Analysis of Variance

Analysis of variance refers to a particular class of statistical model that is used to estimate the magnitude of parameters that account for the effects of multiple independent variables and evaluate their significance based on partitioning of sources of variation. In its simplest form, a general linear model (GLM) formulates a linear relationship between a single dependent variable, Y, and a single independent variable, X. That is, Y is assumed to be a linear function of X with slope β and intercept μ, while the unexplained "residual variance" is represented by ε in the equation $Y = \mu + \beta X + \varepsilon$. This approach can be extended to account for multiple independent variables:

$$Y = \mu + \beta_1 x_1 + \beta_2 x_2 \ldots \ldots \beta_L x_L + \varepsilon.$$

ANOVAs are designed to tease apart the different sources of variation that may contribute to the total experimental variance, including technical and biological factors, in such a way that all of the factors are assessed jointly rather than in a pair-wise manner. Statistical tests of a null hypothesis are formulated using f ratios of the variance due to the treatment of interest to an appropriate measure of the residual variance. The null hypothesis generally stipulates that there is no difference between the means of the populations from which samples are drawn (Ott and Longnecker, 2001). Sokal and Rohlf (1995) provide a comprehensive introduction to the theory, which quickly gets more sophisticated than most nonexperts are comfortable with. Consequently, while it is now possible to implement standard ANOVA algorithms, we recommend close collaboration between biologists and statisticians at all phases of the analysis, starting with experimental design.

Fixed versus Random Effects

There are two general types of effects in a linear model corresponding to fixed and random factors. Operationally, an effect is considered fixed if replication of the experiment would result in resampling of the same population because each of the levels of the factor are represented in the experiment. Examples include Cy3 and Cy5 dyes, the male and female sexes, or high and low lines deliberately chosen to represent the extremes of a trait. In contrast, random factors are considered when the samples are drawn in an unbiased manner from a large population. For example, "array" effects are

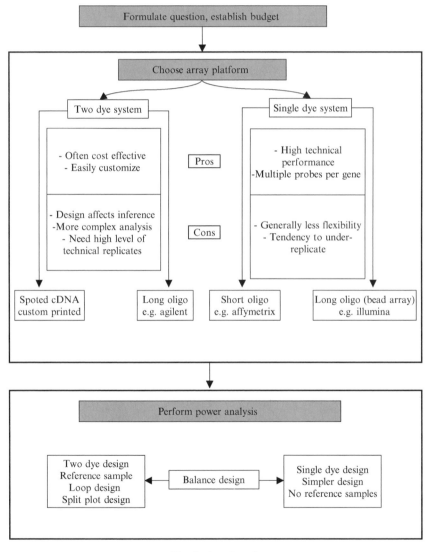

FIG. 1. *(continued)*

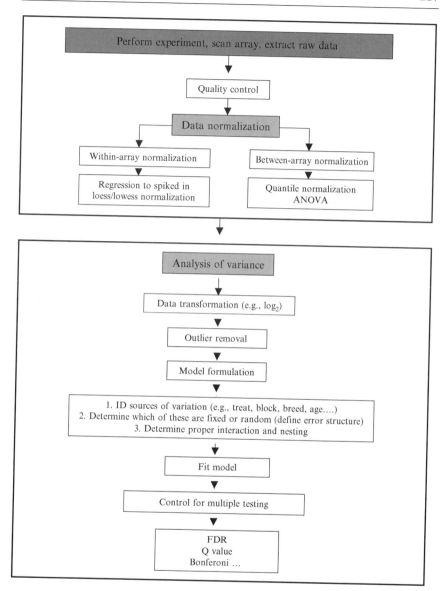

FIG. 1. Flow diagram of steps in design and analysis using ANOVA.

random, as each array used in a given experiment is a unique sample from all of the arrays that could have been used: there is no expected correlation between the measures due to sampling. Some biological factors may be random, for example, individual humans or subpopulations picked by chance to represent a species. If the experiment was to be repeated using a different set of arrays, people or subpopulations, the variance among them should be the same and the parameter estimates should yield the similar results.

A fixed effect model such as one would encounter in Microsoft Excel or by default with PROC GLM in SAS is a model with only fixed terms, except for the residual error, ε, which is always assumed to be distributed randomly with a mean of zero and standard deviation σ. The p values may be inflated artificially (Littell et al., 1996) for several reasons, especially if there is hidden correlation between the errors associated with the measurement of two or more factors as might occur if the experimental design is unbalanced. For example, if two-thirds of the males are labeled with Cy3 and two-thirds of the females with Cy5, these two factors (sex and dye) may be confounded. Wherever a factor is assumed to be random, we recommend the use of mixed model ANOVAs that contain both random and fixed terms. Biological replicates and arrays should be estimated as random effects so that their associated error variance provides the correct error term for the denominator of the f ratio that is tested (Cui and Churchill, 2003). The difference between fixed and random effects is not trivial and serious consideration needs to be paid when designating each effect in the model (for a recent review of fixed and random effects in microarray experiments, see Rosa et al., 2005).

Types of Microarrays

There are several types of microarray technology available, and slightly different analytical approaches are required to deal with the properties of these. The most fundamental distinction is between two-color arrays and single-channel arrays. Two color arrays include spotted cDNA arrays on glass slides or nylon filters (Uhde-Stone et al., 2003), as well as some commercial long oligonucleotide arrays (e.g., Agilent arrays) (Wolber et al., 2006). Microarrays on nylon filters can be probed repeatedly with radioactively labeled samples, in which case the "dye" effect can have multiple levels instead of just two. Short oligonucleotide arrays are typically single channel, notably those produced by Affymetrix (Dalma-Weiszhausz et al., 2006) and Nimblegen (Scacheri et al., 2006), but include multiple probes per gene that can be averaged or treated as individual measurements. Averaging of probe level data can be used to decrease data complexity, assuming high levels of technical replicates; however, probe level variation carries important information useful for outlier removal and subsequent ANOVA. Some long oligonucleotide platforms are only employed with a single dye, notably Illumina bead arrays (Fan et al., 2006), in which transcript abundance is represented as the average (after outlier removal) of the 20 to 30 identical beads per sample.

Two main approaches have been employed with two-color arrays. One is to reserve one dye for a reference sample while the other dye is used to label the treatments. In this case, the ratio of Cy3/Cy5 provides the raw measurement of expression. This approach is intuitive and straightforward and is particularly suitable in cases where there is a large number of treatments of the same factor with low replication, such as a time series or exposure to a compendium of chemicals. Ideally, the reference sample should be chosen to be an approximate average of all of the treatments for each gene. This may be achieved by making a pool of each of the treatments, but doing so compromises the comparability of different experiments. The alternative is to use a common control sample, such as "mouse liver," but this will bias the analysis for genes that have abnormally low or high expression in the reference. Reference sample designs are also wasteful of resources in the sense that there is no biological information of interest in the control. Nevertheless, they remain popular and there is no reason why ANOVA approaches cannot be adopted to analyze ratio data.

The second two-color approach is to use both dyes for samples of interest. This type of design requires considerable forethought in the experimental layout to avoid confounding of factors by unbalanced designs. Because dye effects are prevalent in two-color experiments (Kerr et al., 2000), it is essential to ensure that each sample is represented by technical replicates of both dyes, as far as possible in equal proportions. Loop designs are popular in which sample n is competitively hybridized to sample $n-1$ with one dye, and to $n+1$ with the other dye. With multiple factors in the analysis, it can be difficult to ensure that the factors remain distributed randomly with respect to one another. For example, you do not want all of the liver samples to be labeled with Cy3 and the heart samples with Cy5, as then you cannot tease apart whether the effect is due to the dye or to the tissue. It is not essential that two factors be directly contrasted on an array in order to draw an inference about the effect, but failure to do so will generally reduce the statistical power of the comparison.

An alternative to the loop design is a split-plot design in which factors are deliberately separated (Jin et al., 2001). This has been common practice in agricultural experiments for which mixed model ANOVA has long been used. For example, in an experiment evaluating the joint effects of sex and drug on gene expression in some tissue, if all of the arrays contrast individuals with and without the drug, but none of them contrast males against females (all of the arrays are samples from one sex or the other), the design is a split plot. In this case, there is more statistical power for the factor that is contrasted directly, namely drug, but the sex effect can still be estimated.

If the aim of an experiment is to contrast the contributions of each factor, it is best to employ a loop design as far as possible; if it is to maximize the analysis of one effect across samples that include another effect of lesser interest, then a split-plot is warranted. Many other types of design, including random assignment of treatments to arrays, can be handled with ANOVA, but it must be recognized that the layout of a two-color experiment affects the statistical power and can introduce biases that may go undetected.

Whichever design involving two or more channels (dyes) is employed, it is essential to fit an array term in the model as a random effect. Failure to do so can overestimate effects dramatically because it fails to account for the fact that the two measurements for each spot are correlated. Correlation between measurements from the same spot occurs because spot size and shape, and DNA concentration in each spot, varies from array to array (Minor, 2006). Fitting array as a random effect essentially takes care of the "spot effect" problem that was initially recognized by those who employed ratios and reference sample designs. Note that it is easy to fit ANOVA models without fitting the spot effect, which may lead to the detection of many more significant genes, but it is not valid.

In the case of single-channel arrays, experimental design is much simplified. There is no need to worry about spot and dye effects or confounding of factors because each sample is hybridized onto a different array and is measured independently. Care should be taken to balance the number of replicates of each factor to prevent biases due to overrepresentation of one or more classes. It is also important to avoid hybridizing different types of samples at different times, as batch effects that may be caused by enzyme lots, ozone levels, or other uncontrolled parameters may cause observed differences in fluorescence intensity. The analysis of short oligonucleotide platforms where each gene is represented by multiple probes, each of which has a perfect and a mismatch probe, presents some unique challenges. Affymetrix MAS 5.0 software provides an "average difference" measure that can be taken directly into an ANOVA, but it is also possible, and sometimes more powerful, to perform the analysis at the probe level. Because it turns out that perfect and mismatch hybridization intensity is usually highly correlated (Chu *et al.*, 2002), an assumption of the ANOVA is violated, and it is generally better simply to work with perfect match data only.

Biological versus Technical Replication

Just as the experimental design guides the formulation of the proper linear model (for review, see Rosa *et al.*, 2005), interpretation is also a function of the number of replicates of each level of each factor. Analysis of variance relies on the assumption that the sample size used for each population provides an accurate representation of the variation in the population. Reduction of experimental error requires that a sufficient level of biological replication is employed (Lee *et al.*, 2000), and failure to replicate sufficiently reduces the statistical power to detect differential expression.

An important distinction must be made between technical and biological replicates. Technical replication refers to replicate hybridizations of the same RNA samples (whether from the same or different extractions) that originate from a common biological source. Technical replicates are not truly independent from one another and are designed to validate the accuracy of the transcript-level measurements. They do not provide information about variation in the population. Similarly, replicate probes within an array are designed to limit the need for technical replicates by increasing the confidence of abundance measures for a given target gene. However, each probe cannot be regarded as an independent measurement, and failure to account for the correlation between duplicate probes on an array also inflates the estimate of biological differences. With high-quality commercial arrays, technical error is usually much less than biological variance so there is little point in replicating any sample more than once.

In contrast, biological replication refers to the hybridization of RNA samples originating from independent biological sources under the same conditions, such as samples extracted from different individuals that received the same treatment dosage or two vials or pots of the same fly or plant genotype. These replicates are designed to provide information about variation among individuals. Financial limits will usually affect experimental design, making it infeasible to employ multiple biological replicates, but even if these are not directly of interest, it is a good idea to include biological replication in the study. Simply pooling the replicates to ensure that the individual variation is sampled, even if it is not measured, can do this. It is also good practice wherever possible to duplicate and pool RNA extractions and labeling reactions so as to minimize error due to sampling in each of these steps.

A common question is as follows: how many replicates are required? Unfortunately, it is not possible to precisely answer this question in advance of an experiment because variance components due to different factors are unknown. Computation of the number of samples required to reject a defined percentage of null hypotheses at a particular significance level (i.e., the statistical power) can be performed with a number of online calculators (Table I). These require estimation of the mean difference

TABLE I
ONLINE TOOLS FOR POWER ANALYSIS

From	Web site	By
UCLA Department of Statistics	http://calculators.stat.ucla.edu/powercalc/	Barry Brown et al.
U of Iowa Department of Statistics	http://www.stat.uiowa.edu/~rlenth/Power/	Russ Lenth
York U. Department of Math	http://www.math.yorku.ca/SCS/ Online/power/	Michael Friendly

between samples, and the variance of each sample. Because these values vary greatly among genes on the array, estimates of required sample sizes are by nature approximate only. Furthermore, the experimental design strongly affects power, and as the number of levels of a factor increases, the number of biological replicates required will tend to drop. Power analyses reported by Wolfinger et al. (2001) and Tempelman (2005) generally recommend use of at least four technical replicates per biological replicate for spotted microarrays in order to detect 80% of the differentially expressed genes at experiment-wide thresholds. The higher quality of commercial oligonucleotide arrays allows for just two to three technical replicates, but power is still affected greatly by the mean and variance of the biological samples.

Data Extraction and Normalization

Several programs are currently available to extract spot intensity for two-dye arrays after the slides have been scanned (e.g., Table II). For most applications, raw intensity data from each dye are first converted to a logarithmic scale, usually base 2. This is essentially equivalent, after appropriate adjustment for array effects, to taking the logarithm of the ratio of two dyes, as mathematically the logarithm of a ratio is equal to the difference between the logarithms. Log transformation has two advantages: it makes data more normally distributed and more symmetrical. Although ANOVA assumes that data are normally distributed, it is generally quite robust to departures from normality, but it is always best to work with approximately normal distributions. Because the distribution of raw intensity measurements is always highly skewed with most transcripts showing low expression and maybe 10% of transcripts with high expression, log transformation reduces this bias. It also makes increases and decreases in expression symmetrical: ratios of 2:1 and 0.5:1 become +1 and −1 on the log

TABLE II
LIST OF SOME POPULAR SOFTWARE AVAILABLE TO SCAN SPOTTED ARRAY[a]

Program name	Company/group	Reference	Method
Gene PIX Pro	Axon Instruments Inc.	www.molecular devices.com	Fixed and adaptive circle
ScanAlyze	Lawrence Berkeley Nat Lab	Eisen (1999)	Fixed circle
QuantArray	GSI Lumonics	GSI Lumonics (1999)	Histogram segmentation
Spotfinder	Tigr	Saeed et al. (2006)	Histogram segmentation
Spot	R/Bioconductor	Beare and Buckley (2004)	Adaptive shape
SpotSegmentation	R/Bioconductor	Li et al. (2005)	Adaptive shape

[a] A more complete list can be found at http://www.statsci.org/micrarra/image.html.

base 2 scale. Base 2 is chosen because increments of one correspond to twofold changes, a convenient scale for gene expression comparisons, as most differential expression is of this order of magnitude.

Some data trimming may be in order. Low-intensity spots should be removed from the data set as they may result in unreliable measurements (Wernisch et al., 2003). Nonexpressed genes should also be removed; they can be detected using the presence/absence call provided by MAS 5.0 for oligonucleotide array or by using a nonparametric Wilcoxon sign rank test. Removing these genes is advantageous regarding adjustment for multiple comparisons, but can lead to discarding of differential expression at the low end of transcript abundance. After data quality has been assessed (i.e., checking for artifacts and quality of internal controls on the array), data need to be normalized to remove global effects of arrays and/or channels.

Normalization is a fundamental step in data analysis and should be considered thoroughly. Several methods have been developed and are commonly reviewed in the primary literature. The purpose of normalization is to remove any systematic biases that do not reflect true biological variation within a slide or between slides.

Such biases can be due to the unequal incorporation of dye between samples, variation in the amount of DNA printed on the array, the washing process, or to variation in the ability of the scanner to detect each dye. Several approaches can be used to achieve within-array normalization. Regression to spiked-in controls uses samples of mRNA introduced at known concentrations. Loess normalization is designed to reduce dye and pin effects using a series of local regressions to remove any overall correlation between intensity and ratio, as differential expression should not be a function of transcript abundance (Quackenbush, 2002). All normalization procedures inevitably change data and can introduce artifacts, whereas overfitting can remove true biological signals. There is no one correct approach to normalization and it may be appropriate to compare the results of different approaches.

Between-array normalization adjusts for differences in the intensity level of each slide averaged across all spots. Figure 2 shows how this can affect an analysis. Array intensities vary due to biological and technical factors (especially laser settings during scanning) so it is always important to ensure that the mean transcript abundance for each array and channel is approximately the same. With ratio data, ensuring that the average ratio is unity does this; for log-transformed intensity measurements the simplest adjustment is to subtract the sample mean of the appropriate channel from each measure so as to achieve a channel mean of zero. Considering a gene at the 10th percentile of each expression profile, large differences in the raw data (arrows in Fig. 2A) are largely removed by such a centering process (Fig. 2B). However, because the distributions of variance still have unequal variance, genes at the same percentile may have different relative intensity measurements (compare blue hues and red hues). Several techniques have been developed to ensure that arrays have equal variance and common means (Quackenbush, 2002). The simplest is simply to divide through by the standard deviation of the channel, which reduces the variance to unity, a true normalization procedure. If the distributions are skewed, this may still leave genes at the same percentile with different intensity values so further transformations may be employed to remove such biases. Quantile normalization (Irizarry, 2003) performs a nonlinear transformation that gives each array an equal median, mean, and variance by averaging the intensity of each quantile across arrays (Fig. 2C). Note that there may be good biological reasons for skews in the distributions, with the consequence that such global normalization could artificially remove true biological differences under some circumstances.

Alternatively, variation introduced by arrays and dyes can be removed by modeling these effects in the first of two ANOVA steps (Wolfinger, 2001). In this case, the linear model

$$\text{Log}_2 Y = \mu + \text{Array} + \text{Dye} + \text{Array}^*\text{Dye} + \text{Residual}$$

is fit across all genes on all arrays. The *Residual* is an estimate of the relative fluorescence intensity after accounting for overall array and dye effects, and this value is the input measurement of gene expression used in the gene-specific models described later. It basically expresses the magnitude of fluorescence intensity relative to the sample mean, and the subsequent ANOVA evaluates whether expression relative to the mean

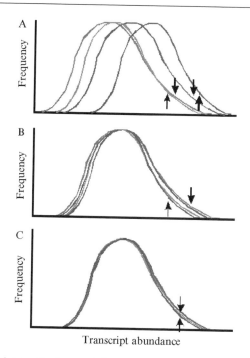

Fig. 2. Effect of normalization on inference of differential expression. (A) Frequency distributions of two arrays with two channels (dyes) each have different means and variance so that a gene at the 10th percentile (arrows) has a different apparent level of transcript abundance on each array. (B) Centralization by subtracting the mean of each channel reduces these effects, but remaining differences in variance still result in apparent differential expression between the red and yellow samples and the blue and green samples. (C) Further normalization to equalize the variance and remove skew may result in similar relative fluorescence intensity for equally ranked genes on each array. (See color insert.)

increases or decreases between factors. This approach gives very similar results to centering by subtraction of the channel mean from each value. More complex models may include the factors of interest at this step, allowing inference of whether there are global effects of each factor on expression.

Another approach is to perform the normalization simultaneously with the assessment of individual gene effects, as proposed by Kerr *et al.* (2000). These authors fit a global model to two-color arrays, of the following form:

$$Y_{ijkg} = \mu + A_i + D_j + (AD)_{ij} + G_g + (AG)_{ig} + (DG)_{jg} + V_k + (VG)_{kg} + \varepsilon_{ijkg}$$

Transcript abundance, Y_{ijk}, on the $\log_2$ scale is expressed as a function of the ith microarray, jth dye, kth variety (or treatment class), and gth gene.

The interaction term of interest is $(VG)_{kg}$, which expresses the effect that condition k has on the expression level of gene g. Arrays and genes, and hence interactions involving these terms, should be regarded as random effects. More complex models involving multiple sources of biological variation can also be employed. Significance of each gene is evaluated by a permutation testing procedure, which effectively assumes that the residual error is the same for each gene. As data sets get larger, though, it is possible to estimate the error variance separately for each gene, which may be biologically appropriate, as there is likely to be considerable variation in the tightness of the regulation of gene expression. This has given rise to the adoption of gene specific-ANOVA models, a few examples of which are described in the next section.

Gene-Specific ANOVA

The generic form of gene-specific ANOVA deals with two-color arrays involving a single class of treatment factor with two or more levels. The treatment may be fixed, such as sex or a comparison of different drug regimens, or random, such as three genotypes picked by chance from a population. A linear model is fit separately for each of the genes on the array:

$$RFI_{ijk} = \mu + A_i + D_j + AD_{ij} + T_k + \varepsilon_{ijk}$$

RFI stands for relative fluorescence intensity, which is the estimated expression level of a transcript following whatever normalization procedure was used to remove global array and dye effects, as discussed earlier. The overall mean is μ, and the residual unexplained error, ε_{ijk}, is different for each gene. The first three terms control for correlations between the two measurements of each spot (A_i), dye effects (D_j), and spot-by-dye interactions $(AD)_{ij}$. T_k is the term of interest, the treatment effect. Most software estimates the least-square mean transcript abundance for each level of the treatment and provides a test statistic by which the significance of differences between these means can be assessed. This does not tell which one of three or more treatments is different from the others, but further test statements can be employed to evaluate the significance of each pair-wise comparison or even of subsets of treatments relative to others.

More complex models can be fit to data by adding terms and interactions representing different treatments. This is particularly useful for simultaneously estimating the relative contributions of two treatments (e.g., sex and age) to expression and for determining whether the effect of one treatment affects the other (do males respond differently than females to aging?). It should be recognized, though, that it is possible to overparameterize models if the number of replicates of each treatment level is small, as there will be insufficient degrees of freedom to evaluate main and/or interaction effects. Furthermore, because little is known about the false-positive rates associated with multiple f tests carried out for thousands of genes, caution must be exercised in interpretation.

Nested designs are used in cases where all the levels of one factor are not represented in the levels of another factor. For example, the five people receiving two different drugs may be different people, so individuals should be nested within drug treatment. More generally, in a randomized block design, if a treatment is applied to different blocks, the treatment effect should be nested within blocks. This type of model effectively stratifies the variance to account for biases that arise because the variance in the block may be confounded with the variance associated with treatment. For very large experiments where different treatments are hybridized at different times, it may be appropriate to nest treatment within the block of hybridizations to account for potentially correlated errors among variance components in the model. A typical model nesting the kth treatment within the lth block is of the form:

$$RFI_{ijk} = \mu + A_i + D_j + AD_{ij} + B_l + T_k(B_l) + \varepsilon_{ijk}$$

ANOVA models can also be applied to the analysis of short oligonucleotide arrays at the probe level (Chu et al, 2002; Dalma-Weiszhausz et al., 2006). In this case, there is no dye effect, but each probe belonging to a probe set is fit as a random factor. Because mismatched data tend to increase the noise and have been shown to be poor indicators of cross hybridization (Chu et al., 2002), use of perfect match probes-only data is recommended (these can be extracted from the .CEL files, which contain the raw intensity reading for all probes, outputted by MAS 5). In the model

$$y_{ijk} = \mu + A_i + P_j + T_k + (PT)_{jk} + \varepsilon_{ijk}$$

y_{ijk} is some measure of probe intensity, such as a $\log_2$ mean-centered intensity value for the jth probe on the ith array. T_k is again the main effect of the treatment, and the probe by treatment effects $(PT)_{jk}$ can be used to determine if there are probes within the probe set for the particular gene that are performing differently according to the treatment. For example, in comparison of expression between two species, polymorphism may cause one species to hybridize less intensely to a particular oligonucleotide (Gilad et al., 2005), which would appear as a probe-by-species interaction (Hsieh et al., 2003). The array should be fit as a random effect, but considering the low level of variance between technical replicates, it may not be necessary to fit array in the model at all. Note that probe effects are often highly significant, which is not surprising considering the variability in annealing temperature and specificity of individual probes.

Significance Thresholds

A convenient tool for visualizing the magnitude and significance of effects is a volcano plot, as shown in Fig. 3 (Wolfinger et al., 2001). Significance on the y axis is plotted as the negative logarithm of the p value, against the magnitude of expression difference on the x axis. Factors that have no effect on gene expression result in plots with small fold-change and significance values (Fig. 3A). If a common error model is employed, all genes with the same expression difference will have the same significance so the significance threshold corresponds to a fold-change cutoff that is determined statistically. With gene-specific error models, significance is only correlated with fold change, as it is a function of the variance in the abundance of each transcript, so a horizontal line defines the significance threshold (Fig. 3B). Some authors have pointed out that error variances are themselves an estimate (e.g., Cui et al., 2005) and will sometimes be unrealistically small, which will tend to reduce p values. Consequently, methods have been developed that attempt to balance gene-specific and common error models, giving rise to the suggestion that significant genes should be chosen from a sector indicated in gray in Fig. 3C.

The raw p value generated by an ANOVA needs to be adjusted for multiple comparison testing because thousands of tests are performed simultaneously. In classical hypothesis testing, the investigator sets an α value defining their willingness to commit a type I error, namely to reject the null hypothesis when it is actually true. Thus, if α is set at 0.05, the probability of detecting a false positive (committing a type I error) is 5%, so for an array of 10,000 genes, 500 are expected to be significant by chance, whether or not they are significant. If only 520 genes are called significant, the majority of them are likely to be false positives, so more stringent approaches are needed.

The most stringent correction is to multiply the observed p value by the number of comparisons to obtain a Bonferroni-corrected p value. This ensures that no genes in the selected sample are false positives at the specified α level. If the most significant gene has a p value of 10^{-5} and there are 10,000 genes on the array, the corrected p value is still 0.1, which is greater than 0.05: it is likely that there is a gene with this significance level by chance in such a large set of comparisons. If the purpose of the experiment is to identify one or two genes with very high confidence, Bonferroni adjustment is appropriate, but for most purposes it is too stringent.

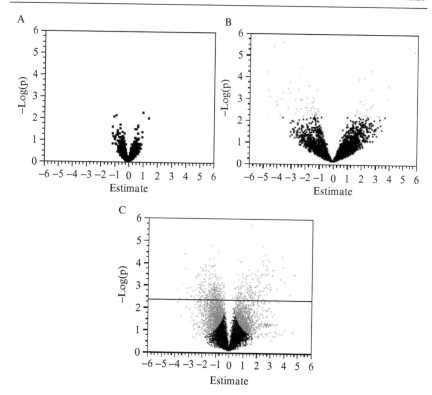

FIG. 3. Volcano plots of significance against magnitude of effect. Significance is represented as the negative logarithm of the p value on the y axis, and magnitude of differential expression on the log base 2 scale on the x axis. Upregulation is to the right, downregulation to the left. (A) A factor that has no effect results in a characteristic plot shown here. (B) A significant factor results in many genes with small p values (toward the top) that tend to have large differential expression. A horizontal cutoff, chosen here at $p = 0.01$ (i.e., negative log $p = 2$), highlights genes (indicated in gray) that are chosen to be significant in the analysis. (C) A more sophisticated selection criterion implemented in R-MANOVA (Wu *et al.*, 2002) reaches a compromise between significance and fold change.

Rather, there is a tradeoff between committing type I and type II errors, the latter being false negatives or cases where the null hypothesis is accepted even though it is false. Several adjustments have been developed to control for multiple testing. Not making any adjustment ensures a high false positive and low false negative rate, whereas Bonferroni adjustment has the opposite effect. A now common compromise is to choose genes based on the false discovery rate (FDR).

Benjamini and Hochberg (1995) developed an intuitive implementation of FDR, which is to adjust the p value such that a specified proportion of the chosen genes is expected to be false positives. In the example given earlier, where 500 genes are expected to be false positives, if 750 genes are called significant at $p < \alpha = 0.05$, then the FDR is $500/750 = 67\%$. In contrast, if 50 genes are significant at $p < = 0.0001$ and because only 1 gene in a sample of 10,000 is expected to be that significant by chance, then the FDR is just 5%. Selection of all 50 genes would be warranted, as all but a handful are likely to be true positives.

A more sophisticated, and in practice slightly more liberal, FDR method from Storey et al. (2003) employs q value cutoffs and is commonly implemented with the $qvalue$ program in $R/Bioconductor$ (Reimers and Carey, 2006). Rather than assuming that all genes are true negatives to begin with, an estimate of the actual fraction of true negatives is made based on the observed distribution of p values. If evidence shows that 30% of genes may be differentially expressed, then only 70% are true negatives, and the expected false-positive rate is adjusted accordingly. A final comment on the adoption of FDR procedures is that the choice of cutoff is arbitrary, but should be specified by the user in advance. For some applications, a 10% FDR may be admissible, for example, where gene ontology class comparisons are made, because a small number of incorrectly identified genes will not affect the conclusions.

Software

The two most popular programs available to perform microarray analysis using linear models are SAS (SAS Institute, Cary, NC) and various implementations in the open-source language R, most of which are available in Bioconductor (Gentleman et al., 2004). SAS offers a high-end Scientific Discovery Solution for handling extremely large data sets and has released a JMP-based version for use by smaller laboratories. Their statistical software has wide application in quantitative genetics and is under license to most academic research institutions. It runs on multiple platforms (Windows, Unix, MacOS), but requires the writing of scripts that nevertheless provide great flexibility, particularly when coupled with the more user-friendly JMP application (Gibson and Wolfinger, 2004). The major procedure use to implement mixed linear models is PROC MIXED, and a detailed manual providing sample code is available on our MMAn-MaDa (mixed model analysis of microarray data) web page (http://statgen.ncsu.edu/ggibson/Manual.html).

R is a freely available mathematical programming language. A large number of statistical packages are available in R that also provide graphical options. Bioconductor is a consortium of statisticians dedicated to providing comprehensive R packages for microarray and other genomic analyses (Reimers and Carey, 2006). Most packages are accompanied with a vignette available on the Bioconductor web site at www.bioconductor.org to help the user. Some of the most popular packages for microarray data analysis are MARRAY (Y. H. Yang), AFFY (R. A. Irizarry), LIMMA (G. Smyth), AFFYPLM (B. Bolstad), and R-MAANOVA (H. Wu and G. Churchill).

References

Beare, R., and Buckley, M. (2004). Spot: cnDA Microarray Image Users Guide. Available from http://spot.cmis.csiro.au/spot/spotmanual.php

Benjamini, Y., and Hochberg, Y. (1995). Controlling the false discovery rate: A practical and powerful approach to multiple testing. *J. R. Stat. Soc. B.* **57**, 289–300.

Chu, T.-M., Weir, B., and Wolfinger, R. (2002). A systematic statistical linear modeling approach to oligonucleotide array experiments. *Math. Biosci.* **176**, 35–51.

Cui, X., and Churchill, G. A. (2003). Statistical tests for differential expression in cDNA microarray experiments. *Genome Biol.* **4**, 210–215.

Cui, X., Hwang, J. T., Qiu, J., Blades, N. J., and Churchill, G. A. (2005). Improved statistical tests for differential gene expression by shrinking variance components estimates. *Biostatistics* **6**, 59–75.

Dalma-Weiszhausz, D. D., Warrington, J., Tanimoto, E. Y., and Miyada, C. G. (2006). The Affymetrix GeneChip platform: An overview. *Methods Enzymol.* **410**, 3–28.

Eisen, M. (1999). ScanAlyze User Manual. Stanford University, US.

Fan, J.-B., Gunderson, K. L., Bibikova, M., Yeakley, J. M., Chen, J., Garcia, E. W., Lebruska, L. L., Laurent, M., Shen, R., and Barker, D. (2006). Illumina universal bead arrays. *Methods Enzymol.* **410**, 57–72.

Gentleman, R. C., Carey, V. J., Bates, D. M., Bolstad, B., Dettling, M., Dudoit, S., Ellis, B., Gautier, L., Ge, Y., Gentry, J., Hornik, K., Hothorn, T., Huber, W., Iacus, S., Irizarry, R., Leisch, F., Li, C., Maechler, M., Rossini, A. J., Sawitzki, G., Smith, C., Smyth, G., Tierney, L, Yang, J. Y., and Zhang, J. (2004). Bioconductor: Open software development for computational biology and bioinformatics. *Genome Biol.* **5**, R80.

Gibson, G., and Wolfinger, R. D. (2004). Gene expression profiling with the SAS microarray solution. *In* "Genetic Analysis of Complex Traits with SAS" (A. M. Saxton, ed.), Chapter 11. Users Press/SAS Publishing, Cary, NC.

Gilad, Y., Rifkin, S. A., Bertone, P., Gerstein, M., and White, K. P. (2005). Multi-species microarrays reveal the effect of sequence divergence on gene expression profiles. *Genome Res.* **15**, 674–680.

GSI Lumonics (1999). Quant Array Analysis Software—Operator's Manual.

Hsieh, W.-P., Chu, T.-M., Wolfinger, R. D., and Gibson, G. (2003). Mixed model reanalysis of primate data suggests tissue and species biases in oligonucleotide-based gene expression profiles. *Genetics* **165**, 747–757.

Irizarry, R. A., Hobbs, B., Collin, F., Beazer-Barclay, Y. D., Antonellis, K. J., Scherf, U., and Speed, T. P. (2003). Exploration, normalization, and summaries of high density oligonucleotide array probe level data. *Biostatistics* **4**, 249–264.

Jin, W., Riley, R. M., Wolfinger, R. D., White, K. P., Passador-Gurgel, G., and Gibson, G. (2001). The contributions of sex, genotype and age to transcriptional variance in *Drosophila melanogaster*. *Nature Genet.* **29**, 389–395.

Kerr, M. K., Martin, M., and Churchill, G. A. (2000). Analysis of variance for gene expression microarray data. *J. Comput. Biol.* **7**, 819–837.

Lee, M. L., Kuo, F. C., Whitmore, G. A., and Sklar, J. (2000). Importance of replication in microarray gene expression studies: Statistical methods and evidence from repetitive cDNA hybridizations. *Proc. Natl Acad. Sci. USA* **97**, 9834–9839.

Li, Q., Fraley, C., Bumgarner, R. E., Yeung, K. Y., and Raftery, A. E. (2005). Donuts, scratches, and blanks: Robust model-based segmentation of microarray images. *Bioinformatics* **21**(12), 2875–2882.

Littell, R. C., Milliken, G. A., Stroup, W. W., and Wolfinger, R. D. (1996). "SAS System for Mixed Models." SAS Institute, Cary, NC.

Minor, J. M. (2006). Microarray quality control. *Methods Enzymol.* **411**, 213–255.

Ott, R. Lyman, and Longnecker, M. (2001). "An Introduction to Statistical Methods and Data Analysis," 5th Ed. Duxbury, Belmont CA.

Quackenbush, J. (2002). Microarray data normalization and transformation. *Nature Genet.* **32**, 496–501.

Reimers, M., and Carey, V. J. (2006). Bioconductor: An open source framework for bioinformatics and computational biology. *Methods Enzymol.* **411**, 119–134.

Rosa, G. J. M., Steibel, J. P., and Tempelman, R. J. (2005). Reassessing design and analysis of two colour microarray experiments using mixed effects models. *Comp. Funct. Genom.* **6**, 123–131.

Saeed, A. I., Bhagabati, N. K., Braisted, J. C., Liang, W., Sharov, V., Howe, E. A., Li, J., Thiagarajan, M., White, J. A., and Quackenbush, J. (2006). TM4 microarray software suite. *Methods Enzymol.* **411**, 134–193.

Scacheri, P. C., Crawford, G. E., and Davis, S. (2006). Statistics for ChIP-chip and DNase hypersensitivity experiments on NimbleGen arrays. *Methods Enzymol.* **411**, 270–282.

Sokal, R. R., and Rohlf, F. J. (1995). "Biometry," 3rd Ed. Freeman, New York.

Tempelman, R. J. (2005). Assessing statistical precision, power, and robustness of alternative experimental designs for two color microarray platforms based on mixed effects models. *Vet. Immunol. Immunopathol.* **105**, 175–186.

Tusher, V. G., Tibshirani, R., and Chu, G. (2001). Significance analysis of microarrays applied to the ionizing radiation response. *Proc. Natl Acad. Sci. USA* **98**, 5116–5121.

Uhde-Stone, C., Zinn, K. E., Ramirez-Yanez, M., Li, A., Vance, C. P., and Allan, D. L. (2003). Nylon filter arrays reveal differential gene expression in proteoid roots of white lupin in response to phosphorus deficiency. *Plant Physiol.* **131**, 1064–1079.

Wernisch, L., Kendall, S. L., Soneji, S., Wietzorrek, A., Parish, T., Hinds, J., Butcher, P. D., and Stoker, N. G. (2003). Analysis of whole-genome microarray replicates using mixed models. *Bioinformatics* **19**, 53–61.

Wolber, P. K., Collins, P. J., Lucas, A. B., De Witte, A., and Shannon, K. W. (2006). The Agilent *in situ*-synthesized microarray platform. *Methods Enzymol.* **410**, 28–57.

Wolfinger, R. D., Gibson, G., Wolfinger, E. D., Bennett, L., Hamadeh, H., Bushel, P., Afshari, C., and Paules, R. S. (2001). Assessing gene significance from cDNA microarray expression data via mixed models. *J. Comp. Biol.* **8**, 625–637.

Wu, H., Kerr, M. K., Cui, X., and Churchill, G. A. (2002). MAANOVA: A software package for the analysis of spotted cDNA microarray experiments. *In* "The Analysis of Gene Expression Data: Methods and Software" (G. Parmigiani, E. S. Garett, R. A. Irizarry, and S. L. Zeger, eds.), Chapter 13. Springer, London.

Further Reading

Churchill, G. A. (2002). Fundamentals of experimental design for cDNA microarrays. *Nature Genet.* **32**, 490–495.

Dudoit, S., Shaffer, J. P., and Boldrick, J. C. (2003). Multiple hypothesis testing in microarray experiments. *Stat. Sci.* **18**, 71–103.

Dudoit, S., Yang, Y. H., Callow, M. J., and Speed, T. P. (2002). Statistical methods for identifying differentially expressed genes in replicated cDNA microarray experiments. *Stat. Sinica* **12**, 111–139.

Kerr, M. K., and Churchill, G. A. (2001). Experimental design for gene expression microarrays. *Biostatistics* **2**, 183–201.

Lee, M. L., Lu, W., Whitmore, G. A., and Beier, D. (2002). Models for microarray gene expression data. *J. Biopharm. Stat.* **12**, 1–19.

Storey, J. D., and Tibshirani, R. (2003). Statistical significance for genome-wide studies. *Proc. Natl Acad. Sci. USA* **100**, 9440–9445.

Yang, Y.-H., and Speed, T. P. (2002). Design issues for cDNA microarray experiments. *Nature Rev. Genet.* **3**, 579–588.

[12] Microarray Quality Control

By JAMES M. MINOR

Abstract

Physically separated groups of specific sequences (probes) provide useful high through put (HTP) measurements for the amount of selected DNA/RNA sequences in a biological target sample. Unfortunately, these measurements are impacted by various technical sources, such as platform production factors, target preparation processes, hybridization method/conditions, and signal-extraction devices and methods. Given the typically huge population of signals, statistical methods are especially effective at estimation and removal of such technical distortions (Churchill, 2002; Kerr *et al.*, 2000; Yue *et al.*, 2001), as well as providing metrics for computer-based quality control (QC), for example, autoQC (Minor *et al.*, 2002a). This chapter reviews statistical procedures that have been validated by successful applications in both large-scale commercial ventures (Ganter *et al.*, 2005) and individual research studies (Parisi *et al.*, 2003, 2004) involving HTP projects. This chapter focuses on methods for spatially distributed probes on a flat medium surface such as glass, collectively known as a microarray.

Introduction

Interest in sequence-expression data is escalating as new applications such as location analysis and comparative genomics hybridization (CGH) are leading to new insights into the biological processes of complex diseases such as cancer. The drug-discovery industry is beginning to develop methods that leverage this new information to find combinatorial compounds that block critical routes of disease pathways. For example, one such method applies design of experiment concepts to a collection of phenotypic and gene expression profiles to effectively identify such combinations for lung cancer (Minor *et al.*, 2003a). The research focus is no longer on the magic bullet "one drug one disease" concept; put another

METHODS IN ENZYMOLOGY, VOL. 411
0076-6879/06 $35.00
DOI: 10.1016/S0076-6879(06)11012-5

way, the pharmaceutical industry is no longer searching for a disease that can be treated with one drug. Hence, proteomics and metabolomics are growing rapidly and are connected closely to genomics and expression patterns. Furthermore, this information could potentially revolutionize clinical medicine supporting concepts of personalized medicine and accurate prognostics. Methods to enable this potential are being developed (Comanor and Minor, 1996; Comanor et al., 2000, 2001; Lau, 1998; Martinot-Peignoux et al., 2006; Minor et al., 2002b). Consequently, the processing and validation of microarray measurements are escalating rapidly in scale and importance.

The microarray procedure begins in manufacturing with a substrate medium such as glass and ends with the processed signals for all probes, ready for the customer's biological applications. Multiple arrays are produced on a block unit of media, for example, a rectangular slide of glass. This procedure is a complex highly integrated sequential process that requires careful monitoring and quality control (QC) at frequent junctures. For example, creating high-quality probes requires both high-quality uniform glass having the right chemical/physical properties and "pens" that create the intended sequences with high probability/reliability with minimal edge effects. Note in this general context that "pen" refers to any device specific to the synthesis of each probe. Examples include clone deposition pens, in situ ink jet nozzles, and lithographic masking, electromagnetic, or optical widows. Each probe or feature is actually a collection of many millions of sequences with typically up to 90% overall being exactly correct, depending on the coupling efficiency of each specific pen. This percentage is typically lower at the edge and middle regions of the probe due to synthesis and hybridization/wash factors.

Although the statistical procedures apply to all phases of the array process, this chapter focuses on the customer experience. Hence, production variations in the glass, slide-substrate batches, slide surface gradients, label kits, and synthesis batches are beyond the scope of this chapter. The focus is on statistical methods and QC metrics useful to the microarray customer to analyze probe signals, surface/sequence gradients, array normalization, and "pen" properties that affect the signal-to-noise performance of expression measurements and expression comparisons from hybridized arrays.

Even with our focus on the customer process, signal noise requires an expanded chapter by itself. Generally, signal noise is a combination from several different sources along the assay process from sample preparation through array scanning. The scanner noise is monotonic to total signal counts, and assay noise is monotonic to the expression component of the signal. These complications typically require advanced statistical methods such as variance components and GLM methods (Nelder and Wedderburn,

1972) to analyze data. However, given the huge and increasing scale of array data, viable simplifying assumptions are necessary to enable the high through put (HTP) process even with fast computers using traditional statistical methods as covered in this chapter.

Despite simplifications, innovative combinations of these methods have proven to be very useful. For example, a profile composed of statistically derived QC metrics relates directly to the performance quality of each array. Through integration of regression analysis with signal networks (Comanor and Minor, 1996), a computational system (Minor *et al.*, 2002a) based on such profiles correctly classifies 95% of arrays as verified by costly manual QC evaluations. The program automatically flags the other 5% for manual inspection only.

The discussion begins with the scanned image, which is stored as a sequence of millions of intensity readings, called pixels. Pixels are grouped into probe regions, and a probe signal is calculated, classified as no expression signal, expression signal, or not reliable.

General statistical models for technical corrections of these probe signals are then derived as well as models for noise. Such models are applied to background evaluations based on probes classed as no expression signal.

The consequent background-corrected signals are then corrected further for technical perturbations due to array gradients. For certain types of arrays the signal can be improved further by a combination of statistics and bioinformatics based on a general stochastic cross-hybridization matrix.

We then look at methods that apply to multiple channel comparisons, for example, dye/labeling corrections and error propagation differences. Given a specific class of sample types, the concept of a perfect array, for example, reference array, is described.

Finally, quality metrics derived at every step of this process become a quality profile (signature) that predicts its quality rating accurately.

Pixel Statistics

The millions of sequence strings created at each probe location produce individual hybridization signals that become partitioned into neighborhoods, called pixels, by a scanner according to its resolution. A typical scanner resolution scale is at most 10 μm per pixel length. Each square micrometer contains many tens of thousands of oligonucleotide strings. Typically, a probe is composed of up to 300 pixels. Each observed pixel signal is the ensemble consequence of its mixed population of correct and defective sequences. This mixed population is a natural consequence of the practical levels of coupling efficiency, which tends to be weaker near the probe edges. Note that we are ignoring the small effects of optical or spectral cross talk among adjacent pixels. As a consequence of nonspecific and/or somewhat specific levels of hybridization (cross-hybridization between nonperfectly base-paired sequences), defective sequences in the pixel population are likely to partially bond to an assortment of target sequences creating a weak signal. The highest expression signals are produced primarily by those pixels with the largest population of the intended sequence as defined for each probe. The complex dynamics of chemistry used in the probe synthesis and wash procedures in conjunction with pen efficiency tend to create either an ellipsoidal or an annulus-like region of best expression pixels that is geometrically complicated and variable (Fig. 1). This implies that a statistical approach is necessary to efficiently combine pixel signals into the optimal composite signal for each probe (Minor, 2003b).

The common procedure involves matching a signal pattern within the identified probe region to an assumed template shape. One then assigns a signal value based on this selection, for example, a mean or median robust against outlier signal levels. However, this selection procedure cannot exclude noisy pixels that impact signal assignment. The method is not robust against nonideal signal patterns, variations in signal coverage (i.e., probe

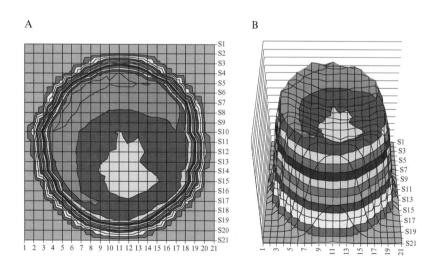

Red = best signals

Green = effect of chemical solutions

Yellow = effect of chemical solutions and pen performance

FIG. 1. Pixel signal patterns. (A) Signal contours. (B) Three-dimensional perspective plot. (See color insert.)

size), or errors in centering the signal pattern. Information that leverages the separation of true signal from true background is lost. Hence, one achieves adequate quantitation only for perfect probes. Clearly, a better more robust efficient tool is desirable that inherently works better for practical probes and scales well for ever larger numbers of even smaller probes.

Similar to the cumulative distribution function (CDF) of probability densities, such a tool is the profile of sorted pixel signals of the region capturing each probe. Note that this profile is not a proper statistical CDF, as data are driven by systematic effects, but the concept is extremely useful for pixel quantitation. The perfect noise-free signal profile of an expressed sequence is a simple step function (Fig. 2) with high levels at the synthesized location of the probe. Given the ubiquitous imperfections of reality, this becomes a smooth sigmoid shape. The shape becomes the simple curve of a uniform random distribution if the sequence is not expressed, that is, is void. Best performing probes driven by optimal binding between perfectly base-paired sequences tend to have a CDF signature near the perfect profile. Hence, the shape of the CDF becomes a useful tool for signal evaluation of each probe (e.g., for testing array designs prior to production and for end users looking to identify poor performing probes that might yield spurious results). The key advantage is robust quantitation without direct reference to the geometry or contour patterns of pixel signals. For each pixel in the CDF profile one may examine its coordinate location for diagnostic purposes. Note that some perspective contained in the native two-dimensional (2D) geometrical image will be lost, but this lost information has little or no importance for signal evaluation. This is the same assumption that applies to our binocular vision system trying to interpret a 3D world.

There must be a sufficient number of pixels to produce at least the upper profile shape of the CDF in order to distinguish good pixels from edge and aberrant pixels. Note that highly dense arrays may have only a few pixels per probe that cannot produce a viable CDF profile. Are these few pixels good or not? In this case all pixels are effectively "edge" pixels. Assuming that one could adequately correct for technical effects such as array location and sequence-based affinity gradients of all oligonucleotides on an array, one could then combine all pixels of probes that measure the same gene to form an ensemble CDF profile to enable a realistic statistical evaluation of expression of that gene. Otherwise, to achieve any sort of quantitation here, one must resort to methods that are designed to use multiple probes that measure the same gene and its noise factors. This is discussed in a later section.

Using pixel profile shapes, one can effectively distinguish between probes measuring primarily signals from random weakly bonded labels, reflecting cross-hybridization between nonperfectly base-paired sequences,

A

Pixel CDF

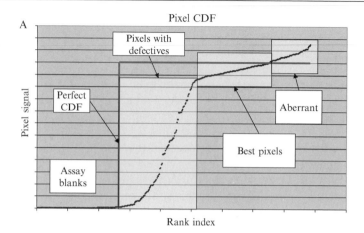

B

CDF'S : from noise to signal dominated

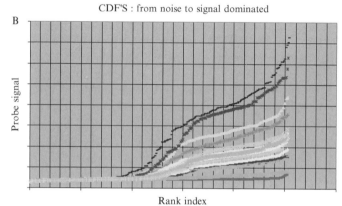

C

Pixel CDF dominated by abberant signals

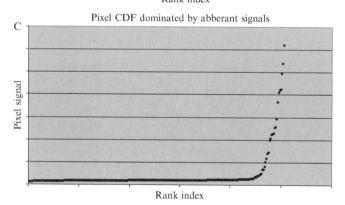

FIG. 2. Pixel signal CDF profiles. (See color insert.)

and probes dominated by expression signal from labels on perfectly base-paired sequences. The identified set of void probes, including negative controls, is useful for processing background noise. The set of best ex-pressed pixel signals is located on the high asymptote of the sigmoid prior to any aberrant signals. The optimal signal in this set tends to be the highest rank signal free of any aberrant impact. Signals in this best set and their ratios between channels tend to be robust with good precision and accuracy across multiple arrays. Precision refers to replicate reproducibility of a measurement, whereas accuracy refers to the truth of a measurement.

Occasionally, aberrantly high signals of localized defects caused by scratches or chemical effects add a distinctive flare at the high end of an expression profile or else they dominate most of the CDF curve, as shown in Fig. 2. The extent of this flare along the CDF measures posthybridization wash residue impact and/or array defect patterns. A connected sequence of aberrant signals across an array implies defects such as scratches, stray deposits, or finger smudges. Cluster analysis on aberrant signals can auto-detect and characterize such connected patterns for QC purposes. Such probes do not provide reliable quantitation.

Pixel Statistics Methods

Assigning correct annotation row and column locations to probes in an image file of an array is the ultimate result of image processing known as the grid-finding problem. This is solved in general by population statistics, time series concepts, and reliable clustering methods (Minor, 2001). For example, one can apply statistical procedures that leverage both global and local information in an image file. The dynamic clustering method (Minor, 2003b) is a very fast effective tool for locating individual signal clusters in an image file. Next, rotation grid-spacing regression models can be applied to a representative sample of such signal clusters to estimate general global grid properties of the scanned image. One can then apply fast predictor-corrector statistical filters to accurately find all individual probes. These methods are robust for all scanners. When applied to high-density arrays, such methods, combined with pixel quantitation/diagnostics, require about a minute of Pentium 4 computation time and provide superior results when compared to other extraction systems. This chapter is not scoped to cover this subject in detail.

Assume that probe locations are reliable. For each probe, select a closed region with geometrical shape and size that captures all probe pixels. Sort the pixels by intensity and remove active pixels from the CDF profile that are at the region edges and likely overlap adjacent probes by coordinate filters. Also, check that the majority of best signals are near the probe center. Otherwise, the region is too big or captures the occasional "black hole" pattern, caused perhaps by noise suppression of a random probe, high surrounding background, and/or high affinity for unlabeled string segments.

If the profile fits a quadratic curve as shown in Fig. 2, then target expression is null. In this case, flag the probe as dominated by background signal. If the profile is dominated by aberrant pixels, flag the probe as aberrant. Otherwise, select pixels adjacent to but avoiding the flare section at the end of the profile. Optionally, this selection can be refined further based on pixel coordinates, that is, pick pixels closer to the probe center. Use these best-pixel expression signals to construct an optimal summary. For example, use mean or median or fit a simple line to these signals and predict the signal at the highest rank index in the profile, which equals the number of pixels in the probe region. Any flare pixels as marked by aberrant signal could also be reported for diagnostic purposes. The relative steepness of the slope section in an expressed signal indicates quality of probe performance. It depends on both the distribution and the cross-hybridization tendency of its realized sequences. Synthesis efficiency and probe design determine such properties.

In summary through image-level statistics, all probes are divided into three signal groups based on the dominating signal source: background, aberrant, or expressed possibly with some aberrant pixels, as depicted in Fig. 2. One can create a "shape" metric for the CDF and then set optimal thresholds that best distinguish these categories. Such groups are important for the next stages of signal processing and metrics.

Smooth Patterns and Block Effects

If probes are located on the array randomly and not organized according to some biological property, then biological information will tend to be stacked in the high spectral domain of a Fourier power spectrum of the signal location pattern. Signal location gradients induced by the technical aspects of assay protocols and conditions, for example, the hybridization process, will tend to be in the low-frequency domain. This spectral separation of information facilitates model-based identification of the biological content in array signals. Block shifts such as "pen" differences are an exception, as they have significant high-frequency content. Therefore, block effects need to be explicitly represented along with smooth gradient effects in any model of technical patterns on an array. Statistically, block effects are analysis of variance terms (ANOVA) and smooth effects are covariates.

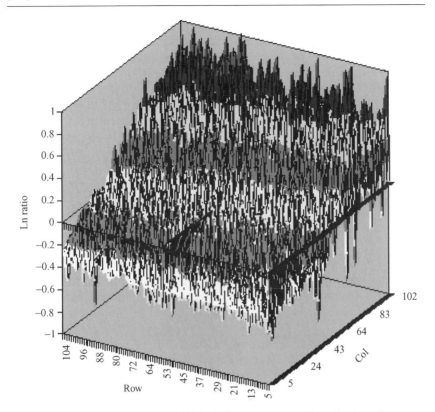

Fɪɢ. 3. Technical patterns in biological measurements. (See color insert.)

Typically, it is difficult to directly see the low-frequency and shift technical patterns because they are hidden within the high-frequency biological measurements and because pen shifts are distributed over the array. However, Fig. 3 portrays an unusual array where these effects are clearly visible (Minor, 1999). Observe the column gradient caused by asymmetry in the hybridization device. The four quadrants each printed by a separate pen are shifted relative to each other. These technical patterns were removed by the statistical models described herein. Better yet, the hybridization chamber was later redesigned to reduce the asymmetry problem. Pen shifts could not be fixed by device redesign. Note that gradients of this magnitude indicate serious quality problems, which should be addressed immediately before continuing with any study.

Other embedded smooth gradient types are also possible. For example, replication of probes measuring the same target sequence can lead to target depletion effects, commonly seen in specialized custom arrays, for example, QC arrays. This reduced signal is caused by the distribution or sharing of a target sequence in a sample over multiple matched probes. The sequence properties of probes can also have smooth technical patterns induced by dye chemistry and physical variables. For example, the signal tends to be proportional to the "G" and "C" content of a probe sequence because of higher melt temperatures. Also, the probe signal tends to increase with its nucleic complement of modified nucleotides in the sample. Hence, sequence-composition gradients are useful for both technical corrections and diagnostics. These gradients are reduced but can still have significant impact on sequence-expression comparisons or ratios; hence, they are important to array applications using intraprobe comparisons such as CGH studies. As stated previously, unusually severe gradients indicate quality problems, in this case with the labeling process or materials.

Models of Array Patterns

One of the earliest commercial applications used to provide the complete model of both ANOVA and covariate terms to ensure better separation of technical from biological patterns was developed by Strategy of Research Company and was marketed by Novation Biosciences, Inc. as Qualifier software (Minor *et al.*, 2002a). Location gradients can be somewhat curvilinear, requiring high-order surface models. There are several options.

1. High-order polynomials (not recommended due to inherent model stability issues, especially near data edges or other data-sparse areas).
2. ANOVA blocks with covariate low-order polynomials and localized Gaussian kernels using Qualifier regression models with SLS terms (Comanor *et al.*, 2001).
3. Locally weighted regression (LWR or Loess methods) (Cleveland, 1979).

LWR is a computation-intensive transform that scales badly. It is prone to overfitting as it tries to model block shifts in data. This reduces intraarray noise at the expense of interarray noise and, typical of transforms, statistical diagnostics are poor. Overfitting occurs when a highly flexible smooth model (LWR) tries to fit nonsmooth shifts in a data set. Hence, some of the intraarray technical variability is removed, improving intraarray noise/signal (CV). However, the high-frequency part of the shift remains mixed in the biological information. Because this shift varies from array to array, the model correction produces higher values of interarray CV. One avoids this problem by including appropriate "ANOVA" shift factors in the smooth model process.

Hence, one could combine Loess patterns as a covariate with ANOVA blocks in a three-step process.

1. Fit data with ANOVA block terms by regression analysis.
2. Fit regression residuals with a Loess transform process.
3. Refit the model with ANOVA block terms and the calculated Loess transform as a term by regression analysis.

This procedure does not resolve the Loess scaling issue. To improve scaling, partition methods are applied such as windowed averages at strategic locations on the array surface. However, Loess cannot inherently interpolate very well nor adapt its effective bandwidth to variations in surface tempo. Consequently, it becomes unstable when actual gradients are small. One can resolve these deficiencies by simply replacing the strategic averages with a set of values optimized to fit the surface better, but this is essentially the SLS process as implemented in the Qualifier system.

The Qualifier model is unique in that it represents all important surface effects with traditional regression diagnostics while scaling much better than transforms. The Qualifier model includes ANOVA block shifts and second-order polynomial terms, known collectively in statistics as a "response surface." Also, the model uses SLS Gaussian kernels to capture localized high-order effects near corners and edges. SLS is a validated method for optimizing Gaussian-kernel designs in general statistical models, including logistic, ordinal, nonlinear, and standard regression models. Related methods are support vector regression and K-nearest-neighbor classifiers (Drucker, 1997; Ripley, 1996).

Conceptual Models for Probe Signals

In general, each expressed probe signal S is proportional to the amount of its specific expressed perfectly base-paired sequence in a fixed volume of target sample, for example, concentration C_j plus background B. The two primary sources of background and signal error are the scanner, index s, and the assay, index a. Hence, error effects are assay proportional to $S\text{-}B$ and scanner proportional to S. A statistical model becomes

$$S_j = \left(A_j C_j^{w_j} e^{\delta_j} + B_{aj} e^{\gamma_j} + \varepsilon_{aj} \right) e^{\eta_j} + B_{sj} e^{\eta_j} + \varepsilon_{sj} \tag{1}$$

$$S_j = \left(A_j C_j^{w_j} e^{\delta_j + \eta_j} \right) + \left(B_{sj} e^{\eta_j} + B_{aj} e^{\gamma_j + \eta_j} \right) + \left(\varepsilon_{sj} + \varepsilon_{aj} e^{\eta_j} \right) \tag{2}$$

where scanner error is a combination of photomultiplier tube multiplicative error η and its optical additive error ε. Note that the assay error is also a combination of chemical errors δ and γ with its additive background ε. The terms are grouped into expression, background B, and additive effects. The equation is typically simplified by assuming that the background and additive effects are represented in form by scanner error.

$$S_j = \left(A_j C_j^{w_j} e^{\Delta_j}\right) + \left(B_j e^{\eta_j}\right) + \left(\varepsilon_j\right) \tag{3}$$

For low expression, an approximation becomes

$$S_j \simeq B_j e^{\eta_j} + \varepsilon_j \tag{4}$$

or, because background is somewhat constant,

$$S_j \simeq B_j + \varepsilon_j \tag{5}$$

For high expression, ignoring the small additive error,

$$S_j - B_j \simeq A_j C_j^{w_j} e^{\Delta_j} \tag{6}$$

Population Models for Probe Signals

The population summary properties of all probe signals of a channel or color can be used to correct or normalize the array to improve comparisons with other signal channels. Such procedures rely on platform protocols and sample assumptions that impact signal properties and their relation to target concentrations. For example, consider the total signal of expression arrays typically used in drug research. Most genes are inert to drug impact. The few that react are important to help identify drug targets. Therefore, the signals of inert genes dominate summary statistics of the signal population.

The molecular basis of summary metrics is interesting. In general, each expressed probe signal S minus background signal B is proportional to the amount of its specific expressed perfectly base-paired sequence in a fixed volume (from Eq. [6]) of target sample, that is, concentration C. A statistical model becomes

$$S_j - B_j = A_j C_j^{w_j} e^{\Delta_j}, \tag{7}$$

where A_j is determined by sequence affinity and hybridization protocol.

For an ideal first-order reaction $w_j = 1$, but given reaction diffusion/ mass transport complications, it becomes effectively less than 1.

Summary total of log signals gives

$$\sum_1^N Ln(S_j - B_j) = \sum_1^N LnA_j + \sum_1^N w_j LnC_j + \sum_1^N \Delta_j. \tag{8}$$

Hence, the log transform provides two advantages:

1. The error model is simplified to be normally distributed uniformly across all probes that are truncated at low signals.
2. The reaction components form a linear model,

$$Ln(S_j - B_j) = LnA_j + w_j LnC_j + \Delta_j, \tag{9}$$

where hyb factors are separated from concentration factors.

The summary total becomes

$$\sum_1^N Ln(S_j - B_j) = \sum_1^N LnA_j + \sum_1^N w_j LnC_j + \sum_1^N \Delta_j. \tag{10}$$

Because N is large the standard error of the summary average is very tight; hence, one has a good metric for normalization. For optimally designed platforms, all w_j are randomly close to one value w' near unity, and hyb factors are well controlled so that the total is essentially

$$\sum_1^N Ln(S_j - B_j) = \mu + w' \sum_1^N LnC_j + \sum_1^N \Delta_j, \tag{11}$$

where μ is a constant of regression.

A significant technical variation in labeling efficiencies and/or channel loading produces a sensitive shift in this total making it an effective metric for array normalization.

Signal Processing Including Control Probes

From Eqs. (3)–(5), if the expression signal is near zero,

$$S_j \simeq B_j e^{\eta_j} + \varepsilon_j$$

or

$$S_j \simeq B_j + \varepsilon_j$$

The background of active expressions may tend to be lower, as probe sequence populations used by expressed target will not be available to ubiquitous noise species in active pixels, a form of noise suppression.

One could use the background portion of the CDF to predict the background signal specific to each probe. Alternatively, one could use the background class of probes determined from CDF metrics to predict background for all other probes using smooth gradients as described previously in the section for array patterns. But if the row–column coverage of the background probes is poor, as indicated by the principles of statistical design of experiments, then add low-signal expressed probes as needed. Subtract the matrix of the estimated background from all viable probes to form the background corrected signals, BC signals.

The logs of all positive BC signals are similarly corrected with smooth gradients to produce the final detrended BC signals, DBC signals. For platforms that exhibit persistent signal correlation between different gene probes, further enhancement of signal to noise is possible through a probe matrix model based on such correlations and bioinformatics.

Matrix Model of Cross-Hybridization Noise (SIAM)

There are a huge variety of sequences in the string population present in a hybridization solution. Each sequence will have a preference profile across all probes on an array determined by complementary sequence similarity or homology, that is, one probe's target is another probe's bias, a form of probe interaction. Likewise, similar probes will have similar noise properties. Furthermore, a probe designed for a specific gene may be correlated to other probes by sequence fragments in the target from that gene. A catalogue platform is designed to minimize such interactions. This improves both precision and accuracy of array results. Other platforms either take advantage of cross-hyb, (e.g., match–mismatch platforms) or accept it (e.g., custom arrays). Statistical methods are available to correct this problem based on multiarray processing (Li and Wong, 2001; Wu and Irizarry, 2004). A unique method that applies to one or more arrays is SIAM (Minor, 2002c).

SIAM: General Method and Simplified Equations

An observed expression signal on probe j is the sum of multiple contributions

$$T_j = \sum_k a_{jk} c_k + \varepsilon_j, \tag{12}$$

where T_j is the signal of probe j, a_{jk} is the bioinformatics parameter for binding of sequence j to probe k including noise factors, c_k is the expression level of sequence k, and ε_j is all other sources of error, for example, optical noise and probe synthesis factors.

Note that summation covers a range of binding properties from specific through degrees of partially specific to nonspecific affinities as determined by sequence similarities and correlations of sequences.

Consider a platform with specially designed pairs of probes,

$$T_j = a_{jj} c_j + \sum_{k \neq j} a_{jk} c_k + \varepsilon_j, \tag{13}$$

$$T_{j+1} = \sum_{k \neq j+1} a_{j+1,k} c_k + \varepsilon_{j+1}, \tag{14}$$

where $j = 1, 3, 5, \ldots$, for example, each paired perfect match "PM" probe j and mismatch "MM" probe $j+1$ of well-known platforms.
Transforming to intensity notation gives

$$P_j = S_j + \sum_{k \neq j} f_{jk} S_k + \varepsilon_j, \tag{15}$$

$$M_{j+1} = f_{j+1,j} S_j + \sum_{k \neq j} f_{j+1,k} S_k + \varepsilon_{j+1}, \tag{16}$$

where P denotes perfect-match T and M denotes its paired mismatch intensity.

Denoting cross-hyb summations $\sum_{k \neq j} f_{j+1,k} S_k$ and $\sum_{k \neq j} f_{jk} S_k$ by factors N_{j+1} and $f'_j N_{j+1}$, respectively, gives

$$P_j = S_j + f'_j N_{j+1} + \varepsilon_j, \tag{17}$$

$$M_{j+1} = f_{j+1} S_j + N_{j+1} + \varepsilon_{j+1}, \tag{18}$$

Relabel these functional forms as bioinformatics factors f_n and f_s, assume all other errors to be small, and reindex to get

$$P_j = S_j + f_n N_j, \tag{19}$$

$$M_j = f_s S_j + N_j. \tag{20}$$

This version of Eq. (12) creates probe-independent bioinformatics factors depending only on the pairing types, mismatched "MM" and perfect matched "PM," and presents a term for ubiquitous cross-hyb bindings as driven by both their large ensemble concentration and a typically mediocre ensemble affinity.

Typical values are $f_s \approx 0.5$ and $f_n \approx 0.95$.

Multichannel Methods

Differential expression information is derived from the comparison of at least two channels of DBC signals produced from two biological samples. Hence, the ratio or log ratio of each probe is evaluated. Variations in incorporation dynamics among samples can bias these ratios. These variations can be abundance (signal intensity) related; as a result, abundance-driven normalization is applied to reduce sample ratio bias. Population statistics is the primary tool for this normalization. It is important that all sources of technical gradients are removed, that is, row–column and

Detrended normalized fly

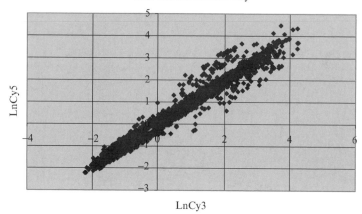

FIG. 4. Detrended normalized fly data from qualifier.

sequence based, prior to any abundance-driven normalization, as such gradients create technical bifurcations in the two-channel plots. Because of the density of points in these plots, such bifurcations may not be obvious. When they are, the consequential patterns are sometimes "affectionately" called "guppies" (Minor, 1999). After all gradients are removed, "guppies" may still be evident as shown in Fig. 4. In the case of unique biological research, such "guppies" are real biological clusters of differential gene expression. In this case the normalization procedure must be especially designed to focus on the primary nondifferential expression cluster of points near the diagonal of the two-channel plot.

In studies where the majority of gene expressions are expected to be invariant among samples, a two-sample plot of the log DBC signal on each axis of every probe should produce an oblong pattern of points with an internal highest density trace on the diagonal (Fig. 5). The pattern in fact may be rotated and curvilinear. This is more likely for multidye systems where chemical differences can impact incoporation dynamics and consequently scanner settings. This density histogram can be calculated rapidly by a new algorithm described in the Appendix.

To correct for this bias, methods are designed to map this pattern onto the diagonal. But which methods effectively reduce such technical shifts?

Because comparable errors impact signals on each axis, one must use statistical methods designed to handle both errors. This section begins with

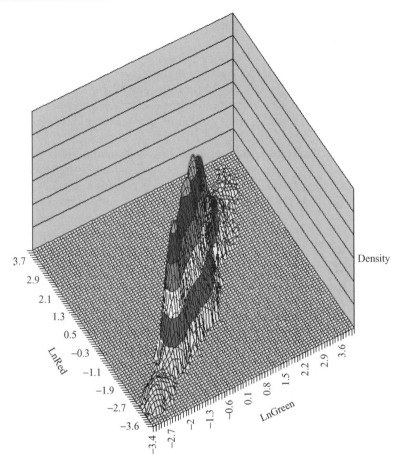

F IG. 5. Two-dimensional histogram of two-channel log signal patterns. (See color insert.)

principal component analysis on the log *DBC* signals, PCA. PCA captures the oblong two-axis pattern as a two-dimensional ellipse with its long axis approximate to the expression amount of the probes and its short axis approximate to the differential expression direction. The important errors are now aligned along the short axis. Random error is reduced on the long axis by signal "averaging" in effect and amplified on the short axis by signal differencing. One can now apply model-fitting procedures to align the pattern correctly onto the long axis, as shown in the next section. The corrected pattern may then be rotated to the plot diagonal.

Model Fitting of Curvilinear Patterns

First calculate the two-dimensional point density (histogram) of all points in the plot pattern. This can be done in a few seconds as described in the Appendix. Locate loci of highest density points within a moving window along the long axis. This trace will tend to have high-frequency jitter by nature. Apply an interpolative low-pass filter to remove this jitter with no phase shift. Subtract this smooth loci from all points to correctly project the pattern onto the long axis.

Another method creates a moving window *CDF* and finds its inflection point, which is the highest density of the population captured by the window. Other dye-normalization methods rely on sorted signals and percentiles to match the two channels, but they require more assumptions and are less robust to random errors and expressed proportion of signals. In general, for all methods the point densities at high abundance are rather sparse and require inertial trace projections to complete the loci.

Reference Pattern Corrections

For a given expression environment a reference is constructed as the summary log intensity profile over many arrays for a given dye, where expression-variant factors are randomized out. For two-color platforms the two reference profiles can then be normalized against each other to produce a dye-independent reference. All arrays measuring expression target from the same environment can now be compared to the reference channels for QC and dye normalization purposes.

Reference patterns are assumed to be very precise. Hence, PCA is not appropriate. Model projections of any channel directly to the diagonal are applied based on the reference log signals. The statistics from such reference corrections become useful metrics.

Two-Channel Error Propagation

Given a reliable estimate of noise for each probe in two channels, one can easily propagate error to the comparison of their signals. A reliable noise estimate can be calculated from replicate arrays corrected for technical variations. Propagation of log ratio error:

$$\mathrm{Var}(\log S1/S2) = \mathrm{Var}(\log S1) + \mathrm{Var}(\log S2) - 2\mathrm{Cov}(\log S1, \log S2),$$

$$(21)$$

where Var and Cov are abbreviations for variance and covariance. The standard deviation is the square root of variance.

For one-color arrays the Cov tends to be small; hence, the Var error of each channel must be small.

For two-color arrays the Cov is inherently significant and positive, thereby effectively canceling much of the single-channel noise. Note that one could gain similar canceling advantage for one-color platforms by matching production and assay conditions as much as possible for all arrays used in a study.

Another technique to identify and reduce two-color error is to run a pair of arrays where each target sample is labeled with both colors, that is, dye-flip arrays (Minor, 1999; Yue *et al.*, 2001). Combining data from the array pair by correctly matching the target samples (known as sense correction) reduces bias error and random error of each probe. But this cannot reduce shifts inherent in the incorporation kinetics specific to each sample.

Metrics from Array Patterns and Reference Channel

Many metrics are possible from all the statistical procedures described. For example, standard dose–response models (four-parameter sigmoid function) can be applied to target "spike-ins" to evaluate the bandwidth, accuracy, and precision performance of each array. The metrics become a profile of quality information for each array. Qualifier was one of the first commercial products to summarize this useful information in a customer QC report, which became very popular.

Furthermore, Qualifier applies an established methodology, SLS, to effectively predict array quality from this profile with high accuracy. The

				Quality scores assigned to each array	Statistical calibration					Negative controls					Complex sample					Concordance			
	RNA prep#	Slide barcode	Hyp date	Quality score	Lower Asymptote	Upper Asymptote	Slope	Ed_{50} calibration curve	Mean square error	Degrees of freedom	Mean signal	Std. dev. signal	CV signal	Efficiency	Degrees of freedom	Mean signal	Std. dev. signal	CV signal	Efficiency	Concordance	Location shift	Scale shift	Precision
Reject	R74490	T00126381	1/18/2002	Reject	−8.12	7.01	0.27	2.55	0.74	20	−2.01	0.45	−0.22	0.41	6549	−0.38	1.54	−4.07	9842	0.82	0.17	1.08	0.84
	R71318	T00126629	1/18/2002	Reject	−1.46	7.04	0.49	4.72	0.42	20	−1.85	0.46	−0.54	0.41	5413	−0.12	1.07	−9.32	9821	0.71	0.00	1.54	0.78
	R72304	T00127044	1/18/2002	Reject	5.23	7.71	0.23	3.91	0.72	22	1.73	0.39	0.23	0.44	6913	−0.14	1.47	−10.18	9830	0.77	0.03	1.13	0.78
	R73501	T00128045	1/18/2002	Reject	−3.28	2.81	0.08	−8.13	1.05	31	0.62	0.92	1.46	0.63	7261	0.24	1.77	8.74	9841	0.25	−0.17	0.94	0.25
	R73261	T00128050	1/18/2002	Reject	−4.43	6.13	0.50	2.82	0.21	29	−2.63	0.39	−0.14	0.60	7320	0.08	1.57	20.82	9831	0.55	−0.06	1.06	0.56
	R74540	T00128511	1/18/2002	Reject	−7.36	6.73	0.33	0.92	0.92	36	−1.74	0.57	−0.33	0.73	6881	−0.19	1.53	−8.11	9817	0.87	0.06	1.09	0.87
Review	R74506	T00126398	1/10/2002	Review	−7.79	7.88	0.00	1.62	0.50	26	−2.31	0.54	−0.20	0.53	6900	−0.37	1.62	−4.37	9040	0.89	0.17	1.02	0.91
	R74545	T00126608	1/18/2002	Review	5.67	2.73	1.63	0.01	1.49	15	2.50	0.61	0.24	0.31	7374	0.03	1.71	51.75	9849	0.98	0.02	0.97	0.98
	R74546	T00126609	1/18/2002	Review	−4.93	1.89	1.23	0.18	1.14	17	−2.53	0.46	−0.18	0.35	7441	−0.09	1.68	−18.82	9830	0.98	0.01	0.98	0.98
	R74482	T00128480	1/17/2002	Review	−4.34	−0.90	0.21	−2.47	0.24	23	−2.36	0.43	−0.18	0.47	7009	−0.43	1.62	−3.80	9819	0.89	0.20	1.03	0.91
	R74484	T00128482	1/17/2002	Review	−4.54	−0.13	0.07	2.56	0.27	20	−2.60	0.36	−0.14	0.40	7335	−0.09	1.69	−19.76	9755	0.99	0.01	0.98	0.99
	R74543	T00128544	1/18/2002	Review	−7.91	6.05	0.34	0.81	0.65	31	−2.30	0.74	−0.32	0.64	7046	−0.24	1.67	−7.05	9841	0.89	0.11	0.99	0.90
Pass	R74539	T00128510	1/18/2002	0.63	−7.16	5.91	0.43	1.01	0.33	22	−2.50	0.60	−0.24	0.45	7419	−0.07	1.69	−24.33	9822	0.98	0.00	0.98	0.98
	R74500	T00120499	1/10/2002	0.60	−7.50	5.74	0.49	0.45	0.59	36	−2.13	0.77	−0.36	0.73	7041	0.02	1.60	106.34	9034	0.96	−0.04	0.99	0.96
	R74529	T00128498	1/18/2002	0.88	7.23	5.91	0.44	0.99	0.35	21	2.55	0.61	0.24	0.42	7373	0.06	1.69	27.16	9823	0.98	0.00	0.98	0.98
	R74507	T00126399	1/18/2002	0.88	−7.43	6.36	0.39	1.12	0.50	31	−2.41	0.70	−0.29	0.64	7269	−0.12	1.69	−13.55	9832	0.97	0.04	0.98	0.97
	R73168	T00127047	1/18/2002	0.98	−7.38	6.33	0.41	1.10	0.26	24	−2.48	0.52	−0.21	0.50	7456	−0.07	1.69	−24.99	9832	0.98	0.00	0.98	0.98
	R74500	T00120395	1/18/2002	0.99	−7.44	6.28	0.42	1.13	0.22	21	−2.62	0.46	−0.17	0.43	7423	−0.06	1.70	−26.23	9842	0.99	0.00	0.98	0.99

FIG. 6. Example of AutoQC report with heat map of metrics profile. (See color insert.)

method also identifies the important metrics in the profile that determine quality as well as the critical profiles that define the boundaries between good and bad arrays. A quality heat map helps visualize critical profiles (Fig. 6). More information on SLS technology is available at www.SLSguy.com.

Appendix: A General Fast Method of Structure Analysis for High-Dimensional Data

Introduction

Structure–analysis problems are defined by

1. A set of profiles encapsulating related information such as DNA sequence strings.
2. A measure of similarity between profiles, for example, a metric.

Biological problems are characterized by large databases of high-dimensional profiles. The similarity metric can be quite complex. To find patterns in data, one typically applies cluster analyses or functional methods such as feature least squares, SLS. To reduce computational burden, one combines and partitions profiles into groups of binned profiles that enable branch-and-bound/divide-and-conquer types of procedures (Bishop, 1998). This in effect is a sorting process on grouped profiles. Another method assigns binary numbers to each profile to facilitate computation (Arya and Mount, 1993). An algorithm is derived that generalizes these techniques via dynamic sorting, for example, adaptive binning, to reduce the N^2 native computational timescale of such problems with N profiles to an improved scale $N\log N$. The key concept is an efficiency parameter that enables the $N\log N$ scaling.

Method

Consider a metric space X with metric D and a finite subset U of X composed of N elements, indexed by set $I = [1, 2, \ldots, N]$.

Define task "C" to be finding for each element j in U the set Hj of all members k of U such that $D(j, k) < h$. This metric associates with a similarity measure between profiles. Note that generalized similarity may not be strictly monotonic but must be bounded by a monotonic bounded envelope, for example, damped oscillations. In this case, metric D associates with the bounding envelope.

The native task C is inherently of minimal complexity N^2, as each profile in turn must be tested against all other profiles to map structural patterns.

A recurrent strategy based on adaptive sorting of profiles in terms of D enables simple efficient processing of required comparisons and reduces the complexity to less than $N\log N$ in computational scale.

For example, consider sorting all profiles by their D values, calculated relative to a specified profile. This sorted list partitions the space of profiles and reduces the search for nearest profiles to a limited set. Strategically repeating this sorting process to assure this efficiency creates a scale for computation time better than $N\log N$.

Basically start with any profile j and sort all other profiles from nearest to farthest neighbors. The problem is solved most efficiently for j, but less efficiently for j's nearest neighbor $j+1$. Efficiency is monotonic nonincreasing with increasing positive integer i form neighbors' index $j+i$. At some threshold of inefficiency occurring at profile $j+I$, it is better to resort to reset efficiency and then continue with the structuring computations. Note that complete sorting of all profiles is not necessary, but only those not yet processed and adequately near profile $j+I$.

Algorithm

1. Set "b" to a positive number <1 and "a" to a positive number. Select an initial element from U, indexed j.
2. For selected j, sort all values of $D(j, k)$ that are less than $H + a$ in increasing order and identified by sorting index $I = 1, 2, \ldots, n$. Note each sorted index value of I specifies a specific element k of U. Denote this function as $k[I]$.
3. Set $I = 1$.
4. If $D(j, k[I])>a$, then go to step 2. After resetting $j = k[I]$ (remaining sorted elements cannot solve the problem for profile j).
5. Set $r = k[I]$.
6. Hr is the subset of all $k[L]$ such that $D(r, k[L]) < H$ for $L = 1, 2, \ldots,$ n of sorted indices. Also, update each such $Hk[L]$ with profile r.
7. Profile r is no longer required. Remove r from U.
8. Efficiency test: if the number of elements in Hr is less than $b*n$, then set $j = k[I+1]$ and go to step 2.
9. Otherwise, update to next sorted index (i.e., set $I = I+1$) and go to step 4.

Enhancement

Note that the efficiency of the algorithm is reduced as the dimension of the profiles increase, that is, more variables. To offset this, one could implement this method recursively to apply to each subproblem as defined within each iteration of the basic algorithm. This would require a simple label be assigned to each profile to dynamically indicate permission to update its neighborhood count or not.

Acknowledgments

Thanks to the dedicated staff of Novation Biosciences, Inc., 2001–2002 for their support of advanced microarray methods: Drew Watson, Hinrich Schuetze, Rob Dickinson, Mikai Illouz, Amitabh Shukla, Srinivas Mangipudi, Atul Kamat, Willie Quinn, Allen Poirson, Damon Horowitz, and Scott Braxton. Thanks to Novation collaborators Maggie Cam and Brian Oliver (NIH), Dov Shiffman and Richard Lawn (CVT), and Virginia Tusher (Stanford U.) Thanks for the support and enlightenment from the staff at Incyte Corp, 1998–2001, especially Rick Johnston, Robert Stack, Scott Eastman, Jeanne Loring, Mike Furness, Ben Cocks, Huibin Yue, and Jeff Seilhamer. Thanks to Kurt Jarnagin, Brigitte Ganter, Alan Roter, and Alex Tolley of Iconix Pharmaceuticals for the application of my advanced technologies to their array projects. Thanks to Barney Saunders and Krishna Ghosh for Agilent support.

References

Arya, S., and Mount, D. M. (1993). Algorithms for fast vector quantization. Technical Report CS-TR-3017, University of Maryland Institute for Advanced Computer Studies (UMIACS).

Bishop, M. J. (ed.) (1998). "Guide to Human Genome Computing," 2nd Ed. Academic Press, New York.

Churchill, G. A. (2002). Fundamentals of experimental design for cDNA microarrays. *Nature Genet.* **32**(Suppl.), 490–495.

Comanor, L., and Minor, J. M. (1996). Method and apparatus for predicting therapeutic outcomes patent U.S. 5,860,917.

Cleveland, W. (1979). Robust locally weighted regression and smoothing scatterplots. *J. Am. Statist. Assn.* **74,** 829–836.

Comanor, L., Minor, J. M., Conjeevaram, H. S., Roberts, E. A., Alvarez, F., Bern, E. M., Goyens, P., Rosenthal, P., Lachaux, A., Shelton, M., Sarles, J., and Sokal, E. M. (2000). Statistical models for predicting response to interferon-alpha and spontaneous seroconversion in children with chronic hepatitis B. *J. Viral Hepat.* **7,** 144–152.

Comanor, L., Minor, J. M., Conjeevaram, H. S., Roberts, E. A., Alvarez, F., Bern, E. M., Goyens, P., Rosenthal, P., Lachaux, A., Shelton, M., Sarles, J., and Sokal, E. M. (2001). Impact of chronic hepatitis B and interferon-alpha therapy on growth of children. *J. Viral Hepat.* **8,** 139–147.

Drucker, H., Burges, C. J., Kaufman, L., Smola, A., and Vapnik, V. (1997). Support vector regression machines. *In* "Advances in Neural Information Processing Systems" (M. C. Mozer, M. I. Jordan, and T. Petsche, eds.), pp. 155–161. MIT Press, Cambridge, MA.

For PTO publications, go to http://www.uspto.gov/patft/index.html.

Ganter, B., Tugendreich, S., Pearson, C. I., Eser, A., Baumhueter, S., Bostian, K. A., Brady, L., Browne, L. J., Calvin, J. T., Day, G., Breckenridge, N., Eynon, B. P., Dunlea, S., Furness, L. M., Ferng, J., Fielden, M. R., Fujimoto, S. Y., Gong, L., Hu, C., Idury, R., Judo,

M. S. B., Kolaja, K. L., Lee, M. D., McSorley, C., Minor, J. M., Nair, R. V., Natsoulis, G., Nguyen, P., Nicholson, S. M., Pham, H., Roter, A. H., Sun, D., Tan, S., Thode, S., Tolley, A. M., Vladimirova, A., Yang, J., Zhou, Z., and Jarnagin, K. (2005). Development of a large-scale chemogenomics database to improve drug candidate selection and to understand mechanisms of chemical toxicity and action. *J. Biotechnol.* **29**, 219–244.

Kerr, M. K., Martin, M., and Churchill, G. A. (2000). Analysis of variance for gene expression microarray data. *J. Comput. Biol.* **7**, 819–837.

Lau, D. T. Y., Comanor, L., Minor, J. M., Everhart, J. E., Wuestehube, L. J., and Hoofnagle, H. (1998). Statistic models for predicting response to interferon-a in patients with chronic hepatitis B. *J. Viral Hepat.* **5**, 105–114.

Li, C., and Wong, W. H. (2001). Model-based analysis of oligonucleotide arrays: Model validation, design issues and standard error application. *Genome Biol.* **2:** research0049.1–0049.12.

Martinot-Peignoux, M., Comanor, L., Minor, J. M., Ripault, M. P., Boyer, N., Castelnau, C., Giuily, N., Hendricks, D., and Marcellin, P. (2006). Accurate model predicting sustained response at week 4 of therapy with pegylated interferon with ribavarin in patients with chronic hepatitis C. *J. Viral Hepat.* **13**.

Minor, J. (1999). Internal memos. Incyte Corp.

Minor, J. (2001). System and method for dynamic data clustering, PTO publication U.S. 20030093411.

Minor, J. (2002c). Analyzing and correcting biological assay data using a signal allocation model, PTO publication U.S. 20050143933.

Minor, J. (2003a). Methods and system for multi-drug treatment discovery. U.S. 20050037363.

Minor, J. (2003b). Feature quantitation methods and system, PTO publication U.S. 20050078860.

Minor, J., Illouz, M., Watson, G., and Dickinson, R. D. (2002a). Microarray performance management system, PTO publication U.S. 20040019466.

Minor, J., and Illouz, M. (2002b). Method and system for predicting multi-variable outcomes, PTO publication U.S. 20040083452.

Nelder, J. A., and Wedderburn, R. W. M. (1972). Generalized linear models. *J. R. Stat. Soc. A* **132**, 107–120.

Parisi, M., Nuttall, R., Naiman, D., Bouffard, G., Malley, J, Andrews, J., Eastman, S., and Oliver, B. (2003). Paucity of genes on the Drosophila X chromosome showing male-biased expression. *Science* **299**, 697–700.

Parisi, M., Nuttall, R., Edwards, P., Minor, J., Naiman, D., Lu, J., Doctolero, M., Vainer, M., Chan, C., Malley, J., Eastman, S., and Oliver, B. (2004). A survey of ovary-, testis-, and soma-biased gene expression in *Drosophila melanogaster* adults. *Genome Biol.* **5**, R40.

Ripley, B. (1996). "Pattern Recognition and Neural Networks." Cambridge University Press, Cambridge.

Wu, Z., and Irizarry, R. A. (2004). Preprocessing of oligonucleotide array data. *Nature Biotechnol.* **22**, 656–658.

Yue, H., Eastman, P. S., Wang, B. B., Minor, J., Doctolero, M. H., Nuttall, R. L., Stack, R., Becker, J. W., Montgomery, .R., Vainer, M., and Johnston, R. (2001). An evaluation of the performance of cDNA microarrays for detecting changes in global mRNA expression. *Nucleic Acids Res.* **29**, e41.

[13] Analysis of a Multifactor Microarray Study Using Partek Genomics Solution

By TOM DOWNEY

Abstract

Partek Genomics Suite (Partek GS) is a powerful statistical analysis and interactive visualization software solution designed to analyze single channel oligonucleotide (Affymetrix) and two-color cDNA microarrays, as well as data from other emerging genomic and proteomic technologies. This chapter takes a simple study on obesity and susceptibility to type 2 diabetes and uses it as an example that demonstrates how Partek GS can be used to analyze data arising from a microarray experiment.

Statistical Analysis of Microarray Data

Experimental design and statistical analysis are powerful scientific tools used to make conclusions about a phenomenon based on empirically measured data. Genomics researchers commonly conduct experiments in which specimens (e.g., animals or cell lines) of different phenotypes are compared or in which specimens are exposed to different treatments whose effects are to be studied. On each sample, the expression level of thousands of genes is measured and then analyzed to identify genes that are correlated to the phenotype or treatment of interest. Because there are so many genes measured by today's microarray technologies, the opportunity to make "false discoveries" is very easy, and properly applied statistical methods provide a means to differentiate random patterns from true signals the biology is sending us. There are many types of statistical methods, and we will cover several, but not all, of them in this chapter. First, we will use "exploratory data analysis" and visualization to see the "big picture" from an experiment. Next we will use "statistical inference" to answer the question "what genes are correlated to the treatment." Finally, we will use explanatory visualization to present the findings from the statistical methods used.

Description of the Experiment

In this experiment, two strains of mice, BTBR and C57BL/6J(B6), are compared in four tissues (adipose tissue, liver, skeletal muscle, and pancreatic islets) to identify genes that are differentially expressed between the strains within each tissue type (Lan *et al.*, 2003). Thus there are two (strains) × four

METHODS IN ENZYMOLOGY, VOL. 411   0076-6879/06 $35.00
          DOI: 10.1016/S0076-6879(06)11013-7

(tissues) = eight treatment combinations. Four B6-ob/ob and 4 BTBR-ob/ob male mice at 14 weeks of age are used. Because the amount of RNA from a single mouse is too small, tissue samples from two mice of each strain are pooled together, producing two pooled RNA samples for each of the eight treatment combinations. In statistical terminology this is a "2 × 4 experiment, replicated 2 times," and the 16 total RNA samples are hybridized to Affymetrix MGU74A mouse GeneChips (Dalma-Weiszhausz *et al.*, 2006). Data from this experiment are available on the NCBI's Gene Expression Omnibus (GEO; Barrett and Edgar, 2006) as experiment GSE2899 (http://www.ncbi.nlm.nih.gov/geo/query/acc.cgi?acc=GSE2899).

Importing and Normalizing GeneChip Data

Both .CEL files (containing raw expression measurements) and .EXP files (containing sample treatment information) are available on GEO, and Partek GS can directly import both of these formats. For this example, we used the popular RMA algorithm (Irizarry *et al.*, 2003) for data normalization upon import of raw expression data from the .CEL files. The sample treatment information (strain, tissue type, and animal pool) was imported from the .EXP files and then merged with summarized expression data (Fig. 1).

Exploratory Data Analysis

Exploratory data analysis is used to identify major effects influencing data, unexpected trends, outliers, and batch effects. We will start by using principal components analysis (PCA) to identify major effects influencing the expression values in this experiment (Hotelling, 1933). The PCA in Partek GS is unique in two important ways.

1. Data do not have to be prefiltered. Partek GS can compute PCA very quickly and with very little memory, even if data contain millions of variables (such as the new Affymetrix Exon arrays containing ~1.4 million exons per array). In this study, each sample contains ~12,500 variables (the number of genes on the Affymetrix MGU74A GeneChip).

2. The graphics used by PCA are true three-dimensional graphic representations, complete with light shading, fogging, and the ability to color, size, and so on by multiple factors simultaneously.

Interpreting the PCA Plot from Fig. 2

1. This is a global analysis of the genome and not an analysis of any gene in particular. Samples that are close together are similar across the whole genome, whereas samples that are far apart are dissimilar across the whole genome.

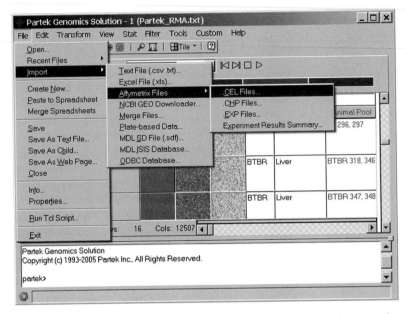

FIG. 1. Importing Affymetrix data into Partek GS. Note that each sample appears in one row and that each gene appears in a single column. The first three columns show thumbnail images for raw chip data, data after RMA normalization, and residuals of the RMA correction, respectively. Double clicking on the thumbnail images will display the full-size image for inspection. (See color insert.)

2. The total variation explained by PCs 1, 2, and 3 is 66.3% (24.9% by PC #1, 22.5% by PC #2, and 19% by PC #3).

3. Tissue is the largest effect in data. As with almost any multitissue experiment, the largest effect in data is due to the different tissue types. This is evidenced by the clustering of samples primarily by tissue.

4. Within each tissue, the small points (B6 strain) are distinct from the large points (BTBR strain). Thus there appears to be a significant overall difference between the two strains, although this effect is much smaller than the tissue effect.

5. To the extent that the connecting lines run relatively parallel to each other, it indicates that samples from each animal pool are more similar to each other than samples from different animals. The lines do not seem to be very parallel, indicating that the difference from one animal pool to the other is relatively small compared to tissue and strain effects.

6. There are no apparent outliers in the PCA plot, thus there is no reason to suspect that any of the samples should be removed for quality reasons. If there was an outlier, it could be inspected easily in Partek by

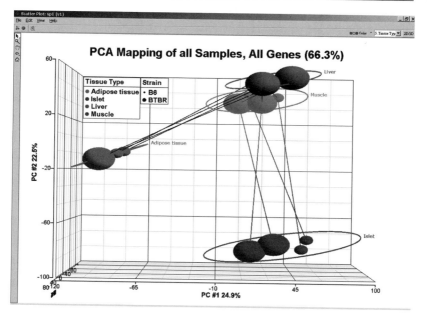

FIG. 2. PCA mapping of the samples from 12,488-dimensional "gene space" to three dimensions for interactive visualization. Samples are colored by tissue type and sized by strain. Lines connect samples from the same animal pool, and ellipsoids are drawn around each tissue group. (See color insert.)

clicking on the corresponding point in the scatterplot and invoking the full size image extracted from the .CEL file.

Multidimensional Scaling (MDS)

Multidimensional scaling is a nonlinear "cousin" of PCA (Torgerson, 1952). Briefly, MDS refers to a family of methods that map high-dimensional data down to a lower dimensionality usually for the purpose of visualization. The important criterion that is optimized by this technique is that objects that are "similar" in high-dimensional space are mapped in such a way that they are "close together" in low-dimensional space. Conversely, objects that are "dissimilar" in high-dimensional space are mapped in such a way that they are "far apart" in low-dimensional space. The user has several choices in defining the meaning of "similarity" in Partek software—most commonly it is a distance measure such as Euclidean distance or other similarity measure such as linear or nonlinear correlation. Figure 3 shows how MDS sees these same data.

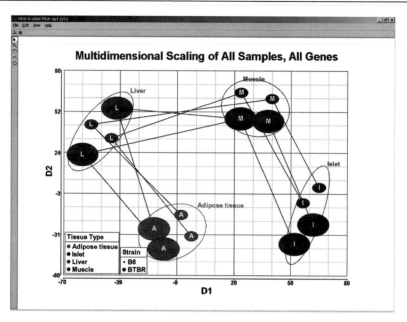

FIG. 3. Multidimensional scaling shows similar patterns as PCA—samples are grouped by tissue type and within each tissue type (except liver), the samples are differentiated by strain. (See color insert.)

Identifying Outliers Using PCA and MDS

Neither PCA nor MDS indicated that any of the samples were extreme outliers, which could be indicative of poor-quality RNA or a flawed hybridization of one or more of the samples. If there were any outliers, the user simply selects the outlier in a graph or on the spreadsheet and invokes an image of the chip (Fig. 4).

Hierarchical Clustering

Hierarchical clustering was one of the first analysis tools used to analyze microarray experiments, and Partek's hierarchical clustering is full featured, allowing dual clustering of genes and samples, interactive branch flipping, and many options for clustering and coloring the resulting dendrograms and heat maps. Figure 5 shows hierarchical clustering of the samples and also shows patterns consistent with PCA and MDS.

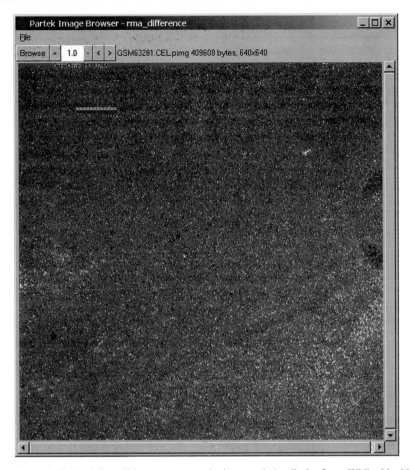

FIG. 4. Individual GeneChips or arrays can be inspected visually for flaws. While this chip shows slight imperfections, the quality is acceptable.

Finding Differentially Expressed Genes Using Analysis of Variance (ANOVA)

Analysis of variance is a very powerful technique for identifying differentially expressed genes in a multifactor experiment such as this one (Fisher, 1925). ANOVA partitions the variability due to treatments from technical and biological noise and then uses signal-to-noise ratios (F ratios) to identify differences that are statistically significant (small p values). In order to identify genes differentially expressed between strains in each of

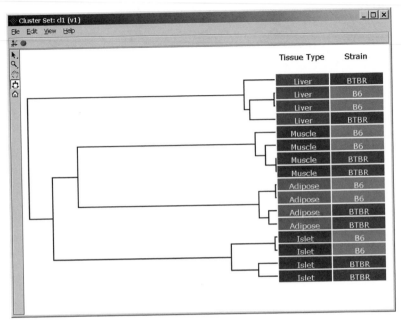

FIG. 5. Hierarchical clustering of samples using Euclidean distance and average linkage. Each branch is annotated with tissue type and strain. Note that samples cluster primarily by tissue and secondarily by strain (with the exception of liver tissues). (See color insert.)

the different tissues, the following mixed model ANOVA was used to analyze these data.

$$y_{gijkl} = \mu_g + S_i + T_j + ST_{ij} + P(S)_{ik} + \varepsilon_{gijk}$$

where y_{gijk} is the expression of the gth gene for ith strain, jth tissue, and kth animal pool. The symbols S, T, ST, and $P(S)$ represent effects due to strain, tissue, strain-by-tissue interaction, and pool-nested-within-strain, respectively. The error for the gth gene for sample ijk is designated as ε_{gijk}.

The aforementioned ANOVA model is dictated by the experiment design and partitions the variability due to strain, tissue, strain by tissue, and animal pool from biological and technical noise. Some of you may be wondering why the animal pool is required in the ANOVA model. Of the three assumptions of ANOVA (normal distribution, equal variance, and independence), the assumption of independence is the most important assumption (for any statistical test) and should not be violated. The assumption of independence requires that all samples within a treatment group be "independent." This means that no two samples are any more like each other than they are like any other within that group. However, four samples from the B6 group come from one pool, and the remaining four samples come from another pool. The same is true for the BTBR strain. Leaving the animal pool out of the ANOVA model would cause us to underestimate the variability within each strain (noise) and thus lead to overoptimistic p values (increased "false discoveries" or type I errors).

Random vs Fixed Effects: Mixed Model ANOVA

There is one more important note about the animal pool—it is referred to as a "random effect." When an ANOVA model contains both random and fixed effects, it is referred to as "mixed model ANOVA."

- Strain is a fixed effect. There are two strains, B6 and BTBR. Because these are the only two strains that we care about for this experiment, it is a fixed effect.
- Tissue is a fixed effect. We care about changes at adipose, islet, liver, and muscle only. We are not trying to make any inferences about any other tissues, thus tissue type is a fixed effect.
- Animal pool, however, is not a fixed effect, as the two animal pools in this experiment represent only a random sample of all the animal pools we wish to make an inference about.

Here is another way to tell if a factor is random or fixed: Imagine repeating the experiment. Would the same levels of each factor be used again?

- Strain. The same strains would be used again—a *fixed* effect.
- Tissue. The same tissues would be used again—a *fixed* effect.
- Animal pool. No, we would use new animals—a *random* effect.

Hierarchical Designs and Nested/Nesting Relationships

In the ANOVA model, we said that "pool is nested in strain." This is a special relationship that results from hierarchical experiment designs such as this one. The multiple samples from the same animal pool are always in the same group for the factor "strain." Thus, knowing the animal pool means that we know the strain, and we say "pool is nested in strain" or "pool (strain)." Pool is the "nested" variable, and strain is the "nesting" variable. The relationship between animal pool and strain is hierarchical, thus this common type of experiment is known as a "hierarchical" or "nested" design. Because Partek detects the nested/nesting relationship automatically, the scientist performing this analysis does not have to be an expert on ANOVA.

Creating Gene Lists of Interest Using ANOVA and Linear Contrasts

The ANOVA model is dictated by the experiment design. The gene lists, however, are dictated by the interests of the researcher. In order to construct the gene lists of interest, we use a technique called "linear contrasts" within the ANOVA model. These linear contrasts are constructed to compare the B6 and BTBR samples within each tissue. Too frequently, researchers who are unfamiliar with mixed model ANOVA and linear contrasts will analyze each tissue separately and use a *t* test (or similar test) to compare the two strains within each tissue. The problem is that this overly simple analysis reduces the sample size from 16 to 4 for each test, which is too small to produce statistically significant results. Sixteen is a small enough sample size to begin with, and we would like to use all 16 samples in our estimates of variance for each gene—linear contrasts and multifactor ANOVA allow us to do so.

Examining the Results

The ANOVA is run on all 12,488 genes, and a table is created allowing the scientist to browse and create gene lists from the results. For each gene, the table includes the following values.

- *p* value for each factor in the experiment
- *p* value for each contrast (e.g., strain *x* vs *y* in each tissue)
- fold change for each contrast.
- gene identification and user-specified annotations
- links to internet databases, genome browsers, etc.

Figure 6 shows one way to summarize results of the ANOVA. It displays the effect sizes for all the factors in the ANOVA model. Consistent with the exploratory analysis (PCA, MDS, and clustering), the tissue effect is by far the largest source of variation in these data. The difference between the strains is also significant with an average signal-to-noise ratio (S/N), or F ratio, of 2.84. There also appears to be a significant strain-by-tissue interaction (average S/N of 1.81), which indicates that the difference between the strains depends on the tissue. The difference between animal pools is not that large, although it is significant (average S/N = 1.15).

Multiple Test Correction

A step-up false discovery rate (FDR) was applied to *p* values from the linear contrasts to determine a cutoff for significantly differentially expressed genes within each tissue (Benjamini and Hochberg, 1995). Table I shows the number of genes that pass an FDR of 10%, meaning that we expected 10% of the genes on each list to be false positives.

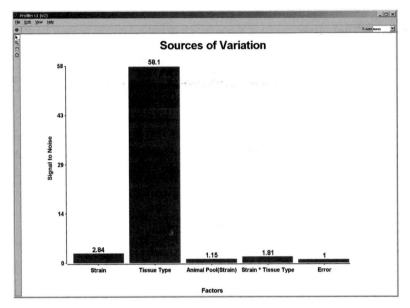

Fig. 6. "Sources of variation" plot. Each bar indicates the average signal for all 12,488 genes. More specifically, the height of the bar is the average "mean square" (ANOVA's name for "variance"). Bars are labeled with the ratio to "error" (noise), thus bars represent an average F ratio (signal-to-noise ratio) for each factor in the model.

TABLE I

NUMBER OF GENES PASSING STATISTICAL SIGNIFICANCE (FDR = 10%)[a]

Factors	Genes passing 10% FDR
Strain	0
Tissue type	9530
Animal pool (strain)	0
Strain × tissue type	45

Linear contrasts	Genes passing 10% FDR
B6/adipose vs BTBR/adipose	29
B6/islet vs BTBR/islet	590
B6/liver vs BTBR/liver	43
B6/muscle vs BTBR/muscle	18

[a] The top of the table lists the number of genes significant for each factor in the ANOVA model, and the bottom of the table lists the number of genes significant for each of the four linear contrasts (one for each tissue).

Examining Results for a Single Gene

Because most people can interpret pictures better than p values, it is important to have effective ways to visualize the patterns detected using ANOVA. Because visualization is a manual and subjective process, it cannot realistically be applied to all of the 12,488 genes on the chip. Thus, we use statistical tests such as ANOVA to identify the interesting genes and then explain those results using appropriate visualizations. Figure 7 shows one such visualization of an interesting gene. This gene is upregulated in the B6 strain in the liver only. The gene was statistically significant in the contrast of B6 to BTBR within the liver tissue. Note that all 16 samples were used to estimate the variance due to noise, which is clearly very small.

Poststatistical Analysis

Genes can be annotated, and gene lists created and compared. Figure 8 shows a Venn diagram created by the Partek list manager. It shows the

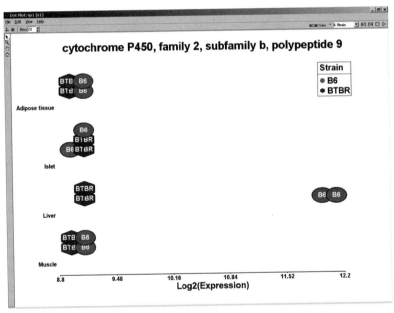

Fig. 7. A "dot plot" is a very effective visualization tool when the sample size is relatively small, such as in this experiment. There are 16 "dots," one for each sample. The x axis represents the $\log_2$ of the expression for this gene. The points are colored (and shaped) by strain and separated on the y axis by tissue. For clarity, dots are also labeled with the strain. Note that there is very little noise in these data and that the 16 samples provide a very good estimate of the noise variance. (See color insert.)

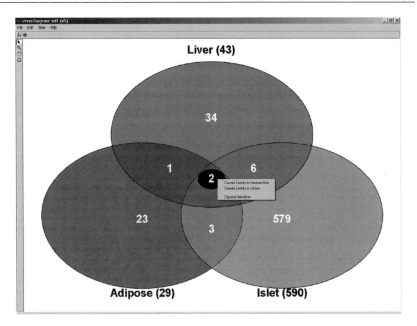

FIG. 8. Often researchers are interested in creating gene lists that are a combination of other lists. The list manager of Partek allows the researcher to combine these lists in several ways. This shows the Venn diagram tool of Partek, which allows the scientist to interactively create gene lists from the intersection or union of two or more lists. (See color insert.)

significant genes for three tissues, and the researcher can look at intersections and unions of genes in each region.

Visualizing Locations of Significant Genes on the Genome

Another useful analysis for many studies is to examine where the differentially expressed genes (or exons, chromosomal copy number) are located on the genome. Figure 9 shows the Partek genome browser, which allows the researcher to find and display interesting patterns based on statistical significance (p value), fold change, etc. Individual chromosomes, regions within chromosomes, cytobands, or individual genes (or SNPs and exons) can be explored easily and interactively. Because the location of the genes on the chromosome for the example experiment used in this chapter is not compelling, the visualizations shown in Fig. 9 are examples from other data sets that show more interesting patterns.

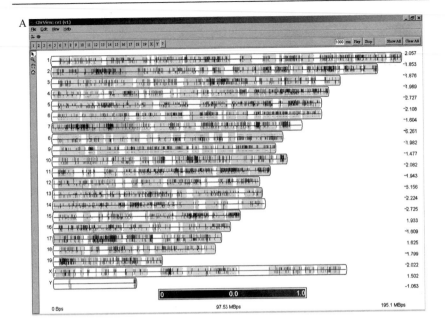

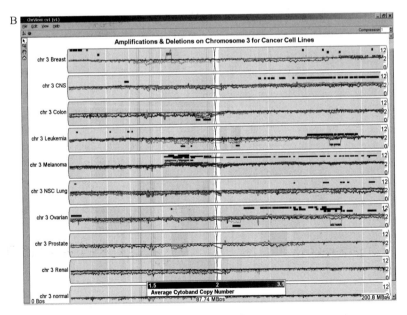

FIG. 9. *(continued)*

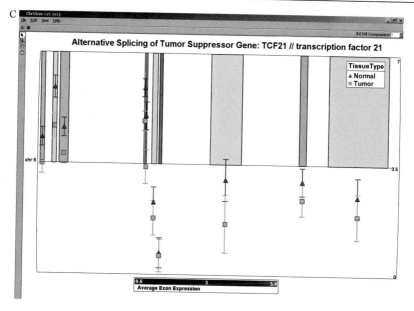

Fig. 9. (A) Genes are displayed in the Partek genome browser. Genes are colored by *p* value, and the heights of the lines indicate fold change. In this screen shot, one line is drawn for each chromosome. (B) In this view, chromosomal copy number amplifications and deletions from a tumor sample can be visualized. Regions of statistically significant alterations are indicated with red (amplification) and blue (deletion) rectangles. In this screen shot, only chromosome 8 is displayed. (C) This view examines individual exons on a single gene. This gene exhibits alternative splicing (detected using analysis of variance). Overall, all genes on the exon are upregulated in the tumor relative to the normal; however, there is a single exon near the center of the gene that is expressed higher in the normal group than the tumor group. This is an indication of alternative splicing. (See color insert.)

Summary

This chapter described the statistical analysis of a microarray experiment using Partek Genomics Solution software. Data were first normalized using RMA. Next exploratory analysis was used for quality assurance and to identify major effects and trends in data, revealing tissue as the largest source of variation, followed by a strain effect. ANOVA was used to partition the variance due to the multiple factors in this experiment, and linear contrasts were used to find the genes of interest to the researcher. Finally, graphical methods were used to display the effects of the experimental treatment(s) on the genes of interest. If you are interested in exploring more example experiments, data and tutorials that accompany them can be found at http://www.partek.com/.

References

Barrett, T., and Edgar, R. (2006). Gene Expression Omnibus (GEO): Microarray data storage, submission, retrieval, and analysis. *Methods Enzymol.* **411,** 352–369.

Benjamini, Y., and Hochberg, Y. (1995). Controlling the false discovery rate: A practical and powerful approach to multiple testing. *JRSS B* **57,** 289–300.

Dalma-Weiszhausz, D. D., Warrington, J., Tanimoto, E. Y., and Miyada, C. G. (2006). The Affymetrix GeneChip platform: An overview. *Methods Enzymol.* **410,** 3–28.

Fisher, R. A. (1925). "Statistical Methods for Research Workers." Oliver & Boyd, Edinburgh.

Hotelling, H. (1933). Analysis of a complex of statistical variables into principal components. *J. Educ. Psychol.*

Irizarry, R. A., Hobbs, B., Collin, F., Beazer-Barclay, Y. D., Antonellis, K. J., Scherf, U., and Speed, T. P. (2003). Exploration, normalization, and summaries of high density oligonucleotide array probe level data. *Biostatistics* **4**(2), 249–264.

Lan, H., Rabaglia, M. E., Stoehr, J. P., Nadler, S. T., Schueler, K. L., Zou, F., Yandell, B. S., and Attie, A. D. (2003). Gene expression profiles of nondiabetic and diabetic obese mice suggest a role of hepatic lipogenic capacity in diabetes susceptibility. *Diabetes* **52**(3), 688–700.

Torgerson, W. S. (1952). Multidimensional scaling. 1. Theory and method. *Psychometrika* **17,** 401–419.

[14] Statistics for ChIP-chip and DNase Hypersensitivity Experiments on NimbleGen Arrays

By PETER C. SCACHERI, GREGORY E. CRAWFORD, and SEAN DAVIS

Abstract

Data obtained from high-density oligonucleotide tiling arrays present new computational challenges for users. This chapter presents ACME (Algorithm for Capturing Microarray Enrichment), a computer program developed for the analysis of data obtained using NimbleGen-tiled microarrays. ACME identifies signals or "peaks" in tiled array data using a simple sliding window and threshold strategy and assigns a probability value (p value) to each and every probe on the array. We present data indicating that this approach can be applied successfully to at least two different genomic applications involving tiled arrays: ChIP-chip and DNase-chip. In addition to highlighting previously described methods for analyzing tiled array data, we provide recommendations for assessing the quality of ChIP-chip and DNase-chip data, suggestions for optimizing the use of ACME, and descriptions of several of ACME features designed to facilitate interpretation of processed tiled array data. ACME is written in R language and is freely available upon request or through Bioconductor.

METHODS IN ENZYMOLOGY, VOL. 411   0076-6879/06 $35.00

Introduction

Advances in microarray technologies have led to unprecedented performance and ultimate flexibility in microarray design. In the production of high-density oligonucleotide microarrays, DNA fragments or "probes" (25–70 nucleotides in length) are designed to span or "tile" across genomic regions of interest with overlapping or nonoverlapping probes. These microarrays, hereafter referred to as "tiled arrays," are technically easier to manufacture than polymerase chain reaction (PCR) amplicon-based probe arrays and allow for higher resolution of analysis due to the large number of measurement points along the region of interest. Given these advantages, it is not surprising that tiled arrays are gaining popularity for applications including array-CGH, chromatin-immunoprecipitation coupled with microarray analysis (ChIP-chip), and genomic expression analysis.

Companies can currently manufacture arrays containing hundreds of thousands to over a million probes on a single slide, and resolution is expected to increase even further. Prior to the development of tiled arrays, virtually all methods for analysis of microarrays were designed to measure the abundance of a given RNA or DNA molecule at a single probe. In contrast, with tiled arrays, increased DNA or RNA abundance often spans several probes, resulting in multiple probes representing a single signal. Due to these fundamental differences, methods for analysis of PCR-amplicon arrays are not well suited for analysis of tiled arrays. New computational methods for tiled arrays are clearly required to maximize detection of true signal from background noise.

This chapter describes ACME (Algorithm for Capturing Microarray Enrichment), a software tool developed for analysis of data obtained from tiled arrays. ACME exploits the "single-tail" and "neighbor effect" characteristics of ChIP-chip and DNase-chip data and is therefore not suitable for analyzing array-CGH or expression data. Briefly, ACME uses a sliding window approach to identify potential sites of enrichment above a user defined threshold and then assigns probability scores (p values) to each probe on the array. ACME is written in R, a free software environment for statistical computing and graphics designed to run on a variety of UNIX platforms, Windows, and MacOS (http://www.r-project.org/). An advantage of R over more commonly used software packages such as Microsoft Excel is that there is virtually no limit to the size of the data file that can be analyzed. This no-limit feature is essential given the massive number of probes on tiled arrays and will also enable ACME to accommodate expected increases in probe quantities.

ChIP-chip: An Overview

Chromatin immunoprecipitation coupled with microarray analysis has become a popular technique for determining the genomic binding sites of transcription factors (Kim *et al.*, 2005; Odom *et al.*, 2004) or the location of histone modifications on chromatin (Bernstein *et al.*, 2005). An overview of the strategy is summarized in Fig. 1A. Cells in culture or tissue samples are treated with a cross-linking agent to covalently link proteins to their cognate DNA sites. The cells are then harvested and chromatin is sheared by sonication. The protein–DNA complexes are then enriched by chromatin immunoprecipitation with antibodies to histone modifications, transcription factors, or other DNA-binding complex members. The resulting ChIP-enriched DNA is purified and the cross-links are heat reversed. Because the yield of factor-bound DNA tends to be low, enriched fragments are usually amplified before labeling and hybridizing to arrays. As a control for hybridization, total genomic DNA that has been cross-linked and sheared is also amplified. For two-color ChIP-chip, the amplified ChIP and total genomic DNA preparations are labeled with distinct fluorophores and cohybridized to arrays. Genomic segments bound by the protein of interest will show enrichment of chromatin-immunoprecipitated DNA over the total genomic reference DNA.

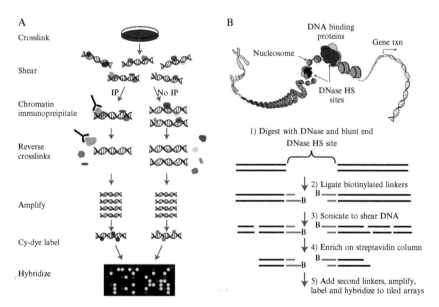

FIG. 1. Overview of ChIP-chip (A) and DNase-chip (B) (see text for details). (See color insert.)

NimbleGen offers completely customizable array designs for ChIP-chip. While tiling across the entire genome is usually most desirable, this option is currently cost prohibitive for most laboratories. Therefore, many users opt for tiling across selected genomic regions, individual chromosomes, or promoter regions (if the protein in question is known to localize to promoters).

DNase-chip: An Overview

Mapping DNase I hypersensitive (HS) sites is an accurate method of identifying the location of active gene regulatory elements, including promoters, enhancers, silencers, and locus control regions (Gross and Garrard, 1988; Wu et al., 1979). While Southern blots have historically been used to identify DNase HS sites one gene at a time, newer strategies have been developed to rapidly identify larger numbers of these regulatory regions (Crawford et al., 2004, 2006; Sabo et al., 2004). One of these high-throughput strategies, called DNase-chip, uses tiled microarrays to accurately identify the location of DNase HS sites (Crawford et al., 2006). This method entails digesting intact chromatin with small amounts of DNase I to preferentially digest open regions of chromatin (Fig. 1B). The DNase-digested ends are blunted, ligated to biotinylated linkers, and sonicated. DNase-digested ends are then captured on a streptavidin column, amplified, labeled, and hybridized to tiled microarrays. As a reference, randomly sheared DNA is cohybridized with DNase-digested DNA.

Properties of ChIP-chip and DNase-chip Data

In a typical two-color ChIP-chip or DNase-chip experiment, two different populations of DNA are labeled with distinct fluorophores and cohybridized to arrays. One population corresponds to enriched genomic fragments captured by either ChIP or treatment with DNase. The other population corresponds to total genomic DNA and serves as a reference for hybridization. A histogram that displays $\log_2$ measurement ratios (ChIP- or DNase-enriched DNA/total genomic DNA) can be used to determine if an experiment was successful. If enrichment is low or completely absent, the distribution of ratio measurements will appear symmetrically bell shaped (Fig. 2A). For ChIP-chip or DNase-chip experiments in which multiple DNA fragments were enriched, the expected distribution will appear asymmetrically bell shaped, with a distinct skew or tail at the positive or right-hand side (Fig. 2B and C). The shape and size of the tail often vary depending on the degree of enrichment. Furthermore, hybridization artifacts such as streaks will often distort the shape of the distribution (Fig. 2D).

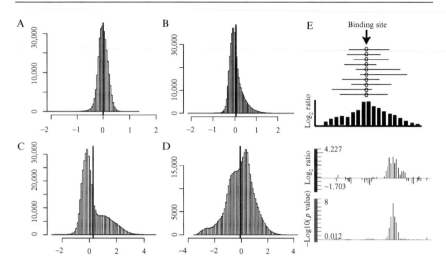

FIG. 2. Examples of "one-tailed" data and "neighbor effect." (A–D) Histograms showing various degrees of enrichment from ChIP-chip experiments using no antibody (no enrichment) (A), antibodies to the MLL1 transcription factor (moderate enrichment) (B), and antibodies to a histone modification (H3K4me3) (high enrichment) (C). (D) Background noise caused by streaks on the array can affect the distribution dramatically, making it difficult to estimate the degree of enrichment. Red lines in A–D denote the mean center of the distribution. Note that this line is shifted to the right in instances where obvious enrichment occurred. Frequency is plotted on the vertical axis; $\log_2$ ratio measurements (ChIP DNA/total genomic DNA) are plotted on the horizontal axis. (E) Theoretical (top) and actual (bottom) examples of the "neighbor effect" principle. (Bottom) $\log_2$ ratio measurements (red) and ACME processed (green) data from a ChIP-chip experiment using antibodies to H3K4me3. Data represent the average of three biological replicates. Probes on the array were spaced at a density of 1 probe per 180 bp. (See color insert.)

The shearing of chromatin by sonication results in multiple fragments of various lengths. Given the length of these fragments (200–1000 bp), we expect true signal to span multiple probes that are genomically located close to one another. The signal from multiple closely spaced probes will often form a peak at or near the binding site or DNase HS site (Fig. 2E), an effect referred to as "neighbor effect." Thus, densely spaces probes that have only single probes yielding spuriously high ratio measurements most likely represent experimental noise.

Previously Developed Methods for Analysis of ChIP-chip Data

Previously described techniques used for analyzing ChIP-chip include the single array error model (SAEM) (Hughes *et al.*, 2000), percentile rank analysis (Lieb *et al.*, 2001), PeakFinder (Glynn *et al.*, 2004), chromatin immunoprecipitation on tiled arrays (ChiPOTle) (Buck *et al.*, 2005), double regression analysis (Kim *et al.*, 2005), and hidden Markov model analysis (HMM) (Li *et al.*, 2005). Algorithms for the SAEM and percentile rank methods were written for analysis of PCR-amplicon arrays, prior to use of high-density tiled arrays. Briefly, SAEM uses control arrays to determine the boundaries outside of which a probe should be considered significantly enriched. SAEM requires multiple control arrays and assumes that the control and experimental arrays are of similar quality. In the percentile rank method, each arrayed probe is assigned a percentile rank based on the degree of ChIP enrichment. This process is repeated for multiple experimental replicates, and the median percentile ranks for each probe are calculated and plotted as a histogram. Provided experimental replicates yield consistent enrichment, the distribution of median rank values will be bimodal. Probes that rank higher than the trough of the bimodal distribution are considered significant. Neither SAEM nor percentile rank utilizes ratio measurements from neighboring probes when determining signal. PeakFinder sorts probes by their genomic location and then smooths data. The derivative of the smoothed line is then used to identify peaks. PeakFinder does not assign a measure of statistical significance to enriched regions. ChIPOTle detects ChIP-enriched regions using a moving-window approach. ChiPOTle assumes that the $\log_2$ ratios are independent and Gaussian distributed (unless the "permutation" option is used, which relaxes the Gaussian distribution assumption). ChiPOTle is available as an Excel macro and is restricted to analyzing data derived from arrays containing less than 65,536 probes due to the limitations of Excel worksheets. Li *et al.* (2005) provided a method for using a two-state hidden Markov model to determine regions of significance in a tiling array experiment. The "double regression" method uses a peak-finding algorithm that predicts binding sites based on hybridization intensity and the triangular-shaped nature of consecutive probes with significant signals. PeakFinder, ChiPOTle, HMM, and double regression utilize information from multiple neighboring probes, and thus take advantage of the neighbor effect.

ACME

ACME can analyze data obtained from any NimbleGen tiled array design. Experimental replicates can be processed separately or, alternatively, the user can average ratio measurements from experimental replicates prior to processing. ACME makes only two assumptions: (1) that data are enriched for signal in the positive direction ("one-tailed") and (2) that the real signal will be represented by multiple probes that are genomically located close to one another (neighbor effect). Unlike other programs

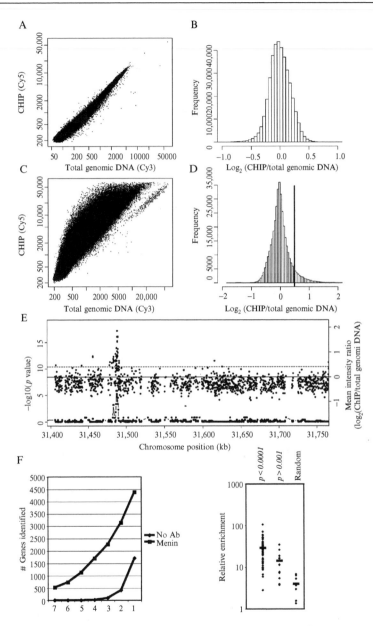

FIG. 3. Analysis of data from tiled arrays using ACME. In this example, promoter-specific arrays (NimbleGen) containing 385,0000 probes tiled at a resolution of 1 probe per 100–180 bp were used to identify genomic-binding sites for the tumor suppressor protein menin. (A and C) Scatter plots of single-array intensities of oligonucleotide probes obtained from chromatin immunoprecipitation with antibodies to menin (C) and a control experiment with no antibody (A). Successful experiments are identified as those that show enrichment of multiple probes in

(Buck *et al.*, 2005; Kim *et al.*, 2005), ACME does not assume that $\log_2$ ratio data are Gaussian distributed or that the signal ratios of consecutive probes form a triangular shape. Despite its simplicity, we have found that ACME is quite robust at detecting true signal from noise (Scacheri *et al.*, 2006).

ACME is designed to process normalized microarray data from NimbleGen arrays that are in standard GFF format (http://genome.ucsc.edu/FAQ/FAQformat), but can be modified easily to analyze data in any format, provided the chromosome coordinates and corresponding signal measurements for each probe are supplied. After loading data into R, ACME automatically sorts probes by their genomic location. The user must then set a threshold within the distribution of the ratio measurements above which true positive signals are expected to be enriched (e.g., 0.9 or 90th percentile). To identify potential sites of enrichment, a window of user-defined size moves stepwise along the tiled region, centering at every probe. Hybridization signals of probes within each window are tested by χ^2 analysis to determine if the window contains a higher than expected number of probes above the defined threshold. ACME then estimates the significance of each probe by assigning probability values to each and every probe on the array. The resulting output contains *p* values with corresponding chromosome coordinates, which can be plotted easily by genomic location in R or imported into the UCSC Genome Browser or the Integrated Genome Browser for visualization. An example of actual ChIP-chip data processed by ACME is depicted in Fig. 3.

---

the Cy5 channel (chromatin immunoprecipitated DNA) over the Cy3 channel (total genomic DNA). (B and D) Histograms of normalized intensity ratios from A and C. Compared to the no antibody control (B), the histogram plotted from the menin ChIP (D) shows a distinct tail at the right-hand end, or positive direction. ACME slides a window of user-defined size along tiled regions and tests statistically whether each window contains a higher than expected number of probes above a user-defined threshold (indicated by the red bar). In this example, the window size is set at 1000 bp and the threshold at 90%. (E) Plots showing data before and after processing with ACME. Normalized $\log_2$ ratio measurements from three experimental replicates were averaged and plotted in black, with corresponding chromosome coordinates on the horizontal axis and the mean intensity ratio on the right vertical axis. Points in red indicate corresponding significance values for each data point following processing by ACME (left vertical axis). The dashed line denotes the 90% threshold level. (F) Number of promoters ACME reported to be bound by menin at various *p* value cutoffs. Compared to the negative control experiment (blue), significant enrichment of multiple promoters was detected for the menin ChIP-chip experiment. Also note that the false-positive rate increases as the *p* values decrease in significance. (G) Real-time PCR validation of promoters determined by ACME to be bound menin. Mean values are indicated by red bars. Compared to randomly selected regions of the genome, promoter regions identified by ACME to be enriched for menin binding at *p* < 0.0001 were enriched more than sevenfold. These data indicate that sites ACME reported to be enriched at *p* <0.0001 reliably represent sites of menin occupancy. Users should empirically determine the optimal *p* value cutoff for their studies using similar methods. (See color insert.)

Due to multiple testing problems, p values reported by ACME should be interpreted with caution. p values can be corrected for multiple comparisons using Bonferroni correction, but even adjusted p values will remain imprecise, as independence between individual data points was violated to an unknown degree. p values reported by ACME must therefore be validated by independent means, such as standard ChIP followed by real-time PCR, to determine true biological significance. In addition, one can estimate empirical p values by permutation testing.

Optimizing Window Size and Threshold

Two parameters require optimization when using ACME: window size and threshold level. The user must choose a window size that is large enough to capture signal from multiple probes located close to one another, but small enough to resolve individual signals or peaks that are close together. Optimal window size is dependent on at least two factors: (1) the distance (bp) between tiled probes on the array and (2) the size of the DNA fragments hybridized. Arrays tiled with little or no space between probes can be analyzed with a shorter window size than arrays containing more sparsely spaced probes. With respect to the size of the DNA fragments hybridized, short fragments will require more closely tiled probes than longer fragments so that signal is not missed. As with window size, the optimal threshold level for a given experiment must also be determined empirically. The user should aim to choose a threshold above background noise, where true signals are likely to be enriched. As a general suggestion, we recommend analyzing most data sets at a threshold level between 0.85 and 0.95 and a window size between 500 and 1500 bp (for arrays tiled at a resolution of 1 probe per 180 bp or less). We have shown that ChIP-chip and DNase-chip data analyzed within these ranges reliably detect significantly enriched genomic regions (Scacheri *et al.*, 2006; Crawford *et al.*, 2006). p values tend to decrease in significance as thresholds are lowered closer to background levels (<0.85) (Fig. 4).

Optimizing Probe Resolution

When designing tiled arrays, the goal is to tile probes that are sufficiently close to one another to maximize sensitivity. Arrays designed with little or no space between probes are usually most desirable, but such resolution is often prohibitive due in part to cost and space limitations on the array. In addition, some regions of the genome are too repetitive to yield useful

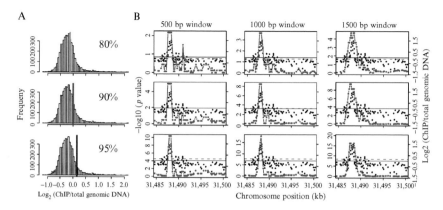

FIG. 4. Examples of data sets processed by ACME at various window sizes and thresholds. (A) Histograms of log₂ ratio measurements. The 80, 90, and 95% thresholds levels at which data were processed are indicated by colored bars. (B) Data at each threshold level indicated in A were processed at window sizes of 500, 1000, and 1500 bp (tiling density = 1 probe per 180 bp). Data analyzed at lower thresholds (80%) yield less significant p values than data processed at higher thresholds (90–95%). Also note that significant p value peak widths increase as window size increases. As discussed in the text, optimal window size and threshold should be determined empirically.

probes for tiling array analysis. To assess the relationship between probe density and sensitivity, we quantified the number of DNase HS sites detectable with variably spaced probes. Data indicate that overlapping probes can detect DNase HS sites with near-perfect accuracy (Fig. 5). Sensitivity decreases as the genomic distance between tiled probes is increased. These data suggest that careful attention should be paid to probe spacing when designing tiled arrays.

Recommendations for Assessing Data Quality

The success of chromatin immunoprecipitation relies heavily on the quality of the antibody. The antibody must be specific for and capable of immunoprecipitating the protein of interest in the context of cross-linked chromatin. When one is performing ChIP-chip for the first time with a new antibody, if is often difficult to assess whether the ChIP was successful, especially when the genomic-binding sites of the protein are not known. Furthermore, the yield of factor-bound DNA in a ChIP experiment is usually low. We recommend assessing the overall degree of enrichment by graphing raw signal intensities as a scatter plot and plotting normalized

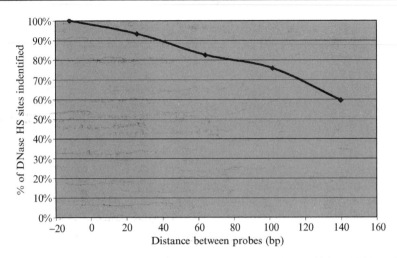

FIG. 5. Sensitivity diminishes as probe resolution decreases. The sensitivity of DNase-chip to detect valid DNase HS sites was calculated for different probe-spacing patterns. Valid DNase HS sites were identified from a previous study (Crawford *et al.*, 2006). DNase-chip was performed on tiled arrays that contained 50-mer probes that overlapped by 12 bases. A significant signal ($p < 0.001$) was detected using ACME for data that included data from all probes, every second probe, every third probe, every fourth probe, and every fifth probe. User-defined window sizes were modified so that different spacing patterns had at least 13 probes per window on average. (See color insert.)

ratio measurements as a histogram (Figs. 2 and 3A–D). If the data points in the scatter plot are distributed along a tight diagonal or if histograms reveal bell-shaped curves that are not single tailed, the degree of enrichment is likely to be low. However, regardless of the distribution, we recommend processing all data with ACME, as even experiments with undetectable amounts of enrichment on scatter plots and histograms can reveal significant regions that are detectable only after processing. Finally, it is often only with repeated experiments on multiple biological replicates that true signal, or lack thereof, becomes evident.

Additional Features of ACME

In addition to providing a measure of significance (p value) for each probe on the array, ACME has several features designed to assist the user view and interpret processed data. A plotting function ("plot") allows the user to view ACME-processed data on the same graph as unprocessed data (Figs. 3E and 4). This plotting feature is particularly useful for optimizing window and threshold parameters, as data analyzed at different window sizes and threshold levels can be compared easily. A function called

"FindRegions" will scan processed data for regions that show significance below a user-defined p value threshold. The chromosome coordinates of significant regions are then returned to the user. "FindClosestGene," an extension of the FindRegions feature, will map significant regions to the nearest transcriptional start site of genes. These features are particularly useful for determining the locations of binding sites or DNase HS sites in relation to known genes. Finally, when analyzing promoter-specific tiled arrays for ChIP-chip, ACME can export the minimum p value associated with each promoter, thereby allowing the user to assemble candidate lists of genes whose promoters are targeted by the protein in question.

Summary

The ACME program presented here provides scientists with a simple but effective set of tools for analyzing the massive data sets generated from ChIP-chip and DNase-chip assays on tiled arrays. Its implementation should facilitate our understanding of chromatin structure and mechanisms underlying gene regulation. The entire ACME software package is available from the authors upon request or through Bioconductor (http://www.bioconductor.org/). Detailed instructions for installation and use of ACME are included.

References

Bernstein, B. E., Kamal, M., Lindblad-Toh, K., Bekiranov, S., Bailey, D. K., Huebert, D. J., McMahon, S., Karlsson, E. K., Kulbokas, E. J., 3rd, Gingeras, T. R., Schreiber, S. L., and Lander, E. S. (2005). Genomic maps and comparative analysis of histone modifications in human and mouse. *Cell* **120**, 169–181.

Crawford, G. E., Davis, S., Scacheri, P. C., Renaud, G., Halawi, M. J., Erdos, M. R., Green, R., Meltzer, P. S., Wofsberg, T. G., Collins, F. S. (2006). DNase-chip: A high resolution method to identify DNaseI hypersensitive sites using tiled microarrays, *Nature Methods*, in press.

Buck, M. J., Nobel, A. B., and Lieb, J. D. (2005). ChIPOTle: A user-friendly tool for the analysis of ChIP-chip data. *Genome Biol.* **6**, R97.

Crawford, G. E., Holt, I. E., Mullikin, J. C., Tai, D., Blakesley, R., Bouffard, G., Young, A., Masiello, C., Green, E. D., Wolfsberg, T. G., and Collins, F. S. (2004). Identifying gene regulatory elements by genome-wide recovery of DNase hypersensitive sites. *Proc. Natl. Acad. Sci. USA* **101**, 992–997.

Crawford, G. E., Holt, I. E., Whittle, J., Webb, B. D., Tai, D., Davis, S., Margulies, E. H., Chen, Y., Bernat, J. A., Ginsburg, D., Zhou, D., Luo, S., Vasicek, T. J., Daly, M. J., Wolfsberg, T. G., and Collins, F. S. (2006). Genome-wide mapping of DNase hypersensitive sites using massively parallel signature sequencing (MPSS). *Genome Res.* **16**, 123–131.

Glynn, E. F., Megee, P. C., Yu, H. G., Mistrot, C., Unal, E., Koshland, D. E., DeRisi, J. L., and Gerton, J. L. (2004). Genome-wide mapping of the cohesin complex in the yeast *Saccharomyces cerevisiae*. *PLoS Biol.* **2**, E259.

Gross, D. S., and Garrard, W. T. (1988). Nuclease hypersensitive sites in chromatin. *Annu. Rev. Biochem.* **57**, 159–197.

Hughes, T. R., Marton, M. J., Jones, A. R., Roberts, C. J., Stoughton, R., Armour, C. D., Bennett, H. A., Coffey, E., Dai, H., He, Y. D., Kidd, M. J., King, A. M., Meyer, M. R., Slade, D., Lum, P. Y., Stepaniants, S. B., Shoemaker, D. D., Gachotte, D., Chakraburtty, K., Simon, J., Bard, M., and Friend, S. H. (2000). Functional discovery via a compendium of expression profiles. *Cell* **102,** 109–126.

Kim, T. H., Barrera, L. O., Zheng, M., Qu, C., Singer, M. A., Richmond, T. A., Wu, Y., Green, R. D., and Ren, B. (2005). A high-resolution map of active promoters in the human genome. *Nature* **436,** 876–880.

Li, W., Meyer, C. A., and Liu, X. S. (2005). A hidden Markov model for analyzing ChIP-chip experiments on genome tiling arrays and its application to p53 binding sequences. *Bioinformatics* **21**(Suppl. 1), i274–i282.

Lieb, J. D., Liu, X., Botstein, D., and Brown, P. O. (2001). Promoter-specific binding of Rap1 revealed by genome-wide maps of protein-DNA association. *Nature Genet.* **28,** 327–334.

Odom, D. T., Zizlsperger, N., Gordon, D. B., Bell, G. W., Rinaldi, N. J., Murray, H. L., Volkert, T. L., Schreiber, J., Rolfe, P. A., Gifford, D. K., Fraenkel, E., Bell, G. I., and Young, R. A. (2004). Control of pancreas and liver gene expression by HNF transcription factors. *Science* **303,** 1378–1381.

Sabo, P. J., Humbert, R., Hawrylycz, M., Wallace, J. C., Dorschner, M. O., McArthur, M., and Stamatoyannopoulos, J. A. (2004). Genome-wide identification of DNaseI hypersensitive sites using active chromatin sequence libraries. *Proc. Natl. Acad. Sci. USA* **101,** 4537–4542.

Scacheri, P. C., Davis, S., Odom, D. T., Crawford, G. E., Perkins, S., Halawi, M. J., Agarwal, S. K., Marx, S. J., Spiegel, A. M., Metzer, P. S., and Collins, F. S. (2006). Genome-wide analysis of menin binding provides insights into MEN1 tumorigenesis. *PLoS Genetics* **2**(4), e51.

Wu, C., Wong, Y. C., and Elgin, S. C. (1979). The chromatin structure of specific genes. II. Disruption of chromatin structure during gene activity. *Cell* **16,** 807–814.

[15] Extrapolating Traditional DNA Microarray Statistics to Tiling and Protein Microarray Technologies

By THOMAS E. ROYCE, JOEL S. ROZOWSKY, NICHOLAS M. LUSCOMBE,
OLOF EMANUELSSON, HAIYUAN YU, XIAOWEI ZHU,
MICHAEL SNYDER, and MARK B. GERSTEIN

Abstract

A credit to microarray technology is its broad application. Two experiments—the tiling microarray experiment and the protein microarray experiment—are exemplars of the versatility of the microarrays. With the technology's expanding list of uses, the corresponding bioinformatics must evolve in step. There currently exists a rich literature developing statistical techniques for analyzing traditional gene-centric DNA microarrays, so the first challenge in analyzing the advanced technologies is to identify which of the existing statistical protocols are relevant and where and when revised methods are needed. A second challenge is making these often very technical ideas accessible to the broader microarray community. The aim of

METHODS IN ENZYMOLOGY, VOL. 411   0076-6879/06 $35.00
                                                    DOI: 10.1016/S0076-6879(06)11015-0

this chapter is to present some of the most widely used statistical techniques for normalizing and scoring traditional microarray data and indicate their potential utility for analyzing the newer protein and tiling microarray experiments. In so doing, we will assume little or no prior training in statistics of the reader. Areas covered include background correction, intensity normalization, spatial normalization, and the testing of statistical significance.

Introduction

Microarray technology (Fodor *et al.*, 1991; Schena *et al.*, 1995) allows for the parallel quantitative assessment of biochemical reactions. On the order of 10^6 measurements can be taken simultaneously with current technology (Cheng *et al.*, 2005). The initial challenge following a microarray experiment is to determine which of these potentially millions of observations are significant and should be studied in more depth. This challenge has been met by hundreds of practitioners in both biomedical and mathematical sciences and literally hundreds of papers have been published on the topic. This chapter aims to illustrate some prevailing ideas and techniques found in the microarray analysis literature. In addition to covering statistics used for traditional microarray experiments, we include those techniques exploited in protein and tiling microarray analyses as well. These latter experiments share some mechanistic aspects with the traditional DNA microarrays, but in several respects, are quite different. Therefore, some of the bioinformatics research done for traditional microarrays is relevant, whereas some of it is not. We will guide our discussion with this as our theme, and focus on two main areas of study: microarray normalization and the assessment of statistical significance.

Prior to delving into the heart of our discussion, we will first introduce some naming conventions, followed by statistical preliminaries. Following these prerequisites, a brief discourse on how microarray data are obtained is given. The first major area of study reviewed is *microarray normalization* or, more concisely, *normalization*. Normalization deals with the technical aspects of the microarray technology that can potentially confound and/or bias the results of the experiment. It does so by correcting measured values so as to remove these effects. Normalization is discussed later. The second area focused on is the assessment of statistical significance. Statistical significance can mean different things for different microarray experiments, depending on their respective goals, and is discussed. In a majority of traditional DNA microarray experiments, significance indicates the presence of differential mRNA expression between two or more biological classes for some gene. An experiment might, for example, assess mRNA concentrations for thousands of genes as cells progress through the cell cycle

(Cho *et al.*, 1998). In such a scenario, we would like to know within each stage those genes that exhibit differential expression (higher or lower concentrations) relative to the other stages. For tiling microarrays, as shown later, significance pertains more loosely to genomic regions. In these experiments, we seek chromosomal regions (consisting of multiple probes) that exhibit higher than expected fluorescent intensities on the microarray. Protein microarrays have two main classes of use: analogous to the DNA microarray, *antibody microarrays* can be used to determine protein abundances, whereas *functional protein microarrays* can be used to detect protein–protein interaction partners *in vitro*. For each of these experiments, significance clearly takes on a different meaning.

Definitions

Some common points of confusion within the microarray literature are how various entities are defined. This section explicitly defines some of these entities so as to minimize the potential for confusion. Herein, we define molecules on the microarray at time of its construction as *probes* and those molecules that are subsequently introduced to the microarray as *targets*. We use the words *spot* and *feature* interchangeably to indicate a collection of probes that have the same sequence and are concentrated at a known position in the microarray design. A collection of targets from a single biological source is called a *sample*. A single event consisting of introducing one or more samples to a microarray is termed *probing*. Finally, a set of probings designed to test certain hypotheses is simply an *experiment*.

Statistical Preliminaries

It is impossible to have a discussion on microarray statistics without any prior knowledge of statistics in general. This section provides some basic concepts that will aid our presentation of microarray analysis. Anyone who has taken an introductory statistics course has seen this material already and can safely skip this section.

Summary Statistics

Assume for the moment that a microarray experiment measures the expression level of just a single gene and that the experiment consists of several technically replicate probings from which a measurement is observed. To generalize the measurements for discussion, let each measurement be denoted by the symbol X_i. Here, the subscript i indicates the ith measurement of the gene. For example, $X_4 = 162$ would indicate that the measurement coming from the fourth microarray is 162.

A first natural question to ask of the experiment is "What is the central tendency of my measurements or, equivalently, how can I best describe my measurements with a single number?" The most commonly used response to this question is to calculate the *arithmetic mean*, or *average*, of the measurements. To calculate the arithmetic mean, we first sum all the measurements and then divide by the total number of measurements observed. If N is the number of measurements taken, then the mean $\bar{X}$ is calculated as

$$\bar{X} = \frac{1}{N} \sum_{i=1}^{N} X_i = \frac{X_1 + X_2 + \ldots + X_N}{N}. \tag{1}$$

We often would like to measure the spread of our measurements in addition to their central tendency. The most commonly used measure of spread is the variance σ^2:

$$\sigma^2 = \frac{\sum_{i=1}^{N}(X_i - \bar{X})^2}{(N-1)}. \tag{2}$$

Note that the numerator consists of N terms, added together. Each term in the summation corresponds to the ith measurement and is the difference between that measurement (X_i) and the mean of all N measurements, $\bar{X}$. Also note that each term is squared. Doing so ensures that the numerator is positive and that measurements less than the mean contribute positively to the variance just as much as those measurements greater than the mean. This measure of spread is roughly the average squared difference from the mean. We say "roughly" here because the denominator in Eq. (2) is $(N - 1)$ rather than the N that we might expect from the definition of arithmetic mean [Eq. (1)]. Why this is so is beyond our scope, but with large N this detail makes little difference. Related to the variance is a quantity called the *standard deviation*. A standard deviation, symbolized as σ, of a group of measurements is simply the square root of those measurements' variance.

We often read or hear the phrase "microarray data are noisy," or some similar (potentially less polite) variant. This can be taken to mean several things, but quite often it is the presence of outliers that is being referred to. An *outlier* is a measurement in large disagreement with other measurements of the same phenomenon. In a microarray experiment, the difference could be due to a biological effect, but more likely the outlier is due to some kind of technical malfunction of the instrument and/or its associated protocol(s). Outliers can have large effects on the aforementioned summary statistics. For an example, consider an experiment where five measurements are taken for the same gene. If these measurements are $X_1 = 12$, $X_2 = 9$, $X_3 = 11$, $X_4 = 507$, and $X_5 = 12$, then

$$\bar{X} = \frac{X_1 + X_2 + X_3 + X_4 + X_5}{N} = \frac{12 + 9 + 11 + 507 + 12}{5} = 110.2. \quad (3)$$

Clearly the quantity 110.2 does not represent the central tendency of data very well. It is not particularly close to any of the measurements. Luckily, there are ways around such pitfalls. One technique is called the trimmed mean. With this approach, some percentage of the most extreme measurements is thrown away prior to calculating the mean. An extreme (and quite common) version of this approach is to calculate the measurements' *median* as a measure of their central tendency. The median is defined as the middle quantity occurring in a sorted list of observations. That is, if N is odd and you first sort your measurements $X_1, X_2, \ldots, X_N$ in either increasing or decreasing order, then the median is the quantity $X_{\frac{N+1}{2}}$. (If N is even, the middle two measurements, $X_{\frac{N}{2}}$ and $X_{\frac{N}{2}+1}$, are averaged.) In our noisy example of five measurements where the mean of 110.2 was obtained, the calculated median is 12. This value intuitively summarizes these data much better.

An analogous calculation can be performed in place of the variance. Recall that the variance is essentially an average squared difference of measurements from the mean [Eq. (2)]. This computation can be made more robust to outliers by first substituting the median for the mean and then computing the median of absolute differences between the measurements and the previously calculated median. This quantity is sometimes referred to as the *median absolute difference* (MAD).

Statistical Significance

The term *p value* comes up frequently in texts about microarray experiments and their analyses. A *p* value is simply the probability of some null hypothesis being true given a set of assumptions and observations. A typical experiment utilizing DNA microarrays might have thousands of such null hypotheses, one for each gene being studied. These null hypotheses would typically claim that the expression level of some gene is not different between two biological samples. As a result, we declare that any gene for which we can compute a low *p* value is *significant* and potentially worthy of further study. To call the gene significant, a *p* value threshold for significance must, of course, be in place. A common interpretation for this threshold is the false-positive rate of the study: the percentage of time replications of the experiment would reject the null hypothesis when it is actually true.

Now, we do not typically know the actual probability of observing something under a given null hypothesis. However, if we know that the numbers being studied follow some known form (e.g., we might know or assume that the gene expression levels are distributed like a bell curve, or a *normal* distribution), then we can use this knowledge to either simulate or directly calculate how likely an average difference between two such groups of measurements would be if there were in fact no difference, for example.

A final note on significance worth noting is that, generally speaking, the more observations we are able to make of some phenomenon, the better is our ability to compute a low p value. To illustrate this point, consider an experiment where we ask, "Is gene A expressed at a higher level in tumors than in healthy tissue?" Let us assume that the answer to this question is, "Yes." If we have one measurement of A from a tumor and one measurement of A from a healthy tissue and the measurement from the tumor is twice as high as its healthy counterpart, we have some limited confidence that the gene is more highly expressed in tumors. This occurrence could be an anomaly, so we still would assign some fairly high probability to the null hypothesis of no difference being. If instead we measure the abundance of gene A in 20 tumors and they are all higher than 20 measurements taken from healthy tissues, we would assign a much lower probability to the null hypothesis because the chance of 20 anomalies is very small.

Multiple Testing

Another issue that comes up frequently in the microarray literature is that of *multiple testing*. Multiple testing simply indicates that more than one statistical test (which generates a p value) is part of the study. For microarray experiments there are thousands and potentially millions of statistical tests being conducted, so clearly we are dealing with multiple testing, but what is our concern when we engage in multiple testing?

In biology, the threshold for considering a p value significant is typically $p < 0.05$ or $p < 0.01$. These criteria arise from a balance between our willingness to accept a 5% or even a 1% false-positive rate and the number of replicate measurements we are able to take. Multiple testing becomes a problem, for example, if we conduct 100 statistical tests and identify that one of them yields a significant p value ($p = 0.04 < 0.05$). It would be tempting to report this seemingly significant finding. The problem here is that within a set of 100 tests, we expect to find 4 of these to yield $p = 0.04$ simply due to random chance (100 tests multiplied by the false-positive rate 0.04 yields 4 tests). This toy example becomes a staggering problem if we are testing, say, 20,000 genes. In this case, at a significance threshold of $p < 0.05$, we will identify roughly 1000 false positives. This number of false positives is potentially more than the actual number of differentially expressed genes that we seek to identify.

The most simple method for dealing with multiple comparisons is to require sufficiently low p values such that the total number of expected false positives is small. The Bonferroni correction (Bonferroni, 1935) does this by controlling the so-called *family-wise error rate* (FWER). The FWER is defined as the probability of detecting a false positive anywhere among the multiple tests. As a result if we want the probability of detecting a false positive among our tests to be less than α, we require that any individual test achieve $p < \frac{\alpha}{N}$ where N is the number of tests. Such corrections pose a problem for microarrays where thousands of genes are being tested for significance and the number of available replicate experiments is small. The problem is more acute for high-density tiling microarrays where the number of tests performed can reach into the millions (see later) and the number of experimental replications is often fewer than five.

Microarray Data

This section reviews briefly how microarray data are obtained.

Data for Traditional, Gene-Centric DNA Microarrays

Each spot on a gene-centric DNA microarray corresponds to a DNA sequence derived from a known or putative gene. That sequence could be the whole spliced form of a gene (such as a cDNA clone) or a tethered 25-bp oligonucleotide sequence, as is the case for Affymetrix GeneChip brand microarrays. Such a microarray typically probes a sample that is derived from a mRNA source.

Subsequent to probing a labeled sample with a microarray, an image representing its surface is generated by subjecting the microarray to a digital-scanning device. Depending on the type of labels used, different scanning technologies are employed. Typically, the samples have been labeled with a fluorescent dye or, alternatively, with radioactive isotopes. For fluorescently labeled samples, the probed microarray is scanned with a laser scanner. There is a wide selection of laser scanners available, including but not limited to ScanArray GX from Perkin-Elmer, GenePix 4200 from Molecular Devices, and DNA microarray scanner from Agilent Technologies. A laser wave length near the absorption maximum of the fluorophor dye used (that was attached to the hybridizing sample) is scanned across the microarray surface from top to bottom and from left to right so that all areas of the microarray are accessed by the laser. The light emitted at each location when laser-excited fluorophors transition to their unexcited state is captured by a detector and translated into a pixel intensity at that location. Microarrays that probe multiple-labeled samples simultaneously must be scanned with a scanner having at least as many unique laser wavelengths as labeled samples. Current scanning resolutions are as high as 1 μm^2 per pixel.

The result of scanning a single probed sample is a monochromatic digital image (usually stored as a TIFF file) of the microarray surface. Bright regions in the image correspond to regions of the microarray with high levels of fluorescence and dim regions likewise correspond to regions devoid of fluorescence. Presumably, the bright regions correspond to spots to which labeled nucleic acid hybridized. If two different samples were labeled with two different dyes and were probed with the same microarray, then the result of scanning is two digital images. There would be one image for each wavelength used.

The microarray images generated by the laser scanner must be further processed with image analysis software. First, the spots of the microarray have to be identified within the image. To do this, rules have to be obtained or assumed that can distinguish between pixels that constitute spots and pixels that belong to background regions. Separating spots from the background is called *segmentation*. Following segmentation, *grid alignment* must be performed. Grid alignment is the process of identifying which spots correspond to which annotation. Basic versions of grid alignment software are usually included with the purchase of a scanner, but there are alternatives, such as TIGR Spotfinder (Saeed *et al.*, 2003), which is freely available under an open-source license, or ScanAlyze (http://rana.lbl.gov/EisenSoftware.htm), which is free for academic and noncommercial use. For most spotted arrays, the grid has to be defined by the user, either manually or semimanually, whereas for many higher density microarrays, such as Affymetrix GeneChip brand microarrays and NimbleGen System's NimbleChips, the alignment of the microarray image to the grid is done automatically by software. This automation is made possible by reserving some spots on the microarray exclusively for grid alignment. Certain labeled cDNA/cRNA molecules that are complementary to the grid alignment probes are spiked into the sample(s), ensuring that the grid alignment probes will appear as bright regions in the scanned image, enabling automatic grid alignment.

After aligning the grid the image analysis software reports back a certain number of key statistics for each of the identified spots. These statistics may include the mean and median pixel intensities within each spot, the standard deviation of those pixels, and sometimes also other information, such as mean intensity ratios in the case of a two-channel experiment. The area of each spot (number of pixels) is also frequently reported. Importantly, if the software considers a particular spot aberrant, for example, irregular in its shape, or if its measured intensity is lower than the surrounding background intensity, the spot may be flagged as irregular. Such flagged spots are often excluded from further statistical analyses.

The end result is a tab-delimited plain text file containing all raw data for each microarray feature within a single row. The tab-delimited, text-based format is easily amenable to further analysis by importing it into a microarray analysis software package such as ExpressYourself (Luscombe et al., 2003) or MIDAS (Saeed et al., 2003) or, of course, into your own microarray analysis pipeline. For simple calculations, a spreadsheet program (e.g., Gnumeric, OpenOffice or Microsoft Excel) could also be used.

For spotted DNA microarrays, it is common that each gene under study is represented by a single spot. An important difference exists for Affymetrix GeneChip brand microarrays. For this technology, each gene is represented by a *probe set*, typically consisting of 10–20 features on the microarray. Within the probe set, each feature contains probes of different sequence. To assess the differential expression of a single gene, multiple spots from each microarray need to be considered.

Data for Tiling Microarrays

The two most widely utilized high-density oligonucleotide platforms are those produced by Affymetrix, using masks to synthesize the oligonucleotides on the microarray (Lipshutz et al., 1999), and those manufactured by NimbleGen Systems, which use a system of mirrors controlled by a digital light processor for synthesis (Nuwaysir et al., 2002). Affymetrix microarrays currently utilize 25-bp oligonucleotide probes for each spot. For every spot corresponding to some 25-bp stretch of genomic DNA (perfect match), there is a corresponding spot (mismatch) where the middle nucleotide of the probe has been substituted with its reverse complement. This perfect match/mismatch setup is also the standard for the Affymetrix GeneChip system as well. The purpose of the mismatch probe in both traditional and tiling applications is to measure the nonspecific binding of the probes within a spot [there is some debate about the usefulness of mismatch probes, however (Irizarry et al., 2003)]. Currently, Affymetrix microarrays are capable of including on the order of 10^6 spots.

Maskless microarrays manufactured by NimbleGen Systems are synthesized such that each microarray can be completely customized with unique probe sequences. These microarrays allow for oligonucleotide lengths of up to 70–80 nucleotides (in fact, isothermal arrays exist where each feature corresponds to a oligonucleotide probe of a different length). Current maskless microarray designs have approximately 390,000 spots per microarray. One important difference between these two high spot density platforms is that Affymetrix brand microarrays can only be hybridized with a single target nucleic acid population, whereas maskless arrays allow the hybridization of two samples simultaneously using different labels, typically

Cy5 (red) and Cy3 (green). This is potentially beneficial when looking for differential expression between samples or for ChIP-chip (Horak and Snyder, 2002; Iyer *et al.*, 2001), where chromatin-immunoprecipitated DNA is labeled differently from some reference DNA.

Tiling microarrays (Bertone *et al.*, 2004; Cawley *et al.*, 2004; Cheng *et al.*, 2005; Kapranov *et al.*, 2002) use high-density capabilities to tile the nonrepetitive sequence of a genome. The word *tile* indicates that probes are selected for inclusion on the microarray at some roughly uniform interval over a potentially large genomic space. In the context of mRNA transcript mapping, this high resolution enables the unbiased detection of individual exons of a spliced transcript. This experiment is not practical on a whole-genome scale in a mammalian species with lower resolution polymerase chain reaction amplicon microarrays due to cost (Bertone *et al.*, 2005).

Tiling microarrays are an evolving medium, and data format standards have not yet materialized. However, several summary statistics about each spot are typically included in a tab-delimited text file. These statistics usually include the mean and/or median pixel intensity of each spot, the number of pixels within each spot, and a standard deviation of the pixel intensities of the spot. It is worth including a cautionary note about tiling microarray data here. Tiling microarrays generate very large data sets. As such, they are difficult or impossible to import into desktop spreadsheets such as Microsoft Excel. Therefore, more robust tools are often needed.

There is one more major difference between traditional DNA microarrays and tiling microarrays to consider. The signal intensity measured at a spot containing short oligonucleotide probes is arguably too unpredictable to score each probe separately. This variability is due to a number of factors, including cross-hybridization and differential binding affinity due to probe sequence and other sequence-based artifacts. In addition, higher standards of statistical significance are typically required for tiling arrays because of the much larger number of spots being queried and therefore require more evidence than that given by a single spot. Thus the methodologies that have been adopted for the analysis of tiling microarrays is to incorporate the intensities of a number of spots that lie within a contiguous genomic region. This methodology is often referred to as a *genomic sliding window* approach.

Data for Protein Microarrays

There are two types of protein, microarrays as defined by their goals (Zhu and Snyder, 2003). One type is protein detection microarrays, or antibody microarrays (Lueking *et al.*, 1999), which use antibodies for its probes and are used to detect and quantify proteins in solution. This design is very similar to its DNA-based counterparts, which quantify mRNA concentrations. The other major class of protein microarrays is functional protein microarrays (Zhu et al., 2001), which aim to identify protein binding or modification capabilities. In such a design, each spot consists of some known protein or protein domain. The target that is introduced will typically consist of a single macromolecule. This target may be labeled so as to detect molecular interaction partners or, as is the case for kinase activity assays, may be probed in the presence of hot ATP to detect phosphorylation events.

An aspect of protein microarrays of note is that the spots therein will usually not contain equal amounts of protein from spot to spot. This discord can cause differences in measured intensity between spots that are not due to molecular activity, but rather to an aspect of the microarray construction.

Regarding software and generated data, protein microarrays utilize the same scanners and scanning software as their DNA-based counterparts and therefore the raw data files they produce are technically very similar. This is an advantage, as some existing computational protocols and interfaces developed for DNA microarrays may be integrated easily with protein microarray analysis.

Microarray Normalization

Once data have been obtained, a usual next step is to perform microarray normalization.

Motivation

Technical aspects of the microarray experiment can cause systematic biases and artifacts to be present in their data. In a two-sample DNA microarray experiment, the probed biological samples may contain different concentrations of RNA, leading to an overall bias in favor of greater measurements in one channel. In addition, the fluorescent dye molecules Cy3 and Cy5 are known to have slightly different properties, leading to a similar problem. Complicating these troubles is that they may be more or less present depending on the intensity of the spot being measured and/or its physical location on the microarray. The following section illustrates an example of how such biases can affect biological conclusions made from microarray data when proper data normalizations are not carried out. We include this example as a cautionary tale and as a motivation for microarray normalization, in general.

Most spotted microarrays are built by depositing solutions of cDNA clones at known locations on a microarray surface. This deposition process is controlled robotically with little human intervention and is therefore completely regular and predictable. Furthermore, the printing process is such that spots close to each other on the microarray surface are printed closely in time as well. Given that a microarray hybridization can be uneven across the surface of the microarray, this leads one to speculate that neighboring spots on the microarray surface might be coordinately affected. An example situation would be if labeled sample were more abundant in one region of the microarray than in others. Spots in that region would have systematically higher observed intensities than those spotted elsewhere.

Indeed, it does appear that such a spatial effect exists. For printed cDNA microarrays, the effect was first reported by examining the relationships between observed spot intensities and the locations of spots in the design of a microarray (Kluger *et al.*, 2003; Qian *et al.*, 2003). Similarities were examined between gene expression profiles (across a large number of probings) for genes that are printed on the microarrays at varying distances. It was found that genes that are close in the microarray design (on average) have higher similarities between their expression profiles than those further away. That is, it might appear that genes that are close on the microarray surface seem more likely to be coexpressed. Note that without knowledge of the microarray design, the genes would be identified as exhibiting coordinated mRNA expression.

It turns out that for the microarray design used in the aforementioned study, genes were printed in an order related to their chromosomal arrangement for organizational convenience. This printing strategy yielded a microarray such that genes located 22 open reading frames (ORFs) away in genomic space are printed as immediate neighbors on the constructed microarrays more often than would occur if they were printed in a random order. Interestingly, by examining the relationships between gene expression and chromosomal localization, a striking similar frequency was found: genes that are approximately 22 ORFs away on the same chromosome are more likely to be coexpressed, whereas genes that are about 11 ORFs away are less likely to be coexpressed (Qian *et al.*, 2003). Furthermore, it was determined that genes on microarrays with a different layout have a different frequency. This last piece of evidence suggests the existence of an artifactual effect related to microarray architecture. One of the aims of microarray normalization is to reduce the effect of such artifactual components of observed data.

Most microarray studies examine the relationship between two biological samples by comparing their relative mRNA expression levels. The idea behind such two-channel experiments is straightforward: labeled (typically red with Cy5 and green with Cy3) nucleic acids in the samples are probed simultaneously with a microarray slide, and relative abundances are derived from comparative fluorescence of the nucleic acid molecules hybridized at each microarray feature. For a given spot i, the relative concentration between the two samples is commonly represented as the log ratio, λ_i, of the measured fluorescence intensities between the two dyes. We summarize the log ratio as

$$\lambda_i = \log\left(\frac{R_i}{G_i}\right) \tag{4}$$

where R_i and G_i denote the observed intensities (mean or median of spot pixels' intensities) for probe i when scanned with red and green lasers, respectively. Note that a log ratio of zero indicates that R_i and G_i are equal. Further, a set of observed log ratios (with measurement error) should center about zero for probes representing genes of equal expression in the two samples. Measurements deviate from this situation proportionately to their degree of up- or downregulation relative to the two samples.

The log ratio measured between a gene in two samples is in itself a normalization technique. Microarray manufacture is not errorfree. Any given spot may be printed poorly on one microarray and printed perfectly on the next. If these two microarrays were used to measure the concentration of the gene corresponding to that spot, the poorly printed spot would likely lead to an artificially low measurement for one sample relative to the sample hybridized to the higher quality spot. If instead both samples were hybridized to both microarrays, then the hybridization of one sample to the poor spot is directly comparable to the hybridization of the other sample to the poor spot and likewise for the higher quality spot. This self-normalization is particularly useful when the two samples hybridized to the microarray are paired in other respects beyond the fact that they were measured with the same instrument. A good example of paired samples is an mRNA sample taken from a tumor biopsy before treatment and an mRNA sample taken after treatment. Regarding log ratios, it should be noted that the Affymetrix GeneChip system only allows hybridization of a single sample to a microarray. Therefore, log ratios are not meaningful as a spot quality normalization. It is believed, however, that Affymetrix microarray construction is much more uniform in terms of quality control than its spotted microarray counterpart so such self-consistency concerns are relatively minor. Log ratios can still be relevant for Affymetrix microarrays in the case of paired samples, such as in the cancer experiment mentioned earlier. Most tiling and protein microarrays yield just a single intensity measurement as well so the log ratio is not always a natural measurement for these experiments either.

Although the log ratio provides an intuitive measure of relative gene expression, it must often be corrected for inconsistencies resulting from the experiment (see earlier discussion). Such corrections are collectively termed normalization. Normalization adjusts the measured intensities for each sample and for each spot as corrective measures. The aim is to compensate for artifactual effects by applying transformations so that equally expressed genes have log ratios approaching zero. (For single-channel experiments, no such baseline generally exists.) Measurements for all spots on the microarray are scaled relative to this baseline. In practice, implementing good normalization has proved challenging; researchers have developed many competing methods, which can lead to divergent results (Hoffmann *et al.*, 2002). The following sections describe some of the more widely implemented strategies.

Background Correction

For many types of microarrays, a measurement of the local background of each spot is recorded in addition to the foreground intensity of the spot. This measurement is, in common practice, the mean or median of all pixels residing in the surrounding regions of the spots (see earlier discussion). It is believed that any measured intensity from this background region is also measured in the foreground pixel intensities of the spot as well. This background fluorescence is attributable, in general, to glass fluorescence and unincorporated label molecules. The background intensities have no biological interpretation so we would ideally like to remove their contribution from spot intensities before proceeding. The easiest way to make this correction is to subtract the mean (median) of all local background pixels measured in the red channel (denoted ρ_i) from the red intensity of each spot, do likewise for the green channel (γ_i), and then compute the background adjusted log ratio as

$$\hat{\lambda}_i = \log\left(\frac{R_i - \rho_i}{G_i - \gamma_i}\right). \tag{5}$$

Equation (5) assumes that $\rho_i < R_i$ and that $\gamma_i < G_i$. Any spot not in agreement with these assumptions should be flagged as a bad spot and subsequently ignored, as it does not make sense for a background region to have a higher intensity than the spot.

We need not rely upon just the background values provided with each spot in a microarray results file. The values of ρ_i and γ_i could actually be computed as the mean or median of all spot background measurements in a localized region before applying Eq. (5). An example of this would be to utilize a spot's eight nearest-neighboring spots' backgrounds to calculate its local background intensity. Such an approach is advisable so as to avoid aberrantly high local background values due to scratches or other artifacts present in a microarray scan. This is of particular importance when dealing with protein microarrays, as these devices can yield spots that smear to bigger sizes due to phosphorylation activity, for example. These smears will often be measured as part of a spot's background, causing it to be erroneously high.

Unfortunately, tiling microarrays will usually seek to maximize feature density in an effort to reduce cost and as such, features are packed immediately next to one another and background calculations may not be possible. In these cases, we can only hope that background intensities are minimal, or at least, constant throughout the microarray.

Normalization via Total Intensity

Following background subtraction, we would like to normalize sample intensities so that their intensity distributions have desirable properties. One commonly desired property within two-sample probings is to have a distribution of log ratios representing nondifferentially expressed genes to center about zero. This is usually reasonable, as in most experiments we do not expect a centering around any other value.

In a differential expression experiment, microarrays should hybridize similar numbers of labeled molecules from each sample, so the total hybridization signals summed over all probes should be the same for both channels. Using these assumptions, we can calculate a scaling factor C_{total} that can be used to correct any observed deviance from this assumption. If M is the total number of features on the microarray, then we have

$$C_{\text{total}} = \log\left(\frac{\sum_{i=1}^{M} R_i}{\sum_{i=1}^{M} G_i}\right). \tag{6}$$

We can then compute the normalized log ratios as

$$\hat{\lambda}_i = \lambda_i - C_{\text{total}}. \tag{7}$$

The result is a distribution of log ratios that are centered somewhere near zero. This method performs well in most standard microarray experiments with sufficiently large numbers of spots ($>20,000$), as in these scenarios, outlier signals make negligible contributions to the total intensities.

A similar approach to Eq. (6) can be used to normalize intensities from one single channel microarray to others. In this application, every intensity in one channel is divided by the summed intensity (e.g., $\sum_{i=1}^{M} R_i$) of spots from the same microarray. Then, these normalized intensities can be used to compare and contrast different samples hybridized to different microarrays. This latter calculation may be useful in experiments where just a single probing is carried out on each microarray. This is always the case for Affymetrix GeneChip technology and is almost always the situation for protein microarrays and for tiling microarrays.

Normalization via Gene Set

The previous method performs fairly well in standard microarray experiments where the number of genes studied is large and overall gene expression differences between the two samples are not excessive. However, the approach must be applied cautiously, as it may mislead researchers into believing that similar numbers of genes are always up- and downregulated. This clearly is not true in some circumstances.

In the following method, sometimes called the *gene set method*, some set of genes is assumed not to be expressed differentially between the samples being studied. This set of genes is typically made up of housekeeping genes. The procedure is analogous to that in Eq. (7), with the only difference being the numbers that are summed are those from the gene set, not all spots. We call this value $C_{geneset}$. Captured in this statistic is the overall deviation that you would expect given no differential expression. Ideally, $C_{geneset}$ is equal to zero, but effects such as unequal RNA concentrations and differences between the fluorescent dyes can cause $C_{geneset}$ to be nonzero. Once $C_{geneset}$ has been calculated, all log ratios (not just those in the gene set) are normalized by $C_{geneset}$ using the relationship

$$\hat{\lambda}_i = \lambda_i - C_{geneset} \tag{8}$$

where $\hat{\lambda}_i$ denotes the normalized log ratio for probe i. Using control spots in this way has an added benefit for sets of microarrays where the spots present on each microarray are not the same. In such a scenario, a common set of control spots can be used to normalize the intensity distributions of the microarrays so that they are similar from microarray to microarray. This is a typical situation for tiling microarrays that require several microarrays having different designs to probe for large fractions of the sequence of a genome.

Normalization via Spiked Controls

A way to guide normalization further is to spike known quantities of external controls into the biological samples prior to fluorescence labeling. Normalization is then based on balancing the signal intensities for those probes corresponding to the control RNA molecules as in Eq. (8).

There are two advantages of this technique. First, the spike-ins are completely controlled—we are sure that they should show no differential expression between two or more samples. Second, different scale factors can be calculated for genes having different expression levels if several different spike-in concentrations are used. A disadvantage, though, is that control probes must be built into the array at the onset. Further, the scaling factor is calculated using a comparatively small number of probes that may be sparsely distributed on the array depending on the design and the correction techniques for spatial microarray biases (discussed later) currently cannot be incorporated easily. A final point of concern is that spiked controls may interact with unintended spots on the microarray in addition to the control spots. For traditional DNA microarrays and tiling microarrays this is manifest as cross-hybridization. For protein microarrays, spiked proteins may interfere with desired protein-binding interactions.

Normalization via Quantiles

Another popular alternative for intensity normalization is so-called *quantile normalization* (Bolstad *et al.*, 2003). In this approach, the first step is to construct a synthetic microarray such that the "measurement of each spot," S_i, is the mean or median of its measurements across all P probings in the experiment. Mathematically, if we use the mean in constructing this synthetic microarray and $X_{i,j}$ is the measurement from the jth probing for spot i, then we have

$$S_i = \frac{\sum_{j=1}^{P} X_{i,j}}{P}$$

The S_i values are then sorted in increasing order, as are the intensities within each probing. The final step in this normalization is to replace each observed intensity by that intensity S_i that occupies the same position within its sorted list. If $X_{1043,2} = 87$ is the third largest observation within probing number two, it is replaced by the third largest value of S. A major advantage of this approach is that it requires no extra probes or spike-ins and yet still can correct for biases that may be present more or less at different intensity levels. This advantage makes this method broadly applicable to any microarray experiment with little concern over experimental nuances.

Correcting Signal Intensity Bias

Numerous reports have indicated that log ratios resulting from a two-sample probing can have a systematic dependency on signal intensity because of differences in the fluorescent properties of the red and green dyes (Quackenbush, 2002; Yang *et al.*, 2002a,b).

Lowess regression (Cleveland, 1981) analysis allows its users to fit a nonlinear curve to a ratio vs intensity distribution. We call the logged product of the measured R_i and G_i intensities ϕ_i and plot each λ_i as a function of its respective ϕ_i. The basic idea of Lowess is then to first find a curve that passes through the "middle" of this ratio versus intensity distribution. The output of Lowess is a value L_i paired with each ϕ_i. Once L_i is calculated, it can be used to correct for intensity biases. The corrected log ratio is

$$\hat{\lambda}_i = \lambda_i - L_i. \qquad (9)$$

The question remains as to how L_i is calculated. This is somewhat beyond the scope of this chapter, but we will sketch the calculation here. For every ϕ_i, a neighborhood of ϕ values is found. The size of this neighborhood is a variable that can be adjusted but is typically set to be 10% of all spots. Once the neighborhood of spots is found, a line is plotted through the values corresponding to the ϕ values in the window. This line is used as a function to compute L_i from ϕ_i. The method can be generalized. In fact, a commonly used variant of this method called Loess (no "w") performs the same functionality but replaces the locally fitted line with a locally fitted quadratic curve.

This technique has no analog for single-channel experiments as in most tiling and protein microarray experiments. The technique can be forced if one microarray is considered a baseline and then all other microarrays are normalized relative to the baseline. This is potentially problematic for tiling microarrays where each microarray may contain different probes and therefore have different expected intensity distributions.

Correcting Array Location Bias

It has become increasingly clear that there are often substantial spatial biases caused by uneven hybridization conditions across a microarray slide. Uncorrected, this can have an effect on results. An example of this is the apparent coexpression of groups of genes, which is actually caused by the proximity of their corresponding spots on the microarray surface (see earlier discussion).

For spotted microarrays, the effect is frequently corrected using subgrid normalization in which local subsets of spots are grouped by their depositing print tip. These groups are then normalized separately using, for example, the method outlined earlier. This approach should be used with caution, as we have observed that most spatial variations do not follow the boundaries of print-tip groups (sometimes referred to as *blocks*).

As an alternative, a variation of the Lowess analysis introduced earlier can be used to correct spatial biases. In this alternative, a surface is fit to the log ratios as a function of their spatial coordinates as opposed to fitting curve to log ratios as a function of total intensity. The corrected intensity is obtained analogously. This procedure can be applied to single-channel intensities as well.

It should be noted here that during the design of a microarray, no regions should be overpopulated with spots that might display coordinated expression level changes. In this unfortunate scenario, the corrective methods will eliminate biologically meaningful variations in the measurements. This limitation can be overcome easily by randomizing spot locations during microarray manufacture.

This procedure may prove difficult for microarrays where a small fraction of spots show measurable signal because there are too few intensities to fit the surface to. Tiling microarrays will usually fall into this category as much of the genomic sequence is inactive at any given time. Functional protein microarrays may fall into this category as well, as a given protein is likely to have just a handful of binding partners.

Normalization by Spot Concentration

Concentrations of probes within each spot will affect measured intensities. For most traditional and tiling microarrays, this is not an issue. However, for protein microarrays, it is difficult to control the amount of protein present at each spot and therefore it is advisable to divide any measurement by the concentration of the spot. The concentration measurements can be obtained by hybridizing a protein microarray with a labeled universal protein marker.

This section briefly described the most common techniques for normalizing microarray data. Many of these methods have been implemented in published software tools that facilitate microarray normalization; examples include Express Yourself (http://bioinfo.mbb.yale.edu/expressyourself/) (Luscombe *et al.*, 2003) and SNOMAD (pevsnerlab.kennedykrieger.org/snomadinput.html) (Colantuoni *et al.*, 2002).

Future improvements in microarray technology may eliminate the need to correct for intensity and spatial bias, or even for normalization all together. However, current technologies still produce substantial artifacts, even if they are not evident from visual inspection of a scanned image.

Scoring for Significance

Following microarray normalization, the intensities are in a more suitable form for statistical testing. This section begins by exploring some of the more common approaches for testing the significance of differences between measured intensities generated from two biological conditions. The discussion is then generalized to the multiple condition case and to tiling and protein microarrays.

Fold Change

Assume for simplicity that we are interested in assessing differential expression for just a single gene between two biological conditions. Call these conditions A and B. Further, assume that we have multiple measurements for the gene within each condition. Let $M > 0$ be the number of measurements obtained for condition A and $N > 0$ be the number of measurements for condition B. Note that M need not be equal to N but ideally they would be equal. To designate the ith measurement from condition A, we will use the notation A_i. We adopt the same convention for measurements of B.

Perhaps the simplest technique for comparing A and B is to compute an average fold change between the two. Call this fold change statistic S_{fold} and define it as

$$S_{\text{fold}} = \max\left\{ \frac{\sum_{i=1}^{M} A_i}{\sum_{i=1}^{N} B_i}, \frac{\sum_{i=1}^{N} B_i}{\sum_{i=1}^{M} A_i} \right\}. \tag{10}$$

In addition to calculating S_{fold}, we also choose cutoff values to deem the statistic potentially interesting. A good way to choose this cutoff is to have control features present on the microarray that are not expected to display differential expression. With enough unique controls, the 95th percentile of their S_{fold} statistics could be a useful cutoff. By the quantity *95th percentile*, we mean that 95% of all S_{fold} values are below this quantity. Such a cutoff would suggest that values above this threshold would occur just 5% of the time for genes not showing differential expression. More commonly, such controls do not exist and an arbitrary cutoff is selected. Often, this cutoff is set at two.

t *Test*

The fold-change method utilizes just a single summary statistic (the sum) for each condition. No information about how widely the measurements vary is considered. In addition, there must be negative control spots in the microarray design to assess how likely an observed fold change would be if the gene was not expressed differentially. Application of the *t* test addresses both of these issues.

The first step in carrying out a t test is to calculate the mean of measurements from A and the mean of measurements from B. We will symbolize these quantities $\bar{A}$ and $\bar{B}$, respectively. We will also need to calculate the conditions' variances, σ_A^2 and σ_B^2. The next value calculated is the standard error, SE

$$SE = \sqrt{\frac{\sigma_A^2(M-1) + \sigma_B^2(N-1)}{(M+N-2)} \times \frac{M+N}{MN}}. \tag{11}$$

The details of what this quantity represents are beyond our scope. For our purposes, it is worthwhile to note, however, that as σ_A and/or σ_B get larger, so does the standard error. SE is large when data are highly variable.

The next calculation we must make is the t statistic. This value is simply the difference between the two cell type means, divided by the standard error calculated in Eq. (11):

$$S_t = \frac{\bar{A} - \bar{B}}{SE}. \tag{12}$$

We note that as the difference between $\bar{A}$ and $\bar{B}$ becomes large, so too does the absolute value of S_t. In addition, as the uncertainty of these means grows (manifest as the variances, σ_A^2 and σ_B^2), the statistic gets smaller. Another way to view this statistic is that it expresses the differences between two means in units of (roughly) standard deviations. This is an advantage over the simpler S_{fold} statistic where variances are not considered. Another nice property about the t statistic is that it is very well studied by statisticians. In fact, we know how likely a given value of S_t is given M, N, and the null hypothesis that there is no real difference between the two means. Therefore, we can assign a p value for any value of S_t without the requirement of negative control spots.

The corresponding p values of the t statistic should be interpreted carefully, however. The knowledge we have about these probabilities assumes that the observations from each cell type are distributed normally (bell curve shaped). Unfortunately, replicate measurements coming from a microarray experiment do not always behave this way (Thomas *et al.*, 2001) and should be considered when utilizing the t test.

Another potentially troublesome aspect of the t test is that two quantities can lead to large values of S_t. The first is the value we are chiefly interested in, the difference between two conditions. The second quantity that can lead to large S_t is a small SE term. A problem with most microarray experiments is that there are few replicates available from which to calculate the standard error. This leads to the situation where SE can be quite small just by chance, resulting in high S_t values regardless of differences between the two groups of measurements. This is a situation we may not want to deem significant and worthy of further study. A useful guard against this situation is to require low p values computed with the t test *and* some fold-change criterion to consider genes for further study (Rinn *et al.*, 2004).

Significance Analysis of Microarrays (SAM)

The statistic used in SAM (Tusher *et al.*, 2001) is a slight variant of the one given in Eq. (12). The only difference is the so-called "fudge factor" f:

$$S_{sam} = \frac{\bar{A} - \bar{B}}{SE + f}. \tag{13}$$

The purpose of f is to disallow inflated test statistics solely due to standard errors close to zero. Effectively, it sets a lower bound on the denominator of Eq. (13). This factor gives an advantage over the t statistic but it is arguably not the greatest contribution of SAM.

In SAM, the concept of permutation testing was introduced as a means to calculate a *false discovery rate* (FDR). To perform this technique, we first fix a p value threshold T. Next, we identify those genes that have p values less than T. These are our positives. Then, for each gene, the class associations are randomized, that is, we randomly assign measurements for that gene to one of the two classes being compared. Using Eq. (13), S_{sam} is computed for each of these randomized genes. Once the S_{sam} statistics are computed along with their associated p values, the number of these p values less than T is counted. The randomization procedure is repeated a number (100 or 1000) of times and a count is made for each repetition. The median of these counts divided by the total number of genes in the study is then the reported FDR. The intuition of this is that the randomized genes represent genes that do not experience differential expression; therefore, any time one of their p values falls below T, this event can be considered a false discovery.

The notion of a FDR is an important one for microarray experiments having thousands of genes that need to be tested. It helps interpret results of an experiment in light of the multiple testing problem.

Cyber T

Equation (13) introduced the fudge factor f. The purpose of adding this factor was to guard against selecting genes that have a low mean difference and unusually low variances. Another way to protect against such situations is to apply another variation of the *t* test, called Cyber T (Baldi and Long, 2001). In this test, standard error is replaced by an expression that is a function of both the standard error of the gene and the standard error computed over all genes. The assumption here is that most genes should have similar standard errors; by utilizing this assumption, we can lessen the degree to which unexpectedly low or high gene-specific standard errors affect the *t* statistic. This method has been demonstrated to be quite powerful for detecting differences between two samples in experiments using Affymetrix GeneChip brand microarrays (Choe *et al.*, 2005).

Wilcoxon Rank Sum Test

An alternative method for computing significance levels when *t* test assumptions do not hold is the Wilcoxon rank sum test. This test, like many other so-called nonparametric tests, transforms measurements to their magnitude ranks and calculates probabilities based on rank-based statistics. This test was introduced in the microarray literature in Troyanskaya *et al.* (2002). (As an aside, it should be noted that when the assumptions of the *t* test hold, that test should be used, as it is more likely to detect a difference if it exists.)

The basic idea of the Wilcoxon rank sum test is to count the number of times a measurement from one group is greater than a measurement from a second group. The properties of how this value behaves under the null hypothesis of no difference between the groups' medians are well known so we can directly calculate a *p* value from this number. The actual computations of the procedure are not straightforward and lie beyond the scope of this chapter.

Wilcoxon Signed Rank Test

The previously described Wilcoxon rank sum test is generally applicable for comparing two sets of numbers. When the two sets of numbers are paired in some way (such as gene expression levels before and after a treatment), a more powerful nonparametric test is available. This test is called the Wilcoxon signed rank test. To begin, the difference D_i for the ith spot is calculated for each pair in a set of N measurements:

$$D_i = X_i - Y_i \qquad (14)$$

where X_i and Y_i are the paired measurements. Next, each D_i is assigned a rank value R_i of its absolute value

$$R_i = \sum \text{Rank of } |D_i| \tag{15}$$

Next, we sum the ranks of those D_i values that are positive

$$R_+ = \sum R_i \text{ with } D_i > 0 \tag{16}$$

and do the same summation for ranks that have negative D_i values

$$R_- = \sum R_i \text{ with } D_i < 0. \tag{17}$$

Now if we sum all ranks regardless of whether D_i is negative or positive, we will obtain the quantity $1 + 2 + \ldots + N = \frac{N(N+1)}{2}$. If there is no difference between the paired values being compared, then both R_+ and R_- should be roughly half of this previous quantity: $\frac{N(N+1)}{4}$ Therefore, if we take one of the R values as in

$$S = \min(R_+, R_-), \tag{18}$$

we known that under the null hypothesis of zero difference between the two groups, S is expected to be $\frac{N(N+1)}{4}$. We then determine how far away S is from this expected value. Again, the statistic is well studied, and given S and the number of measurements N, we can compute a corresponding p value.

The Wilcoxon signed rank test has utility in experimental designs having perfect match and mismatch probes. In fact, this a commonly used statistic for Affymetrix tiling microarray analysis.

Analysis of Variance (ANOVA)

Previous sections showed how to test for the differential spot intensities measured between two conditions. Frequently, however, a study consists of three or more conditions and the researcher would still like to deduce which genes differ in expression levels between the conditions under study. The standard statistical tool for solving such problems is the ANOVA.

To begin, we need a null hypothesis. For ANOVA, our null hypothesis will be that for all conditions, the gene under study has the same expression level. It may seem strange that a model for assessing equality of means is called analysis of variance. However, the basic idea of ANOVA is to compare the variance of within-condition means to the variance calculated within each condition. (The variance of within-condition means will hereafter be called the between condition variance, and the variance within the samples as the within condition variance.)

Consider measurements $X_{i,j}$ for a single gene where subscript i indicates that the measurement is from the ith biological condition being studied and

j denotes the jth measurement from this condition. If we symbolize the average intensity within condition i as $\bar{X}_i$ and the average of all measurements as $\bar{X}$, we can compute the between condition variance as

$$\sigma^2_{\text{between}} = \frac{\sum_{i=1}^{K} N_i(\bar{X}_i - \bar{X})^2}{K - 1} \tag{19}$$

where K is the number of conditions being studied, N is the total number of measurements, and N_i is the number of measurements taken for condition i. Note that if there are no differences among the conditions, then the variance of their means is small. Likewise, if there are differences the terms $(\bar{X}_i - \bar{X})^2$ become larger. We would like to compare this number to the amount of variation we expect if there are no differences. We can estimate this level of variation by calculating the within condition variance:

$$\sigma^2_{\text{within}} = \frac{\sum_{i,j}(X_{i,j} - \bar{X}_i)^2}{N - K}. \tag{20}$$

We can then compare these two variances [Eqs. (19) and (20)] via a ratio:

$$S_{\text{anova}} = \frac{\sigma^2_{\text{between}}}{\sigma^2_{\text{within}}}. \tag{21}$$

Like previous statistics, we know how this statistic behaves under the null hypothesis of no differential expression and we use this information to calculate its corresponding p value.

The aforementioned discussion on ANOVA is intended to provide a basic feel of the technique and is useful in the case where just one factor (such as biological condition) is expected to affect the measured intensities. Clearly, it can easily be the case that several factors affect microarray measurements. As an example, let us assume that our microarray measurements are expected to vary due to two independent factors in a cancer study. First, we might expect to see differences based on which of several tissue types the measured mRNA came from. Example tissues might include "healthy tissue," "localized cancer," and "metastatic cancer." Second, we could also expect that expression measurements are affected by the race of the individual from which the tissue was obtained. The goal of the study is to identify whether the expression level of some gene changes among the healthy, localized, and metastatic samples.

Given the stated goal of the study, it is tempting to simply apply Eqs. (19), (20), and (21) to elucidate an answer. The problem with doing so is that σ^2_{within} is large when there are unaccounted sources of variation. This translates into lower values of S_{anova} and higher p values.

Why would this higher σ^2_{within} be the case? Recall that the two factors are independent. Therefore, when we bin data by a single factor (e.g., tissue), each bin contains a number of measurements from each class of the other factor (race). Now if there are differences among the classes of the second factor, this will lead to some spread within each tissue bin. This spread leads to higher values of σ^2_{within}. To give us the best chance of detecting a difference among the factor we care about, we need to do some additional work.

First, accounting for two sources of variation requires a little more notation. Previously, we used $X_{i,j}$ to indicate the jth measurement of the ith condition. Now because we have an additional source of variation we wish to model, we must extend this to the term $X_{i,j,k}$, which symbolizes the kth measurement of those belonging to both the ith class of one factor and the jth class of the second. For example, $X_{3,1,7}$ could symbolize the seventh measurement taken of those of the third tissue type (e.g., metastatic tissue) and the first race (e.g., African). In addition, we previously used the variable K to indicate the number of classes we were testing between. Now in addition to K, we also need a variable that denotes the number of classes of the other factor we are studying. Let this variable be B. In our example we might have $K = 3$ ("healthy," "localized," and "metastatic") and $B = 4$ ("African," "Asian," "Caucasian," and "Latino").

In studying the differences between the different stages of cancer, we calculate $\sigma^2_{between}$ as before using Eq. (19), where we use tissue labels as the different classes. The main difference in our analysis lies in how σ^2_{within} is calculated. If we let $\bar{X}_{i,j}$ be the mean of all measurements where factor 1 (e.g., tissue type) is i and factor 2 (e.g., race) is j, then σ^2_{within} is calculated as

$$\sigma^2_{within} = \frac{\sum^{i,j,k}(X_{i,j,k} - \bar{X}_{i,j})^2}{N - BK}. \tag{22}$$

We can then use Eq. (21) as before and use knowledge of its distribution under the null hypothesis to obtain a corresponding p value. Intuitively, all we have done in moving from one factor to two is to adjust the within condition variance so that it does not include potential variation from known sources such as age, race, or gender. Accounting for these sources of variation gives us an enhanced ability to detect differences between some conditions of interest. This increase in sensitivity comes at a cost, however. To calculate σ^2_{within} accurately, there must be a number of measurements available for each combination of the factors we wish to model. As the number of factors in our model increases, so too does the number of replicate experiments needed to estimate σ^2_{within}.

Given that ANOVA can account for different sources of variability, it is also capable of merging microarray normalization with differential expression detection. To do this, sources of variation within the model are not only those of biological interest (such as cancerous vs healthy tissue), but also those of technical concern (such as microarray used and dye used for labeling) (Kerr et al., 2000). The application of ANOVA to microarray data in this context is reviewed nicely in Kerr (2003).

Extensions to Tiling Microarrays

The tests described earlier can be applied to tiling microarrays as well. Recall that in a tiling microarray, we are looking for regions of consecutive probes (in genomic space) that exhibit intensities higher than some background level. To assess this, a windowing approach is often taken where we do not simply assess a single feature by itself, but rather we assess that feature along with a window of neighboring features. To apply the t statistic, for example, we may test the intensities of each window to a random sampling of intensities from any genomic region, to intensities from within putative promoters (which are not expected to be transcribed), or to a control set of features. For Affymetrix tiling microarrays that contain a mismatch probe for every perfect match probe on the microarray, the mismatch probes can serve as this control set to which the comparison can be made. The extension of this approach to fold change, SAM, etc. is straightforward.

Following scoring each window in this manner, the resulting statistics are thresholded by some criteria (set by negative and positive control probes or theoretical considerations). The result is a set of putatively "on" and "off" probes. Spots that meet the threshold criterion and that are within a short distance of each other in genomic space are combined (the spacing between probes above threshold must be less than *maxgap* bp apart) to form larger continuous regions. These combined fragments are then filtered to remove short fragments (require a length longer than *minrun* bp) that are likely to be spurious results.

Extensions to Protein Microarrays

For antibody microarrays that assess concentrations of proteins in solution, the methods described in this section can be applied directly to testing abundance differences between two or more biological conditions. For functional protein microarrays, however, the question is usually one of event detection. In these cases, control experiments must be designed so that they represent the activities of proteins in some baseline state. Once a suitable control is identified, then the methods described here are suitable as well.

Summary

The microarray platform is emerging as a standard tool in biological and biomedical research. This is partly because of its ever-expanding utility, as evidenced by both tiling and protein microarray applications. As is true for any standard tool, it is important that the microarray technology be well understood by its practitioners. For microarrays, part of this technological understanding is resident in the understanding of microarray statistics. Here, in this chapter, widely used methods for microarray normalization and significance testing are presented with the aim of providing this understanding in at least a broad sense. We have indicated where and when gene-based microarray statistics can be useful for tiling and protein microarrays in our discussion. The information conveyed was intended to provide at least a motivation and intuition for what happens to microarray data after it leaves the bench.

Acknowledgment

This work was supported by NIH Grant HG02357. Microsoft and Excel are registered trademarks in the United States and/or other countries.

References

Baldi, P., and Long, A. (2001). A Bayesian framework for the analysis of microarray expression data: Regularized t-test and statistical inferences of gene changes. *Bioinformatics* **17**(6), 509–519.

Bertone, P., Gerstein, M., and Synder, M. (2005). Applications of DNA tiling arrays to experimental genome annotation and regulatory pathway discovery. *Chromosome Res.* **13**(3), 259–274.

Bertone, P., Stolc, V., Royce, T., Rozowsky, J., Urban, A., Zhu, X., Rinn, J., Tongprasit, W., Samanta, M., Weissman, S., Gerstein, M., and Snyder, M. (2004). Global identification of human transcribed sequences with genome tiling arrays. *Science* **306**(5705), 2242–2246.

Bolstad, B., Irizarry, R., Astrand, M., and Speed, T. (2003). A comparison of normalization methods for high density oligonucleotide array data based on variance and bias. *Bioinformatics* **19**(2), 185–193.

Bonferroni, C. E. (1935). Il calcolo delle assicurazioni su gruppi di teste. *In* "Studi in Onore del Professore Salvatore Ortu Carboni. Rome," pp. 13–60.

Cawley, S., Bekiranov, S., Ng, H., Kapranov, P., Sekinger, E., Kampa, D., Piccolboni, A., Sementchenko, V., Cheng, J., Williams, A., Wheeler, R., Wong, B., Drenkow, J., Yamanaka, M., Patel, S., Brubaker, S., Tammana, H., Helt, G., Struhl, K., and Gingeras, T.

(2004). Unbiased mapping of transcription factor binding sites along human chromosomes 21 and 22 points to widespread regulation of noncoding RNAs. *Cell* **116**(4), 499–509.

Cheng, J., Kapranov, P., Drenkow, J., Dike, S., Brubaker, S., Patel, S., Long, J., Stern, D., Tammana, H., Helt, G., Sementchenko, V., Piccolboni, A., Bekiranov, S., Bailey, D., Ganesh, M., Ghosh, S., Bell, I., Gerhard, D., and Gingeras, T. (2005). Transcriptional maps of 10 human chromosomes at 5-nucleotide resolution. *Science* **308**(5725), 1149–1154.

Cho, R., Campbell, M., Winzeler, E., Steinmetz, L., Conway, A., Wodicka, L., Wolfsberg, T., Gabrielian, A., Landsman, D., Lockhart, D., and Davis, R. (1998). A genome-wide transcriptional analysis of the mitotic cell cycle. *Mol. Cell* **2**(1), 65–73.

Choe, S., Boutros, M., Michelson, A., Church, G., and Halfon, M. (2005). Preferred analysis methods for Affymetrix GeneChips revealed by a wholly defined control dataset. *Genome Biol.* **6**(2), R16.

Cleveland, W. S. (1981). Lowess: A program for smoothing scatterplots by robust locally weighted regression. *Am. Stat.* **35**, 54.

Colantuoni, C., Henry, G., Zeger, S., and Pevsner, J. (2002). SNOMAD (Standardization and NOrmalization of MicroArray Data): Web-accessible gene expression data analysis. *Bioinformatics* **18**(11), 1540–1541.

Fodor, S., Read, J., Pirrung, M., Stryer, L., Lu, A., and Solas, D. (1991). Light-directed, spatially addressable parallel chemical synthesis. *Science* **251**(4995), 767–773.

Hoffmann, R., Seidl, T., and Dugas, M. (2002). Profound effect of normalization on detection of differentially expressed genes in oligonucleotide microarray data analysis. *Genome Biol.* **3**(7), RESEARCH0033.

Horak, C., and Snyder, M. (2002). ChIP-chip: A genomic approach for identifying transcription factor binding sites. *Methods Enzymol.* **350,** 469–483.

Irizarry, R., Hobbs, B., Collin, F., Beazer-Barclay, Y., Antonellis, K., Scherf, U., and Speed, T. (2003). Exploration, normalization, and summaries of high density oligonucleotide array probe level data. *Biostatistics* **4**(2), 249–264.

Iyer, V., Horak, C., Scafe, C., Botstein, D., Snyder, M., and Brown, P. (2001). Genomic binding sites of the yeast cell-cycle transcription factors SBF and MBF. *Nature* **409**(6819), 533–538.

Kapranov, P., Cawley, S., Drenkow, J., Bekiranov, S., Strausberg, R., Fodor, S., and Gingeras, T. (2002). Large-scale transcriptional activity in chromosomes 21 and 22. *Science* **296** (5569), 916–919.

Kerr, M. (2003). Linear models for microarray data analysis: Hidden similarities and differences. *J. Comput. Biol.* **10**(6), 891–901.

Kerr, M., Martin, M., and Churchill, G. (2000). Analysis of variance for gene expression microarray data. *J. Comput. Biol.* **7**(6), 819–837.

Kluger, Y., Yu, H., Qian, J., and Gerstein, M. (2003). Relationship between gene co-expression and probe localization on microarray slides. *BMC Genom.* **4**(1), 49.

Lipshutz, R., Fodor, S., Gingeras, T., and Lockhart, D. (1999). High density synthetic oligonucleotide arrays. *Nature Genet.* **21**(Suppl. 1), 20–24.

Lueking, A., Horn, M., Eickhoff, H., Bussow, K., Lehrach, H., and Walter, G. (1999). Protein microarrays for gene expression and antibody screening. *Anal. Biochem.* **270**(1), 103–111.

Luscombe, N., Royce, T., Bertone, P., Echols, N., Horak, C., Chang, J., Snyder, M., and Gerstein, M. (2003). Express Yourself: A modular platform for processing and visualizing microarray data. *Nucleic Acids Res.* **31**(13), 3477–3482.

Nuwaysir, E., Huang, W., Albert, T., Singh, J., Nuwaysir, K., Pitas, A., Richmond, T., Gorski, T., Berg, J., Ballin, J., McCormick, M., Norton, J., Pollock, T., Sumwalt, T., Butcher, L., Porter, D., Molla, M., Hall, C., Blattner, F., Sussman, M., Wallace, R., Cerrina, F., and

Green, R. (2002). Gene expression analysis using oligonucleotide arrays produced by maskless photolithography. *Genome Res.* **12**(11), 1749–1755.

Qian, J., Kluger, Y., Yu, H., and Gerstein, M. (2003). Identification and correction of spurious spatial correlations in microarray data. *Biotechniques* **35**(1), 42–44, 46, 48.

Quackenbush, J. (2002). Microarray data normalization and transformation. *Nature Genet.* **32** (Suppl), 496–501.

Rinn, J., Rozowsky, J., Laurenzi, I., Petersen, P., Zou, K., Zhong, W., Gerstein, M., and Snyder, M. (2004). Major molecular differences between mammalian sexes are involved in drug metabolism and renal function. *Dev. Cell* **6**(6), 791–800.

Saeed, A., Sharov, V., White, J., Li, J., Liang, W., Bhagabati, N., Braisted, J., Klapa, M., Currier, T., Thiagarajan, M., Sturn, A., Snuffin, M., Rezantsev, A., Popov, D., Ryltsov, A., Kostukovich, E., Borisovsky, I., Liu, Z., Vinsavich, A., Trush, V., and Quackenbush, J. (2003). TM4: A free, open-source system for microarray data management and analysis. *Biotechniques* **34**(2), 374–378.

Schena, M., Shalon, D., Davis, R., and Brown, P. (1995). Quantitative monitoring of gene expression patterns with a complementary DNA microarray. *Science* **270**(5235), 467–470.

Thomas, J., Olson, J., Tapscott, S., and Zhao, L. (2001). An efficient and robust statistical modeling approach to discover differentially expressed genes using genomic expression profiles. *Genome Res.* **11**(7), 1227–1236.

Troyanskaya, O., Garber, M., Brown, P., Botstein, D., and Altman, R. (2002). Nonparametric methods for identifying differentially expressed genes in microarray data. *Bioinformatics* **18**(11), 1454–1461.

Tusher, V., Tibshirani, R., and Chu, G. (2001). Significance analysis of microarrays applied to the ionizing radiation response. *Proc. Natl. Acad. Sci. USA* **98**(9), 5116–5121.

Yang, I., Chen, E., Hasseman, J., Liang, W., Frank, B., Wang, S., Sharov, V., Saeed, A., White, J., Li, J., Lee, N., Yeatman, T., and Quackenbush, J. (2002a). Within the fold: Assessing differential expression measures and reproducibility in microarray assays. *Genome Biol.* **3**(11), research0062.

Yang, Y., Dudoit, S., Luu, P., Lin, D., Peng, V., Ngai, J., and Speed, T. (2002b). Normalization for cDNA microarray data: A robust composite method addressing single and multiple slide systematic variation. *Nucleic Acids Res.* **30**(4), e15.

Zhu, H., Bilgin, M., Bangham, R., Hall, D., Casamayor, A., Bertone, P., Lan, N., Jansen, R., Bidlingmaier, S., Houfek, T., Mitchell, T., Miller, P., Dean, R., Gerstein, M., and Snyder, M. (2001). Global analysis of protein activities using proteome chips. *Science* **293**(5537), 2101–2105.

Zhu, H., and Snyder, M. (2003). Protein chip technology. *Curr. Opin. Chem. Biol.* **7**(1), 55–63.

[16]   Random Data Set Generation to Support Microarray Analysis

By DANIEL Q. NAIMAN

Abstract

As microarray analyses become increasingly routine, involving the simultaneous investigation of huge numbers of genes, researchers can easily search for and uncover what appear to be promising patterns in their data. In such circumstances tools are needed to help decide the extent to which these patterns are meaningful or can be explained by chance alone. The purpose of this chapter is to describe examples of the use of microarray analysis for inferential purposes and how validation of inference is addressed by Monte-Carlo techniques, which essentially amounts to investigation of statistical methods on *synthetic* or *random* data sets.

Introduction

The organization of this chapter is as follows. After a discussion of random permutations, three different scenarios in which microarray data lead to inferential questions are described: tests of gene/phenotype association, significance of gene clusters, and classification using gene expression. This purpose of this chapter is not to present every instance of the use of the Monte-Carlo method to microarray analysis, but rather to demonstrate how it is useful in a small sample of instances and to give the reader some feeling for directions in which the field is evolving.

Random Permutations

Random *permutations* form a basic building block for creating random data sets. A permutation of a finite set of objects is a vector in which the objects all appear in some particular order. For example, the set {1, 2, 3}, has six permutations:

$$(1, 2, 3), (1, 3, 2), (2, 1, 3), (2, 3, 1), (3, 1, 2), (3, 2, 1).$$

More generally, a set consisting of N objects $\{O_1, \ldots, O_N\}$ has $N! = N(N-1)$ $\ldots 1$ permutations.

Permuting the objects does not necessarily involve moving them around, but can be accomplished by permuting their identifiers. This comment is

METHODS IN ENZYMOLOGY, VOL. 411
0076-6879/06 $35.00
DOI: 10.1016/S0076-6879(06)11016-2

quite a practical one, as for our purposes each object might occupy a large portion of computer memory (think of the case in which each object constitutes data from a single microarray experiment), and rather than moving them around in memory, we need only manipulate an array consisting of *locations* or *pointers* (Fig. 1): When we consider permutations of N objects, there is no loss of generality if we focus on permuting the set of indices $\{1, \ldots, N\}$.

We will need a procedure for generating a random permutation $\Pi = (\Pi_1, \ldots, \Pi_N)$ of $\{1, \ldots, N\}$ with all $N!$ being equally likely. While many computing platforms already provide this capability, one may encounter situations in which it becomes necessary to write one's own code for the task. Such a procedure is easily built using the more primitive *random number generator*, which produces a sequence of random numbers $U_1, U_2, \ldots$ that are independent and distributed uniformly in the interval $(0, 1)$. The following is pseudo code for making use of this generator to give a random permutation $\Pi_1, \ldots, \Pi_N$ of $1, \ldots, N$.

Step (1) Set $\Pi_i = i$ for $i = 1, \ldots, N$
Step (2) For i from 1 to $N - 1$ do
    Step (a) Generate U uniformly distributed in $(0, 1)$
    Step (b) Set $J = \lfloor (N + 1 - i)U \rfloor + i$
    Step (c) Set $t = \Pi_i$
    Step (d) Set $\Pi_i = \Pi_J$
    Step (e) Set $\Pi_J = t$
Step (3) Return $\Pi = (\Pi_1, \ldots, \Pi_N)$

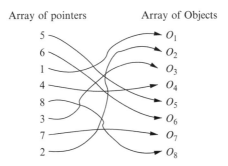

Array of pointers   Array of Objects

FIG. 1. Permuting a list of objects stored in memory. The objects $O_1, \ldots, O_8$ are stored in an array. Permuting the objects is accomplished by permuting the indices $1, 2, \ldots, 8$ giving an array $\pi = (\pi_1, \pi_2, \ldots, \pi_8) = (5, 6, 1, 4, 8, 3, 7, 2)$. The list of permuted objects becomes $(O_{\pi_1}, O_{\pi_2}, \ldots, O_{\pi_8}) = (O_5, O_6, O_1, O_4, O_8, O_3, O_7, O_2)$.

Here the notation $\lfloor x \rfloor$ in Step (b) means remove the fractional part of x, for example, $\lfloor 2.71828 \ldots x \rfloor = 2$. In the ith interation of Step (2), the index J is selected at random from $\{i, \ldots, N\}$, and the value Π_i is swapped with the value Π_J. A simple induction argument can be used to verify that all permutations are equally likely to be generated by this algorithm.

Random permutations can be used to accomplish many of the kinds of tasks required for the statistical validation methods described later. In particular, n-fold cross-validation requires random partitioning of a set of N data points into n groups of sizes $q_1, \ldots, q_n$, where $q_1 + \ldots + q_n = N$. This can be accomplished by generating a random permutation $\Pi_1, \ldots, \Pi_N$ of $\{1, \ldots, N\}$ and extracting the groups in order so that the first group is the first q_1 indices in the permutation, the second group is the next q_2, and so on. As a special case, a random permutation can be used to draw a random subset of size q_1 from a data set of size N, as this is equivalent to breaking the set into groups of sizes q_1 and $q_2 = N - q_1$.

Tests of Genetic Association

In a genome-wide association study (Cardon and Bell, 2001), cases and controls are genotyped at a number of markers with the goal of determining a marker or markers giving a high degree of association between genotype and disease. *Pools* of cases and controls can be analyzed as well, with various benefits (Sham *et al.*, 2002), including that of reducing the number of microarray chips required and providing a degree of privacy that would presumably increase study participation. Gene chips are now available for genotyping at 100,000 SNPs (Kreiner and Buck, 2005). Once a marker is found to have significant association, a search can be carried out for genes causing the disease near such a putative marker.

Consider an experiment in which n sampled cases and $\tilde{n}$ sampled controls are genotyped at each of N marker loci. Data from such an experiment can be arranged in the form of a matrix

$$
\begin{bmatrix}
X_1^{(1)} & \cdots & X_1^{(n)} & \tilde{X}_1^{(1)} & \cdots & \tilde{X}^{(\tilde{n})} \\
\vdots & \vdots & \vdots & \vdots & \vdots & \vdots \\
X_N^{(1)} & \cdots & X_N^{(n)} & \tilde{X}_N^{(1)} & \cdots & \tilde{X}_N^{(\tilde{n})}
\end{bmatrix}
\tag{1}
$$

where each row corresponds to a marker, each column represents a microarray chip, and each entry could be a genotype.

For each marker i, a statistic T_i can be calculated to measure disease association, and we can usually arrange it so that these statistics have a common cumulative distribution function (cdf) F_T under the assumption of no association between the marker and the disease. For example, if each matrix entry $X_j^{(i)}$ or $\tilde{X}_j^{(i)}$ is one of three possible genotypes aa, aA, or AA, genotype counts could be entered in the form of a table

	cases	controls
aa	n_{aa}	$\tilde{n}_{aa}$
aA	n_{aA}	$\tilde{n}_{aA}$
AA	n_{AA}	$\tilde{n}_{AA}$

then T_i could be the χ^2 statistic for testing for association (Ott, 1991) and F_T would be the cdf for the χ^2 distribution with two degrees of freedom.

We need a threshold for deciding that a particular T_i is large enough to conclude an apparent association to be actually significant. Typically, one introduces a null hypothesis, which says that the microarray columns, both cases and controls, constitute a random sample from some multivariate distribution F_X so that none of the genes are associated with disease. Then, one chooses a threshold so that the probability of no false marker associations takes some nominal value α (usually $\alpha = 0.05$).

If a single marker were selected for analysis *a priori*, the desired false detection probability would be obtained by using the upper α critical point of F, i.e., $F_T^{-1}(1 - \alpha)$. However, when N markers are scanned, it is more likely to find *some* marker i with $T_i > F_T^{-1}(1 - \alpha)$, and in fact we would obtain an average of $N\alpha$ markers with this property, even under the assumption that no markers are associated with disease.

The threshold needs to be *adjusted for multiple testing*, which means determining the cdf F_M of $M = \max_{i=1, \ldots, N} T_i$ under the null hypothesis. Using information about the multivariate distribution F_M various tools would become available for setting a threshold for M. For example, under the assumption of linkage equilibrium, the rows of the matrix and hence the T_i could be taken to be independent. Then $F_M(x) = F_T(x)^N$ so the desired threshold takes the form $F_T^{-1}(\{1 - \alpha\}^{1/N})$. As marker sets become increasingly dense, linkage disequilibrium between nearby markers cannot be ignored, and modeling the dependence realistically becomes essential, as the appropriate threshold will be very sensitive to degree of dependence assumed.

Were it the case that plausible choices for F_M describing the nature of gene–gene dependence were available, one could use ordinary Monte-Carlo simulation (Fishman, 1996) or some variant (see, e.g., Naiman and Priebe, 2001). In ordinary Monte-Carlo simulation, we would sample F_M to produce a large number m of realizations of random data sets as in (1), which requires $(n + \tilde{n}) \times N \times m$ sample columns. For each of these data sets, the maximum statistic M could be calculated to yield a sample $M^{(1)}, \ldots, M^{(m)}$

from F_M. From this sample, the threshold $F^{-1}(1 - \alpha)$ can be estimated by the $1 - \alpha$ sample quantile.

The problem with this approach is that specifications of F_M tend to be speculative as the sample sizes are too small to measure the complex dependencies reliably. What is needed is a *model-free* or *nonparametric* approach to setting a threshold. One commonly employed technique is to use a Bonferroni (Naiman and Priebe, 2001) threshold of $F_T^{-1}(1 - \alpha/N)$. This choice guarantees that the probability of at least one false detection is no greater than α, no matter what the nature of the dependence between markers is, but one pays a price for using such a procedure in that it can be overly conservative: the probability of at least one false detection will typically be much smaller than α and the power of the test is unacceptably low.

A standard nonparametric technique to use in this situation is the *permutation test*. The idea is the same as in ordinary Monte-Carlo simulation in that we create m random data sets, calculate a sample $M^{(1)}, \ldots, M^{(m)}$, and use the $1 - \alpha$ quantile to estimate a threshold for M. The difference here is that the data sets are created by permuting the columns in the original data set so that effectively n random columns are taken as controls and the remaining $\tilde{n}$ columns are taken as cases. The resulting test has the desirable property that the probability of false detection takes the nominal value of α no matter what null distribution is being sampled. The discussion here has been focused on what is referred to as the *population model*, where two different populations are sampled (cases/controls), but the method also applies in the *randomization model* in which subjects are assigned randomly to treatment and control groups. This distinction and its ramifications are discussed in Ernst (2004), which provides a comprehensive overview of permutation tests.

Gene Clustering

Statistical analyses are required frequently to determine whether apparent spatial clustering of disease occurrences is real or due to chance (Kulldorff and Nagarwalla, 1995). Similarly, one might classify a collection of genes as "interesting" by some criterion, and observe that the interesting genes tend to appear in clusters on the genome. Generally, criteria for establishing which genes are interesting can be developed using microarray experiments. For example, a gene could be declared interesting if it is expressed differentially when samples are analyzed under a pair of distinct experimental conditions. An example of such an analysis appears in Parisi *et al.* (2003).

Significance of apparent clustering can be assessed using spatial scan statistics (Naiman and Priebe, 2001). To keep the discussion simple, consider genes $i = 1, \ldots, m$ linearly ordered along a chromosomal segment. Let $X_i = 0$ or 1, where $X_i = 1$ is interpreted as saying that gene i is classified as interesting. A *cluster* can be thought of as a sequence of L contiguous genes $j, j + 1, \ldots, j + L - 1$ (a *window*) in which the number $\mathcal{N}_{j,L} = \sum_{i=j}^{j+L} X_i$ of interesting ones is unusually high, say $\mathcal{N}_{j,L} \geq c_L$, where c_L is a constant chosen so that the probability of exceeding a threshold is roughly the same for all window sizes, assuming the interesting genes are distributed randomly about the genome.

Here the search for clusters can be carried out over a predetermined collection $\mathcal{W} = \{L_1, \ldots, L_q\}$ of window sizes, and as a special case, one can take $\mathcal{W} = \{L\}$ a single window size decided upon a priori. It is important to keep in mind that the specific search procedure used for finding clusters is a critical factor in determining the significance level of a discovered cluster. The significance level of a cluster is a function of the size m of the chromosomal segment in which the search is carried out, as well as of the collection of window sizes $\mathcal{W}$. The larger the search space is, the more likely we are to find large clusters by chance alone. By ignoring this very fundamental fact (and reporting a finding without properly accounting for the method used), it is quite easy to produce results that cannot be validated upon further investigation.

Permutation tests provide both a formal way to compute the significance level of a cluster and an informal way to contemplate the discovery process (see Fig. 2). At the formal level, having found that $\mathcal{N}_{j,L} \geq c_L$ for some choice of j and $L \in \mathcal{W}$, we can calculate a significance level as the proportion of random permutations $\Pi = (\Pi_1, \ldots, \Pi_m)$, out of a large number, say $N = 100,000$ of them, for which the permuted data $X_{\Pi_1}, \ldots, X_{\Pi_m}$ give some window of some size $L \in \mathcal{W}$, with $\mathcal{N}_{j,L} \geq c_L$. On an informal level, one can generate random permutations of data, display them graphically, and obtain a sense for the kind of clustering to be expected even when the interesting genes are distributed randomly throughout the chromosomal segment.

Apparent cluster

FIG. 2. An apparent cluster of 3 *interesting* genes in a window consisting of 5 genes. The entire genomic segment consists of 312 equispaced genes of which 17 are interesting. Interesting genes are shown as solid lines, and noninteresting genes appear dashed (when shown). When the interesting genes are permuted randomly, there is about a one in five chance of such a cluster appearing somewhere.

Supervised Classification

A variety of methods have been developed for supervised classification of biological samples using gene expression. Supervised classification provides a framework in which many different problems are solved routinely using a variety of random data set approaches. The framework for this can be described as follows. Several samples are collected from some population to be used as training data. Each sample is assumed to possess one of various possible phenotypes so there is a finite set of p categories $\{C_1, \ldots, C_p\}$ into which each sample can be *classified*, and we use Y to denote the category of a particular sample. It is assumed throughout this discussion that all classification of training samples is done without error. In addition, gene expression levels for a large collection of genes are obtained using a microarray so that we have a column vector $X = [X_1 \ldots X_G]$ of gene expression levels for each sample, where G denotes the number of genes.

Assuming there are N samples, a training data set $\mathcal{T}$ will then consist of expression profile/class label pairs $(X^{(i)}, Y^{(i)})$ for $i = 1, \ldots, N$ and these data are conveniently combined into a matrix having G rows and N columns

$$[X^{(1)}, \ldots, X^{(N)}] = \begin{bmatrix} X_1^{(1)} & \cdots & X_1^{(N)} \\ \vdots & \vdots & \vdots \\ X_G^{(1)} & \cdots & X_G^{(N)} \end{bmatrix},$$

where the columns are assigned labels $Y^{(1)}, \ldots, Y^{(N)}$. Each row of this matrix corresponds to a specific gene, and each column, together with its label, corresponds to a sample microarray profile.

The goal is to find a way to use only gene expression values of samples to classify them, that is, to build a *classifier* or *class predictor* (Fig. 3), a function f that takes as input an observed microarray profile and outputs a class label.

Training data are used to build such a function, and the process of going from training data to a classifier is referred to as a *learning algorithm*. A learning algorithm $\mathcal{L}$ is a function whose input is a training data set $\mathcal{T}$ with column labels as defined earlier and whose output is a classifier $f = \mathcal{L}_\mathcal{T}$ (Fig. 4).

For a concrete example, as in Singh *et al.* (2002), the population consists of patients with a high level of prostate-specific antigen in their blood, and the samples are prostate tissue samples. The pathologist's examination of the tissue sample reveals whether it is cancerous or not, making the set of categories $\{N, C\}$, or the pathologist assigns a *Gleason's score* to the sample, in which case the set of categories is $\{2, \ldots, 10\}$.

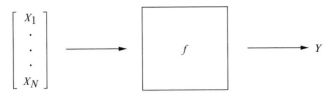

FIG. 3. A classifier is a function that assigns a class Y to each input profile X.

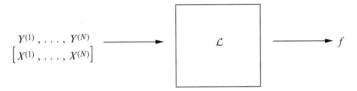

FIG. 4. A learning algorithm $\mathcal{L}$ is a function whose input is expression profiles $[X^{(1)}, \ldots, X^{(N)}]$ with class labels $Y^{(1)}, \ldots, Y^{(N)}$ and whose output is a classifier f.

There are several available methods for classification, including k-nearest neighbors (Cover and Hart, 1967), prediction analysis for microarrays (PAM) (Tibshirani *et al.*, 2002), variants of Fisher's discriminant analysis, including *diagonal quadratic discriminant analysis* and *diagonal linear discriminant analysis* (Dudoit *et al.*, 2002), and support vector machines (Furey *et al.*, 2000; Statnikov *et al.*, 2005), many of which are described in Dudoit and Fridlyand (2003). A recent new approach to classification is the *TSP* classifier described in Geman *et al.* (2004).

For concreteness, we describe the *k-nearest neighbors* classification. To simplify the discussion, we assume there are only two classes (the dichotomous classification problem). The k-nearest neighbors classifier $f = f_{d,k}$ is built from a training set (so it is inherently a learning algorithm) and is based on a distance measure d between pairs of profiles, for example, $d(X,X') = \sum_{i=1}^{N} |X_i - X_i'|^2$, and on a choice of an odd integer k. Once the choice of d and k are made, a new profile X is classified by determining the k *sample* profiles $X^{(i_1)}, \ldots, X^{(i_k)}$ closest in distance from X and picking the class Y that appears most often among $Y^{(i_1)}, \ldots, Y^{(i_k)}$. In other words, the closest samples *vote* for a class and the class decision is based on majority rule.

For the classifier just described, even assuming the distance function d is fixed, the integer k may be fixed *a priori* or can be viewed as a *parameter* to be *learned* from the training set. Parameter selection is a feature common to many learning algorithms.

Estimating Classification Error

Usually the researcher not only wants to construct a good classifier f but also wants an estimate $\hat{e}$ of its misclassification rate, for example, the probability of misclassification for a newly sampled profile. The following discussion uses the abbreviated notation Z for a sampled gene expression/class pair (X, Y) and assumes these are sampled from a distribution denoted by F_Z.

The assumption that a single population is sampled can be relaxed to allow for the situation when each *stratum* (defined by the class variable Y) is sampled, and most of the methods discussed here can be modified easily to address that situation, provided that the probabilities of the strata $\pi_j = P[Y = j], j = 1, \ldots, p$ are assumed to be known.

We define the misclassification error for a newly sampled profile $Z = (X, Y)$ (not part of the training set) as $e(f) = P[f(X) \neq Y] = E[\delta(f(X), Y)]$, where

$$\delta(Y', Y) = \begin{cases} 0 & \text{if } Y = Y' \\ 1 & \text{if } Y \neq Y'. \end{cases}$$

The process of arriving at the classifier and an estimate of its misclassification rate are usually intertwined, and it is appropriate to view a learning algorithm as a procedure that not only produces a classifier from training data, but also an estimate of its misclassification rate. It is important to recognize that without exercising care, use of the same data to build a classifier, as well as estimate the error rate, can lead to misleading results. In particular, the so-called *apparent error rate*, defined as the proportion of misclassified training samples $\hat{e}_{app} = \frac{1}{N}\sum_{i=1}^{N} \delta(\mathcal{L}_T(X^{(i)}), Y^{(i)})$, tends to underestimate the actual error rate. For example, when a k-nearest neighbors classifier is used to classify a training sample $Z^{(i)}$, this same sample will always get to vote being the nearest one, which results in a tendency for the class $Y^{(i)}$ itself to be favored in the voting process.

Far better methods for estimating the misclassification are readily available that avoid the pitfalls of over fitting. The following is a brief description of some of the available methods for estimating the misclassification rate for a classifier.

Separating Training from Testing

One of the simplest ideas for estimating the misclassification rate is to set aside a random subset $\{Z^{(i)}, i \in I\}$, use the remaining data as a training set, so that $T = \{X^{(i)}, i \notin I\}$, to learn a classifier $f = \mathcal{L}_T$, and once this is done, the misclassification rate is estimated to be the proportion of errors made on the left-out samples $\hat{e} = \frac{1}{|I|} \sum_{i \in I} \delta(f(X^{(i)}), Y^{(i)})$. While this method leads to unbiased estimates of misclassification error with a variance that can be controlled by manipulating the size of I, the samples in I are not available at the training stage when they would be more useful.

n-*fold Cross-Validation* (n-*fold CV*)

In n-fold cross-validation, the available data set $T = \{Z^{(i)}, i = 1, \ldots, N\}$ is broken up randomly into n roughly equal-sized pieces $T_i, i = 1, \ldots, n$, where n is an integer, for example, $n = 10$. For convenience, let us assume that the size of the training sample N is evenly divisible by n and let $q = N/n$ so that each piece T_i is of size q. Random partitioning was described earlier.

To estimate the classification error, we proceed as follows. For each $i = 1, \ldots, n$ set aside T_i, and combine the remaining $n - 1$ data sets T_j, $j \neq i$ into a single data set $\tilde{T}_i$ of size $(n - 1)q = N - q$. Train a classifier using the learning algorithm $\mathcal{L}$ on $\tilde{T}_i$ to give a classification rule $\tilde{f}_i = \mathcal{L}_{\tilde{T}_i}$. Estimate the classification error rate for $\tilde{f}_i$ on each labeled profile $Z = (X, Y) \in T_i$ in the data set aside as the proportion of misclassified profiles $\hat{e}_i = \frac{1}{q} \sum_{(X,Y) \in T_i} \delta\left(\tilde{f}_i(X), Y\right)$. Finally, estimate the classification error to be $\hat{e}_{CV} = \frac{1}{n} \sum_{i=1}^{n} \hat{e}_i$. Multiple estimates obtained in this way using different random partitions can be averaged.

Conveniently, we can use the same data set T to arrive at a classifier, as well as to calculate an estimate of the misclassification rate. Because samples used to train the classifier are separated from those used to estimate the misclassification rate, the estimated rate is unbiased for the misclassification rate based on the use of training sets of size $N - q$ rather than N. Because a larger sample size gives rise to improved performance, these estimates tend to overestimate the misclassification rate.

From the point of view of *bias* minimization, one should use as large a value of n as possible. In the most extreme case, we can take $n = N$ to minimize the bias. Here, in each of the N iterations we leave out a single observation, use the remaining $N - 1$ training samples to build a classifier, and use the classifiers on the left-out samples to estimate the classification rate. This procedure, known as *leave-one-out cross-validation*, does not involve randomization, as every one of the N profiles is left out in the analysis.

However, bias is not the only source of statistical error in estimating the misclassification rate. There is also variance, and the optimal choice of n involves a bias/variance tradeoff that is so common in modern statistical inference (Friedman, 1996; Geman *et al.*, 1992; Kohavi and Wolpert, 1996). The larger we take n to be, the more the profiles the training sets will tend to share in common. The resulting dependence between classifiers leads to a greater variance in the estimator $\hat{e}$.

Bootstrap Error Estimates

It is important to keep in mind that the classification error $e(f)$ is actually a random variable, as $f = \mathcal{L}_T$ depends on the training data $T = \{Z^{(i)} = (X^{(i)}, Y^{(i)}), i = 1, \ldots, N\}$. It is helpful to recast the problem of estimating classification error as that of estimating the *expected* classification error $E[e(f)] = E[\delta(\mathcal{L}_T(X^{(N+1)}, Y^{(N+1)}))]$. Clearly, this quantity takes the form of the expected value of a known function of a sample $Z^{(i)}$, $i = 1, \ldots,$ $N + 1$ of size $N + 1$ from F_Z, the function being determined by the learning algorithm.

The bootstrap (Efron, 1979) is a general nonparametric technique for estimating such an expectation when we only get to observe a random sample $Z^{(i)}$, $i = 1, \ldots, N$ of size N from F. The idea of the bootstrap is to approximate this quantity by the expected value of the same function applied to a *bootstrap* sample $Z^{(i)*}$, $i = 1, \ldots, N + 1$, defined as a random data set formed by sampling *with* replacement N times from the set of observed data $\{Z^{(i)}, i = 1, \ldots, N\}$. A large body of articles and books have been written about the bootstrap. It is recommended that the interested reader consult Hall (1992), LePage and Billard (1992), Shao and Tu (1995), Efron and Tibshirani (1993), and Davison and Hinkly (1997).

The bootstrap idea is implemented using Monte-Carlo simulation: independent samples of random variables having the target expectation value are averaged. In fact, one can take the view that the bootstrap is, in essence, model-free Monte-Carlo simulation where the fact that the underlying cdf F_Z is unknown is addressed by replacing it by the empirical cdf. A simple approach to bootstrapping this quantity proceeds as follows. In the ith of m iterations, a bootstrap sample $Z^{(j)*}$, $j = 1, \ldots, N + 1$ is drawn. The training set $T^{(i)}$ is taken to be $\{Z^{(j)*}, j = 1, \ldots, N\}$, and the learning algorithm is used to produce a classifier $f^{(i)} = f_{T_{(i)}}$. Then we determine the performance of this classifier on the left-out bootstrap sample $Z^{(N+1)*}$ by taking $\delta_i = \delta(f^{(i)}(X^{(N+1)*}), Y^{(N+1)*})$. These values are averaged over the m iterations to give $\hat{e}_{boot_1} = \frac{1}{m}\sum_{i=1}^{N}\delta_i$.

A simple improvement on this approximation is obtained by observing that in any given iteration, once the classifier $f^{(i)}$ is trained on the bootstrap sample T^i the random variable $\hat{e}_{boot}^{(i)} = \frac{1}{N}\sum_{j=1}^{N}\delta(f^{(i)}(X^{(j)}), Y^{(j)})$, which represents the average classification error over *all* of the original samples is typically relatively easy to calculate, has the same expectation as δ_i but has a variance that is smaller by a factor of $1/N$. Thus, the bootstrap expectation can be approximated more effectively by taking $\hat{e}_{boot_2} = \frac{1}{m}\sum_{i=1}^{m}\hat{e}_{boot}^{(i)}$.

It turns out that this way of using the bootstrap idea is somewhat naive. Efron (1983) described an alternative bootstrap approach that leads to improved performance. The idea is to use the bootstrap to estimate expected "optimism" of the apparent error rate $w = E[e(f) - \hat{e}_{app}] = E[\delta(\mathcal{L}_T(X), Y) - \frac{1}{N}\sum_{i=1}^{N}\delta(\mathcal{L}_T(X^{(i)}), Y^{(i)})]$ calling this estimate $\hat{w}_{boot}$, and *correct* the apparent error rate by taking as an estimated classification error $\hat{e}_{boot_3} = \hat{e}_{app} + \hat{w}_{boot}$.

As in the naive bootstrap approximation, the quantity w to be approximated is the expectation of a known function $\Delta = \delta(\mathcal{L}_T(X^{(N+1)}), Y^{(N+1)}) - \frac{1}{N}\sum_{i=1}^{N}\delta(\mathcal{L}_T(X^{(i)}), Y^{(i)})$ of a sample $Z^{(i)}, i = 1, \ldots, N+1$ from F_Z.

A Monte-Carlo algorithm for approximating $\hat{w}_{boot}$ can be described as follows. As before, in the ith of m iterations a bootstrap sample $Z^{(j)*}, j = 1, \ldots, N+1$ is drawn, we can take as the training set $T^{(i)} = \{Z^{(j)*}, j = 1, \ldots, N\}$, which can be used to produce a classifier $f^{(i)} = f_{T^{(i)}}$. Then we take $\Delta_i = \delta(f^{(i)}(X^{(N+1)}), Y^{(N+1)}) - \frac{1}{N}\sum_{j=1}^{N}\delta(f^{(i)}(X^{(j)}), Y^{(j)})$ and average these to give $\hat{w}_{boot_1} = \frac{1}{m}\sum_{i=1}^{m}\Delta_i$.

As was the case for the *naive* bootstrap, it is easy to find a random variable with the same expectation as Δ_i but with smaller variance. We can take as an alternative $\Delta_i' = \frac{1}{N}\sum_{j=1}^{N}\delta(f_{T^*}(X_j), Y_j) - \frac{1}{N}\sum_{j=1}^{N}\delta(f^{(i)}(X^{(j)}), Y^{(j)}) = \frac{1}{N}\sum_{j=1}^{N}(1 - r_j)\delta(f_{T^*}(X_j), Y_j)$, where r_i is the number of times Z_i appears in T^*, and take $\hat{w}_{boot_2} = \frac{1}{m}\sum_{i=1}^{m}\Delta_i'$.

Permutation Tests

Having chosen an algorithm $\mathcal{L}$ for learning a classifier and estimating its misclassification rate $\hat{e}$ (hopefully low), it is natural to wonder how likely it is that one would obtain such a rate, even using data for which the gene expression profiles do not contain useful information about class labels.

In standard statistical parlance, we would like to estimate a *p-value*

$$P_{H_0}[\hat{e} \leq \hat{e}_{\text{obs}}]$$

This is the probability of achieving such a low misclassification rate as the one just observed $\hat{e}_{\text{obs}}$ from real data, when we use the same learning algorithm $\mathcal{L}$ on data collected under a null hypothesis of profile noninformativeness. Here noninformativeness means that the distribution of a profile X is independent of its class label, that is, all labeled profiles can be seen as sampled from a common distribution.

One can use the permutation test approach to compute a *p*-value: repeatedly modify the training set by permuting the row of class labels (making the class labels noninformative), and estimate a *p*-value by the proportion of cases in which the error rate is no greater than $\hat{e}_{\text{obs}}$.

Conclusions

The Monte-Carlo approach to statistical validation, namely repeatedly performing statistical analyses on synthetic data sets, has been with us for a long time. A key goal of this chapter was to describe some common situations in which the Monte-Carlo approach to statistical validation is useful in microarray analysis and to give the reader a sense for how new methodology in this area is still being developed. Some relevant work has not been described here at all. In particular, Bayesian techniques (Chen *et al.*, 2001), which in the past have been viewed as computationally intractable, are now becoming more routine. Also, various improvements on standard Monte-Carlo via the technique of importance sampling (Naiman and Priebe, 2001) are being discovered somewhat regularly. As computational power increases, more varieties of analyses requiring these techniques are likely to appear and become standard tools in every microarray analysts tool box.

References

Cardon, L., and Bell, J. (2001). Association study designs for complex diseases. *Nature Rev. Genet.* **2**, 91–99.

Chen, M.-H., Shao, Q.-M., and Ibrahim, J. G. (2001). "Monte Carlo Methods in Bayesian Computation." Springer, New York.

Cover, T., and Hart, P. (1967). Nearest neighbor pattern classification. *IEEE Trans. Inform. Theory* **13**, 21–27.

Davison, A., and Hinkly, D. (1997). "Bootstrap Methods and Their Application." Cambridge University Press, Cambridge.

Dudoit, S., and Fridlyand, J. (2003). Classification in microarray experiments. *In* "Statistical Analysis of Gene Expression Microarray Data" (T. Speed, ed.), pp. 93–158. Chapman and Hall.

Dudoit, S., Fridlyand, J., and Speed, T. (2002). Comparison of discrimination methods for the classification of tumors using gene expression data. *J. Am. Stat. Assoc.* **97**, 77–87.

Efron, B. (1979). "The Jacknife, the Bootstrap, and Other Resampling Plans." SIAM NSF-CBMS, Monograph No. 38.

Efron, B. (1983). Estimating the error rate of a prediction rule: Improvement on cross-validation. *J. Am. Stat. Assoc.* **78**(382), 316–331.

Efron, B., and Tibshirani, R. (1993). "An Introduction to the Bootstrap." Chapman and Hall.

Ernst, M. (2004). Permutation methods: A basis for exact inference. *Stat. Sci.* **19**, 676–685.

Fishman, G. (1996). "Monte Carlo: Concepts, Algorithms and Applications." Springer, New York.

Friedman, J. (1996). On bias, variance, 0/1 loss and the curse of dimensionality. *J. Data Mining Knowledge Disc.* **1**, 55–97.

Furey, T., Cristianini, N., Duffy, N., Bednarski, D., Schummer, M., and Haussler, D. (2000). Support vector machine classification and validation of cancer tissue samples using microarray expression data. *Bioinformatics* **16**, 906–914.

Geman, D., d'Avignon, C., Naiman, D., and Winslow, R. (2004). Classifying gene expression profiles from pairwise mrna comparisons. *Stat. Appl. Genet. Microbiol.* **3**.

Geman, S., Bienenstock, E., and Doursat, R. (1992). Neural networks and the bias/variance dilemma. *Neural Comput.* **4,** 1–58.

Hall, P. (1992). "The Bootstrap and Edgeworth Expansion." Springer-Verlag, New York.

Kohavi, R., and Wolpert, D. (1996). Bias plus variance decompositions for zero-one loss functions. *In* "Machine Learning: Proceedings of the 13th International Conference."

Kreiner, T., and Buck, K. (2005). Moving toward whole-genome analysis: A technology perspective. *Am. J. Health Syst. Pharm.* **62,** 296–305.

Kulldorff, M., and Nagarwalla, N. (1995). Spatial disease clusters: Detection and inference. *Stat. Med.* **14,** 799–810.

LePage, R., and Billard, L. (1992). "Exploring the Limits of Bootstrap." Wiley-Interscience, New York.

Naiman, D., and Priebe, C. (2001). Computing scan statistic p-values using importance sampling with applications to genetics and medical image analysis. *J. Comput. Graph. Stat.* **10,** 296–328.

Ott, J. (1991). "Analysis of Human Genetic Linkage." Johns Hopkins.

Parisi, M., Nuttal, R., N., Naiman, D., Bouffard, G., Malley, J., Andrews, J., Eastman, S., and Oliver, B. (2003). Paucity of genes on the drosophila x chromosome showing male-biased expression. *Science* **299,** 697–700.

Sham, P., Bader, J., Craig, J., O'Donovan, M., and Owen, M. (2002). DNA pooling: A tool for large scale association studies. *Nature Rev. Genet.* **3,** 862–871.

Shao, J., and Tu, D. (1995). "The Jacknife and Bootstrap." Springer, New York.

Singh, D., Febbo, P., Jackson, D., Manola, J., Ladd, C., Tamayo, P., Renshaw, A., D'Amico, A., Richie, J., Lander, E., Loda, M., Kantoff, P., Golub, T., and Sellers, W. (2002). Gene expression correlates of clinical prostate cancer behavior. *Cancer Cell* **1**(2), 203–209.

Statnikov, A., Aliferis, C., Tsamardinos, I., Hardin, D., and Levy, S. (2005). A comprehensive evaluation of multicategory classification methods for microarray gene expression cancer diagnosis. *Bioinformatics* **21,** 631–643.

Tibshirani, R., Hastie, T., Narasimhan, B., and Chu, G. (2002). Diagnosis of multiple cancer types by shrunken centroids of gene expression. *Proc. Natl. Acad. Sci. USA* **99**(10), 6567–6572.

[17] Using Ontologies to Annotate Microarray Experiments

By PATRICIA L. WHETZEL, HELEN PARKINSON, and
CHRISTIAN J. STOECKERT, JR

Abstract

Consistent annotation of studies using microarrays is critical to optimal management and use of microarray data. Ontologies provide defined and structured terminology suited for this purpose. The Gene Ontology (GO) has aided the analysis of expression studies greatly by providing consistent functional annotation of array sequence features. The intent of the MGED Ontology (MO) is to provide consistent experimental annotation. The MO has been developed as a community effort in support of the Minimum Information

METHODS IN ENZYMOLOGY, VOL. 411   0076-6879/06 $35.00
                          DOI: 10.1016/S0076-6879(06)11017-4

About a Microarray Experiment standard and is tied to the Microarray Gene Expression object model. The MO is freely available and has been incorporated into the annotation systems of several public microarray database systems.

Introduction

The use of high-throughput technologies, such as microarrays, has increased rapidly in the last few years as improvements in the technology have made their use accessible and affordable to many laboratories. Consequently, bench biologists have faced the problem of data management. Due to the complexity and enormity of data, the microarray community has developed standards regarding the type of information to collect and the format of data. The Minimum Information about a Microarray Experiment (MIAME) describes the minimum information needed in order for the experiment to be repeated, and the Microarray Gene Expression Markup Language (MAGE-ML) provides a standard format for reporting data to facilitate data exchange (Brazma *et al.*, 2001; Spellman *et al.*, 2002). The community has accepted these standards, and many journals now require that MIAME-compliant data be deposited in a public microarray repository as a prerequisite for publication. However, text annotation and deposition of ill-defined data do not allow us to manage it well. Therefore, in addition to the content standard MIAME and the data exchange format MAGE-ML, well-defined terms are required to unambiguously annotate the experiment. These terms can exist as an ontology, terminology, or controlled vocabulary. The MGED Ontology (MO) provides terms or descriptors required to unambiguously annotate a microarray experiment as described in the MIAME document.

This chapter

- explains what an ontology is
- explores the Gene Ontology (GO)
- outlines the scope and purpose of the MGED Ontology
- describes the "use cases" for the MGED Ontology
- explains where a biologist might see and use terms from the MO
- gives examples of software applications that use the MO and briefly explain their use
- outlines the future development of the MGED Ontology.

What Is an Ontology?

The word ontology is Greek in origin and dates back to the time of Aristotle meaning the science of being. In more recent years, this word has been used in research on artificial intelligence and knowledge representation with the definition of "An ontology is an explicit specification of a conceptualization" (Gruber, 1993). This rather cryptic definition can be simplified to "an ontology is a set of concepts and the relationships between these concepts in a domain(s)." The usage of the term ontology in the biological world is less stringent, and often controlled vocabularies such as Swissprot (Boeckmann et al., 2003) key word and terminologies such as SNOMED (http://www.snomed.org/) are grouped under the ontology umbrella. An ontology is not just an annotation aid, as ontologies are used in many fields as a method to model information about a domain. By modeling the domain using an ontology, various types of relationships between the concepts can be represented explicitly. An example would be the use of the relationship types "is-a" and "part-of," which can be found in many ontologies. Such complex relationship types are rarely found in controlled vocabularies. Representing ontologies is a science in itself and therefore ontologies can be available in multiple formats and can be constructed in various tools. The usual starting places for an ontology is a piece of paper, but tools exist to manage and develop ontologies as they become more complex. Biologists may be familiar with the GO (Ashburner et al., 2000), which is created using the DAG-Edit editor (http://www.godatabase.org/dev/java/dagedit/docs/index.html); Protégé (http://protege.stanford.edu/; Noy et al., 2003) is also widely used by projects such as PharmGKB (Hewett et al., 2002). Both DAG-Edit and Protégé have their own ontology representation format (DAG-Edit OBO and CLIPS, respectively). Efforts to develop a standard representation for the semantic web have produced OWL—the Web Ontology Language (http://www.w3.org/TR/owl-ref/), which is supported by Protégé, replacing DAML as the standard (Horrocks, 2002). Both Protégé and DAG-Edit also support the construction of machine-readable ontologies. This is explored in detail in Stevens et al. (2002). Ontologies in OWL can be developed using Protégé with the OWL plug-in and can be reasoned over using Racer (Haarslev and Moller, 2001, http://www.racer-systems.com). One advantage of developing an ontology in a language such as OWL is that concepts can be asserted or defined by specifying restrictions as either necessary or necessary and sufficient. This is useful in ontology development, as concepts can be placed as subclasses of new parent classes after classification of the ontology when using a reasoner, such as Racer or FaCT, therefore avoiding the use of multiple parents (Horrocks, 1999). In constructing an ontology in this way, different views of the ontology can be developed and used for different purposes, as discussed in the next section.

Gene Ontology

The GO is worth exploring in some detail, as it is now considered a paradigm for using and building biological ontologies and is extremely successful; over 50 tools using the GO are now available (http://www.

geneontology.org/GO.tools.shtml). The GO was developed to provide terms for the annotation of gene products, initially by the model organism databases, and exists as a directed acyclic graph (DAG), which means that one term can have multiple parents, each link is directional, and there are no cyclic relationships. The GO consists of three branches—biological process, function, and cellular compartment—and is used to annotate gene products based on what they do (function), what processes they are part of, for example, metabolism, and where they are found (cellular compartment). The GO uses the relationship types "is-a" and "part-of"; for example, the nucleolus is "part-of" the nucleus and DNA binding "is-a" nucleic acid binding. The GO terms and definitions are essentially based on natural language, and the DAG structure means that is it an attractive tool for biologists. However, the GO is not fully machine readable and therefore efforts are underway to develop a workflow to convert GO into a more expressive ontology language (Wroe et al., 2003). The GO developers are also developing their own formalism OBOL, which is more expressive than the standard OBO format. A critical factor in the success and evolution of the GO has been the community authoring methodology, which was also adopted by the developers of the MO. The advantages of community authoring are that a solution can be developed that addresses the needs of multiple users, and although compromises are made in the development of the ontology, it is usable by a wider community.

What Is the MO Used For?

The MO was developed in concert with the MIAME standard and the MAGE-OM, which are standards for describing and modeling a microarray experiment, respectively. The MO therefore describes the process of performing a microarray experiment and some of the components used in such an experiment. The MO contains terms for the annotation of experiment designs, sample processing, and protocols. The MO does not provide terms for the annotation of samples from a specific species, as these terms are typically provided in other ontologies. For example, the Jackson Laboratory provides several ontologies for mouse strains, developmental stages, and anatomy, and the MO references these resources as sources of appropriate terms but does not contain these terms directly. It does, however, model the concepts about a sample that could be annotated. For example, the age, developmental stage, and strain attributes of a sample are all concepts present in the MO.

How Was the MO Built?

The MO was designed to facilitate queries such as "show me all expression values for gene x in experiments using organism y of strain z" and "show me all human samples in tissue comparison experiments that are from nondiseased samples." Execution of such queries requires that gene x is annotated, that the samples are annotated consistently with species, strain, and disease state, and that experiment types are consistent. The rules adopted in construction of the MO can be summarized as follow.

Concepts must be orthogonal (nonoverlapping)

Concepts must not be species specific

Where terms exist in other ontologies, the MO will refer to that ontology rather than reinvent the terms locally

The MO supports the MAGE-OM and is organized in the same way

Terms are not deleted from the MO but made obsolete, with a reason

The MO addresses only microarray experiments

The MO core is stable and the underlying structure cannot be changed.

Where Can I Get the MO?

Documents describing the MO and versions of the MO exist in various formats (DAML+OIL, OWL etc.) and can be found on the sourceforge site at http://mged.sourceforge.net/ontologies/index.php. It is possible to obtain terms from the MGED Ontology for use in annotating experiments directly from on-line resources such as the NCI DTS browser (http://nciterms.nci.nih.gov/NCIBrowser/Dictionary.do).

The ontology itself is located at http://mged.sourceforge.net/ontologies/MGEDontology.php. The web site provides easy access to descriptions for each term and enables browsing. Terms for the MO are organized into classes, properties, and individuals. Information on the ontology and links to notes, files, and sites for submitting terms (sourceforge tracker) are also provided on the web site. Classes are listed alphabetically, and there are hypertext links to definitions, superclasses and subclasses, constraints (i.e., properties of the class), and individuals (i.e., instances of the class). Listed classes, properties, and individuals are all hypertext linked to their own definitions. Properties and individuals are similarly listed alphabetically. Classes, properties and individuals can be accessed directly on the web as http://mged.sourceforge.net/ontologies/MGEDontology.php#[term]. For example, to access the class "Action," add "#Action" to the end of the URL for the MGED Ontology (i.e., http://mged.sourceforge.net/ontologies/MGEDontology.php#Action). Figure 1 illustrates the html page that is returned with a description of the class Action. Also shown are examples for a MO property (hasID) and individual (Adult_Mouse_Anatomy). In order to see the class hierarchy or search the MO the ontology is best downloaded and viewed in the ontology editor Protégé or viewed in the NCI DTS browser.

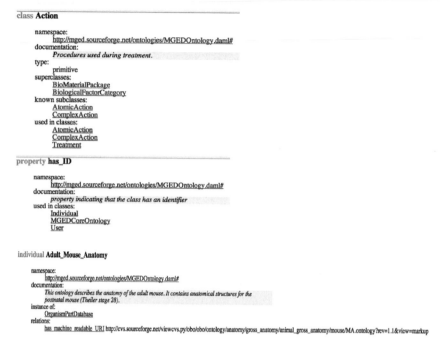

FIG. 1. Examples of a class, property, and individual in the MGED ontology. Shown are screen shots from the MGED ontology web site obtained by adding "Action," "has_ID," or "Adult_Mouse_Anatomy" to the end of the URL: http://mged.sourceforge.net/ontologies/MGEDontology.php#.

MO in Detail

The various parts of a microarray experiment are divided into "packages" by MIAME and the MAGE-OM. For example, there are packages that describe the samples, the experimental design, and the annotation of the genes on the array. As the ontology supports the MAGE-OM, it uses the same organizing principles. The MO is intended primarily as an annotation resource that is used in other applications; consequently, the user will rarely see the entire structure, only the parts of relevance to the task at hand. The GO is used in a similar fashion; expert curators have an overview of the whole structure, while the biologist will likely see only the GO terms associated with a gene product(s) of interest. Detailed knowledge of the MO structure is useful for those building applications that deal with microarray data for biologists and here we provide an overview of the MO structure and design principles. The MO contains two parts: MGEDCoreOntology (MCO), which contains the stable core, and the MGEDExtendedOntology (MEO), which is used for terms outside the MAGE object model. The class hierarchy of the MCO is directly parallel with the MAGE-OM structure (Fig. 2). In other words, for each package in the MAGE object model, a similar named package exits in the MCO. Within each high-level package grouping of the MCO, the ontology classes mimic the structure of the MAGE object model. The MCO does not contain all the classes that exist in the MAGE object model, only those that provide a path to the various references to the MAGE-OM OntologyEntry class. In addition to following the structure of the MAGE object model, the MCO class names are also similar to the MAGE object model association name that is used to point to the OntologyEntry class in order to facilitate mapping from the object model to the MO. Annotations in MAGE-OM are provided using two mechanisms: inclusion of terms within the MAGE object model and by using a reference to the OntologyEntry class. Terms that exist within the model are provided as an enumerated list in the MO, and the definition of the term is included in the MAGE object model documentation. Terms that are referenced via the OntologyEntry class exist within the MO. In some cases the MCO uses specialized property names where the MAGE object model used a generic association name such as types. For example, the association from ExperimentDesign to Ontology Entry in the MAGE object model is named types; in the MO the property that represents this association is named has_experiment_design_type. In cases where associations in the MAGE object model are to existing ontologies, for example, NCBI taxonomy, the MO provides references to these external semantic resources. This is done in a similar way within the MCO by providing these resources as a subclass of the MCO class OntologyEntry. In this way, terms from these external semantic resources can be used to annotate the experiment and be referenced within the MCO. The MEO contains classes from previous versions of the MO that represent concepts to maintain, but did not mirror concepts in the MAGE-OM and therefore did not fit in the MO after it was restructured to support MAGE-OM. This is the result of early parallel development of the MO and MAGE-OM and the subsequent restructuring of the MO to support the MAGE-OM. All classes, properties, and individuals in the MO are defined, and exact or nonexact synonyms are included in the definition of an individual. This is a consequence of the original ontology editor tool used; OilEd (Bechhofer et al., 2001) had limited synonym handling.

Who Uses the MO?

Several institutions have developed databases and applications to manage, annotate, and analyze microarray data that use the MO, including the European Bioinformatics Institute, the University of Pennsylvania,

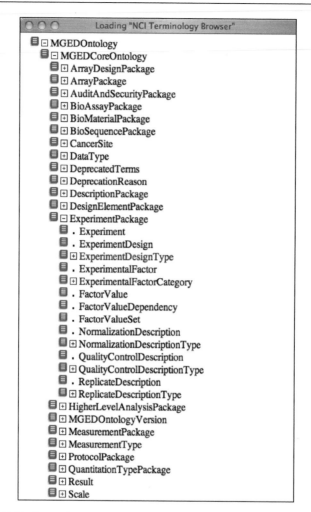

FIG. 2. NCI view of the class hierarchy of MGED ontology. The top class, MGEDOntology, is the top-level container of descriptors for microarray experiments. The subclass, MGEDCoreOntology, is the container class for descriptors of microarray experiments that are tied to the MAGE object model. The ExperimentPackage is a class of descriptors for microarray experiments tied to the experiment package of the MAGE object model. ExperimentDesignType is a class of descriptors for microarray experiments that provide terms for the MAGE object model association from ExperimentDesign to OntologyEntry in order to describe the types of experimental designs. The figure is from an image generated by the NCI DTS browser.

TABLE I
DATABASES AND APPLICATIONS THAT USE THE MO

Application	URL	Institution	Data annotation	Data query	Ontology server
ArrayExpress	www.ebi.ac.uk/arrayexpress	EBI[a]	Yes	Yes	No
caArray	caarraydb.nci.nih.gov/	NCICB[b]	Yes	Yes	No
Enterprise Vocabulary Service	http://nciterms.nci.nih.gov/ NCIBrowser/Dictionary.do	NCICB	No	No	Yes
maxD	Bioinf.man.ac.uk/microarray/maxd/	University of Manchester	Yes	Yes	No
MIAMExpress	www.ebi.ac.uk/MIAMExpress	EBI	Yes	Yes	No
MiMiR	microarray.csc.mrc.ac.uk/	CSC/IC[c]	Yes	Yes	Yes
RAD Study Annotator	www.cbil.upenn.edu/RAD/	University of Pennsylvania	Yes	Yes	No
SMD	genome-www5.stanford.edu/	Stanford University	Yes	Yes	No

[a] European Bioinformatics Institute.
[b] National Cancer Institute Center for Bioinformatics.
[c] Clinical Sciences Centre/Imperial College.

A

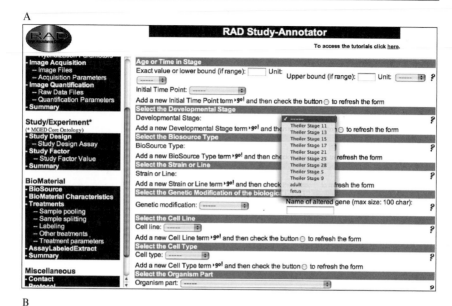

B

Stanford, the University of Manchester, the National Center for Biotechnology Information, and the National Center for Toxicogenomics. Although the tools developed by these institutions provide microarray data annotated with the MO, these institutions have developed different implementations of the MO, as discussed later, in the same way that there are many tools that use the GO. Table I summarizes some of these tools and describes their functions.

How Is the MO Presented to Users?

This section examines two applications in detail: (1) The RAD Study Annotator (RADSA) and (2) the Enterprise Vocabulary Service.

The RAD Study Annotator

The RADSA was developed at the University of Pennsylvania (Manduchi *et al.*, 2004) for biologists to annotate their microarray experiments. The RADSA is a suite of web annotation forms that provide MO terms in select menus. These terms are stored in a cache table and therefore only necessary terms from the MO are presented to the user (Fig. 3). This cache table points to the entire set of MO terms in case new terms need to be added (Fig. 4). This system also has a query interface that allows the user to retrieve data based on MO terms for the ExperimentDesignType, Organism, OrganismPart, ExperimentFactorType, and PlatformType.

The Enterprise Vocabulary Service

The Enterprise Vocabulary Service at the National Cancer Institute Center for Bioinformatics (NCICB) is a set of services and resources that address the National Cancer Institutes needs for controlled vocabulary (Covitz *et al.*, 2003). The Enterprise Vocabulary Service project provides the cancer centric MetaThesaurus, which includes terms from the Unified Medical Language System and other vocabularies used in NCICB applications. The NCI has

---

FIG. 3. (A) RAD Study Annotator–BioMaterialCharacteristics form. This form shows the various concepts within the MGED ontology that are applicable to describe human or mouse samples. Terms for the developmental stage were obtained from external resources listed in the MGED ontology and are stored in a cache table within the schema, which allows for terms to be displayed in a project-specific manner. Note that this form displays terms for the MGED ontology concept DevelopmentalStage that are relevant to human or mouse samples. (B) RAD Study Annotator–BioMaterialCharacteristics form. This form shows the BioMaterialCharacteristics form for plasmodium samples. This form displays fewer concepts as they are not applicable to plasmodium, such as cell type. Also, note that terms for the concept DevelopmentalStage display only those that are relevant to plasmodium.

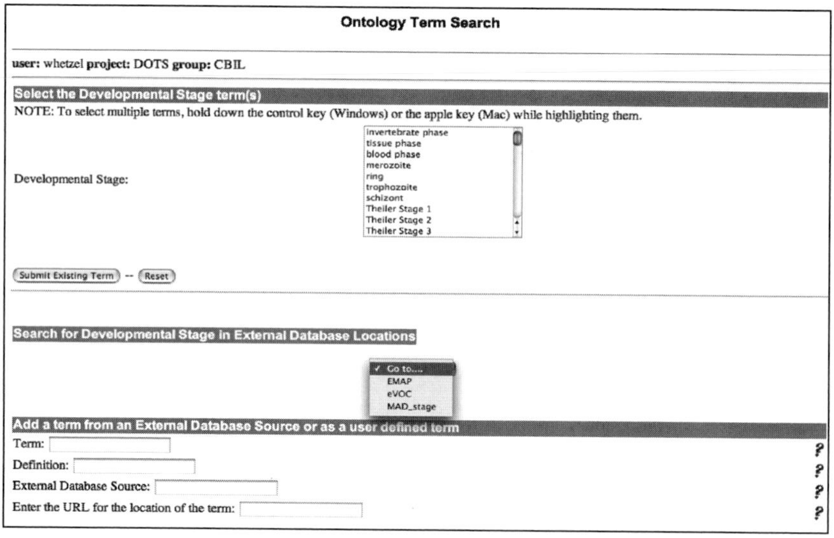

Fig. 4. RAD Study Annotator–Ontology Term Search Form. This form is accessed by clicking on the "GO" link in forms that allow the user to add new terms for concepts. This form shows how the user can add new terms to be displayed in the BioMaterialCharacteristics form for the concept DevelopmentalStage. The top of the form displays all terms stored in the cache table for the selected concept. The bottom of the form displays the external resources that the MGED ontology lists as resources for DevelopmentalStage terms. The user can browse these resources and then add the information into the text boxes.

provided a terminology browser, which can be used to view the MO and other ontologies as a tree structure. The MO can be searched for a term via the user interface and the user can choose to view the details of the term and its position in the hierarchy as a collapsible tree or presented on the screen with clickable links to super- and subclasses (Fig. 2). An example of a MO concept displayed in the NCI browser is shown in Fig. 5.

Releases and Management of the MO

The MO is an evolving ontology and new terms can be suggested by anyone via a tracker (http://sourceforge.net/tracker/?atid=603031&group_id=16076&func=browse). New terms are gleaned from submissions to the participating databases, submitted via the tracker, and discussed during conference calls where the definition of term and the placement of the term in the ontology are determined. The changes in the MO are reflected in the version number; the current version is 1.2.0, and a complete set of release

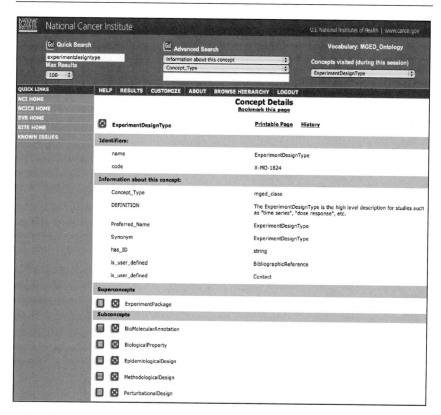

FIG. 5. NCI DTS browser view of a MO class. This view shows the class ExperimentDesignType, the definition of the class, properties of this class, and the superclass and subclasses of this term.

notes is provided with each successive release. Typically changes between releases are due to the addition of new terms, for example, terms describing comparative genomic hybridization experiments have been added recently, or obsoletion of existing terms, most frequently as they are split into finer grained terms. The MO underwent a major change between versions 1.1.9 and 1.2.0 as it was migrated from DAML+OIL format to OWL. Version 1.1.9 is the most recent DAML+OIL version, and all future development will be done in OWL. This change was made as OWL is now a W3C standard and DAML+OIL has become a legacy format. There have also been considerable improvements in the editing tools and the MO will be developed exclusively in Protégé in the future.

Future of the MO

The MO in its existing form will continue to be maintained and new terms will be added on request. However, the world has changed since the MAGE-OM and the MO were conceived. The proteomics, environmental biology, and metabol/nomics communities have a need for an ontology to describe their experiments. As all these disciplines require the annotation of biological samples, the MO developers have joined forces with these communities to develop a new ontology that meets the requirements of the wider community. A new model called FuGE (http://fuge.sourceforge.net/) is being developed to represent these new domains. The MO is being used as the foundation to develop the Functional Genomics Investigation Ontology (FuGO), which is designed to model functional genomics domains such as proteomics and metabol/nomics, as well as transcriptomics, and will be extended to additional communities working together under Reporting Structure for Biological Investigations (RSBI).

Conclusion

In summary, the MO provides descriptors to unambiguously annotate microarray data. The MO was originally developed in DAML+OIL; however, future versions will be developed in OWL. Annotation applications and databases that implement the MO provide high-quality, well-annotated data that are available publicly.

The MO will serve as the foundation to develop an ontology to describe functional genomics, transcriptomics, proteomics, toxicogenomics, environmental genomics, and metabol/nomics, which will allow comparison across these technologies and biological domains using a common terminology.

References

Ashburner, M., Ball, C. A., Blake, J. A., Botstein, D., Butler, H., Cherry, J. M., Davis, A. P., Dolinski, K., Dwight, S. S., Eppig, J. T., Harris, M. A., Hill, D. P., Issel-Tarver, L., Kasarskis, A., Lewis, S., Matese, J. C., Richardson, J. E., Ringwald, M., Rubin, G. M., and Sherlock, G. (2000). Gene ontology: Tool for the unification of biology. *Nature Genet.* **25,** 25–29.

Bechhofer, S., Horrocks, I., Goble, C., and Stevens, R. (2001). OilEd: A reasonable ontology editor for the semantic web. *In* "Proceedings of the Joint German Austrian Conference on Artificial Intelligence, LNAI 2174," pp. 396–408.

Boeckmann, B., Bairoch, A., Apweiler, R., Blatter, M. C., Estreicher, A., Gasteiger, E., Martin, M. J., Michoud, K., O'Donovan, C., Phan, I., Pilbout, S., and Schneider, M. (2003). The SWISS-PROT protein knowledgebase and its supplement TrEMBL in 2003. *Nucleic Acids Res.* **31,** 365–370.

Brazma, A., Hingamp, P., Quackenbush, J., Sherlock, G., Spellman, P., Stoeckert, C., Aach, J., Ansorge, W., Ball, C. A., Causton, H. C., Gaasterland, T., Glenisson, P., Holstege, F. C., Kim, I. F., Markowitz, V., Matese, J. C., Parkinson, H., Robinson, A., Sarkans, U., Schulze-Kremer, S., Stewart, J., Taylor, R., Vilo, J., and Vingron, M. (2001). Minimum information about a microarray experiment (MIAME)-toward standards for microarray data. *Nature Genet.* **29**, 365–371.

Covitz, P. A., Hartel, F., Schaefer, C., De Coronado, S., Fragoso, G., Sahni, H., Gustafson, S., and Buetow, K. H. (2003). caCORE: A common infrastructure for cancer informatics. *Bioinformatics* **19**, 2404–2412.

Gruber, T. R. (1993). A translation approach to portable ontologies. *Knowledge Acquisition* **5**, 199–220.

Haarslev, V., and Moller, R. (2001). Racer system description. *In* "International Joint Conference on Automated Reasoning," June 18–23, Siena, Italy.

Hewett, M., Oliver, D. E., Rubin, D. L., Easton, K. L., Stuart, J. M., Altman, R. B., and Klein, T. E. (2002). PharmGKB: The pharmacogenetics knowledge base. *Nucleic Acids Res.* **30**, 163–165.

Horrocks, I. (2002). DAML+OIL: A reasonable web ontology language. *In* "Proceedings of the Advances in Database Technology-EDBT," pp. 2–13, Prague, Czech Republic.

Horrocks, I. (1999). FaCT and iFaCT. *In* "Proceedings of the International Workshop on Description Logics," pp. 133–135. Linkoping, Sweden.

Manduchi, E., Grant, G. R., He, H., Liu, J., Mailman, M. D., Pizarro, A. D., Whetzel, P. L., and Stoeckert, C. J., Jr. (2004). RAD and the RAD Study-Annotator: An approach to collection, organization and exchange of all relevant information for high-throughput gene expression studies. *Bioinformatics* **20**, 452–459.

Noy, N. F., Crubezy, M., Fergerson, R. W., Knublauch, H., Tu, S. W., Vendetti, J., and Musen, M. A. (2003). Protege-2000: An open-source ontology-development and knowledge-acquisition environment. *AMIA Annu Symp Proc.* 953.

Spellman, P. T., Miller, M., Stewart, J., Troup, C., Sarkans, U., Chervitz, S., Bernhart, D., Sherlock, G., Ball, C., Lepage, M., Swiatek, M., Marks, W. L., Goncalves, J., Markel, S., Iordan, D., Shojatalab, M., Pizarro, A., White, J., Hubley, R., Deutsch, E., Senger, M., Aronow, B. J., Robinson, A., Bassett, D., Stoeckert, C. J., Jr., and Brazma, A. (2002). Design and implementation of microarray gene expression markup language (MAGE-ML). *Genome Biol.* **3**, RESEARCH0046.

Stevens, R., Goble, C., Horrocks, I., and Bechhofer, S. (2002). Building a bioinformatics ontology using OIL. *IEEE Trans. Inf. Technol. Biomed.* **6**, 135–141.

Wroe, C. J., Stevens, R., Goble, C. A., and Ashburner, M. (2003). A methodology to migrate the gene ontology to a description logic environment using DAML+OIL. *Pac Symp. Biocomput.* 624–635.

[18] Interpreting Experimental Results Using Gene Ontologies

By TIM BEISSBARTH

Abstract

High-throughput experimental techniques, such as microarrays, produce large amounts of data and knowledge about gene expression levels. However, interpretation of these data and turning it into biologically meaningful knowledge can be challenging. Frequently the output of such an analysis is a list of significant genes or a ranked list of genes. In the case of DNA microarray studies, data analysis often leads to lists of hundreds of differentially expressed genes. Also, clustering of gene expression data may lead to clusters of tens to hundreds of genes. These data are of little use if one is not able to interpret the results in a biological context. The Gene Ontology Consortium provides a controlled vocabulary to annotate the biological knowledge we have or that is predicted for a given gene. The Gene Ontologies (GOs) are organized as a hierarchy of annotation terms that facilitate an analysis and interpretation at different levels. The top-level ontologies are molecular function, biological process, and cellular component. Several annotation databases for genes of different organisms exist. This chapter describes how to use GO in order to help biologically interpret the lists of genes resulting from high-throughput experiments. It describes some statistical methods to find significantly over- or under-represented GO terms within a list of genes and describes some tools and how to use them in order to do such an analysis. This chapter focuses primarily on the tool *GOstat* (http://gostat.wehi.edu.au). Other tools exist that enable similar analyses, but are not described in detail here.

Introduction

DNA microarrays are a high-throughput experimental technique used for gene expression analysis (Lockhart *et al.*, 1996; Schena *et al.*, 1995). Such high-throughput techniques have greatly facilitated the discovery of new biological knowledge. However, this kind of knowledge is often difficult to grasp, and turning raw microarray data into biological understanding is by no means a simple task. Classical analysis of microarray data is based critically on comparing gene expression at the level of single genes and determining the significantly differentially expressed genes (Ayroles

METHODS IN ENZYMOLOGY, VOL. 411
0076-6879/06 $35.00
DOI: 10.1016/S0076-6879(06)11018-6

and Gibson, 2006; Beissbarth *et al.*, 2000; Downey, 2006; Saeed *et al.*, 2006; Smyth, 2004) or on clustering techniques used to determine genes with a similar expression pattern in a series of experiments (Eisen *et al.*, 2000; Gollub and Sherlock, 2006; Saeed *et al.*, 2006). In other popular techniques, such as serial analysis of gene expression (SAGE), the focus is often just to find the transcripts expressed in a given tissue (Beissbarth *et al.*, 2004). In each case, results can be expressed as a list of genes or as a sorted list of all genes on the array ranked by a score (see Fig. 1).

In the more recent literature, many groups came up with methods that use prior biological knowledge in order to help interpret these lists of genes (e.g., Al-Shahrour *et al.*, 2004; Beissbarth and Speed, 2004; Dennis *et al.*, 2003; Doniger *et al.*, 2003; Yue and Reisdorf, 2005). Together with the application of clustering techniques to microarray data the term "guilt by association" has been used, indicating that by using prior biological knowledge it may be possible to infer the function of coexpressed genes (Clare, 2002; Quackenbush, 2003). Furthermore, with a single experiment producing a list of hundreds of differentially expressed genes, automatic annotation and grouping of genes are necessary merely to make interpretation of the results possible (Beissbarth *et al.*, 2003; Boon *et al.*, 2004). In addition,

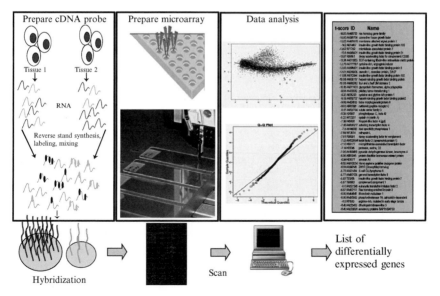

FIG. 1. Overview of the process of performing and analyzing microarray experiments. Frequently the output of such an analysis is a list of significantly differentially expressed genes or a list of genes ranked by a score or *p* value.

the grouping of genes and subsequent testing for the differential expression of these groups of genes may help find significant biological processes even in cases where single gene analysis fails to produce significant results.

The main problem is how to incorporate prior biological knowledge and where to get it from. The type of knowledge that comes to mind first that would be useful to incorporate is that of biological pathways. However, the information on this as stored in pathway databases such as the Kyoto Encyclopedia of Genes and Genomes (KEGG) (Kanehisa et al., 2000; Mao et al., 2005) is quite sparse. Other sources of information could be SWISS-Prot key words, presence of transcription factor-binding sites, protein domains, etc. Further, it is important to organize this information in a structured way. Ontologies are a commonly used concept in computer science (Draghici et al., 2003). They are used to define terms in a controlled vocabulary as well as relations between those terms. They can also be viewed as a directed acyclic graph, which defines a hierarchy of terms. The Gene Ontology Consortium (http://www.geneontology.org) is a worldwide consortium that defines an ontology that can be used to annotate genes (Ashburner et al., 2000). Furthermore, there are a number of Gene Ontology (GO) databases, where annotation for genes of a certain organism or several organisms are stored (e.g., Blake et al., 2003; Camon et al., 2004). The hierarchical structure of GO annotations is illustrated in Fig. 2.

A biological annotation can be expressed as a gene set. Gene ontology annotations can be cut at different levels, allowing the definition of gene sets for each GO term, leading to a hierarchy of gene sets. The subsequent statistical test for whether a certain biological function (or gene set) is

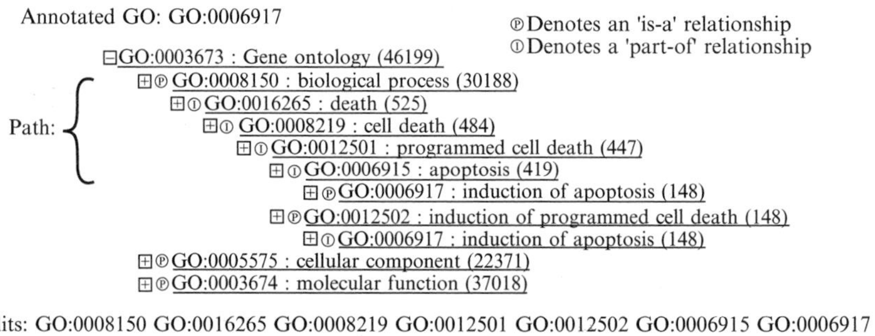

FIG. 2. Structure of a GO annotation. Gene ontologies can be represented as a hierarchy or as a directed acyclicgraph. There are two types of relationships: "is a" and "part of." Top-level ontologies are "molecular function," "biological process," and "cellular localization." Each GO term can be reached by a path, and the GO terms that lead to a certain GO annotation are called "splits."

associated with the experimental outcome of a microarray is often referred to as gene set enrichment analysis (Lamb *et al.*, 2003; Mootha *et al.*, 2003), and analysis with GO groups has become widely used for the analysis of microarray data (also see Gollub and Sherlock, 2006; Hennetin and Bellis, 2006; Whetzel *et al.*, 2006). This analysis allows testing for gene sets (or functional groups) that are represented significantly more frequently in a list of genes that appears interesting in a microarray experiment than in a comparable list of genes that would be selected randomly from all genes on the microarray. This then allows the association of functions to a list of genes.

Find Statistically Overrepresented GO Terms within a Group of Genes

The most commonly used methods to test for enrichment of a gene set in a list of genes selected from an experiment are based on hypergeometric distribution and use either Fisher's exact test or the χ^2 test. These methods work in a similar way: a list of genes is selected from a microarray first, for example, by choosing all significantly differentially expressed genes using a cutoff value. The test for enrichment involves counting how many genes in the gene set occur in the list of selected genes and how many occur on the microarray, and then estimating what proportion of genes from the gene set would be expected to appear in randomly selected lists. The Fisher's exact test gives optimal results for small counts; with larger counts (e.g., counts > 5) the χ^2 test is accurate and much faster to compute (for introduction into statistical testing, see Dalgaard, 2002; Freedman *et al.*, 1997).

A commonly used argument against these methods is that the initial selection of the list of interesting genes from the microarray is arbitrary. The test could possibly be much more sensitive by including genes just below the selection cutoff or by taking all of the genes on the microarray into account. There are several different approaches to do this, a few of which are outlined.

Scores or statistics computed from microarray data for each gene can be accumulated in the gene set. These scores could be, for example, *p* values, *t* statistics, or ranks. It is then possible to create a summary score and carry out tests, for example, for differential expression of the whole gene set rather then for the individual genes. Permutation testing can be used to compute the significance.

Furthermore, the Kolmogorov–Smirnov (KS) test has been used in order to test for significant deviations of the ranks that are collected in a gene set. In the KS test it is assumed that the ranks of genes that are collected under a certain gene set should be drawn from a uniform distribution 1 to N (N being the number of genes on the microarray).

For example, if the ranks 1 through 10 are collected in a certain gene set with 10 genes and the microarray has 10,000 genes, this would be significant. See Fig. 3 for an illustration.

It is worth noting that the KS test checks for any deviation from uniformity, for example, a gene set collecting ranks 5001 through to 5010 would also be significant. Because this is usually not desirable, we recommend using Wilcoxon's signed rank test. The Wilcoxon test checks whether the ranks collected in a gene set are consistently high or consistently low. A significant result indicates that the gene set is enriched in genes with low or high ranks from the experiment.

In most cases there are a multitude of gene sets that one wants to test enrichment for. For example, when defining gene sets from GO annotation, it is common to test all GO terms annotated in the list of interesting genes as well as all GO terms that are higher in the hierarchy than GO terms annotated directly. This means there are usually a few hundred to a thousand gene sets that are tested independently. This leads, however, to a multiple testing problem, as with this many tests it becomes very likely that some tests will show a significant result by chance. Therefore, multiple testing correction is necessary. Many methods for multiple testing correction exist (Dudoit, 2002; Shaffer, 1995). The most simple and most conservative method is to multiply the p values by the number of tests carried out (Bonferroni, 1936). We recommend using a less stringent method controlling the false discovery rate (FDR). The FDR gives the expected rate of false discoveries when selecting below a certain cutoff (Benjamini and Hochberg, 1995), for example, when selecting 100 GO terms with a FDR cutoff of 0.05 we expect that 5 of these resulted due to random chance. This is a less stringent criteria than a usual p value, which controls the family wise error rate, that is, the probability of any error.

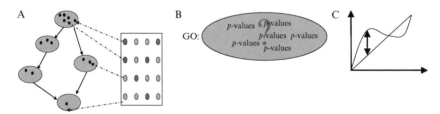

FIG. 3. Scores, p values, or ranks are collected within a gene set. Gene sets can be defined from GO terms. Genes of interest from a microarray analysis may link to different GO terms in the GO hierarchy (A). This leads to a collection of p values, scores, or ranks for each GO term (B). The Kolmogorov–Smirnov (KS) test can be used to test whether the distribution of ranks is significantly different from uniform. It uses the KS distance, which is defined as the maximum distance between two cumulative distributions (C).

It is important to keep in mind, however, that the p values and FDRs computed here are merely indications for significance and give a ranking of the most significant gene sets. Many of the methods described make assumptions, some of which (e.g., independence of tests, normal distribution) are clearly violated. The same is true for the p values or FDRs computed for differential gene expression from microarray data. A common assumption that is clearly violated is that of independence. For multiple testing corrections, there are a few methods that allow arbitrary dependence structures in data. However, these methods are often too conservative.

Another important consideration is the power of the tests, that is, that if the list of genes of interest is small or does not have many annotations or if the gene set that is tested has only a few genes, the p values that can be reached are not infinitely small, and with thousands of tests performed it is quite likely that a p value smaller than 1 can never be reached. Therefore, when testing for hundreds of GO terms, the list of interesting genes studied needs to be sufficiently large (>50). Even when choosing not to use a cutoff and using the KS or Wilcoxon test this is not necessarily a more powerful method. If there are clear factors that determine whether a gene is interesting or not as opposed to a smooth ranking over all the genes, then methods that rank the complete list will be less powerful. It might be better in these cases to choose a higher cutoff to get to a larger list of genes allowing for some false positives and to perform the Fisher's exact test. An apparent idea to incorporate a smooth score for each gene rather than just a rank or than choosing a fixed cutoff would be to use the p values indicating the differential expression of a gene. These p values should ideally follow a uniform distribution between 0 and 1 and KS tests could be performed, similarly as on the ranks, to find gene sets where the distribution of p values differs from uniformity. However, as the p values that result out of significance testing for differential gene expression in microarray experiments are usually strongly skewed (e.g., biased toward 0 or 1) and far from uniformly distributed, it is not usually advisable to do this. Currently these issues are not fully resolved and there is no clear answer on which method to use. Therefore, it is best to try several methods.

GOstat: A Tool to Find Statistically Overrepresented Gene Ontologies

Description

The program *GOstat* helps in the analysis of lists of genes and will compute statistics about the GO terms contained in data and sort the GO annotations giving the most representative GO terms first. The program also allows the computation of statistics for GO terms based on ranked gene lists from microarrays. It is available via the web site http://gostat. wehi.edu.au. Figure 4 shows a view of the *GOstat* input form and an example of a *GOstat* output.

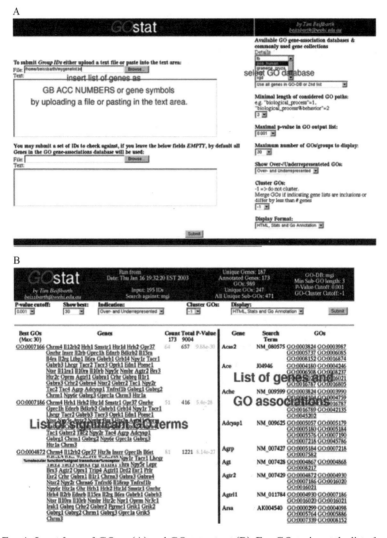

FIG. 4. Input form of *GOstat* (A) and *GOstat* output (B). For *GOstat* input the list of genes of interest, as well as complete list of genes on the microarray, has to be provided. The output is a sorted list of significant GO terms as well as the GO annotation of the genes specified.

Using GOstat

The program requires as input a list of *gene identifiers,* which specify the group of genes of interest. As the different GO databases for different organisms use different standard IDs, the type of gene identifiers that have to be put in are different based on the used GO gene associations database (e.g., MGI numbers for mouse, Swissprot accession numbers for human). Details about the different GO databases, as well as complete annotation files, can be downloaded from http://www.geneontology.org. For some of the supplied GO databases, for example, human, mouse, and rat, *GOstat* searches several synonyms. The links for GenBank and EST accession numbers are provided based on the EST clusters of Unigene (Boguski and Schuler, 1995; Haas *et al.*, 2000). Many of the Affymetrix array IDs (Dalma-Weiszhausz *et al.*, 2006) can be used. Another useful tool that can be used for ID conversion and to preprocess the IDs put into *GOstat* is SOURCE (http://source.stanford.edu).

The default version of *GOstat* requires two lists of genes as input: the list of genes of interest and the complete list of genes from the microarray that these genes were selected from. If the second list does not contain all genes in the first list, *GOstat* merges them. All duplicates as well as non-annotated genes are removed automatically. The lists of genes can be either uploaded as plain text files or pasted into the text fields. An alternative to specifying the second list of genes is to select gene lists of predefined commonly used microarrays as the second input. If the second list is left blank and no microarray is selected, then all of the annotated genes in the GO gene associations database are used as a reference list.

A second version of *GOstat* does rank-based statistics. Here the input is the complete list of genes on the microarray with each gene ID followed by a number. The numbers are converted automatically into ranks by first randomly scrambling and then sorting the gene list. Duplicates as well as non-annotated genes are removed automatically. *RankGOstat* has an option that the numbers specified can be treated as p values rather than converted into ranks; this is, however, not recommended as most p values from microarray studies are biased.

At this stage the method of multiple testing correction has to be selected. Multiple testing correction has to be applied in order to get a more realistic idea of the p value. *GOstat* offers several different options to adjust for multiple testing: Benjamini and Hochberg (1995) correction controls the "false discovery rate" but assumes independence; Benjamini and Yekutieli (2001) drop the assumption of independence but are more conservative; and the Holm (1979) method controls the "family wise error rate," which is even more conservative. The default option FDR should be adequate in most cases. Further, a cutoff for the very top levels of ontologies such as "biological process" is recommended, as these levels contain little information. Also the testing can be restricted to GO terms containing a particular annotation by specifying a search term. All other options can be changed later.

GOstat *Algorithm*

The program will determine all annotated GO terms and all GO terms that are associated (i.e., lower in the hierarchy) with these for all the genes analyzed. The default version of *GOstat* will then count the number of appearances of each GO term for the genes in the list of interesting genes and for genes in the reference list. Fisher's exact tests are performed to judge whether the observed differences are significant or not; in cases where abundances are high, χ^2 tests are performed instead. This will result in a *p* value for each GO term, estimating the probability that the observed counts could have occurred by chance. *RankGOstat* uses Wilcoxon tests or KS tests, which test if the ranks for the genes of a particular GO term are distributed randomly within the complete list of genes on the microarray or if they are biased toward low or high ranks. The KS test checks whether the ranks of all genes associated with a GO follow the uniform distribution. The primary difference from using Wilcoxon tests is that KS will test for any deviations from uniformity (i.e., also those GO terms with ranks in the middle) while the Wilcoxon test only considers high or low ranks.

GOstat *Output*

The program will result in a list of *p* values (or FDRs) that state how specific the GO terms associated with the provided list of genes are, either through direct annotation or through the GO hierarchy. The resulting list of GO terms is sorted by the *p* value and can be limited by the number of terms, as well as by a *p* value cutoff. *p* values of GO terms that are overrepresented in the data set are typeset in green, while *p* values of underrepresented GO terms are colored red. The counts of genes associated to a GO term in the list of genes of interest are given as well as the total count for all the genes on the microarray. The output of *RankGOstat* is similar, except that the column "Total" represents the mean rank for the genes in the gene set. It is possible to display only the over- or underrepresented terms (or high or low ranks). Frequently, the most significant GO terms all represent similar functions or the same subset of genes due to the hierarchical structure of GO and because the genes may each have several GO annotations that are similar. To make the resulting list of GO terms easier to read, GO terms corresponding to similar subsets of genes can be grouped together. The path structure of the GO terms can be displayed with AmiGO (http://www. godatabase.org), a visualization tool for the GO hierarchy. Also, a list of all the annotated associations of genes in the input list to GO terms is provided. It is also possible to download the output as a tab-delimited text file or to save the complete output of the program as a gzipped text file.

Visualization and Further Analysis

GOstat does not as yet provide any tools for visualization of the results. This is due to the fact that creating a good visualization requires interpretation of the results. Determining the significant GO terms in a list of genes is only a first step. The *p* values provide an indication of what processes or functions might be relevant. However, these results must be scrutinized further to determine whether they are reasonable or not. Careful evaluation of the results is necessary in order to produce a good graphical representation. There are several ways to display the output from a GO analysis and a few tools that have options for visualization. Most commonly, a pie chart display of the distribution of GO terms at a certain level of the GO hierarchy is used. However, this display does not give an indication of which GO terms are significant and it is often hard to decide which level of the hierarchy should be used. Another option is to select some of the significant GO terms from the *GOstat* output and display the log-odds ratios in a bar chart. The *GOstat* output can be exported and loaded into a spreadsheet, such as Excel.

Discussion

There are many other tools besides *GOstat* that can help scrutinize the gene ontology structure and annotation to get more out of your microarray data. A more or less complete list of these is provided at http://www. geneontology.org/GO.tools.shtml. Tools that allow more options to display and analyze the graph structure of the annotations, as well as a more automated usage of such analyses, are the Bioconductor (Reimers and Carey, 2006) package *gostats* (http://www.bioconductor.org) and the package *OntologyTraverser* (Young, 2005), which are available as part of the statistical computing environment of the R software. R is a software package and scripting language for statistical data analysis (http://www. r-project.org). Another easy-to-use tool that is similar to *GOstat*, which has more graphical display options, is *FatiGO* (http://fatigo.bioinfo.cnio.es). We also plan to add some graphical features into *GOstat*.

One should keep in mind that the value of such an analysis is highly dependent on the quality of the GO structure and annotation. Also the size of the input list of genes plays an important role; with very small lists of interesting genes and many possible GO categories or pathways to test, there is likely to be no significant result. Here it might make sense to use prior biological knowledge or hypotheses and test only for the most relevant GO categories or pathways. However, very large input lists almost certainly yield some significant GO terms, as in these, very small over- or underrepresentation leads to significant results. It is important to use multiple testing correction to get realistic p values and to carefully consider the biological relevance of the results. After all, the analysis of gene lists using GO terms and pathways is an explorative way to interpret data, which can lead to easier to comprehend sets of results than the analysis focused on single differentially expressed genes. However, the hypotheses generated should still be tested by different means.

Acknowledgments

Thanks to Terry Speed, Hamish Scott, and Annemarie Poustka for giving the author the opportunity to develop and maintain *GOstat*. Thanks to Nick Tan and Dirk Ledwinka for IT support. Thanks to James Wettenhall, Lavinia Hyde, and Matthew Ritchie for proofreading and help with the manuscript. Tim Beissbarth and *GOstat* were supported by the Deutsche Forschungsgemeinschaft, WEHI, NHMRC, and NGFN.

References

Al-Shahrour, F., Diaz-Uriarte, R., and Dopazo, J. (2004). FatiGO: A web tool for finding significant associations of Gene Ontology terms with groups of genes. *Bioinformatics* **20,** 578–580.

Ashburner, M., Ball, C., Blake, J., Botstein, D., Butler, H., Cherry, J., Davis, A., Dolinski, K., Dwight, S., and Eppig, J. (2000). Gene ontology: Tool for the unification of biology. *Nature Genet.* **25,** 25–29.

Ayroles, J. F., and Gibson, G. (2006). Analysis of variance of microarray data. *Methods Enzymol.* **411,** 214–233.

Beissbarth, T., Borisevich, I., Hoerlein, A., Kenzelmann, M., Hergenhahn, M., Klewe-Nebenius, A., Klaeren, R., Korn, B., Schmid, W., Vingron, M., and Schuetz, G. (2003). Analysis of CREM-dependent gene expression during mouse spermatogenesis. *Mol. Cell. Endocrinol.* **212,** 29–39.

Beissbarth, T., Fellenberg, K., Brors, B., Arribas-Prat, R., Boer, J. M., Hauser, N. C., Scheideler, M., Hoheisel, J. D., Schuetz, G., Poustka, A., and Vingron, M. (2000). Processing and quality control of DNA array hybridization data. *Bioinformatics* **16,** 1014–1022.

Beissbarth, T., Hyde, L., Smyth, G. K., Job, C., Boon, W. M., Tan, S. S., Scott, H. S., and Speed, T. P. (2004). Statistical modeling of sequencing errors in SAGE libraries. *Bioinformatics.* **20,** I31–I39.

Beissbarth, T., and Speed, T. P. (2004). GOstat: Find statistically overrepresented Gene Ontologies within a group of genes. *Bioinformatics* **20,** 1464–1465.

Benjamini, Y., and Hochberg, Y. (1995). Controlling the false discovery rate: A practical and powerful approach to multiple testing. *JRSS-B* **57,** 289–300.

Benjamini, Y., and Yekutieli, D. (2001). The control of the false discovery rate in multiple testing under dependencies. *Anal. Stat.* **29,** 1165–1188.

Blake, J. A., Richardson, J. E., Bult, C. J., Kadin, J. A., Eppig, J. T., and the members of the Mouse Genome Database Group (2003). MGD: The mouse genome database. *Nucleic Acids Res.* **31,** 193–195.

Boguski, M., and Schuler, G. (1995). Establishing a human transcript map. *Nature Genet.* **10,** 369–371.

Boon, W. M., Beissbarth, T., Hyde, L., Smyth, G. K., Gunnersen, J., Denton, D. A., Scott, H. S., and Tan, S. S. (2004). A comparative analysis of transcribed genes in the mouse hypothalamus and neocortex reveals chromosomal clustering. *Proc. Natl. Acad. Sci. USA* **101,** 14972–14977.

Camon, E., Magrane, M., Barrell, D., Lee, V., Dimmer, E., Maslen, J., Binns, D., Harte, N., Lopez, R., and Apweiler, R. (2004). The Gene Ontology Annotation (GOA) database: Sharing knowledge in Uniprot with Gene Ontology. *Nucleic Acids Res.* **32,** D262–D266.

Clare, A., and King, R. D. (2002). How well do we understand the clusters found in microarray data? *In Silico Biol.* **2,** 511–522.

Dalgaard, P. (2002). "Introductory Statistics with R." Springer, New York.

Dalma-Weiszhausz, D. D., Warrington, J., Tanimoto, E. Y., and Miyada, C. G. (2006). The Affymetrix GeneChip Platform: An overview. *Methods Enzymol.* **410,** 3–28.

Dennis, G., Sherman, B., Hosack, D., Yang, J., Gao, W., Lane, H., and Lempicki, R. (2003). DAVID: Database for annotation, visualization, and integrated discovery. *Genome Biol.* **4,** P3.

Doniger, S., Salomonis, N., Dahlquist, K., Vranizan, K., Lawlor, S., and Conklin, B. (2003). MAPPFinder: Using Gene Ontology and GenMAPP to create a global gene-expression profile from microarray data. *Genome Biol.* **4,** R7.

Downey, T. (2006). Analysis of a multifactor microarray study using Partek genomics solution. *Methods Enzymol.* **411,** 256–270.

Draghici, S., Khatri, P., Bhavsar, P., Shah, A., Krawetz, S., and Tainsky, M. (2003). Onto-Tools, the toolkit of the modern biologist: Onto-Express, Onto-Compare, Onto-Design and Onto-Translate. *Nucleic Acids Res.* **31,** 3775–3781.

Dudoit, S., Shaffer, J., and Boldrick, J. (2002). Multiple hypothesis testing in microarray experiments Technical Report 110. Division of Biostatistics, UC Berkeley.

Freedman, D., Pisani, R., and Purves, R. (1997). "Statistics," 3rd Ed. Norton.

Gollub, J., and Sherlock, G. (2006). Clustering microarray data. *Methods Enzymol.* **411,** 194–213.

Haas, S., Beissbarth, T., Rivals, E., Krause, A., and Vingron, M. (2000). GeneNest: Automated generation and visualization of gene indices. *Trends Genet.* **16,** 521–523.

Hennetin, J., and Bellis, M. (2006). Clustering methods for analyzing large data sets: Gonad development, a study case. *Methods Enzymol.* **411,** 387–407.

Holm, S. (1979). A simple sequentially rejective multiple test procedure. *Scand. J. Stat.* **6,** 65–70.

Kanehisa, M., Goto, S., and Goto, S. (2000). KEGG: Kyoto encyclopedia of genes and genomes. *Nucleic Acids Res.* **28,** 27–30.

Lamb, J., Ramaswamy, S., Ford, H. L., Contreras, B., Martinez, R. V., Kittrell, F. S., Zahnow, C. A., Patterson, N., Golub, T. R., and Ewen, M. E. (2003). A mechanism of cyclin D1 action encoded in the patterns of gene expression in human cancer. *Cell* **114,** 323–334.

Lockhart, D. J., Dong, H., Byrne, M. C., Follettie, M. T., Gallo, M. V., Chee, M. S., Mittmann, M., Wang, C., Kobayashi, M., Horton, H., and Brown, E. L. (1996). Expression monitoring by hybridization to high-density oligonucleotide arrays. *Nature Biotechnol.* **14,** 1675–1680.

Mao, X., Cai, T., Olyarchuk, J. G., and Wei, L. (2005). Automated genome annotation and pathway identification using the KEGG Orthology (KO) as a controlled vocabulary. *Bioinformatics* **21**, 3787–3793.

Mootha, V. K., Lindgren, C. M., Eriksson, K. F., Subramanian, A., Sihag, S., Lehar, J., Puigserver, P., Carlsson, E., Ridderstraele, M., Laurila, E., Houstis, N., Daly, M. J., Patterson, N., Mesirov, J. P., Golub, T. R., Tamayo, P., Spiegelman, B., Lander, E. S., Hirschhorn, J. N., Altshuler, D., and Groop, L. C. (2003). PGC-1alpha-responsive genes involved in oxidative phosphorylation are coordinately downregulated in human diabetes. *Nature Genet.* **34**, 267–273.

Reimers, M., and Carey, V. J. (2006). Bioconductor: An open source framework for bioinformatics and computational biology. *Methods Enzymol.* **411**, 119–134.

Quackenbush, J. (2003). Genomics. Microarrays: Guilt by association. *Science* **302**, 240–241.

Saeed, A. I., Bhagabati, N. K., Braisted, J. C., Liang, W., Sharov, V., Howe, E. A., Li, J., Thiagarajan, M., White, J. A., and Quackenbush, J. (2006). TM4 microarray software suite. *Methods Enzymol.* **411**, 134–193.

Schena, M., Shalon, D., Davis, R. W., and Brown, P. O. (1995). Quantitative monitoring of gene expression patterns with a complementary DNA microarray. *Science* **270**, 467–470.

Shaffer, J. (1995). Multiple hypothesis testing. *Annu. Rev. Psychol.* **46**, 561–584.

Smyth, G. K. (2004). Linear models and empirical Bayes methods for assessing differential expression in microarray experiments. *SAGMB.* **3**, Article 3.

Whetzel, P. L., Parkinson, H., and Stoecket, C. (2006). Using ontologies to annotate microarray experiments. *Methods Enzymol.* **411**, 325–339.

Young, A., Whitehouse, N., Cho, J., and Shaw, C. (2005). OntologyTraverser: An R package for GO analysis. *Bioinformatics* **21**, 275–276.

Yue, L., and Reisdorf, W. C. (2005). Pathway and ontology analysis: Emerging approaches connecting transcriptome data and clinical endpoints. *Curr. Mol. Med.* **5**, 11–20.

Further Reading

Eisen, M. B., Spellman, P. T., Browm, P. O., and Botstein, D. (1998). Cluster analysis and display of genome-wide expression patterns. *Proc. Natl. Acad. Sci. USA* **95**, 14863–14868.

Masys, D., Welsh, J., Fink, J. L., Gribskov, M., Klacansky, I., and Corbeil, J. (2001). Use of keyword hierarchies to interpret gene expression patterns. *Bioinformatics* **17**, 319–326.

[19]   Gene Expression Omnibus: Microarray Data Storage, Submission, Retrieval, and Analysis

By Tanya Barrett and Ron Edgar

Abstract

The Gene Expression Omnibus (GEO) repository at the National Center for Biotechnology Information archives and freely distributes high-throughput molecular abundance data, predominantly gene expression data generated by DNA microarray technology. The database has a

METHODS IN ENZYMOLOGY, VOL. 411   0076-6879/06 $35.00
                                                   DOI: 10.1016/S0076-6879(06)11019-8

flexible design that can handle diverse styles of both unprocessed and processed data in a Minimum Information About a Microarray Experiment-supportive infrastructure that promotes fully annotated submissions. GEO currently stores about a billion individual gene expression measurements, derived from over 100 organisms, submitted by over 1500 laboratories, addressing a wide range of biological phenomena. To maximize the utility of these data, several user-friendly web-based interfaces and applications have been implemented that enable effective exploration, query, and visualization of these data at the level of individual genes or entire studies. This chapter describes how data are stored, submission procedures, and mechanisms for data retrieval and query. GEO is publicly accessible at http://www.ncbi.nlm.nih.gov/projects/geo/.

Purpose and Scope of the Gene Expression Omnibus (GEO)

The postgenomic era has led to a multitude of high-throughput methodologies that generate massive volumes of gene expression data. The GEO repository was established by National Center for Biotechnology Information (NCBI) in 2000 to house and distribute these data to the public with no restrictions or login requirements (for more information, please read the GEO data disclaimer[1]). The primary role of GEO is data archiving, functioning as a hub for data deposit, and retrieval (Barrett et al., 2005; Edgar et al., 2002). ArrayExpress (Brazma et al., 2006) serves a similar function.

GEO is currently the largest, fully public gene expression resource. At the time of writing, the database holds over 80,000 samples, comprising approximately a billion individual expression measurements, 13 million gene expression profiles, for over 100 organisms, submitted by almost 1500 laboratories. These data address a very broad diversity of biological themes, including disease, development, evolution, metabolics, toxicology, immunity, ecology, and transgenesis. Most data are provided by the research community in compliance with grant or journal provisos that require microarray data to be made available in a public repository, with the objective being to facilitate independent evaluation of results, reanalysis, and full access to all parts of the study (Ball et al., 2004).

Data types currently stored include, but are not limited to, cDNA and oligonucleotide microarrays that examine gene expression, serial analysis of gene expression (SAGE), massively parallel signature sequencing, array comparative genomic hybridization, chromatin–immunoprecipitation on arrays studies, and peptide profiling techniques such as tandem mass

---

[1] http://www.ncbi.nlm.nih.gov/projects/geo/info/disclaimer.html.

spectrometry (MS/MS). In keeping with the theme of the book, this chapter focuses on gene expression data generated by DNA microarrays.

Although primarily a data storage and retrieval facility, it was clear early on that the resource must also enable effective searching and data mining as means to identify entries of interest. Consequently, several user-friendly web-based query tools have been developed to assist even those unfamiliar with microarray technology to effectively explore and analyze GEO data. However, it is important to realize that GEO is not intended to be used as a laboratory information management system or a pre-/first-analysis environment, as data submitted to GEO are generally processed data that form the basis for discussion in accompanying manuscripts.

This chapter explains the database design for storage of microarray information, how to submit data, and how to effectively retrieve and examine information in the GEO database.

Structure

The GEO database architecture is designed for the efficient capture, storage, and retrieval of heterogeneous sets of high-throughput molecular abundance data. The structure is sufficiently flexible to accommodate evolving state-of-the-art technologies. There are many different varieties of microarray technology, and researchers use a wide assortment of hardware and software packages to generate and process data. Consequently, data have many different styles and comprise varying content. For example, the sequences on an array may be described by multiple attributes, including gene symbols, GenBank accession numbers, clone identifiers, ontology categories, and feature coordinates, to name a few. Similarly, hybridization data may contain many types of supporting measurements and calculations that supplement final expression values. Importantly, expression data are worthless unless complemented with comprehensive contextual biological details and data analysis methodologies under which they were generated. GEO was built with all these considerations in mind and has an open, adaptable design that can handle variety and a Minimum Information About a Microarray Experiment (MIAME)-supportive (Brazma *et al.*, 2001) infrastructure that promotes fully annotated submissions. Extensive technical details regarding database design and data flow are beyond the scope of this chapter, but it helps to understand that data and metadata are stored separately within the database. The versatility of GEO is largely attributed to the fact that tabular data are not fully granulated in the core database but instead are treated as "blobs," that is, compressed text tab-delimited tables that may contain any number of rows or columns. Data in selected columns are extracted to a secondary database and used in subsequent indexing and query applications. Descriptive or informative metadata are fully normalized in the schema as needed.

Submitter-Supplied Data

Data supplied by submitters are stored as three main entities in a MSSQL server relational database.

Platform: Includes a summary description of the array and a data table defining the array template. Each row in the table corresponds to a single element and includes sequence annotation and tracking information as provided by the submitter.

Sample: Includes a description of the biological source and the experimental protocols to which it was subjected and a data table containing hybridization measurements for each element on the corresponding platform.

Series: Defines a set of related samples considered to be part of a study and describes the overall study aim and design.

Each of these three objects is assigned an accession number that may be used to cite and retrieve the records. In addition to sample data tables and descriptive information, accompanying supplementary files such as original microarray scan images or preprocessed quantification data are accepted and stored on an FTP site with database links.

GEO-Constructed Data Sets

Despite the variability in the style and content of incoming data, a common set of salient information is submitted:

- sequence identity tracking information for each feature on the array
- normalized hybridization measurements
- a description of the biological source used in each hybridization.

Using a combination of automated data extraction and manual curation, this information is rendered into an upper-level unit called a GEO DataSet (Fig. 1). A DataSet represents a collection of similarly processed, experimentally related sample hybridizations and provides a coherent synopsis for a study. Samples within a DataSet are further categorized according to experimental variables, for example, they are organized by gender and disease state.

A DataSet provides two separate perspectives of data.

1. An *experiment-centered* rendering that encapsulates the whole study. This information is presented as a "DataSet record." DataSet records comprise a synopsis of the experiment, a breakdown of the experimental

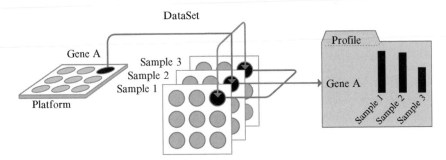

FIG. 1. Schematic diagram of relationships among GEO platform, sample, DataSet, and profiles. For each gene on a platform, multiple sample measurement values are generated. Related samples constitute a DataSet from which multiple gene expression profile entities are generated.

variables, access to auxiliary objects, several data display and analysis tools, and download options (Fig. 2).

2. A *gene-centered* rendering that presents quantitative gene expression measurements for one gene across a DataSet. This information is presented as a "GEO Profile." A GEO Profile comprises gene identity annotation, the DataSet title, links to auxiliary information, and a chart depicting the expression level of that gene across each sample in the DataSet (Fig. 3). The following section describes more information on interpreting GEO profile charts.

DataSets enable transformation of diverse styles of submitted data such that they are readily accessible in a uniform format upon which to base downstream data analysis tools.

Interpreting GEO Profiles Charts

GEO profile charts track the expression behavior of one gene across all samples in a DataSet. Several categories of information are presented in GEO profile charts: expression measurement values, expression measurement rankings, and an outline of the experimental design and variables (Fig. 3).

The *value* data (red bars, scale at the left side of the chart shown in Fig. 3) are extracted from the "VALUE" column of corresponding sample records from which the DataSet is composed. All sample data tables include this column, which contains the final normalized expression level measurements as supplied by the submitter. Other than to log transform single-channel expression counts for graphic visualization, no additional processing is applied by GEO to *value* data.

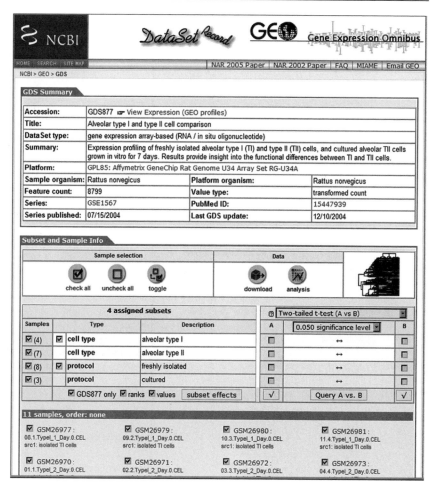

FIG. 2. Screen shot of a typical DataSet record GDS877 (Gonzalez *et al.*, 2005). The record includes a summary of the experiment, links to related records and publications, subset designations and classifications, download options, and access to mining features such as cluster heat maps and "Query group A vs B" tool.

An important point to consider is that there is no standard unit for gene expression; because a very wide variety of technologies, software packages, and algorithms generate these data, the *values* should be considered arbitrary units. Consequently, it is inadvisable to attempt to draw direct comparisons between expression values in unrelated DataSets. However, it can be assumed that the *value* measurements of each sample within a DataSet

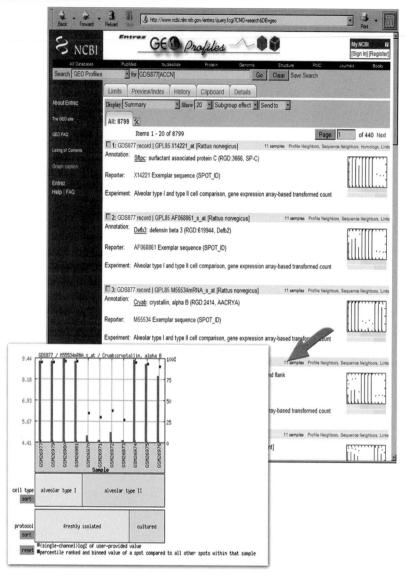

FIG. 3. Screen shot of Entrez GEO profile retrieval results; each entity includes sequence identifier and DataSet information and a thumbnail profile image. Links to other Entrez databases or related profiles are provided above the thumbnail image. The expanded profile chart depicts values (bars) and rank (squares) information for the crystallin gene across each sample in GEO DataSet GDS877 (Gonzalez et al., 2005). Experimental subset groupings are reflected in labels at the foot of the chart.

are comparable and have been calculated in an equivalent manner, that is, considerations such as background processing and normalization/scaling are consistent across the DataSet. The "Value distribution" box and whisker plots available on DataSet records allow users to easily evaluate how well distributed, and thus comparable, the sample *values* within a DataSet are.

In addition to the *value* profile display for individual genes, most DataSets also provide a *rank* percentile view (blue squares, scale on the right side of the chart shown in Fig. 3). Ranks provide an indication of the expression level of that gene compared to all other genes on that array. Ranks are calculated as follows: (i) the total number of genes in the sample is divided to 100 bins such that there are n genes per bin; (ii) genes are sorted by *value*, and (iii) the lowest n genes are assigned to the first bin, subsequent n genes to the next bin, and so on. Binning is rather sensitive to local (sample) distribution and global (DataSet) normalization. It is therefore useful to note if a gene displays the same pattern of behavior in both value and rank space, as a disparity in trends can indicate that data are not normalized or the existence of other effects, such as nonspecific hybridization.

Currently, faded data points are specific to Affymetrix technology (this mode of display will likely be applied to other technology types in the future). They indicate where the Affymetrix algorithms have assigned a "Detection call = absent" to an expression signal. An absent call can be assigned for two reasons: either the detected signal was so low that the transcript was deemed not to be present or stray cross-hybridization was detected, in which case the signal is deemed unreliable for that transcript.

Bars at the horizontal foot of the chart provide experiment annotation and contextual information about the gene expression profile under review. The "sort" button allows users to resort the samples in the DataSet according to a particular experimental parameter, thus assisting visualization of expression trends in experiments with complex design.

Submission

The GEO database is a MIAME-supportive infrastructure; the MIAME guidelines outline the minimal information that should be provided to allow unambiguous interpretation of microarray experiment data (Brazma *et al.*, 2001). While the submission procedures promote MIAME compliance, ultimately it is the submitters' responsibility to ensure that their data are sufficiently well annotated. Large volumes of contextual information may be provided, including the cell or tissue type, characteristics of the organism (e.g., species, age, sex, disease state) from which the sample was isolated, comprehensive explanations of the perturbations that the cells or organisms were subjected to, sample isolation and preparation protocols, data processing and normalization strategies, and more.

There are several ways in which data may be deposited with GEO. Deciding which method to use depends on the amount of data to be submitted, what format data are in already, and the level of computational expertise of the submitter. Regardless of the submission method, the final GEO records look the same and contain equivalent information.

> Web deposit: The web submission process is designed for the quick and easy deposit of individual records by occasional submitters. This route comprises a set of interactive web forms that provide a simple step-by-step procedure for deposit of data tables and accompanying descriptive information.
>
> Batch direct deposit using Simple Omnibus Format in Text (SOFT) format: SOFT is a simple line-based format designed for rapid batch submission (and retrieval) of data. A single SOFT file can hold both data tables and accompanying descriptive information for multiple platforms, samples, and/or series records. SOFT files may be produced readily from common database and spreadsheet applications and can be uploaded directly to the database.
>
> FTP deposit: If data are already in matrix format (e.g., Affymetrix pivot file), submission via a SOFT-formatted spreadsheet is recommended. Valid MAGE-ML-formatted (Spellman *et al.*, 2002) reports are also acceptable. These file types are transferred to GEO via FTP.

Full instructions and examples of these various submission routes and formats are provided on the GEO web site. All submissions are reviewed and checked by a GEO curator, ensuring that records contain meaningful information and are organized correctly. If no structural or content problems are identified the submissions are approved and assigned GEO accession numbers. If problems are identified, the curator will work with the submitter to make any modifications necessary to achieve successful deposit. The GEO accession numbers are unique and stable and may be quoted in corresponding manuscripts. The records may remain private for several months, typically pending manuscript publication. Submitters may generate a secondary account that enables collaborators or reviewers read-only, confidential access to prepublication data. Submitters retain full editorial control over their records and may perform updates and edits at any time.

Navigating GEO and Finding What You Need

Browsing

Original submitter-supplied platform, sample, and series records may be browsed using the repository browser at http://www.ncbi.nlm.nih.gov/geo/query/browse.cgi. These browser pages allow data to be sorted by various categories, such as submitter, organism, platform and sample type, titles, release dates, and supplementary file type. DataSet records may be browsed at http://www.ncbi.nlm.nih.gov/projects/geo/gds/gds_browse.cgi and may be sorted by title, organism, type, creation date, and platform. Within records, reciprocal links are provided to all related records for easy, uninterrupted browsing.

Downloading

Several download options are available.

- Each platform, sample and series record has a mechanism at the head of the page that enables download (SOFT format) or viewing (HTML) of that record and/or related records, with the option to restrict to only descriptive data or tabular data.
- DataSet records include a link for download of a text tab-delimited value matrix and associated platform element gene annotation.
- All platform, sample, series, DataSet, and supplementary data are available for bulk download via FTP at ftp://ftp.ncbi.nih.gov/pub/geo/.

Query and Analysis

GEO provides a variety of strategies for locating and visualizing information of interest. Query approaches include standard and Boolean text-based searches, sequence-based searches, mining based on expression behavior characteristics, or combinations of these parameters. Figure 4 depicts a schematic overview of the query workflow and how the various features and tools are interlinked. A summary of where these features are located, their purpose, and methodology is provided.

Deciding where to begin a search generally depends on what type of information one needs to retrieve. Often, there is more than one way to identify relevant data. Users should always keep in mind that the features provided on the GEO site are not intended for robust systematic analyses. The heterogeneous nature of GEO data, coupled with the limitations of web browsing, limits to some extent the statistical tools that can be developed. Diverse data are treated similarly; criteria such as sample size, number of

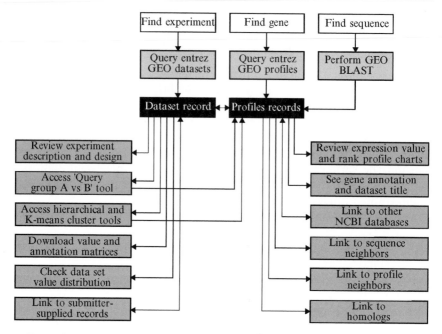

Fig. 4. Schematic overview of the query workflow and how the various features and tools are interlinked.

repeats, prior filtering, and normalization factors are not considered. That said, these tools are extremely useful for the quick and easy identification of relevant and noteworthy data.

Entrez GEO DataSets

> Where: From the GEO home page or at http://www.ncbi.nlm.nih.gov/entrez/query.fcgi?db=gds.
> Purpose: A query interface that facilitates identification of DataSets relevant to a particular area of study.
> Method: Effective query and mining is achieved using keywords or Boolean phrases restricted to supported attribute fields (Table I). Retrievals display the DataSet titles, a brief experiment description, and a link to the complete DataSet record (Fig. 2), as well as links to publications and other databases.

Entrez GEO Profiles

> Where: From the GEO home page or at http://www.ncbi.nlm.nih.gov/entrez/query.fcgi?db=geo.

TABLE I
ENTREZ QUALIFIER FIELDS[a]

Field name	Field description
GEO DataSets	
Author	Authors associated with the experiment
Experiment type	Experiment type, e.g., cDNA, genomic, protein, SAGE
GDS text	DataSet description text
GEO accession	GEO accession number
GEO description/title text	Text provided in the description/title of original records
Number of samples	Number of samples in the DataSet[b]
Number of platform probes	Number of platform reporters in the DataSet[b]
Organism	Organism from which the reporters on the array were derived/designed
Reporter identifier	Identifier for the array reporter (GenBank accession, gene name, etc.).
Sample source	Source biological material of the sample
Sample title	Sample title
Submitter institute	Submitter institute
Subset description	Description of the experimental variable
Subset variable type	Type of experimental variable, e.g., age, strain, gender
GEO profiles	
Experiment type	Experiment type, e.g., cDNA, genomic, protein, SAGE
Flag information	Specific experimental variable flags, e.g., age, strain, gender
Flag type	Flag types, e.g., rank and value subset effects
GDS text	DataSet description text
GEO accession	GEO accession number
GEO description/title text	Text provided in the description/title of original records
GI	Mapped GenBank identifier
Gene description	Gene description, symbol, alias
ID_REF	Unique identifier for a reporter as given on the array
Max value rank	Maximum value rank[b]
Max value in profile	Maximum value in profile[b]
Median value in GDS	Median value in DataSet[b]
Median value in profile	Median value in profile[b]
Min value rank	Minimum value rank[b]
Min value in profile	Minimum value in profile[b]
Number of samples	Number of samples in the DataSet[b]
Organism	Organism from which the samples were derived
Ranked standard deviation	Ranked standard deviation
Reporter identifier	Identifier for a reporter
Sample source	Source biological material of the sample

[a] Useful qualifier fields for performing restricted GEO DataSets and GEO profile queries.
[b] Possible range operation, e.g., 20:50[number of samples] will find DataSets containing 20 to 50 samples.

Purpose: A query interface that facilitates identification of gene expression profiles of interest.

Method: Effective query and mining is achieved using keywords or Boolean phrases restricted to supported attribute fields (Table I). Retrievals display the mapped gene name, the DataSet title, a thumbnail image of the gene expression profile, as well as links to publications and other databases. Clicking on the thumbnail image will enlarge the chart to display the full profile details and Sample subset partitions that reflect experimental design (Fig. 3).

Advanced Entrez Features

Where: The tool bar at the head of all NCBI Entrez query and retrieval pages.

Purpose: Facilitates powerful mining and linking across many NBCI databases (Schuler *et al.*, 1996; Wheeler *et al.*, 2005).

Method: The "Preview/Limits" link assists greatly in the construction of complex queries. Users employ indices to browse and/or select the terms by which data are described and build multipart queries. The "History" tab stores previous queries, which can be combined to form a new search query, enabling sophisticated mining that traverses DataSets and platforms. The "Display" pull-down menu enables users to find related data in other Entrez resources in batch mode.

DataSet Clusters

Where: On the DataSet record under the "analysis" button.

Purpose: Clustering is a popular method used to visualize and examine high-dimensional DataSets. Typically, the goal of a microarray cluster analysis is to organize genes so that those with similar expression patterns are grouped together. It can be hypothesized that genes that behave similarly might have a coordinated transcriptional response, possibly inferring a common function or regulatory elements.

Method: Many different clustering algorithms exist (see Gollub and Sherlock, 2006); all employ various combinations of mathematical distance metrics and linkages (Eisen *et al.*, 1998). Nine varieties of precomputed hierarchical clusters are available on GEO DataSet records, as well as user-defined K-means or K-median clustering. Results are depicted as a color-coded "heat map" image, where rows represent individual elements on the array (genes) and columns represent individual samples (hybridizations), and color A (high expression level) transitions into color B (low expression level). Users can scan these images visually for cluster "hot spots" that represent a group of genes with similar expression. The heat maps are interactive; after selecting a region, or regions, of interest using a movable box, corresponding data may be downloaded as a text file or linked to genes in Entrez GEO profiles. Care must be taken not to over-interpret cluster output. Different clustering algorithms may yield different clustering solutions using the same data. Clustering provides suggestions for possible relationships between data, but does not prove them.

Profile Neighbors

Where: The "Profile Neighbors" link on the top right side of Entrez GEO profile retrievals.

Purpose: Connects groups of genes that show a similar or reversed profile shape within a DataSet. It can be hypothesized that genes that behave similarly might be coregulated or have related functionality.

Method: Profile neighbors are precalculated using an adjusted Pearson linear correlation. The user need only click the "Profile Neighbors" link to retrieve related genes. Currently, Profile neighbors are subject to a GEO-defined arbitrary cutoff limit imposed in order to restrict the number of links that can be managed effectively.

Sequence Neighbors

Where: The "Sequence Neighbors" link on the top right side of Entrez GEO profile retrievals.

Purpose: Connects groups of genes related by nucleotide sequence similarity across all DataSets. Genes related by sequence similarity can provide insights into the possible function of the original sequence if it has not yet been characterized or can identify related gene family members.

Method: Sequence neighbors are precalculated using standard BLAST (Altschul *et al.*, 1990). The user need only click the "Sequence Neighbors" link to retrieve related genes. Currently, Sequence neighbors are subject to a GEO-defined arbitrary cutoff limit imposed in order to restrict the number of links that can be managed effectively.

Links

Where: The "Links" link on the top right side of Entrez GEO profiles and Entrez GEO DataSets retrievals.

Purpose: Connects GEO data to related data in other NCBI resources, facilitating seamless navigation and cross-referencing between multiple data domains.

Method: Where possible, reciprocal links are provided to and from GenBank, PubMed, Gene, UniGene, OMIM, Homologene, Taxonomy, SAGEMap, and MapViewer databases. The user need only click the "Links" link and select the relevant resource from the pull-down menu to link to retrieve related data.

Geo Blast

Where: The GEO BLAST link on the GEO home page.

Purpose: Retrieves gene profiles that are related to a user-defined nucleotide sequence of interest.

Method: This tool performs a BLAST (Altschul *et al.*, 1990) search of a user-provided nucleotide sequence against all GenBank identifiers represented on microarray platforms or SAGE libraries in GEO. Retrievals resemble conventional BLAST output with each alignment receiving a score and expected value and a link to corresponding GEO profiles. This interface is helpful in locating expression data for specified nucleotide sequences, for identifying sequence homologs, for example, related gene family members or for cross-species comparisons, or for providing insight into potential roles of the original sequence if it has not yet been characterized functionally.

Sorting and Limit Options Using Subset Effects Flags

Where: Intrinsic to standard Entrez GEO profiles retrievals, which are default ordered according to subset effect flags and specifiable using [Flag Type] and [Flag information] qualifiers (Table I) in Entrez GEO profiles.

Purpose: Attempts to identify genes that display marked differences in expression level according to experimental variables.

Method: Genes whose values or ranks pass a threshold of statistical difference between any nonsingle experimental variable subset and another are flagged in the database. This allows users to search across all GEO for genes that show an interesting effect with respect to particular experimental variable types, such as "age." The fact that standard Entrez GEO profile retrievals are default ordered according to these flags makes potentially interesting results more visible (alternative sorting options include profile deviation and mean value). It is important to realize that subset effects are calculated with arbitrarily defined thresholds with no consideration of data type and processing and merely provide suggestions of what could be interesting profiles.

Query Group A vs B Tool

Where: On the DataSet record on the right side of the subset assignment section.

Purpose: Assists filtering and identification of gene profiles that display marked differences in the expression level between two specified sets of samples within a DataSet.

Method: Using checkboxes, the user assigns one or more samples to group A and other samples to group B. Samples are selected/deselected on the basis of their experimental subset designations. The user then chooses from several varieties of filtering procedures and stringency parameters by which to compare the two groups, including one-tailed or two-tailed *t* tests or a mean log values or ranks fold difference. Genes that meet the user-defined criteria are presented in Entrez GEO profiles. Note that this tool uses rudimentary means of filtering data, as retrievals may have no statistical significance; the compared subsets may be too small to provide any statistic value.

Conclusion

DNA microarray technology has led to a rapid accumulation of gene expression data. GEO serves as a unifying resource for these data, operating primarily as a public archive, but also providing flexible data mining strategies and tools that allow users to query, filter, select, and inspect data in the context of their specific interests. Many of these features use traditional data reduction techniques designed to filter inherently noisy data and concise displays that allow human scanning. The integration of GEO data with extensive sequence, mapping, and bibliographic resources via the Entrez system of linked databases offers additional ancillary information that can assist in the interpretation of biological data and evaluate the relevance of microarray results.

Examination of published gene expression data can help researchers prioritize candidates for further study and direct the design of new experiments. The literature reveals that researchers are using GEO data to complement and support their own studies (e.g., Brockington *et al.*, 2005; Nakai *et al.*, 2005; Ozyildirim *et al.*, 2005; Rico-Bautista *et al.*, 2005; Yant *et al.*, 2005).

Compiling large volumes of diverse gene expression data into one location and making them accessible through common integrated interfaces impart a powerful investigative factor not attainable when considering solitary experiments. This large compendium of data affords more opportunity to gather corroboratory evidence for global metabolic and regulatory networks, to investigate what the majority of evidence implies about the behavior and function of a gene or group of genes, and to generate hypotheses on functional models and themes (e.g., Jordan *et al.*, 2004; Ott *et al.*, 2005; Zhou *et al.*, 2005). This macro approach to discovery will only strengthen as the database continues to grow.

Because the GEO database and tools continue to undergo intensive development and modification, the features and data presentation strategies discussed in this chapter will evolve over time. To receive announcements of site developments, subscribe to the GEO-announce list at geo@ncbi.nlm.nih.gov.

Acknowledgments

The authors unreservedly acknowledge the efforts of the GEO curation and programming staff, including Tugba Suzek, Dennis Troup, Steve Wilhite, Pierre Ledoux, Dmitry Rudnev, Carlos Evangelista, and Alexandra Soboleva. Also, Todd Groesbeck is thanked for assistance with manuscript figures. This chapter is an official contribution of the National Institutes of Health; not subject to copyright in the United States.

References

Altschul, S. F., Gish, W., Miller, W., Myers, E. W., and Lipman, D. J. (1990). Basic local alignment search tool. *J. Mol. Biol.* **215,** 403–410.

Ball, C. A., Brazma, A., Causton, H., Chervitz, S., Edgar, R., Hingamp, P., Matese, J. C., Parkinson, H., Quackenbush, J., Ringwald, M., Sansone, S. A., Sherlock, G., Spellman, P., Stoeckert, C., Tateno, Y., Taylor, R., White, J., and Winegarden, N. (2004). Submission of microarray data to public repositories. *PLoS Biol.* **2**(9), e317.

Barrett, T., Suzek, T. O., Troup, D. B., Wilhite, S. E., Ngau, W. C., Ledoux, P., Rudnev, D., Lash, A. E., Fujibuchi, W., and Edgar, R. (2005). NCBI GEO: Mining millions of expression profiles—database and tools. *Nucleic Acids Res.* **33,** D562–D566.

Brazma, A., Hingamp, P., Quackenbush, J., Sherlock, G., Spellman, P., Stoeckert, C., Aach, J., Ansorge, W., Ball, C. A., Causton, H. C., Gaasterland, T., Glenisson, P., Holstege, F. C., Kim, I. F., Markowitz, V., Matese, J. C., Parkinson, H., Robinson, A., Sarkans, U., Schulze-Kremer, S., Stewart, J., Taylor, R., Vilo, J., and Vingron, M. (2001). Minimum information about a microarray experiment (MIAME)-toward standards for microarray data. *Nature Genet.* **29,** 365–371.

Brazma, A., Kapushesky, M., Parkinson, H., Sarkins, U., and Shojatalab, M. (2006). Data storage and analysis in ArrayExpress. *Methods Enzymol.* **411,** 370–386.

Brockington, M., Torelli, S., Prandini, P., Boito, C., Dolatshad, N. F., Longman, C., Brown, S. C., and Muntoni, F. (2005). Localization and functional analysis of the LARGE family of glycosyltransferases: Significance for muscular dystrophy. *Hum. Mol. Genet.* **14**(5), 657–665.

Edgar, R., Domrachev, M., and Lash, A. E. (2002). Gene Expression Omnibus: NCBI gene expression and hybridization array data repository. *Nucleic Acids Res.* **30,** 207–210.

Eisen, M. B., Spellman, P. T., Brown, P. O., and Botstein, D. (1998). Cluster analysis and display of genome-wide expression patterns. *Proc. Natl. Acad. Sci. USA* **95,** 14863–14868.

Gollub, J., and Sherlock, G. (2006). Clustering microarray data. *Methods Enzymol.* **411** (this volume).

Gonzalez, R., Yang, Y. H., Griffin, C., Allen, L., Tigue, Z., and Dobbs, L. (2005). Freshly isolated rat alveolar type I cells, type II cells, and cultured type II cells have distinct molecular phenotypes. *Am. J. Physiol. Lung Cell Mol. Physiol.* **288**(1), L179–L189.

Jordan, I. K., Marino-Ramirez, L., Wolf, Y. I., and Koonin, E. V. (2004). Conservation and coevolution in the scale-free human gene coexpression network. *Mol. Biol. Evol.* **21**(11), 2058–2070.

Nakai, H., Wu, X., Fuess, S., Storm, T. A., Munroe, D., Montini, E., Burgess, S. M., Grompe, M., and Kay, M. A. (2005). Large-scale molecular characterization of adeno-associated virus vector integration in mouse liver. *J. Virol.* **79**(6), 3606–3614.

Ott, S., Hansen, A., Kim, S. Y., and Miyano, S. (2005). Superiority of network motifs over optimal networks and an application to the revelation of gene network evolution. *Bioinformatics* **21**(2), 227–238.

Ozyildirim, A. M., Wistow, G. J., Gao, J., Wang, J., Dickinson, D. P., Frierson, H. F., Jr., and Laurie, G. W. (2005). The lacrimal gland transcriptome is an unusually rich source of rare and poorly characterized gene transcripts. *Invest. Ophthalmol. Vis. Sci.* **46**(5), 1572–1580.

Rico-Bautista, E., Greenhalgh, C. J., Tollet-Egnell, P., Hilton, D. J., Alexander, W. S., Norstedt, G., and Flores-Morales, A. (2005). Suppressor of cytokine signaling-2 deficiency induces molecular and metabolic changes that partially overlap with growth hormone-dependent effects. *Mol. Endocrinol.* **19**(3), 781–793.

Schuler, G. D., Epstein, J. A., Ohkawa, H., and Kans, J. A. (1996). Entrez: Molecular biology database and retrieval system. *Methods Enzymol.* **266**, 141–162.

Spellman, P. T., Miller, M., Stewart, J., Troup, C., Sarkans, U., Chervitz, S., Bernhart, D., Sherlock, G., Ball, C., Lepage, M., Swiatek, M., Marks, W. L., Goncalves, J., Markel, S., Iordan, D., Shojatalab, M., Pizarro, A., White, J., Hubley, R., Deutsch, E., Senger, M., Aronow, B. J., Robinson, A., Bassett, D., Stoeckert, C. J., Jr., and Brazma, A. (2002). Design and implementation of microarray gene expression markup language (MAGE-ML). *Genome Biol.* **3**, RESEARCH0046.

Wheeler, D. L., Barrett, T., Benson, D. A., Bryant, S. H., Canese, K., Church, D. M., DiCuccio, M., Edgar, R., Federhen, S., Helmberg, W., Kenton, D. L., Khovayko, O., Lipman, D. J., Madden, T. L., Maglott, D. R., Ostell, J., Pontius, J. U., Pruitt, K. D., Schuler, G. D., Schriml, L. M., Sequeira, E., Sherry, S. T., Sirotkin, K., Starchenko, G., Suzek, T. O., Tatusov, R., Tatusova, T. A., Wagner, L., and Yaschenko, E. (2005). Database resources of the National Center for Biotechnology Information. *Nucleic Acids Res.* **33**, D39–D45.

Yant, S. R., Wu, X., Huang, Y., Garrison, B., Burgess, S. M., and Kay, M. A. (2005). High-resolution genome-wide mapping of transposon integration in mammals. *Mol Cell. Biol.* **25**(6), 2085–2094.

Zhou, X. J., Kao, M. C., Huang, H., Wong, A., Nunez-Iglesias, J., Primig, M., Aparicio, O. M., Finch, C. E., Morgan, T. E., and Wong, W. H. (2005). Functional annotation and network reconstruction through cross-platform integration of microarray data. *Nature Biotechnol.* **23**(2), 238–243.

[20] Data Storage and Analysis in ArrayExpress

By ALVIS BRAZMA, MISHA KAPUSHESKY, HELEN PARKINSON, UGIS SARKANS, and MOHAMMAD SHOJATALAB

Abstract

ArrayExpress is a public resource for microarray data that has two major goals: to serve as an archive providing access to microarray data supporting publications and to build a knowledge base of gene expression profiles. ArrayExpress consists of two tightly integrated databases: ArrayExpress repository, which is an archive, and ArrayExpress data warehouse, which contains reannotated data and is optimized for queries. As of December 2005, ArrayExpress contains gene expression and other microarray data from almost 35,000 hybridizations, comprising over 1200 studies, covering 70 different species. Most data are related to peer-reviewed publications. Password-protected access to prepublication data is provided for reviewers and authors. Data in the repository can be queried by various parameters such as species, authors, or words used in the experiment description. The data warehouse provides a wide range of queries, including ones based on gene and sample properties, and provides capabilities to retrieve data combined from different studies. The ArrayExpress resource also includes Expression Profiler (EP)—a microarray data mining, analysis, and visualization tool—and MIAMExpress—an online data submission tool. This chapter describes all major ArrayExpress components from the user perspective: how to submit to, retrieve from, and analyze data in ArrayExpress.

Introduction

Since the first genome-wide microarray gene expression studies were published in 1997 (e.g., De Risi *et al.*, 1997), microarrays have become a standard technology in life sciences research. The amounts of data generated in a single microarray experiment considerably exceed that generated by any traditional technology, or by DNA sequencing. Not only do microarrays produce large amounts of data, but these data are complex. A series of non-trivial data processing steps have to be applied to raw microarray data to obtain biologically meaningful results. It has been widely acknowledged that to interpret microarray experiment results both raw and processed data

METHODS IN ENZYMOLOGY, VOL. 411   0076-6879/06 $35.00

are needed, as well as metadata describing the biological samples and experimental and data transformation procedures. These requirements are summarized in the MIAME guidelines (Ball *et al.*, 2004a), which have been adopted by a growing number of scientific journals. However, publishing, maintaining, and providing access to MIAME-compliant microarray data on an author's or a journal's web site is not a trivial task—professionally developed and maintained public repositories are more appropriate for it (Brazma *et al.*, 2000). Storing these data centrally also allows for access to all data on the same web site using a standard interface. The European Bioinformatics Institute (EBI) established a MIAME supportive public repository for microarray data ArrayExpress in 2002 (Brazma *et al.*, 2003; Parkinson *et al.*, 2005; Sarkans *et al.*, 2005).

As the numbers of laboratories using microarrays are increasing, data submission tools are improving, and journals are becoming more forceful in requiring submission to public repositories, the volume of data in Array-Express is growing rapidly. Its size has tripled during the last 12 months, and as of December 2005, the repository contains almost 35,000 hybridizations comprising over 800 studies related to 70 species (Fig. 1). The available studies cover a wide variety of experiment types, such as gene

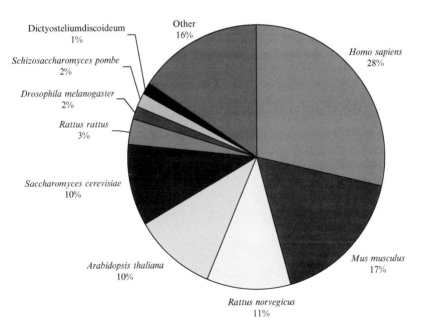

FIG. 1. Data in ArrayExpress by organism (December 2005). (See color insert.)

expression related to compound treatments, disease states, organism part comparisons, or developmental studies (Fig. 2). For instance, the experiment with accession number E-TOXM-16 investigates whether genotoxic carcinogens at doses known to induce liver tumours in rat bioassay deregulate a common set of genes in a short-term *in vivo* study. Raw and normalized data are provided. The experiment uses 137 hybridizations on 126 different samples on Affymetrix array RG_U34A. It combines experimental factors compound, dose, and time. The experiment E-UMCU-12 studies 9-day glucose starvation stationary phase culture in yeast *Saccharomyces cerevisiae* exit and entry from quiescence. It provides time series data for 34 time points and provides raw, normalized, and normalized smoothed data. Among other gene expression data sets in the database are human and mouse tissue expression data (e.g., E-AFMX-4, E-AFMX-5) and *Arabidopsis thaliana* development and differentiation expression data (e.g., E-AFMX-8). Slightly over 20% of the gene expression experiments provide time course data. Roughly a third of the experiments have been performed on the Affymetrix platform.

Although most data relate to gene expression, number of experiments used array comparative genomic hybridization (Erickson and Spana, 2006) or DNA-binding site identification (so-called ChIP-on-chip experiments)

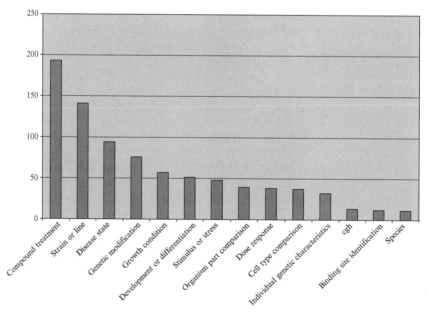

FIG. 2. Data in ArrayExpress by experiment type (MGED ontology terms, December 2005).

(Negre *et al.*., 2006; Scacheri *et al.*. 2006). An example of nongene expression data sets is ChIP-chip data for most yeast transcription factors (E-WMIT-1,2,10).

ArrayExpress is one of the three international repositories recommended by the Microarray Gene Expression Data (MGED) society (Ball *et al.*, 2004b) for storing microarray data related to publications [the other two being GEO (Barrett and Edgar, 2006; Edgar *et al.*, 2002) and CIBEX (Ikeo *et al.*, 2003)]. This defines the role of ArrayExpress as a primary archive and obliges it to accept all microarray data related to peer-reviewed publications without any changes unless approved by the submitter. The second goal of ArrayExpress is to build a knowledge base of gene expression providing easy access to high-quality, well-annotated data characterizing expression profiles of all genes in different organisms under different conditions. To meet these two goals, a separate database from the ArrayExpress repository, namely the ArrayExpress data warehouse, has been developed. It contains a subset of MIAME-compliant reannotated data and provides more powerful queries—ones based on gene names and properties. It also allows retrieval of data combined from different studies. As of December 2005 the ArrayExpress warehouse contained about 5% of data from the repository, but this percentage is expected to grow substantially in 2006.

Two additional tools are available to the user as a part of ArrayExpress: Expression Profiler is an online microarray data analysis tool linked to the database and MIAMExpress is a web-based microarray data annotation and submission tool (Fig. 3).

The target user community for ArrayExpress includes three major groups.

- Microarray experimentalists who are interested in experimental designs, array designs and protocols, and data from published experiments. This group will primarily use the repository, as well as data submission tools for submitting their own data.
- Biologists who are interested in expression patterns of particular genes. This group is primarily served by the data warehouse through gene attribute-based queries.
- Biologists and bioinformaticians who are interested in genome-wide studies. They can use the repository to upload published data sets in their own analysis tools or analyze them online using Expression Profiler, as well as use the data warehouse to retrieve gene expression data matrices combined from different experiments.

The next two sections describe (1) how data can be queried and retrieved from the ArrayExpress repository and warehouse, respectively,

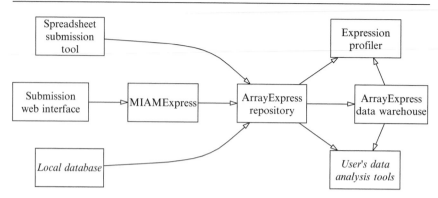

FIG. 3. ArrayExpress components.

(2) how data can be analyzed in Expression Profiler, and (3) how data can be submitted to ArrayExpress.

How to Query and Retrieve Data from the ArrayExpress Repository

Reflecting its archival role, the ArrayExpress repository organizes data by *experiments*, that is, a collection of hybridizations related to a particular study, often related to a publication. Each experiment can be retrieved by its accession number. Additionally, *array designs* and *protocols* have their own accession numbers, which enable the experimentalists to reuse the arrays and protocols submitted earlier (possibly by a different submitter), thus facilitating standardization.

The repository query interface provides queries for experiments, protocols, and array designs by a variety of attributes, such as species, experiment types, words, or phrases used in experiment descriptions or array platforms used in the experiment (see http://www.ebi.ac.uk/arrayexpress/query/entry). For instance, one can query for all experiments performed on Affymetrix arrays, the description of which contains the word "leukaemia." To access proprietary data (e.g., data related to a publication that is under review), the user needs to log in and provide the user name and password.

The result of a query for each experiment is a summary page containing a short description and links to data matrices and more detailed metadata, such as experimental or data processing protocols or sample properties. The experiment structure can be visualized as a block diagram (see Fig. 4) or as a set of spreadsheets, depicting how samples, labeled RNA extracts, different arrays, and data files relate. Users can examine the description of

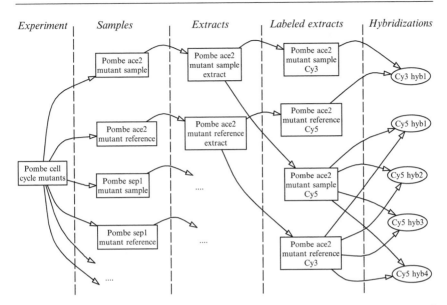

FIG. 4. Microarray experiment structure visualization in ArrayExpress repository and MIAMExpress submission tool.

the samples and protocols by navigating through the hyperlinks provided and download data for analysis locally or in Expression Profiler (see later).

In the data retrieval page the user can choose experimental conditions (effectively the columns in the data matrix; default option is all) and which measurements they want to retrieve (e.g., raw Cy3 or Cy5 signal, or normalized log ratios), as well as which annotation to export in the matrix (e.g., RefSeq IDs and various database accession numbers; note that in the repository the choice of array annotation is limited to that provided by the original submitter). The exported data matrix can be analyzed in Expression Profiler or downloaded for analysis in the user's own tools.

Array designs are provided in a generic format known as array description format (ADF)—a spreadsheet with each array feature (spot) on a separate row; the set of columns in ADF reflects the annotation provided by the data submitters.

How to Query Data in the ArrayExpress Data Warehouse

The ArrayExpress Data warehouse is a separate database that contains a subset of reannotated data from the ArrayExpress repository. Experiments in the repository are reviewed carefully by the ArrayExpress curators and selected for the warehouse on the basis of the quality of annotation, presence of raw and normalized data, and MIAME compliance. The array annotation is improved and updated using the Ensembl genome database if the respective array features have been mapped in Ensembl. For genomes or arrays not present in Ensembl, the UniProt database is used. Annotation, such as Gene Ontology (GO) (Harris *et al.*, 2004) terms, gene names, synonyms, and InterPro IDs (Mulder *et al.*, 2003), are added from the latest release of the respective databases.

The ArrayExpress data warehouse supports queries on gene attributes, for example, gene names, GO terms, InterPro terms, and on sample and experiment properties, for example, anatomical terms, disease states, or array platforms used in the experiment (see http://www.ebi.ac.uk/arrayexpress). The user can retrieve, combine, and visualize the gene expression values for multiple experiments. For example, a query of gene name "jun" and sample property "leukemia" results in retrieval of all experiments that contain data for one or several genes matching this name (e.g., jun, junb, jund) and that appear in experiments where one or more samples are annotated as "leukemia." First a list of genes that match the query, that is, jun, junb, and jund, is retrieved, from which the user can make a subselection by ticking the respective check boxes. Data for the selected genes are visualized using line plots and can be selected for further analysis. Links are also provided back to the repository where users can access the full annotation and supporting raw data.

Instead of querying for individual genes by names or properties, *all* genes of an organism can be selected. In this case the query retrieves full gene expression data matrix in a tab-delimited format, and matrices from different experiments can be selected (using tick boxes) and combined in a single matrix. Figure 5 shows the data selection path in the data warehouse *simple* interface, with boxes corresponding to sets of data objects and arrows describing filtering and data retrieval happening as a result of query parameters supplied by the user.

The *advanced* data warehouse query option enables users to choose the query path in a more complex way. A variety of data fields can be used to select subsets of experiments, hybridizations, samples, and genes and at any point export the data matrix with the chosen gene, sample, and hybridization annotation. For instance, the following query scenarios are possible:

- Select all experiments performed in a given laboratory.
- Select all hybridizations that belong to these experiments, and where the sample has been treated with a certain compound.
- Export all data available for these hybridizations.
- Select genes belonging to a certain GO category.

- Export data for selected genes and selected hybridizations.
- Export data for selected genes and all hybridizations.

The data flow corresponding to this scenario is shown in Fig. 6.

Data Analysis with Expression Profiler

Expression Profiler is a web-based tool that provides access to many basic exploratory analysis and visualization modules for microarray data (Kapushesky *et al.*, 2004). Data can be loaded into EP (http://www.ebi.ac. uk/expressionprofiler/) from ArrayExpress or from any source, including the user's own desktop PC. EP presents a graphical user interface to the most popular components of BioConductor (Gentleman *et al.*, 2004; Reimers and Carey, 2006), in addition to providing several unique tools implemented within the EBI. The platform also supports easy integration of novel algorithms into the uniform interface presented to the user.

Expression Profiler provides a set of functionalities responding to three basic use-case groups: identification of differentially expressed genes, cluster

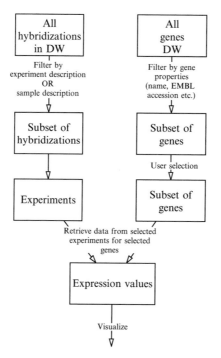

Fig. 5. Example workflow for the ArrayExpress data warehouse simple query interface.

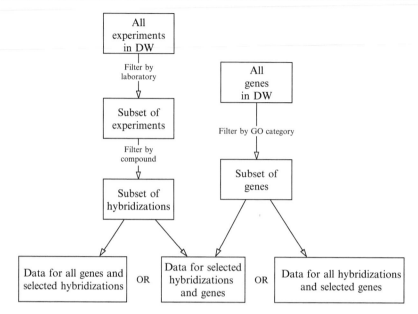

FIG. 6. Example workflow for the advanced data warehouse interface.

analysis of large data sets, and comparative analysis of several groups of biological samples. EP also contains a range of modules providing data subselection, normalization, and various transformations and data publishing. To use EP the user only needs a web browser, as all the computations are done on the server side at the EBI.

All the EP components are presented in a set of drop-down menus running across the top of the screen; this menu is referred to in the following sections. On selecting an item from the menu, the respective component will appear on screen, with horizontal sections containing various input fields and subsection tabs, as well as the Execute button in the bottom right corner.

Data Selection, Normalization, and Transformations

After data have been imported into Expression Profiler, descriptive statistics and basic data visualization graphics are provided—a histogram of the overall distribution density and a line plot of all rows in the data set or specialized plots for Affymetrix CEL files (Fig. 7). The *Data Selection* component can be used as a first step for identifying differentially expressed genes. It also provides a way to subselect a "slice" of the gene expression matrix by row or column names via the *Select rows* and *Select*

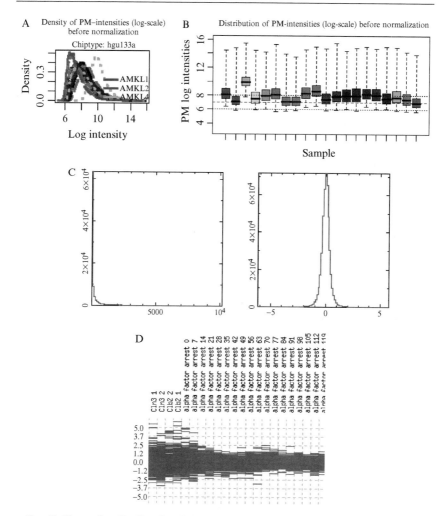

FIG. 7. Expression Profiler descriptive statistics visualizations. (A) PM intensity plot. (B) PM intensity box plot. (C) Distribution density histograms (one- and two-channel experiments, absolute and log-ratio data). (D) Multigene line plot. (See color insert.)

columns tabs. The *Missing values* subsection filters out rows of the matrix with more than a specified percentage of the values marked as NA (not available). The *Select by similarity* option provides an option to supply a list of genes and, for each of those, select a specified number of most similarly expressed ones in the same data set.

Expression Profiler provides a graphical interface to three commonly used BioConductor data normalization routines for Affymetrix and other microarray data, namely GCRMA, RMA, and VSN (Huber *et al.*, 2002; Irizarry *et al.*, 2003; Wu *et al.*, 2003) through the *Data Normalization* component (from the *Data Transformation* menu). In EP, GCRMA and RMA can only be applied to Affymetrix CEL file imports, whereas VSN can be applied universally to all types of data.

The *Data Transformation* component is used where data need to be transformed to make it suitable for some specific analysis. For instance, with the *Absolute-to-Relative* transformation, absolute expression values for each gene can be transformed into relative ones: either relative to a specified column of the data set or relative to the mean value of the gene.

Identification of Differentially Expressed Genes

The *Data Selection* component provides several basic mechanisms to filter out the highly variable genes in the data set as ones likely to be expressed differentially. The *Value ranges* option under *Subselection* takes two parameters and can be used to filter genes that are above or below their mean at least a chosen number of standard deviations, in at least a certain percentage of the conditions. This is similar to the commonly applied fold change criterion, with the difference that it considers the variability of each gene across multiple conditions.

The *t test* component in the *Statistics* menu provides a way to apply this basic statistics test for comparing the means from two distributions in the following differentially expressed gene identification situations: looking for genes expressed significantly above background/control or looking for genes expressed differentially between two sets of conditions. In the first case, the user specifies either the background level to compare against or selects the genes in the data set that are to be used as controls. In the second case, the user specifies which columns in the data set represent the first group of conditions and which represent the second group. The *t* test calculates the mean in both groups being tested for each gene. When testing against controls, the mean over all control genes is taken as the second group mean, and in both cases the difference between the two means is compared to a theoretical *t* statistic.

Clustering Analysis

Expression Profiler provides fast implementations of two classes of clustering algorithms (clustering menu): hierarchical clustering and flat partitioning in the *Hierarchical Clustering* and *K-means/K-medoids Clustering* components respectively. A range of distance measures such as Euclidean or correlation-based distance measures can be used. The *Signature Algorithm* component is an alternative approach to clustering-like analysis, based on the method by Bergmann *et al.* (2003).

A common problem with hierarchical clustering is that it is difficult to identify branches within the hierarchy that form optimally tight clusters. Similarly, in the case of flat partitioning, determination of the number of desired clusters is often arbitrary and unguided. The *Clustering Comparison* component aims to alleviate these difficulties by providing an algorithm and a visual depiction of a mapping between a dendrogram and a set of flat clusters. The implementation is based on Torrente *et al.* (2005) and provides a choice of two comparison functions based on mutual information in the two clusterings and a heuristic measure for maximizing overlap while minimizing visual complexity (Fig. 8). The clustering comparison component highlights the tree branches that best correspond to one or more flat clusters from the partitioning and is also useful when comparing hierarchical clustering to a predefined functionally meaningful grouping of the genes. This component can also be used to compare a pair of flat partitioning clusterings to help establish the optimal parameter K by starting with a high number of clusters and letting the comparison algorithm identify the appropriate number of superclusters, that is, groups made up of overlaps between the flat ones.

Comparative Group Analysis

When more than two groups of samples are to be compared and more than one factor is being studied (e.g., multiple cell lines with several different chemical treatments), there is a need to identify both the distinguished groups of samples and, simultaneously, the differentiating genes between these groups. Although the questions asked in this scenario are similar to those for identifying differentially expressed genes between two samples, the statistical mechanisms need to be more powerful and robust for this type of analysis, which is provided the *Between Group Analysis* component (from the *Ordination* menu).

Between-group analysis (BGA) is a multiple discriminant approach that is performed by ordinating the specified groups of samples and projecting individual sample locations on the resulting axes (Culhane *et al.*, 2002). In Expression Profiler the ordination step involved in BGA can be either principal components analysis (PCA) or correspondence analysis (COA). Both of these are standard statistical tools for reducing the dimensionality of the data set being analyzed. This is done by a calculation of an ordered set of values that correspond to greatest sources of variation in data and using these values to "reorder" the genes and samples of the matrix. BGA combined with COA is especially powerful because it provides a simultaneous view of the grouped samples and the genes that most facilitate the discrimination between them. Algorithms of the BGA component are provided through an interface to the

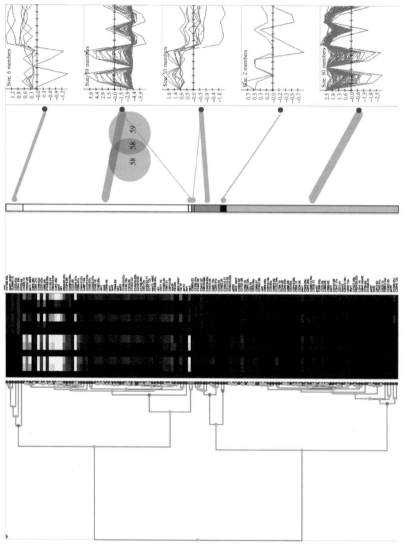

Fig. 8. Expression Profiler clustering comparison visualization. A hierarchical flat comparison, matching a K-means clustering ($K = 5$) to a dendrogram. Overlaps between matching clusters are shown with Venn diagrams. (See color insert.)

BioConductor package "made4," which, in turn, refers to the R multivariate data analysis package "ade4" (Thioulouse *et al.*, 1997).

In addition to the various plots, BGA produces several numerical tables, including tables of gene and array coordinates. The gene coordinates table is of special interest because it provides a set of numbers that provide a measure of how variable each gene is in each of the identified strong sources of variation. The sources of variation (principal axes/components) are ordered from left to right. Thus genes that have the highest or lowest values in the first column of the gene coordinates table are the likeliest candidates for differential expression. When BGA is run within the COA framework, since the first column in this table corresponds to the first principal component, looking at the plot we can examine which groups of samples are best discriminated by the corresponding (first) axis so the identified genes are most distinguishing between these groups.

Exporting Graphics and Results

Expression Profiler stores all parameters, results, and graphics files for every analysis step. These can be retrieved at any stage in the analysis by clicking the *Display output* button in the history display in the top section of any component.

Scalable vector graphics are one of the native formats Expression Profiler employs. These graphics are standard publication quality images that can be easily imported into, for instance, Adobe Illustrator software. For those requiring alternative image formats, PNG files can also be exported.

Expression Profiler is a developing platform: more components are being added, both for providing comprehensive visualizations for basic techniques and for enabling the analysis of data from complex experiment designs with specially developed methods.

How to Submit Data to ArrayExpress

Data can be submitted to ArrayExpress either online using the submission tool MIAMExpress (http://www.ebi.ac.uk/miamexpress/) or as MAGE-ML files (Spellman *et al.*, 2002) from external databases or applications. To use MIAMExpress, one needs only a web browser. No prior knowledge of MIAME guidelines is required, as the tool guides the submitter through a series of web forms and context-sensitive help is provided. Large submissions can be made via a spreadsheet upload submission system tab2mage. A visualization module provides a graphical representation of experiment structure submitted up to a given moment (as in Fig. 4). Data files are uploaded from the user's desktop PC and linked to the experiment annotation.

To submit MAGE-ML files directly from a local database or application, a MAGE-ML pipeline can be established. To establish such a pipeline,

ArrayExpress curators will work with the local database developers to make sure that the exported MAGE-ML files are consistent with the best practice recommendations (see http://www.mged.org/Workgroups/MIAME/miame_ mage-om.html), as well as that the use of identifiers is consistent with those in ArrayExpress. Once such a pipeline is established, submitting data to ArrayExpress becomes a simple task and there is no need to use MIAM-Express. Pipelines from more than 10 different laboratory database have been established, including Stanford Microarray Database (SMD) (Sherlock, 2001), MIDAS at TIGR (Saeed *et al.*, 2003, 2006), and University Medical Center Utrecht microarray database, as well as the array manufacturers Affymetrix and Agilent, and more are under construction.

Details of the submission process, including when and how to build a pipeline and what is needed for submission, can be found at www.ebi.ac.uk/ arrayexpress.

Future

One of the immediate future goals of ArrayExpress is to populate the data warehouse with more data from the repository. We estimate that 50% of data submitted to the repository will eventually be loaded into the warehouse. At the same time we are working to extend the functionality of the data warehouse, and new features, such as selecting genes by similar expression profiles (which is already possible in Expression Profiler), will be added. The data warehouse will be closely integrated with Expression Profiler data analysis tools. Links from other EBI resources, including Ensembl, will be expanded so that users querying, for instance, genomic information, can easily get the related expression information. We are also exploring a possibility that data loaded in the data warehouse can be renormalized for consistency (note that currently we use normalized data provided by the authors).

Another major development task is making data submissions to Array Express easier. The batch upload tool facilitates the submission of large experiments. This tool will be improved further to simplify submissions of standard experimental designs, such as experiments on one-channel arrays. The batch upload tool will also be used to establish direct submission routes to ArrayExpress from laboratory databases.

One of the guiding principles of ArrayExpress development has been the support of community data standards, such as MIAME, MAGE-ML, and MGED ontology. As the development of MAGE-ML is continuing, ArrayExpress will be updated constantly to support these developments.

Acknowledgments

EMBL, the European Commission (FELICS), International Life Sciences Institute (ILSI), and the National Institutes of Health (NIH) support ArrayExpress development.

References

Ball, C. A., Brazma, A., Causton, H., Chervitz, S., Edgar, R., Hingamp, P., Matese, J. C., Parkinson, H., Quackenbush, J., Ringwald, M., Sansone, S. A., Sherlock, G., Spellman, P., Stoeckert, C., Tateno, Y., Taylor, R., White, J., and Winegarden, N. (2004). Submissions of microarray data to public repositories. *PLOS Biol.* **2,** e317.

Ball, C. A., Sherlock, G., Parkinson, H., Rocca-Sera, P., Brooksbank, C., Causton, H. C., Cavalieri, D., Gaasterland, T., Hingamp, P., Holstege, F., Ringwald, M., Spellman, P., Stoeckert, C. J., Jr., Stewart, J. E., Taylor, R., Brazma, A., and Quackenbush, J., Microarray Gene Expression Data Society (2004). Standards for microarray data. *Science* **298,** 539.

Barrett, T., and Edgar, R. (2006). Gene Expression Omnibus: Microarray data storage, submission, retrieval, and analysis. *Methods Enzymol.* **411,** 352–369.

Bergmann, S., Ihmels, J., and Barkai, N. (2003). Iterative signature algorithm for the analysis of large-scale gene expression data. *Phys. Rev. E. Stat. Nonlin. Soft Matter Phys.* **67**(3 Pt 1), 031902.

Brazma, A., Parkinson, H., Sarkans, U., Shojatalab, M., Vilo, J., Abeygunawardena, N., Holloway, E., Kapushesky, M., Kemmeren, P., Lara, G. G., Oezcimen, A., Rocca-Sera, P., and Sansone, S. A. (2003). ArrayExpress: A public repository for microarray gene expression data at the EBI. *Nucleic Acids Res.* **31**(1), 68–71.

Brazma, A., Robinson, A., Cameron, G., and Ashbumer, M. (2000). "One-stop shop for microarray data." *Nature* **403**(6771), 699–700.

Culhane, A. C., Perriere, G., Considine, E. C., Cotter, T. G., and Higgins, D. G. (2002). Between-group analysis of microarray data. *Bioinformatics* **18**(12), 1600–1608.

DeRisi, J. L., Iyer, V. R., and Brown, P. O. (1997). Exploring the metabolic and genetic control of gene expression on a genomic scale. *Science* **278**(5338), 680–686.

Edgar, R., Domrachev, M., and Lash, A. E. (2002). Gene Expression Omnibus: NCBI gene expression and hybridization array data repository. *Nucleic Acids Res.* **30**(1), 207–210.

Erickson, J. N., and Spana, E. P. (2006). Mapping *Drosophila* genomic aberration breakpoints with comparative genome hybridization on microarrays. *Methods Enzymol* **410,** 377–386.

Gentleman, R. C., Carey, V. J., Bates, D. M., Bolstad, B., Dettling, M., Dudoit, S., Ellis, B., Gautier, L., Ge, Y., Gentry, J., Hornik, K., Hothorn, T., Huber, W., Iacus, S., Irizarry, R., Li, F. L. C., Maechler, M., Rossini, A. J., Sawitzki, G., Smith, C., Smyth, G., Tierney, L., Yang, J. Y. H., and Zhang, J. (2004). Bioconductor: Open software development for computational biology and bioinformatics. *Genome Biol.* **5,** R80.

Harris, M. A., Clark, J., Ireland, A., Lomax, J., Ashburner, M., Foulger, R., Eilbeck, K., Lewis, S., Marshall, B., Mungall, C., Richter, J., Rubin, G. M., Blake, J. A., Bult, C., Dolan, M., Drabkin, H., Eppig, J. T., Hill, D. P., Ni, L., Ringwald, M., Balakrishnan, R., Cherry, J. M., Christie, K. R., Costanzo, M. C., Dwight, S. S., Engel, S., Fisk, D. G., Hirschman, J. E., Hong, E. L., Nash, R. S., Sethuraman, A., Theesfeld, C. L., Botstein, D., Dolinski, K., Feierbach, B., Berardini, T., Mundodi, S., Rhee, S. Y., Apweiler, R., Barrell, D., Camon, E., Dimmer, E., Lee, V., Chisholm, R., Gaudet, P., Kibbe, W., Kishore, R., Schwarz, E. M., Sternberg, P., Gwinn, M., Hannick, L., Wortman, J., Berriman, M., Wood, V., de la Cruz, N., Tonellato, P., Jaiswal, P., Seigfried, T., and White, R., and Gene Ontology Consortium (2004). The Gene Ontology (GO) database and informatics resource. *Nucleic Acids Res.* **31,** 236–261.

Huber, W., von Heydebreck, A., Sultmann, H., Poustka, A., and Vingron, M. (2002). Variance stabilization applied to microarray data calibration and to the quantification of differential expression. *Bioinformatics* **18**(Suppl. 1), S96–S104.

Ikeo, K., Ishi-i, J., Tamura, T., Gojobori, T., and Tateno, Y. (2003). CIBEX: Center for information biology gene expression database. *C. R. Biol.* **326**(10–11), 1079–1082.

Irizarry, R. A., Bolstad, B. M., Collin, F., Cope, L. M., Hobbs, B., and Speed, T. P. (2003). Summaries of affymetrix GeneChip probe level data. *Nucleic Acids Res.* **31**(4), e15.

Kapushesky, M., Kemmeren, P., Culhane, A. C., Durinck, S., Ihmels, J., Korner, C., Kull, M., Torrente, A., Sarkans, U., Vilo, J., and Brazma, A. (2004). Expression Profiler: Next generation—an online platform for analysis of microarray data. *Nucleic Acids Res.* **32,** W465–W470.

Mulder, N. J., Apweiler, R., Attwood, T. K., Bairoch, A., Barrell, D., Bateman, A., Binns, D., Biswas, M., Bradley, P., Bork, P., Bucher, P., Copley, R. R., Courcelle, E., Das, U., Durbin, R., Falquet, L., Fleischmann, W., Griffiths-Jones, S., Haft, D., Harte, N., Hulo, N., Kahn, D., Kanapin, A., Krestyaninova, M., Lopez, R., Letunic, I., Lonsdale, D., Silventoinen, V., Orchard, S. E., Pagni, M., Peyruc, D., Ponting, C. P., Selengut, J. D., Servant, F., Sigrist, C. J. A., Vaughan, R., and Zdobnov, E. J. (2003). The InterPro Database, 2003 brings increased coverage and new features. *Nucleic Acids Res.* **31,** 315–318.

Negre, N., Lavrov, S., Hennetin, J., Bellis, M., and Cavalli, G. (2006). Mapping the distribution of chromatin proteins by ChIP on chip. *Methods Enzymol.* **410,** 316–341.

Parkinson, H., Sarkans, U., Shojatalab, M., Abeygunawardena, N., Contrino, S., Couison, R., Fame, A., Lara, G. G., Holloway, E., Kapushesky, M., Lija, P., Mukherjee, G., Oezcimen, A., Rayner, T., Rocca-Serra, P., Sharma, A., Sansone, S., and Brazma, A. (2005). "Array-Express—A public repository for microarray gene expression data at the EBI." *Nucleic Acids Res.* **33,** Database Issue: D553–D555.

Reimers, M., and Carey, V. J. (2006). Bioconductor: An open source framework for bioinformatics and computational biology. *Methods Enzymol.* **411,** 119–134.

Saeed, A. I., Bhagabati, N. K., Braisted, J. C., Liang, W., Sharov, V., Howe, E. A., Li, J., Thiagarajan, M., White, J. A., and Quackenbush, J. (2006). TM4 microarray software suite. *Methods Enzymol.* **411,** 134–193.

Saeed, A. I., Sharov, V., White, J., Li, J., Liang, W., Bhagabati, N., Braisted, J., Klapa, M., Currier, T., Thiagarajan, M., Sturn, A., Snuffin, M., Rezantsev, A., Popov, D., Ryltsov, A., Kostukovich, E., Borisovsky, I., Liu, Z., Vinsavich, A., Trush, V., and Quackenbush, J. (2003). TM4: A free, open-source system for microarray data management and analysis. *Biotechniques* **34**(2), 374–378.

Sarkans, U. P. H., Lara, G. G., Dezcimen, A., Sharma, A., Abeygunawardena, N., Contrino, S., Holloway, E., Rocca-Serra, P., Mukherjee, G., Shojatalab, M., Kapushesky, M., Sansone, S. A., Fame, A., Rayner, T., and Brazma, A. (2005). "The ArrayExpress gene expression database: A software engineering and implementation perspective." *Bioinformatics* **21**(8), 1495–1501.

Scacheri, P. C., Crawford, G. E., and Davis, D. (2006). Statistics for ChIP-chip and DNase hypersensitivity experiments on NimbleGen arrays. *Methods Enzymol.* **411,** 270–282.

Sherlock, G., Hernandez-Boussard, T., Kasarskis, A., Binkley, G., Matese, J. C., Dwight, S. S., Kaloper, M., Weng, S., Jin, H., Ball, C. A., Eisen, M. B., Spellman, P. T., Brown, P. O., Botstein, D., and Cherry, J. M. (2001). The Stanford Microarray Database. *Nucleic Acids Res.* **29**(1), 152–155.

Spellman, P. T., Miller, M., Stewart, J., Troup, C., Sarkans, U., Chervitz, S., Bernhart, D., Sherlock, G., Ball, C., Lepage, M., Swiatek, M., Marks, W. L., Goncalves, J., Markel, S., Iordan, D., Shojatalab, M., Pizarro, A., White, J., Hubley, R., Deutsch, E., Senger, M., Aronow, B. J., Robinson, A., Bassett, D., Stoeckert, C. J., Jr., and Brazma, A. (2002). Design and implementation of microarray gene expression markup language (MAGE-ML). *Genome Biol.* **3**(9), RESEARCH0046.

Thioulouse, J., Chessel, D., Dolédec, S., and Olivier, J. M. (1997). ADE-4: A multivariate analysis and graphical display software. *Stat. Comput.* **7**(1), 75–83.

Torrente, A., Kapushesky, M., and Brazma, A. (2005). A new algorithm for comparing and visualizing relationships between hierarchical and flat gene expression data clusterings. *Bioinformatics* **21,** 3993–3999.

Wu, Z., Irizarry, R., Gentleman, R., Murillo, F., and Spencer, F. (2003). "A Model Based Background Adjustment for Oligonucleotide Expression Arrays." John Hopkins University, Baltimore, MD.

[21] Clustering Methods for Analyzing Large Data Sets: Gonad Development, A Study Case

By JÉRÔME HENNETIN and MICHEL BELLIS

Abstract

With the development of data set repositories, it is now possible to collate high numbers of related results by gathering data from experiments carried out in different laboratories and addressing similar questions or using a single type of biological material under different conditions. To address the challenge posed by the heterogeneous nature of multiple data sources, this chapter presents several methods used routinely for assessing the quality of data (i.e., reproducibility of replicates and similarity between experimental points obtained under identical or similar biological conditions). As gene clustering on large data sets is not straightforward, this chapter also presents a rapid gene clustering method that involves translating variation profiles from an ordered set of comparisons into chains of symbols. In addition, it shows that lists of genes assembled based on the presence of a common term in their functional description can be used to find the most informative comparisons and to construct from them exemplar chains of symbols that are useful for clustering similar genes. Finally, this symbolic approach is extended to the overall set of biological conditions under study and shows how the resultant collection of variation profiles can be used to construct transcriptional networks, which in turn can be used as powerful tools for gene clustering.

Introduction

The advent of microarray technology (Pease *et al.*, 1994; Schena *et al.*, 1995) has highly stimulated the field of transcriptome studies over the last decade, and results are now accumulating rapidly in public repositories [Gene Expression Omnibus (GEO, http://www.ncbi.nlm.nih.gov/geo/), European Bioinformatics Institute (EBI, http://www.ebi.ac.uk/arrayexpress)]. Although a great number of methods have been described for coping with simple situations in which a small number of biological conditions are compared in order to identify genes varying at statistically significant levels (Breitling *et al.*, 2004; Neuhauser and Senske, 2004), to treat more complex configurations in which numerous experimental points are used to cluster genes (Gao *et al.*, 2004), or to find gene expression signatures (Golub *et al.*, 1999), far fewer

METHODS IN ENZYMOLOGY, VOL. 411
0076-6879/06 $35.00
DOI: 10.1016/S0076-6879(06)11021-6

reports have addressed methods for helping users deal with the specific issues that arise when working with such large sets of data. For example, when analyzing data coming from heterogeneous sources it is difficult to assess the reproducibility of replicates and to measure the similarity of points obtained under identical or similar biological conditions. This chapter presents several methods devised to treat this kind of problem. Another key issue relates to the clustering of genes, in which is imperative to simply and succinctly summarize the main class of genes defined by their variation profile across the entire set of biological conditions examined. Unfortunately, while a plethora of clustering methods are available (Datta and Datta, 2003), results delivered by this type of analysis are strongly affected not only by the particular method used, but also by the setup of the parameters, if any (Fred and Jain, 2005).

Another problem relates to the size of data, as huge data sets cannot be treated in an acceptable amount of time. This chapter presents a clustering method that is rapid, very simple to implement and use, and is robust with regard to setting up the necessary parameters. Finally, it explains how this clustering method can be extended to an entire set of data and used to construct a transcriptional network, which in turn can be used as a powerful tool for gene clustering. To demonstrate all of these methods, we have collected a set of transcriptome experiments related to gonad development in mice [Schultz et al., 2003 (T2); Shima et al., 2004 (T3); Small et al., 2005 (OV and T1); Hamra et al., 2004 (LA); and Costoya et al., 2004 (ZP); see Table I for abbreviations used]. These experiments allow one to assemble 76 experimental points and to consider 33 distinct biological conditions represented by duplicates.

Selection of Data Sets

GEO stores a huge number of microarray data sets in different formats. We limited our search to Affymetrix chipset models for two main reasons: (1) it is a single channel technology, which means that experiments carried out in different laboratories can be compared easily, and (2) as it is an industrial technology, the genes probed in a data set are determined by the particular chipset used and are independent of the experimenter. We note that in this technology, a single gene is represented by one or more collections of probes called a probe set.

We interrogated version June 2005 of GEO and searched for experiments (GSE items in GEO vocabulary) related to meiosis and using Affymetrix technology. We found the GSE described in Table I. Experiments and biological conditions are designated by their abbreviations (columns 4 and 5 of Table I). Experimental points are referred to by a name constructed

TABLE I
GEO Affymetrix Experiment Related to Meiosis in Mouse

Rank	Experimentation	GSE[a]	Exp[b]	Biological conditions[c]
1	Ovary postcoitum	GSE1359	OV	11d5, 12d5, 14d5, 16d5, 18d5
2	Testis postcoitum	GSE1358	T1	11d5, 12d5, 14d5, 16d5, 18d5
3	Testis 1 day to adult	GSE640	T2	*01d*, 04d, 08d, 11d, **14d, 18d,** **26d, 29d, Ad,** 14/18d, 26/29d
4	Testis postpartum	GSE926	T3	00d, 03d, 06d, 08d, 10d, 14d, 18d, 20d, 30d, 35d, 56d
5	Laminin binding	GSE829	LA	LA, NL, TU, **INT**
6	Pure germinal cells	GSE 2736	PC	MY, SE, CSC, **SGA, SGB, SC, PSC, RST**
7	Zp145 spermatogonia	GSE 1399	ZP	**WT, KO**

[a] GSE reference [some data sets related to pure germinal cells (PC, rank = 6) were obtained from the following address: http://www.wsu.edu/~griswold/microarray/].
[b] Abbreviations used in the text for designating the experiments.
[c] Abbreviations used in the text for designating biological conditions. All biological conditions marked in bold are represented by only one experimental point. All others have exactly two replicates, except for the one in italics, which has three (T2 01d). T2 14d and T2 18d have been grouped into 14/18d to allow comparisons with other conditions. T2 26d and 29d were grouped into 26/29d for the same reason. All numerical abbreviations refer to days either postcoitum (OV and T1) or postpartum (T2 and T3). Ad, adult; LA, laminin binding; NL, laminin nonbonding; TU, tubular cells; INT, interstitial cells; MY, myoid cells; SE, Sertoli cells; SEC, Sertoli cell line (MSC-1); SGA, spermatogonia of type A; SGB, spermatogonia of type B; SC, spermatocyte; PSC, pachytene spermatocyte; RST, round spermatid; WT, wild type; KO, knock-out of *Zfp145*.

by the concatenation of the experimental and biological abbreviations and a replicate rank: T2 04d R1, T2 04d R2, T2 14d R0 (R0 stands for no replicated points).

Detection of Statistically Significant Variations by the Rank Difference Analysis of Microarray Method (RDAM)

RDAM (Martin *et al.*, 2004) allows the identification of statistically significant signal variations between two biological conditions when each contains at least two replicates. One of the first steps of this method consists of calculating an absolute rank (AR) for each gene by ordering their signals from 0 to N. In earlier versions of their analysis software (MAS3 or MAS4), Affymetrix allowed negative values as a measure of expression. While all experiments used in this study were analyzed by MAS5 (Hubbell *et al.*, 2002), which delivers only positive values, in general usage the situation could be encountered in which the results were analyzed by a mixture of MAS4 and MAS5 or where all the results are of the MAS4 type. In such cases, negative signals can be treated as follows: in the case of MAS4 results only, all negative signals have their rank set to zero; in the case of a mixture of MAS4 and MAS5 results, all signals must be ranked with different values irrespective of whether they are positive or negative.

Variations are expressed as standardized rank difference (zRD). The following quantities are estimated by RDAM: the total variation (TV), the p value of a variation, the false discovery rate (FDR), that is, the percentage of false positives in a selection, and the sensitivity (S), that is, the percentage of the total variation that is found in a given selection. Subsets of genes are selected using one of the three latter parameters (p value, FDR, or S) or a combination of them.

Reproducibility of Replicates

Before any analysis designed to detect expression variation can be carried out, the quality of the assembled experiments must be assessed. While this quality can be envisaged in different ways, we consider the reproducibility between replicates in each experiment to be a priority because the detection of statistically significant variations by the RDAM algorithm largely depends on this property. Indeed, in order to assign a p value to each standardized variation, we use replicates to empirically construct the variation distribution observed in the case of the null hypothesis. If the reproducibility of these replicates is low, the p value will be high, and the power of the test will be low as well.

We devised a method, illustrated in Fig. 1, which allows the reproducibility to be examined both qualitatively and quantitatively.

Table II shows that the mean reproducibility score (RS) for each experiment is consistently higher with chip C than with chip B, which is itself higher than with chip A.

Figure 2 confirms this observation for each of the pairs of replicates studied (listed in Table I) and indicates that most of the RS increase can be explained by an elevation in the curve corresponding to a rank smaller than 50.

Figures 3 and 5 show the profound effect of this shift on the power of detection: total variation TV is inversely related to RS and is consistently lower with chip C than with chip B, which is lower than with chip A. In view of the form of the RC curves (Fig. 2), we predict that this effect should be more pronounced for probe sets having their rank smaller than the 50th

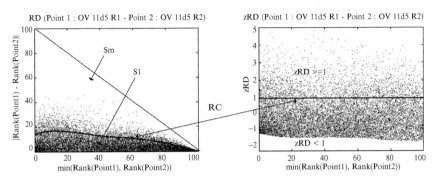

FIG. 1. Construction principle of the reproducibility curve (RC) and the reproducibility score (RS). (Right) The standardized rank difference of all probe sets (zRD) is plotted against the minimum of the ranks of the two replicated points under consideration (OV 11d5 R1 and OV 11d5 R2). The horizontal line zRD = 1, which represents the boundary between points that are less than one standard deviation unit from those that are equal to or higher than one unit, is mapped (left) onto the plot of RD against the minimum of the ranks of the two replicated points to give the reproducibility curve RC. This mapping used the inverse of the standardization function of RD, i.e., RDi = (zRD * std(RD(Ri))) − μ(RD(Ri)). In this formula, the sample mean and standard deviation (μRD and stdRD) were calculated for all probe sets having a rank within a given neighborhood of Ri, with the rank of points i localized on the line of equation zRD = 1. This notation reflects the fact that the RD distribution (left) is not gene specific but rank dependent. The surface S1 under the curve SC is measured and normalized by the surface Sm under the diagonal, which is the maximal possible value of S1: RS = S1/Sm.

TABLE II
MEAN AND STD VALUES OF THE REPRODUCIBILITY SCORE IN DIFFERENT EXPERIMENTS

	Chip A		Chip B		Chip C	
Experiment	Mean	STD	Mean	STD	Mean	STD
Ovary postcoitum (OV)	19.4	1.52	27.6	2.19	35.6	2.51
Testis postcoitum (T1)	19.2	2.39	25.2	2.59	32.6	1.52
Testis 1 day to adult (T2)	18.83	2.64	21	2.71	30	2.16
Testis postpartum (T3)	22.73	3.64	30.18	4.49	34.64	5.24
Laminin binding (LA)	27	1	27.5	2.12	30.5	0.71
Pure germinal cells (PC)	19.67	7.09	26.33	8.33	33	7

percentile. It is striking that the same effect was observed in all of the experiments, and we deem the most plausible explanation to be that the same target preparation was used successively in the three chips starting with chip A and finishing with chip C.

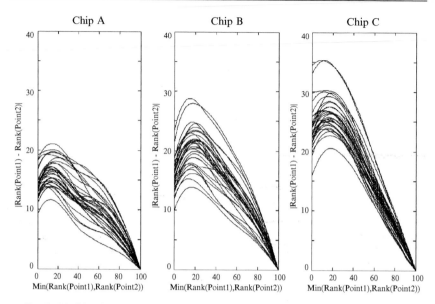

FIG. 2. Modification of the form of the reproducibility curves from chip A to chip C. All the biological conditions with replicated points have their RC curves traced, as explained in Fig. 1. Thirty-three (30+3) curves are displayed (three RC curves are traced for T2 01d because this biological condition has three replicated experimental points).

Relationships between Experimental Points

The second point to be verified is the existence of consistent relationships among the whole set of experimental points. It is expected, for example, that replicates will be highly similar, as will points obtained under similar or identical conditions. An elegant way to visualize the quality of the entire data set is to trace a dendrogram, which is a two-dimensional "tree-like" diagram that summarizes the distances between experimental points by joining points according to their level of similarity. We proceed as follows.

1. We construct a median sample in which the rank of a gene represents the median of its ranks over all experimental points.
2. We compare each experimental point with the median sample by applying the RDAM algorithm.
3. We select all of the probe sets having a p value smaller than, for example, 0.005, in at least one comparison.
4. We calculate the distance between each pair of experimental points in the probe set space, with $\log_2$ of the signal being used as the unit

and with each axis representing one of the probe sets selected in the previous step.

5. We trace the corresponding dendrogram, as shown in Fig. 4.

The principal conclusion to be drawn from this procedure is that the grouping of experimental points is mainly determined by whether they originate from the same experiment. For example, corresponding points between the two time courses T2 and T3 were not grouped together as they would have been if there had been high levels of reproducibility between identical or similar biological conditions. At this step, it might have been opportune to conduct a complementary analysis to possibly eliminate one of these experiments. We did not explore this possibility, however, and decided to keep both of them [we show elsewhere (Negre *et al.*, 2006) how a dendrogram, constructed using a slightly different method, can inform decisions concerning whether to keep or to eliminate experimental points in the presence of poor reproducibility]. This anomaly aside, the experimental points are well grouped: within each time course series, the biological

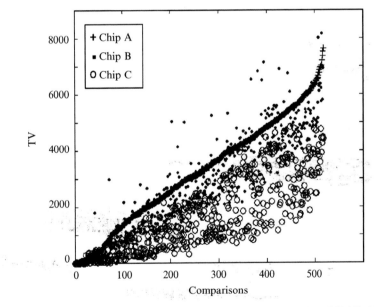

FIG. 3. Total variation (TV) observed in 528 comparisons in chips A, B, and C. All of the 528 possible comparisons among the 31 biological conditions having replicates, plus T2 14/18d and T2 26/29d, were carried out. TV for each comparison in chips A, B, and C were plotted versus the rank of the ordered TV values of chip A. Chip A, B, and C results are represented by plus signs, points, and circles, respectively.

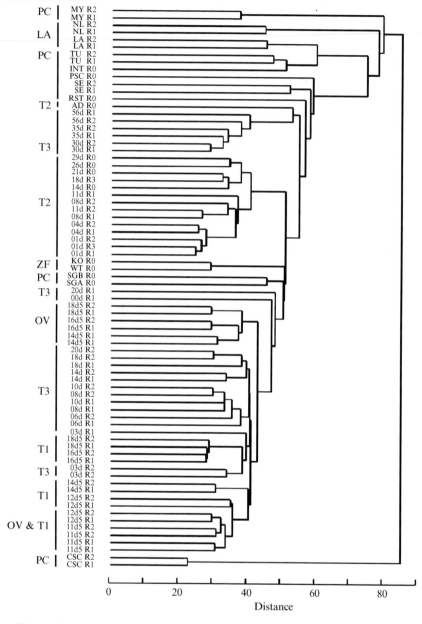

FIG. 4. Dendrogram showing the relationships among the experimental points. The selection at *p* value 0.005 filtered 3541 probe sets.

conditions are arranged according to the experiment's date, most of the replicates are adjacent to each other, and in OV and T1, the points for 11d5 and 12d5, in which the gonads are still undifferentiated, are grouped. It is worth noting that trees constructed in this way are very stable and are largely independent of the p value limit used in step 3.

Evolution of Total Variation across Ordered Comparisons in
   Each Experiment

Having verified the overall quality of data, the following step was then carried out in order to obtain a general overview of the variation in gene expression. We looked at the evolution of the total variation estimate (TV) in a subset of ordered comparisons for each experiment. In an experiment with n biological conditions, $n*(n-1)/2$ different comparisons can be analyzed. In reality, we restricted the number of comparisons to three types, as exemplified in Fig. 5. Important conclusions can be drawn from a visual inspection of the resulting profiles. For example, it can be seen that the second comparison (14d5 vs 12d5) from the OV incremental profile (group 1 in Fig. 5A) has a TV of roughly 1000 probe sets. In contrast, the second comparison (14d5 vs 11d5) from the OV cumulative profile (group 1 in Fig. 5B) has a TV close to zero. This apparent discrepancy can be explained by the observation that the probe sets that were increased between 14d5 and 12d5 were decreased systematically between 12d5 and 11d5. This decrease is not statistically significant, however, when tested individually on each gene, explaining why no variation was detected between OV 12d5 and OV 11d5 (first comparison of incremental and cumulative OV profiles). This illustrates that methods aimed at detecting significant variation in individual comparisons, which are based on the repetition of a statistical test separately for each gene, cannot take into account the "collective" behavior of genes and are less powerful than clustering techniques that are sensitive to this information.

We can generally state that incremental comparisons are less powerful for detecting genes than cumulative comparisons, particularly with this type of kinetics in which the expression of a particular gene increases steadily over several time points. This effect is particularly visible with T3: the 6th (14d vs 10d) and 10th (56d vs 35d) comparisons are the only ones that show a significant amount of TV in the incremental comparisons. In contrast, all cumulative comparisons starting from the 6th up to the 10th have a significant amount of TV. As expected from the substantial distance separating the two testis time courses T2 and T3 in the dendrogram of Fig. 4, the differential comparisons between T2 and T3 (group 6 in Fig. 5B) show a high number of varying probe sets, confirming that these two experiments cannot be considered as similar.

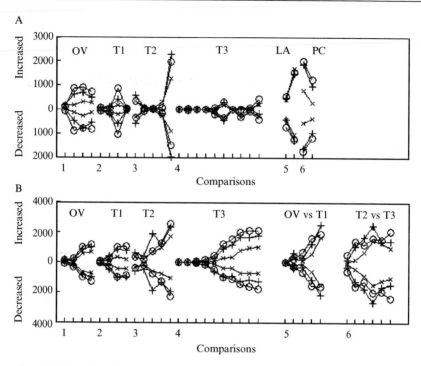

FIG. 5. Estimation of total variation (TV) in several comparisons. Comparisons are grouped, and the first comparison of each group is indicated by its group number on the abscissa. The identity of the group is indicated above it. Chip A, B, and C results are represented by circles, plus signs, and crosses, respectively. Three types of comparisons are displayed: (1) incremental comparisons [groups 1, 2, 3, and 4 (top)]; each comparison is between time $tn+1$ and time tn ($tn+1$ vs tn) or between biological conditions $n+1$ and n (the fourth and fifth comparisons of T2 use 14/18d and 26/29d); (2) cumulative comparisons [groups 5 and 6 (top) and groups 1, 2, 3, and 4 (bottom)]; each comparison is between time tn and $t1$ (tn vs $t1$) or between biological conditions n and 1 (the first comparison of LA is TU vs LA and the second is NL vs LA; the first comparison of PC is MY vs SC and the second is SE vs SC, the fourth and fifth comparisons of T2 use 14/18d and 26/29d); and (3) differential comparisons [groups 5 and 6 (bottom)]; each comparison is between two identical or similar biological conditions (the successive comparisons of T2 vs T3 are 01d vs 00d, 04d vs 03d, 08d vs 08d, 11d vs 10d, 14/18d vs 14d, and 26/29d vs 30d).

Similarity of Comparison Results

Another way to gain general insight into the relationships between different biological conditions is to count the number of statistically significant variations that any two comparisons have in common. To compare two conditions with two replicates, we applied a selection criterion of 20% FDR. Concerning comparisons between conditions without replicates,

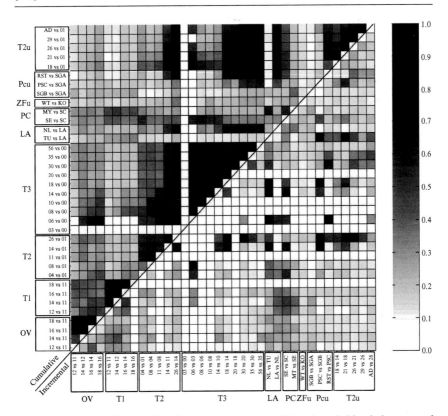

FIG. 6. Fraction of increased probe sets common among lists selected either in incremental comparisons or in cumulative comparisons in chip A. The fraction of probe sets common to lists 1 and 2 is the number of common probe sets normalized by the minimum number of probe sets in the two lists. Cumulative comparisons are represented above the diagonal, and the incremental analysis is shown below. The selection was carried out at 20% FDR except in experiments T2u, Pcu, and ZFu, where the 400 most increased probe sets were selected. The fraction of identity is displayed on a gray scale, as indicated on the right.

we proceed as follows. Because the RDAM method is unable, in this particular case, to reliably estimate TV, FDR, and S, instead of using two replicates to construct the empirical noise distribution, we used the two points under comparison. This means that true variations were treated as insignificant noise, explaining why the procedure was ineffective at detecting variations. In contrast, the standardization procedure, which transforms RD into zRD and allows p values to be assigned to each variation, preserves the ranking of the variations. In other words, the smallest p values are assigned to the most significant variations. Using this type of comparison, we retained the 400 probe sets showing the greatest increase. Figure 6 presents a systematic comparison of these lists concerning increased probe sets (II overlapping type) and obtained using either incremental or cumulative analysis. Obviously, overlapping of results is less important in incremental analysis, which is in part explained by the lower statistical power of this type of analysis.

Another conclusion is that existing overlaps generally agree with expectations: with cumulative analysis, for example, all the results from the comparisons in T2u, which concern the late stages of development (18d to adult), overlapped extensively with their corresponding comparisons in T3. For early stages, while the results overlapped correctly at 11d and 14d, the agreement was less obvious at the very beginning of development at 8d and 4d. These observations suggest that despite the aforementioned lack of reproducibility between T2 and T3, the RDAM method is able to establish lists with many genes in common. Another group, with a different method, has also established the presence of a common expression pattern between T2 and T3 (Wrobel and Primig, 2005). Similarly, with incremental analysis, we observed an expected overlap between pachytene spermatocytes and type B spermatogonia (PSC vs SGB) and T2 14/18d vs 11d, T2 26/29d vs 14/18d, T3 14d vs 10d, and T3 18d vs 14d. We note, however, that the comparison between round spermatids and pachytene spermatocytes showed more overlap with an a priori unrelated early stage (T3 08d vs 06d) than with an appropriate later stage (T3 35d vs 30d). A similar analysis, when conducted for the three other overlapping types (ID, DI, and DD, results not shown), detected several other overlaps, which may merit further investigation (e.g., overlaps between increased in T1 18d5 vs 16d5 and decreased in T1 16d5 vs 14d5, increased in OV 16d5 vs 14d5 and decreased in T3 30d vs 20d, increased in OV 18d5 vs 14d5 and decreased in T3 14d vs 10d, decreased in PCS vs SGB and increased in T2 26/29d vs 14/18d, decreased in Zfu WT vs KO and decreased in T3 14d vs 10d).

Combinatorial Clustering

Plotting TV against time or condition (Fig. 5) is not informative regarding the course of individual genes and does not allow genes with a similar variation profile to be grouped into disjoint classes. We have developed an approach, called combinatorial clustering, which is based on the systematic construction of chains of symbols and which allows the infinite number of variation profiles to be partitioned into a finite and determined number of equivalence classes. The basic symbols we manipulate are I, D, and N, which stand for increased, decreased, and not significant variation, respectively. In an experiment arranged into n ordered comparisons, 3^n such chains of symbols can be constructed, each representing a basic variation profile: for example, in the OV experiment analyzed by increment, the class NIIN would contain all the genes detected as unchanged in comparisons 12d5 vs 11d5 and 18d5 vs 16d5 and as increased in comparisons 14d5 vs 12d5 and 16d5 vs 14d5 (in a cumulative analysis, the corresponding genes would have been classified as NIII). This method is very easy to implement and allows rapid detection of the major variation trends in each experiment.

For each comparison, the assignment of a gene to one of the three symbols depends on the criterion used to select varying genes, and the population of each class will increase or decrease according to the stringency of the selection. For example, the class NINN in the OV experiment analyzed by increment contains 459, 275 and 482 probe sets for selection conducted at FDR 20%, S 50%, and S 99%, respectively, on chip A (Table III). Using the same selection conditions, the class NDDN would contain 42, 28, and 74 probe sets, respectively. The values of the selection parameters also influence the number of nonempty classes, and we observe 51, 56, and 78 nonempty classes (among the $3^4 = 81$ possible classes), respectively, with the same selection conditions. Despite this fluctuation, the classification can be considered to be robust because the relative weights of the most populated classes are stable (see the frequencies f in the last line of Table III) and because most of the variations occur at a single comparison even if very different selection conditions are applied (see section of Table III marked "One variation" at FDR 20%, S 50%, or S 99%). The results observed here for the OV experiment can be generalized to T1, T2, and T3 (results not shown).

Boolean Clustering

Each chain of symbols generated by combinatorial clustering is de facto a Boolean expression, for example, a series of intersections (&) or unions (|) between several subsets, and the NIIN variation profile discussed earlier corresponds to the simple expression N&I&I&N. More complex expressions can be used to answer particular biological questions. For example, based on prior knowledge, we could say that meiotic genes should be particularly expressed in OV (at 12d5 vs 11d5, T2 at 14/18d vs 10d, T3 14d vs 10d), and that variation profile III, if it exists in this ordered series of comparisons, should contain a high fraction of meiotic genes. Additional information could impose supplementary conditions, for example, a possible delay in expression (T3 18d vs 14d) or a stabilization of expression in the final stages of testis development (T2 26/29d vs 14/18d, T3 20d vs 18d, T3 30d vs 20d, T3 56d vs 30d), and we would thus be interested in the cluster of genes verifying that expression: ((IIINNNN)|(IININNN)). In the presence of a

TABLE III

MOST POPULATED CLASSES OF OV EXPERIMENT ANALYZED BY INCREMENT (CHIP A)

| Symbol | One variation[a] | | | | | | Symbol | Two successive variations[a] | | | | | | Symbol | Two disjoint variations[a] | | | | | |
| | FDR 20% | | S 50% | | S 99% | | | FDR 20% | | S 50% | | S 99% | | | FDR 20% | | S 50% | | S 99% | |
	R	#	R	#	R	#		R	#	R	#	R	#		R	#	R	#	R	#
nInn	2	459	6	275	5	482	nIDn	8	124	11	62	11	212	nInI	12	55	12	46	16	147
nnDn	3	401	5	302	4	545	nDIn	9	77	10	68	10	266	nInD	15	35	14	35	17	130
nDnn	4	342	3	336	2	702	nnDI	10	66	9	69	8	314	nDnD	17	25	13	41	14	168
nnIn	5	323	7	256	6	465	nnID	11	55	8	70	9	270	nDnI	19	14	19	18	12	178
nnnI	6	251	2	354	1	849	nIIn	13	42	17	28	21	74	InIn	35	2	40	2	57	5
nnnD	7	153	4	314	3	682	nDDn	14	39	16	29	19	93	DnDn	40	1	25	10	30	33
Innn	21	11	21	15	18	16	nnII	19	15	17	16	20	83	InnI	44	1	50	1	60	4
Dnnn	22	10	15	33	7	100	IDnn	24	9	20	11	28	37	InDn	45	1	51	1	54	6
							nnDD	27	4	26	8	25	50							
							DInn	30	3	24	9	24	51							
							DDnn	33	2	27	6	41	15							
							IInn	37	2	29	2	75	1							
#		1950		1885		3841	#		436		376		1465	#		134		154		671
f		0.75		0.72		0.56	f		0.17		0.14		0.21	f		0.05		0.06		0.11

[a] Columns marked R and # indicate the rank of the cluster and the number of probe sets it contains, respectively.

large number of results, the number of possible Boolean expressions is enormous, and the effectiveness and fertility of this approach depend on a judicious choice of comparisons and an ability to adequately translate prior knowledge or a testing hypothesis into a relevant Boolean expression.

One possible source of prior knowledge that is rarely used as such is Gene Ontology (GO) (Ashburner *et al.*, 2000). Indeed, GO classification is used very often simply to assess, a posteriori, the pertinence of a particular selection by searching for significantly overrepresented terms. In our search scheme, we started by observing the variation profile of a subset of genes that were selected based on a particular set of GO terms. Figure 7A presents variation profiles of 45 probe sets (26, 6, and 13 for chips A, B, and C, respectively) whose corresponding genes have a GO biological process explicitly related to meiosis. We observed a significant increase for these 45 probe sets at OV for 12d5 vs 11d5, T2 for 14/18d vs 10d, T3 for 14d vs 10d, LA for LA vs NL, OV vs T1 for 12d5 vs 12d5, and PC for SC vs SE. We recovered 38 probe sets (18, 11, and 9 in chips A, B, and C, respectively) in cluster IIIIII by interrogating these six comparisons at 99% S. Figure 7B shows that the variation profile of this selection is more homogeneous than that of the first selection based on GO terms. In order to increase the number of selected probe sets, it is possible to admit some degenerancy in the cluster [e.g., 95 supplementary probe sets are selected with one degenerated position (NIIIII, INIIII,...)].

Gene Clustering of Transcriptional Networks

While only intraexperimental comparisons have been discussed here, there is no fundamental reason why analysis must be restricted in this way. It is thus tempting to apply the combinatorial clustering method at the level of the overall set of experiments and to construct the variation profiles of the $33*32/2 = 528$ possible comparisons, even if variation profiles with 528 symbols are too complex to be interpreted directly. The information content of the resulting profiles is very high and can be used to construct a transcriptional network using the following method. If we align the variation profiles of two probe sets, we can observe, at each position corresponding to a particular comparison, three informative types of correlation between the probe sets: the two symbols are II or DD and the two probe sets are linked by a positive correlation, the symbols are ID or DI and the two probe sets are linked by a negative correlation, or the symbols are IN, NI, DN, or DD and the two probe sets are said to be linked by an unknown correlation (#Corr, #Anti, and #Quest are the respective number of positions in which these three correlation types are encountered in two aligned variation profiles). We then calculate for the given pair of probe sets a positive correlation score, CORR

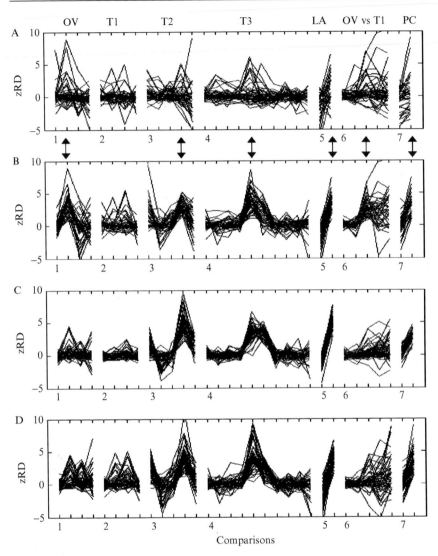

FIG. 7. Variation profiles of meiotic genes (zRD). Comparisons are grouped, and the first comparison of each group is indicated by its group number on the abscissa. The identity of each group is indicated at the top of the figure. (A) Forty-five probe sets selected using GO based on having the term "meiosis" in the definition of biological function. (B) Thirty-eight probe sets selected by Boolean clustering of the comparisons marked by a double-headed arrow (cluster IIIIII). (C) The top 50 probe sets from a list of 1400 probe sets from cluster 1 in Fig. 9. (D) The top 50 probe sets from a list of 160 probe sets from cluster 10 in Fig. 9.

= #Corr/(#Corr + #Anti + #Quest), and a negative correlation score, ANTI
= #Anti/(#Corr+#Anti+#Quest). This calculus is repeated for all possible
probe set pairs, insignificant scores are set to zero [scores below the surface
MinQuest = min(#Quest = f(#Corr,#Anti)), constructed from the #Corr,
#Anti, and #Quest tables calculated after randomization of the variation
profiles, are considered to be significant], and we obtain two tables of
36,899 × 36,899 CORR and ANTI scores. Figure 8 shows part of the resulting
transcriptional network obtained by applying a selection level of 1% FDR to
each of the 528 comparisons.

These transcriptional networks can be used to clusterize probe sets by
extracting a matrix of N × 36,899 CORR values, corresponding to the
correlation of N probe sets from a list of genes (either a list extracted with
GO terms or a list generated by Boolean clustering). This matrix can then be
transposed and treated by the CLICK clustering algorithm [available in the
EXPANDER suite, http://www.cs.tau.ac.il/~rshamir/expander/expander.
html (Sharan *et al.*, 2003)], which arranges the 36,899 probe sets into several

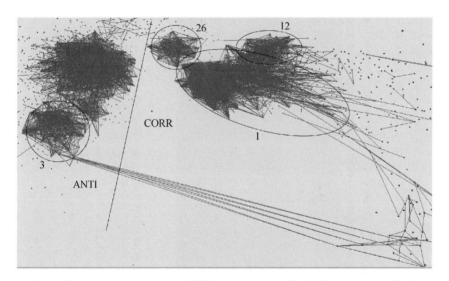

FIG. 8. Transcriptional network of 6135 probe sets (see Fig. 9). Probe sets are displayed
that had a positive or negative correlation score greater than 0% with at least one probe set of
the meiosis cluster IIIIII discussed in the text. Cluster numbers refer to clusters of Fig. 9.
Probe sets correlated positively (negatively) to the meiosis cluster IIIIII are plotted to the
right (left) of the oblique line. The geometry of the network is determined by a physical model
in which CORR and ANTI values are treated as attractive and repulsive forces, respectively.
Starting from a configuration in which all probe sets are distributed evenly on a sphere, the
free interplay of these two types of forces makes the system evolve toward a stable
configuration. An early step of this accretion process is displayed here.

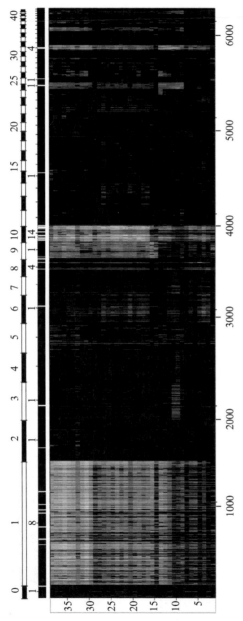

FIG. 9. Gene clustering on the transcriptional network with an interrogation list of 38 probe sets. Represented on the *X* axis are 6135 probe sets with a positive or negative correlation score greater than 0% with at least one probe set of the interrogation list plotted on the *Y* axis. Each pair of probe sets (*x,y*) has its CORR value displayed on a heat map (from black for CORR = 0% to white for CORR = 100%). The upper bar delimits the clusters found by CLICK (cluster zero contains the probe sets that could not be clusterized adequately).The second upper bar displays the position and the number of probe sets of the interrogation list found in each cluster. (See color insert.)

clusters. Finally, the N lines of the $N \times 36,899$ CORR matrix are clusterized using an annealing cluster (Alon *et al.*, 1999). Figure 9 shows the results of this procedure on the list of 38 probe sets from the meiosis cluster IIIIII discussed earlier

It can be seen that there are roughly 1400 probe sets that are highly correlated with this list of 38 probe sets (cluster nb1). Fifty of them, the most highly correlated, are displayed in Fig. 7C. There were also two populated clusters that were strongly correlated with a high number (cluster 9) or the quasi totality (cluster 10) of interrogation probe sets. We observed that clusters 24, 31, and 35 had relatively few probe sets (around 50 each), but were also highly correlated with the interrogation genes. Cluster 10 is represented in Fig. 7D. Comparing the different profiles displayed in Fig. 7 shows that while their overall trends are very similar, each profile has at least one particularity. For example, if we focus on comparisons T3 14d vs 10d and T3 18d vs 14d (points 5 and 6 in group 4), we can see that the Boolean cluster IIIIII (Fig. 7B) has a superposition of two peaks, that cluster 1 (Fig. 7C) has only one peak encompassing the two comparisons, and that cluster 10 (Fig. 7D) has only one peak corresponding to the first comparison. We conclude that CLICK clustering on CORR values is capable of differentiating subtle variations and that each cluster has its own specific characteristics.

Two other properties of this approach deserve attention. First, the method is very efficient in detecting the "collective" behavior of genes. Indeed, the network was constructed using a very stringent selection level (1% FDR), and most of the probe sets clusterized in clusters 1 and 10 would not have been detected as increased at this level of selection in many of the discriminative comparisons indicated on Fig. 7 by double-headed arrows. This means that comparisons other than those used in the Boolean clustering of Fig. 7 are also very discriminative for detecting meiotic genes and that our method of constructing transcriptional networks is able to use all the information contained in data. Second, we have seen that the low reproducibility score of chip C had a profound effect on the detection power of individual comparisons (Fig. 3). This effect did not exist at the level of the transcriptional networks, and we observed that roughly one-third of the probe sets in clusters 1, 9, 10, and 31 originated from each chip.

Finally, we looked at the distribution of the GO Biological Process annotations related to meiosis or gonad development for each of the four clusters discussed earlier. The first cluster was found to be linked to male gamete generation (GO:0048232, 17 genes), fertilization (GO:0007338, 1 gene), and sperm motility (GO:0030317, 3 genes). For cluster number 9, we found the following annotations: male gamete generation (GO:0048232, 1 gene) and germ cell development (GO:0007281, 1 gene). The third cluster was recorded as being involved in male gamete generation (GO:0048232, 4 genes), M phase of meiotic cell cycle (GO:0051327, 3 genes), female gonad development (GO:0008585, 2 genes), and female gamete generation (GO:0007292, 1 gene). Finally, cluster number 31 contained 1 gene marked as being involved in the M phase of the meiotic cell cycle (GO:0051327) and 1 gene related to male gamete generation (GO:0048232).

Not surprisingly, there were also differences in the distribution of GO terms that are unrelated to gonad development. For example, cluster 9 had many more terms related to negative regulation of cellular metabolism (GO:0031324), negative regulation of nucleobases (GO:0045934), and cell migration (GO:0016477) than cluster 1. Similarly, cluster 10 contained many genes that are related to DNA repair (GO:0006281) and response to radiation (GO:0009314), which were represented far less in cluster 1.

Conclusion

As the number of microarray data sets deposited in public repositories grows, the number of biological conditions studied in multiple independent experiments will increase as well. We have shown, using the particular example of gonad development, that the application of methods adapted to such composite data sets with a high information content is promising. Representing variation profiles by chains of symbols, an approach that had previously been proposed for the study of time series (Phang *et al.*, 2003), revealed itself to be very fruitful when applied to en masse analysis. First, this method is efficient for clustering genes rapidly and getting an overview of the main variation profiles across ordered comparisons. Second, it is easy to construct chains of symbols designed to clusterize genes that carry out a particular biological function. Finally, we have shown that a natural extension of this method is the construction of transcriptional networks, which can in turn be used as powerful tools for gene clustering.

Acknowledgments

We thank Bernard de Massy and Patrick Chaussepied for incentive discussions and N. Schultz, D. Garbers, J. Shima, and M. Griswold for providing access to unpublished data (GSE 2736).

References

Alon, U., Barkai, N., Notterman, D. A., Gish, K., Ybarra, S., Mack, D., and Levine, A. J. (1999). Broad patterns of gene expression revealed by clustering analysis of tumor and normal colon tissues probed by oligonucleotide arrays. *Proc. Natl. Acad. Sci. USA* **96,** 6745–6750.

Ashburner, M., Ball, C. A., Blake, J. A., Botstein, D., Butler, H., Cherry, J. M., Davis, A. P., Dolinski, K., Dwight, S. S., Eppig, J. T., Harris, M. A., Hill, D. P., Issel-Tarver, L., Kasarskis, A., Lewis, S., Matese, J. C., Richardson, J. E., and Ring, J. E. (2000). Gene ontology: Tool for the unification of biology. *Nature Genet.* **25,** 25–29.

Breitling, R., Armengaud, P., Amtmann, A., and Herzyk, P. (2004). Rank products: A simple, yet powerful, new method to detect differentially regulated genes in replicated microarray experiments. *FEBS Lett.* **573**, 83–92.

Costoya, J. A., Hobbs, R. M., Barna, M., Cattoretti, G., Manova, K., Sukhwani, M., Orwig, K. E., Wolgemuth, D. J., and Pandolfi, P. P. (2004). Essential role of Plzf in maintenance of spermatogonial stem cells. *Nature Genet.* **36**, 653–659.

Datta, S., and Datta, S. (2003). Comparisons and validation of statistical clustering techniques for microarray gene expression data. *Bioinformatics* **19**, 459–466.

Fred, A. L., and Jain, A. K. (2005). Combining multiple clusterings using evidence accumulation. *IEEE Trans. Pattern Anal. Mach. Intell.* **27**, 835–850.

Gao, F., Foat, B. C., and Bussemaker, H. J. (2004). Defining transcriptional networks through integrative modeling of mRNA expression and transcription factor binding data. *BMC Bioinformat.* **5**, 31.

Golub, T. R., Slonim, D. K., Tamayo, P., Huard, C., Gaasenbeek, M., Mesirov, J. P., Coller, H., Loh, M. L., Downing, J. R., Caligiuri, M. A., Bloomfield, C. D., and Lander, E. S. (1999). Molecular classification of cancer: Class discovery and class prediction by gene expression monitoring. *Science* **286**, 531–537.

Hamra, F. K., Schultz, N., Chapman, K. M., Grellhesl, D. M., Cronkhite, J. T., Hammer, R. E., and Garbers, D. L. (2004). Defining the spermatogonial stem cell. *Dev. Biol.* **269**, 393–410.

Hubbell, E., Liu, W. M., and Mei, R. (2002). Robust estimators for expression analysis. *Bioinformatics* **18**, 1585–1592.

Martin, D. E., Demougin, P., Hall, M. N., and Bellis, M. (2004). Rank Difference Analysis of Microarrays (RDAM), a novel approach to statistical analysis of microarray expression profiling data. *BMC Bioinformat.* **5**, 148.

Negre, N., Lavrov, S., Hennetin, J., Bellis, M., and Cavalli, G. (2006). Mapping the distribution of chromatin proteins by ChIP on chip. *Methods Enzymol.* **410**, 316–341.

Neuhauser, M., and Senske, R. (2004). The Baumgartner-Weibeta-Schindler test for the detection of differentially expressed genes in replicated microarray experiments. *Bioinformatics* **20**, 3553–3564.

Pease, A. C., Solas, D., Sullivan, E. J., Cronin, M. T., Holmes, C. P., and Fodor, S. P. (1994). Light-generated oligonucleotide arrays for rapid DNA sequence analysis. *Proc. Natl. Acad. Sci. USA* **91**, 5022–5026.

Phang, T. L., Neville, M. C., Rudolph, M., and Hunter, L. (2003). Trajectory clustering: A non-parametric method for grouping gene expression time courses, with applications to mammary development. *Pac. Symp. Biocomput.* **35**, 1–362.

Schena, M., Shalon, D., Davis, R. W., and Brown, P. O. (1995). Quantitative monitoring of gene expression patterns with a complementary DNA microarray. *Science* **270**, 467–470.

Schultz, N., Hamra, F. K., and Garbers, D. L. (2003). A multitude of genes expressed solely in meiotic or postmeiotic spermatogenic cells offers a myriad of contraceptive targets. *Proc. Natl. Acad. Sci. USA* **100**, 12201–12206.

Sharan, R., Maron-Katz, A., and Shamir, R. (2003). CLICK and EXPANDER: A system for clustering and visualizing gene expression data. *Bioinformatics* **19**, 1787–1799.

Shima, J. E., McLean, D. J., McCarrey, J. R., and Griswold, M. D. (2004). The murine testicular transcriptome: Characterizing gene expression in the testis during the progression of spermatogenesis. *Biol. Reprod.* **71**, 319–330.

Small, C. L., Shima, J. E., Uzumcu, M., Skinner, M. K., and Griswold, M. D. (2005). Profiling gene expression during the differentiation and development of the murine embryonic gonad. *Biol. Reprod.* **72**, 492–501.

Wrobel, G., and Primig, M. (2005). Mammalian male germ cells are fertile ground for expression profiling of sexual reproduction. *Reproduction* **129**, 1–7.

[22] Visualizing Networks

By GEORGE W. BELL and FRAN LEWITTER

Abstract

An interrelated set of genes or proteins can be represented effectively as a network that describes physical interactions, regulatory relationships, or metabolic pathways. Visualizing a network can be a helpful method to extract biological meaning and to generate testable hypotheses about large-scale biological data. This chapter describes some potential rationales for visualizing networks of microarray and other data types, which can be integrated and filtered to show potentially significant relationships. It also presents a practical introduction to Osprey and Cytoscape, two software platforms that are powerful tools for visualizing, integrating, and manipulating networks.

Introduction

The study of genomics requires large sets of complex data describing the behavior of genes and gene products functioning not in isolation, but as interconnected complexes. Making and describing biologically significant conclusions about these data can be aided greatly by visualizing this multi-dimensional interrelated information. Computing with large-scale expression and other microarray data can be enhanced by the use of network structures, which can describe relationships between entities in ways not possible with the typical representation by a matrix of genes by conditions or cell types. Extracting biological meaning from a microarray experiment can be enhanced by network visualization of array data alone, but the real power of network representations comes from the ability to add other perspectives from additional data sources on the interactions between proteins and other molecules.

Networks (or, more precisely, "graphs" in a mathematical sense) have been an essential mathematical representation of data even before the arrival of computers, and biologists have often used them to show metabolic relationships. With high-throughput biological data sets, networks are becoming more common as ways to represent relationships among genes, RNA, proteins, and/or other biomolecules.

We can think of a network as simply a set of points ("nodes" such as genes, RNAs, or proteins), with pairs joined with lines ("edges") describing some type of interaction. From a geometrical perspective, the two main

METHODS IN ENZYMOLOGY, VOL. 411
0076-6879/06 $35.00
DOI: 10.1016/S0076-6879(06)11022-8

types of networks are undirected (such as a pair of interacting proteins) or directed (such as a transcription factor binding a gene or the flow of metabolic energy). Both types of networks can be visualized just as easily and both can contain loops ("cycles"), but interpretation is generally dependent on the biological significance of the edge. Because most proteins function together in complexes, clues to the role of a novel protein may be predicted from the functions of proteins with which it interacts, a case of "guilt by association" (Oliver, 2000). High-throughput methods can identify interactions that occur *in vivo* at only specific developmental times, tissue types, or cytoplasmic compartments so a corresponding network may include information that is physiologically relevant at only a specific time and place. Topologically, most biological networks have a power law distribution in which most proteins have only a few interactions and only a few interact with many others (Bray, 2003). Not surprisingly, the more connected proteins tend to be those that are the most indispensable (Estrada, 2006).

It is common for a microarray analysis to begin with thousands of or "all" genes and, after filtering, to generate a subset (using differential expression or selection by function or other properties) containing a much smaller number of genes. Depending on the goal of the experiments, interpretation of array data itself may be aided by a network analysis of either all genes or these final gene sets. The practicalities of network visualization and analysis are influenced greatly by whether we wish to start with a population of data or a specific sample. Because visualizing a genome's worth of data can be computationally challenging and visually difficult to interpret, one theme of the tutorials will be ways to reduce the size of the network to a meaningful amount.

The following are some applications of networks in the context of microarray analysis, with the potential for integration of other types of networks.

• Using a set of expression profiles for some or all genes, a distance matrix can be calculated (as a prelude to clustering) to assay for correlated genes. Then instead of using the genes by genes distance matrix to generate a dendrogram, the distances between correlated genes can be visualized as a network showing coexpressed genes in close proximity (Zhang *et al.*, 2005). Expression networks from different time points or tissue types can also be compared for specificity.

• Using any network of expression data for one's species of study, layering on a similar network of orthologous genes or proteins can indicate phylogenetically conserved interactions. One can use multiple species interaction data to predict functional orthologs (Bandyopadhyay *et al.*, 2006) or, conversely, one can use one species' network, together with sequence-based orthology data, to predict interactions in another species (Kelley *et al.*, 2004).

• The representation of protein–protein interactions is the most obvious genome-scale application of networks. Combined with expression ratios or profiles, such a network can extract groups of coexpressed genes that interact physically (Ge *et al.*, 2003; Ideker *et al.*, 2002; LaCount *et al.*, 2005; Rual *et al.*, 2005).

• Starting with a network of protein–DNA interactions, such as those derived from large-scale chromatin immunoprecipitation (ChIP chips), adding another network of expression ratios or profiles can indicate, among other interpretations, how binding by a transcription factor influences transcript abundance (Boyer *et al.*, 2005).

• Using a set of known metabolic pathways, highlighting one's gene set(s) can help elucidate which pathways may describe any metabolic themes of the gene set(s).

• Starting with one of the three Gene Ontology (GO) networks, overlaying enrichment data (such as from the cumulative hypergeometric distribution) for a gene set can help determine any GO themes of the gene set (Maere *et al.*, 2005).

• Combining a network of expression ratios or profiles with a directed network of microRNAs and their gene targets (Lewis *et al.*, 2005) can help determine the relationship (if any) between miRNA activity and gene expression level (Farh *et al.*, 2005).

For most biologists, a key requirement for the use of networks is an effective environment for their visualization. This section presents some software tools for network visualization that can effectively display and manipulate sets of networks.

Osprey

One choice of software for visualizing networks is Osprey. It is a Java-based software platform and runs on desktop operating systems and Linux. It is freely available to academic scientists after registering at the site (http://biodata.mshri.on.ca/osprey). For-profit users can purchase a license for the software.

There is a very thorough user's manual that describes the many features and ways to customize the displays you create. The software comes with published interaction data or you can enter your own data easily. Several formats are available and are well documented. Currently there is extensive information for yeast, worm, and fly available with the software. Data are created for the General Repository for Interaction Datasets (GRID). The BioGRID (Stark *et al.*, 2006) is the next-generation version of the interaction database and contains information on human interactions as well as the model organisms available through the GRID. The information is culled from the literature and BioGRID contains more than 116,000 interactions. Of potential interest is the BioGRID web site (http://www. thebiogrid.org), which allows the visualization of networks for one gene at a time (Fig. 1).

An Example. In this example, we start with a list of 37 genes identified in a microarray experiment of yeast as being repressed in response to neutrophils (Rubin-Bejerano *et al.*, 2003). After opening Osprey, choose Yeast as the default database. Then from the Insert menu or the tool bar,

FIG. 1. The BioGRID web site (http://www.thebiogrid.org) contains downloadable interaction data. In addition, one can visualize interaction networks for one gene at a time by entering the appropriate identifier. (See color insert.)

select "Add Nodes." You can copy and paste your list of genes into the text box. In our example, one gene does not get added because its name, HAP2, is ambiguous in the Saccharomyces Genome Database (Hirschman *et al.*, 2006).

The first display you will see after adding this gene list is color coded according to GO annotation (Ashburner *et al.*, 2000), and the genes/edges are distributed randomly in the window. Note that to see a list of GO categories or to change the color of the categories, select Colour Indexes and then "GO Process" from the View menu. To get a better view of the genes, select all genes (Edit menu or typing control-z). Then select "One Circle" from the Circular menu item under Layout. This organizes the display by GO annotation. To get information on a particular gene, click on the gene of interest and view the information in the upper left window. Another way to get information about the genes is to get a Node Report. First select the nodes/genes of interest and then choose "Selected Node Report" from the View menu.

Reducing the Data Set. For the rest of the example, we will be working with the seven genes classified as Cell Organization and Biogenesis according to the GO annotations. If necessary, move a gene that you are not interested in by dragging it to another position in the window. Then, select the genes of interest by dragging and holding the mouse around them. Now select "Invert Selected Nodes" from the Edit menu. Next select "Remove Nodes" from the Edit menu. To make the display more readable, again select all and then choose One Circle from the Circular menu item under Layout. Reselect the remaining genes and then, either by right clicking on the mouse or from the Insert menu, select "All Interactions for Selected Nodes."

The edges (lines connecting genes) are color coded to identify the source of interaction data. If an interaction was observed in more than one experiment, it will be multicolored. If you select an edge, information about the interaction will be displayed in the upper left window, including a link to the PubMed references reporting the interactions. To select interactions identified by a particular method (e.g., two-hybrid, synthetic lethality), you can filter the networks by Experimental System(s), selecting more than one method if you prefer. To do this, use the Network Filters window in the lower left window.

You can continue building networks by selecting any node and then clicking on the Get New Interactions button in the Gene Info window in the upper left-hand window. Be careful not to add too many interactions as it will become difficult to view a large network. If you do this and want to locate a particular gene of interest you can select Find from the Edit menu.

Once you have displayed the networks to your liking, you will want to export them in a format that can be imported into presentations and publications (Fig. 2).Your choices include standard graphics formats, such as PNG, SVG, and JPEG.

Summary of Osprey. We have shown how to get started in Osprey using interaction data supplied by the GRID. It is worth reading through the well-written online program documentation to understand the many features and ways to customize the layout of your network. For those scientists interested in exploring the interactions of one or a few genes, the BioGRID web interface is a good alternative.

Osprey is under continuing development (personal communication) and many enhancements are planned. To make sure you have the latest update, click on "Check for Updates" under the Help menu.

Cytoscape

Another choice of software for visualizing networks is Cytoscape [cytoscape.org; (Shannon *et al.*, 2003)], a freely available open source Java-based software platform that works on all major operating systems.

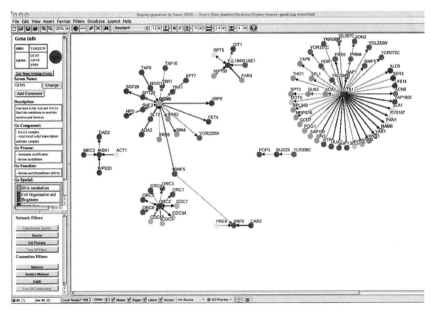

FIG. 2. A display of an interaction network in Osprey for seven genes of interest. The nodes (circles) are color coded by GO category. The edges (lines) are color coded by the publication source of interaction data. The window on the upper left gives annotation about a highlighted node. Filtering of networks is available from the window on the lower left. (See color insert.)

It can be used to visualize networks of any type, as long as data are formatted in Simple Interaction Format (SIF; three columns indicating interacting molecules and the type of interaction) or GML format (which indicates layout, as well as nodes and edges). Start Cytoscape by running "cytoscape.sh" or "cytoscape.bat."

Viewing and Filtering a Network. To load a network, go to the File menu, select Load, and then Network. We will start with some recent protein–protein interaction data. As an example, we can use some human interactome data (Rual *et al.*, 2005) showing interactions from yeast two-hybrid, coaffinity purification, and literature-based association data. After downloading (http://www.cytoscape.org/cgi-bin/moin.cgi/Data_Sets) and then loading data, note that the network has been created but not visualized. Visualizing a large network of thousands of nodes is of questionable help unless one adds a z axis (Stuart *et al.*, 2003) or other measure of node density that can help organize the network into highly connected subnetworks.

To annotate our nodes, we can start by linking the NCBI Entrez Gene IDs (in the network file) to gene symbols. The Cytoscape data download site includes these data in a file format of lines such as "4089 = SMAD4." To import these data, go to the File menu, select Load, and then "Node Attributes."

To concentrate on a subregion of the network, we may wish to select node subsets such as the following:

- A favorite protein and the neighbors with which it interacts
- A set of proteins that interact according to specified types of experimental evidence
- A set of genes, such as those with a similar response or profiles from an expression array.

To select a node by ID, highlight the network (Rual.sif). Then go to the Select menu, select Nodes, and "By Name." Entering 4089 should select SMAD4. If we would like to look at the proteins with which SMAD4 interacts, go to the Select menu, Nodes, and "First neighbors of selected nodes" (Fig. 3). These interactions can then be used to create another (much smaller) network by going to Select, To New Network, and "Selected nodes, all edges." CytoPanel 1 will show this new network and its size. Right clicking and selecting "Create View" will finally create a visualization. To create a layout that is more helpful, go to the Layout menu, Apply Spring Embedded Network, and "All Nodes" (Fig. 4). To select a set of nodes by ID, we can create a text file with a list of desired Entrez Gene IDs and select those nodes by going to the Select menu, followed by Nodes and "From File."

One powerful side of Cytoscape is the flexibility of its display. Going to the Visualization menu, selecting Set Visual Style, and clicking on "Duplicate"

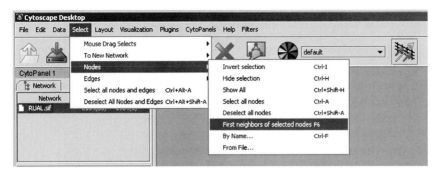

FIG. 3. Selection of first neighbors of selected nodes on Cytoscape. The network "RUAL. sif" has been highlighted and at least one node has been selected previously.

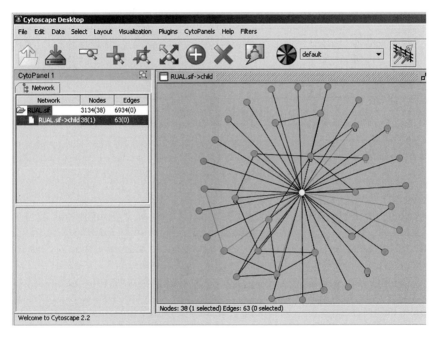

FIG. 4. Example of a Spring Embedded Network view in Cytoscape. Note that this is a child network (subnetwork) of the "RUAL.sif" network containing 38 nodes (with one selected) and 63 edges.

will copy the current style as a starting point for further customization. Then clicking on Define opens a Set Visual Style window (Fig. 5), with options for node and edge attributes, including edge coloring by type of interaction. Another global setting is the selection of algorithm for the network layout.

FIG. 5. Set Visual Style window in Cytoscape. After clicking on the Node Attributes button, the Node Label option was selected. Under Mapping, we have chosen to label each node with its official name. We can also specify a variety of other node parameters.

In addition to the Spring Embedded Network layout, Cytoscape has a series of yFiles layouts (derived from Java implementations), such as Hierarchic and Circular. Certain layouts may be much more effective than others for representing biological relationships of a certain type.

Combining Interaction and Expression Data. After a brief introduction to the loading, filtering, and visualization of a network with Cytoscape, we may be interested in overlaying expression information to identify any patterns in these two complementary high-throughput data types. To import expression information, a key requirement is the use of the same nomenclature system for interactions and expression. This is an important practicality for integrating data from multiple sources; all interaction data must use the same system for naming nodes, whether it be gene symbol, name, or an identifier from a reference database (such as NCBI Gene IDs or yeast systematic ORFs). In the simplest case, Cytoscape can import a space-delimited matrix of expression ratios or values, with each column for an experimental condition and each row a gene. Cytoscape also has a plug-in that can import files from the NCBI Gene Expression Omnibus repository using their "SOFT" file format, as long as the network uses the same set of identifiers.

To load an expression matrix, go to the Load menu and select "Expression Matrix File" (Fig. 6). Then once expression data are linked to the nodes, we have several choices about how to use these data. Using a numeric filter, nodes above or below a threshold expression value/ratio can be selected. Otherwise, nodes can be colored one of several colors in a spectrum depending on user-defined cutoffs. To set up this coloring by expression value, go to the Visualization menu and select Visual Styles. Click on Define, then Node Attributes, and select Node Color. Under Mapping, select RedGreen, and an expression condition can be selected via Map Attribute from the list of headers in one's expression file. Then a range of colors can be chosen, such as bright green to bright red, to indicate ranges of expression values. Finally, clicking on "Apply to Network" will color the nodes of the network (Fig. 7).

Visualizing Gene Ontology Annotations. Gene Ontology (Ashburner *et al.*, 2000) annotations can be applied to expression data in several different ways, including systems of network-wide (or array-wide) analysis and filtering, or overrepresentation analysis of a predefined gene set. Analyzing GO annotations first requires loading of both GO ontologies (the list of GO terms and the relationships between them, described in an OBO file) and a species-specific association (annotation) file, linking node identifiers to GO terms. Cytoscape comes with the ontology file and several association files, but one may need to modify or create an association file to obtain annotation information for protein identifiers matching those of the network.

Before loading any GO information, go to the Edit menu and select Preferences. Under Properties, modify the field defaultSpeciesName, if

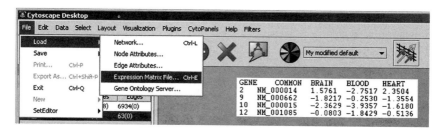

FIG. 6. Loading an Expression Matrix File in Cytoscape. The white matrix shows a few lines of this file, with one or two columns of node identifiers, followed by one or more columns of expression ratios (commonly log transformed). The first column (Entrez Gene IDs) must use the same identifiers as the corresponding nodes of the network (RUAL.sif).

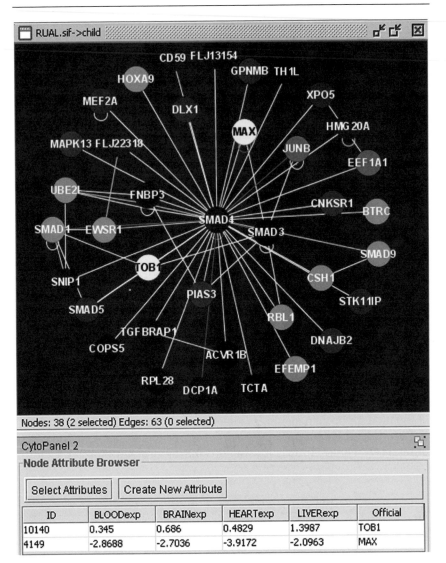

FIG. 7. An interaction network with nodes colored to indicate expression ratios in a selected tissue. The two selected nodes are listed in CytoPanel 2 (below), with corresponding expression ratios and other information shown (for those fields chosen by going to the Select Attributes button). (See color insert.)

necessary, to reflect the species of the network, as this species and that of the GO association file must match. Next, go to the File menu and select Load and then Gene Ontology Server. After selecting the Gene Ontology file format, you can set the location of the two GO files and then load GO data.

To begin selecting or coloring nodes by GO annotation, click on the "A" (annotation) icon. The Annotation window that pops up should show the three GO ontologies in the Available box. Expanding an ontology by clicking on the "+" will show a series of levels, from most general (1) to most specific. After selecting a level, click on Apply Annotation to All Nodes, and the level of that ontology will appear in the Current Annotations box. After expanding the level in Current Annotations, selecting a GO term will also select the corresponding nodes.

Enrichment analysis is another common way to use GO data to help extract biological themes from a gene set derived from microarray data. The BiNGO plug-in (Maere *et al.*, 2005) is an effective way to perform both under- and overrepresentation analysis in the context of Cytoscape. After having installed BiNGO from the Cytoscape web site, it can be selected from the Plugins menu. The options to select are mostly similar to other GO enrichment tools, but the ontology of enriched GO terms can be visualized in Cytoscape, with the degree of enrichment indicated by node colors. After generating the initial figure, Visual Styles of nodes and edges can be configured further as with any Cytoscape network.

Summary of Cytoscape. We have shown just a few ways to visualize networks in combination with or derived from expression analysis. The Cytoscape web site links to a series of tutorials with detailed step-by-step instructions. Ideally an integrated approach combining expression and other data will encourage the generation of better, more readily testable, hypotheses. Extending the core functionality of Cytoscape, developers have created more than 20 plug-ins to perform specific tasks or to fetch data from public sources so potential applications to microarray analysis go well beyond some common tasks described earlier. When an analysis is complete, images of networks created in Cytoscape can be exported in a wide variety of graphic formats, including publication-quality vector formats such as PostScript, SVG, and PDF. The Cytoscape development team continues to improve and expand the capabilities of Cytoscape, so it is expected that this resource will remain an effective tool for visualizing biological networks.

Summary

Analysis of biological networks is an effective way of describing themes of a highly interrelated ensemble of genes, proteins, or other molecules. Extracting biological information from a network can be aided greatly by intuitive network visualization tools. Software applications such as Osprey and Cytoscape are powerful tools for the experimental biologist or the bioinformatics scientist to display, filter, and explore genome-scale coexpression data, either alone or with other sources of biological data represented as networks. Both allow simple data input and manipulation, flexibility in the representation of nodes and edges according to their properties, and series of algorithms to create different graphical layouts. Together with other tools for expression analysis, these network visualization tools can help generate hypotheses describing functional relationships between genes or proteins present in very complex sets of data.

Acknowledgments

We thank H. Ge, J. Rodriguez, K. Walker, and V. Vyas for critical reading and discussion of the manuscript.

References

Ashburner, M., Ball, C. A., Blake, J. A., Botstein, D., Butler, H., Cherry, J. M., Davis, A. P., Dolinski, K., Dwight, S. S., Eppig, J. T., Harris, M. A., Hill, D. P., Issel-Tarver, L., Kasarskis, A., Lewis, S., Matese, J. C., Richardson, J. E., Ringwald, M., Rubin, G. M., and Sherlock, G. (2000). Gene ontology: Tool for the unification of biology. *Nature Genet.* **25,** 25–29.

Bandyopadhyay, S., Sharan, R., and Ideker, T. (2006). Systematic identification of functional orthologs based on protein network comparison. *Genome Res.* **16,** 428–435.

Boyer, L. A., Lee, T. I., Cole, M. F., Johnstone, S. E., Levine, S. S., Zucker, J. P., Guenther, M. G., Kumar, R. M., Murray, H. L., Jenner, R. G., Gifford, D. K., Melton, D. A., Jaenisch, R., and Young, R. A. (2005). Core transcriptional regulatory circuitry in human embryonic stem cells. *Cell* **122,** 947–956.

Bray, D. (2003). Molecular networks: the top-down view. *Science* **301,** 1864–1865.

Estrada, E. (2006). Virtual identification of essential proteins within the protein interaction network of yeast. *Proteomics* **6,** 35–40.

Farh, K. K., Grimson, A., Jan, C., Lewis, B. P., Johnston, W. K., Lim, L. P., Burge, C. B., and Bartel, D. P. (2005). The widespread impact of mammalian microRNAs on mRNA repression and evolution. *Science* **310,** 1817–1821.

Ge, H., Walhout, A. J., and Vidal, M. (2003). Integrating 'omic' information: A bridge between genomics and systems biology. *Trends Genet.* **19,** 551–560.

Hirschman, J. E., Balakrishnan, R., Christie, K. R., Costanzo, M. C., Dwight, S. S., Engel, S. R., Fisk, D. G., Hong, E. L., Livstone, M. S., Nash, R., Park, J., Oughtred, R., Skrzypek, M., Starr, B., Theesfeld, C. L., Williams, J., Andrada, R., Binkley, G., Dong, Q., Lane, C., Miyasato, S., Sethuraman, A., Schroeder, M., Thanawala, M. K., Weng, S., Dolinski, K., Botstein, D., and Cherry, J. M. (2006). Genome Snapshot: A new resource at the Saccharomyces Genome Database (SGD) presenting an overview of the *Saccharomyces cerevisiae* genome. *Nucleic Acids Res.* **34,** D442–D445.

Ideker, T., Ozier, O., Schwikowski, B., and Siegel, A. F. (2002). Discovering regulatory and signalling circuits in molecular interaction networks. *Bioinformatics* **18**(Suppl. 1), S233–S240.

Kelley, B. P., Yuan, B., Lewitter, F., Sharan, R., Stockwell, B. R., and Ideker, T. (2004). PathBLAST: A tool for alignment of protein interaction networks. *Nucleic Acids Res.* **32**, W83–W88.

LaCount, D. J., Vignali, M., Chettier, R., Phansalkar, A., Bell, R., Hesselberth, J. R., Schoenfeld, L. W., Ota, I., Sahasrabudhe, S., Kurschner, C., Fields, S., and Hughes, R. E. (2005). A protein interaction network of the malaria parasite *Plasmodium falciparum*. *Nature* **438**, 103–107.

Lewis, B. P., Burge, C. B., and Bartel, D. P. (2005). Conserved seed pairing, often flanked by adenosines, indicates that thousands of human genes are microRNA targets. *Cell* **120**, 15–20.

Maere, S., Heymans, K., and Kuiper, M. (2005). BiNGO: A Cytoscape plugin to assess overrepresentation of gene ontology categories in biological networks. *Bioinformatics* **21**, 3448–3449.

Oliver, S. (2000). Guilt-by-association goes global. *Nature* **403**, 601–603.

Rual, J. F., Venkatesan, K., Hao, T., Hirozane-Kishikawa, T., Dricot, A., Li, N., Berriz, G. F., Gibbons, F. D., Dreze, M., Ayivi-Guedehoussou, N., Klitgord, N., Simon, C., Boxem, M., Milstein, S., Rosenberg, J., Goldberg, D. S., Zhang, L. V., Wong, S. L., Franklin, G., Li, S., Albala, J. S., Lim, J., Fraughton, C., Llamosas, E., Cevik, S., Bex, C., Lamesch, P., Sikorski, R. S., Vandenhaute, J., Zoghbi, H. Y., Smolyar, A., Bosak, S., Sequerra, R., Doucette-Stamm, L., Cusick, M. E., Hill, D. E., Roth, F. P., and Vidal, M. (2005). Towards a proteome-scale map of the human protein-protein interaction network. *Nature* **437**, 1173–1178.

Rubin-Bejerano, I., Fraser, I., Grisafi, P., and Fink, G. R. (2003). Phagocytosis by neutrophils induces an amino acid deprivation response in *Saccharomyces cerevisiae* and *Candida albicans*. *Proc. Natl. Acad. Sci. USA* **100**, 11007–11012.

Shannon, P., Markiel, A., Ozier, O., Baliga, N. S., Wang, J. T., Ramage, D., Amin, N., Schwikowski, B., and Ideker, T. (2003). Cytoscape: A software environment for integrated models of biomolecular interaction networks. *Genome Res.* **13**, 2498–2504.

Stark, C., Breitkreutz, B. J., Reguly, T., Boucher, L., Breitkreutz, A., and Tyers, M. (2006). BioGRID: A general repository for interaction datasets. *Nucleic Acids Res.* **34**, D535–D539.

Stuart, J. M., Segal, E., Koller, D., and Kim, S. K. (2003). A gene-coexpression network for global discovery of conserved genetic modules. *Science* **302**, 249–255.

Zhang, L. V., King, O. D., Wong, S. L., Goldberg, D. S., Tong, A. H., Lesage, G., Andrews, G., Bussey, H., Boone, C., and Roth, F. P. (2005). Motifs, themes and thematic maps of an integrated *Saccharomyces cerevisiae* interaction network. *J. Biol.* **4**, 6.

[23] Random Forests for Microarrays

By ADELE CUTLER and JOHN R. STEVENS

Abstract

Random Forests is a powerful multipurpose tool for predicting and understanding data. If gene expression data come from known groups or classes (e.g., tumor patients and controls), Random Forests can rank the genes in terms of their usefulness in separating the groups. When the groups are unknown, Random Forests uses an intrinsic measure of the similarity of the genes to extract useful multivariate structure, including clusters. This chapter summarizes the Random Forests methodology and illustrates its use on freely available data sets.

Introduction

Microarrays present new challenges for statistical methods because of the large numbers of genes and relatively small numbers of microarrays. Random Forests (Breiman, 2001; Breiman and Cutler, 2005) provide a general-purpose tool for predicting and understanding data. They are becoming popular for analyzing microarray data (see, e.g., Díaz-Uriarte and Alvarez de Andrés, 2006) in part because they can handle large numbers of genes without formal variable selection, they are robust to outliers, do not require data to follow the normal (or any other) distribution, can be used for badly unbalanced data sets, and can impute missing values intelligently. This chapter refers to gene expression microarrays, although the ideas transfer directly to tissue arrays or even mass spectrometry data. We assume that all gene expression data have been normalized appropriately using a preprocessing method such as RMA (Irizarry et al., 2003). Because the Random Forest methods discussed are invariant under monotone transformations, data do not need to be log transformed, although it may be advisable for numerical reasons.

A Random Forest is a collection of classification trees generated by bootstrap sampling from data and randomly sampling predictor variables at each node. This chapter describes trees and forests in more detail and intuitively shows how the trees in a Random Forest combine to give more accurate results.

Random Forests do not perform formal statistical inference and do not do significance tests or give p values. They are not intended for small,

METHODS IN ENZYMOLOGY, VOL. 411
Copyright 2006, Elsevier Inc. All rights reserved.   0076-6879/06 $35.00
DOI: 10.1016/S0076-6879(06)11023-X

carefully controlled experiments. However, results of a Random Forests analysis might *suggest* interesting experiments and might give insight that could be missed by formal procedures.

Two quite general applications are considered. In the first situation, we have two or more labeled groups of microarrays (e.g., tumor versus control) and want to classify new microarrays or determine which genes would be useful in classifying new microarrays. We refer to this first situation as *classification* and note that it is a form of *supervised learning* (see Gollub and Sherlock, 2006). The second situation in which Random Forests may be useful is when we have unlabeled microarrays and want to find clusters or other interesting multivariate structure. This situation is referred to as *unsupervised learning*. Classical statistical clustering methods (Gollub and Sherlock, 2006) are popular in this situation and can be used in conjunction with a Random Forests analysis.

This chapter focuses on classification, although the unsupervised learning approach is also discussed.

Classification

Classification deals with data comprising a number of observations, each of which is known to come from one of a number of distinct groups or classes. For each observation, we have a number of predictors. Our goal is to use the predictors to classify unlabeled observations and to learn which predictors are important or useful in the classification.

In the microarray context, the observations usually represent the microarrays themselves (or the observational units, such as patients, from which they are obtained) and the predictors represent the genes. For example, we may have microarrays for cancer patients and controls and we may want to classify a new person into one of these two groups based on their microarray results. Perhaps more importantly, we may also want to determine which genes on the microarray are useful in classifying the new person, with the idea of developing a more efficient diagnostic tool or giving useful information about the genetic basis of the disease itself.

One common approach to data like these is to treat the genes individually and perform something similar to a *t* test to decide which genes are "significantly different" between the two groups, presumably with an adjustment for the number of comparisons. Methods such as significance analysis of microarrays (Tusher *et al.*, 2001) have this flavor. If there are more than two groups, an ANOVA approach might be used (Ayroles and Gibson, 2006). We refer to these procedures as "significance testing."

Classification differs from the significance testing approach in several important ways. Perhaps the most fundamental difference is that t tests and ANOVAs test whether the population means for the groups are different. For overlapping groups, we may conclude that the population means differ, but the groups may not be distinct enough to make accurate predictions of group membership. A second difference is that classification is inherently multivariate, so instead of asking whether each gene is *individually* good at separating the groups, we are asking whether the gene expression information from *all* the genes is useful. The collective expression levels of groups of genes may capture higher order terms such as interactions between genes, which may allow us to separate the groups better than any single gene.

Traditional statistical methods for classification include linear discriminant analysis and logistic regression (see, e.g., Hastie *et al.*, 2001). For microarray data, these methods are not directly applicable because the number of genes is too large. One approach is to do some sort of gene filtering. For example, the significance testing approach may be used to determine a small set of genes that can then be used in a classification. However, as well as the distributional assumptions, this form of gene filtering ignores the multivariate structure of data and it is not clear whether valuable information may be lost. Another common approach is to use principal components analysis to reduce the dimensionality of data. One problem with this approach is that principal components analysis concentrates on finding combinations of genes with large variance and ignores the class labels. Genes with large variance dominate, and outliers can have a huge impact.

Random Forests can handle gene expression data sets without gene filtering and without assuming normality.

Random Forests for Classification

A Random Forest, as the name suggests, is made up of a collection of classification trees. This section briefly describes classification trees and the particular type of "random" classification tree used in the Random Forests method. It also explains how the trees are combined and how measures of variable importance and other useful quantities are obtained.

Classification Trees

Classification trees (Breiman *et al.*, 1984) are a binary decision. An example of a classification tree is given in Fig. 1. This tree was fit to data

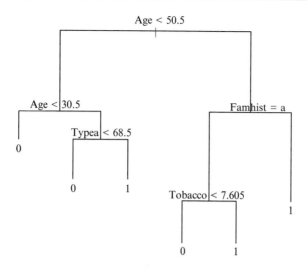

FIG. 1. Tree diagram for heart disease data.

from a South African study on coronary heart disease in men (Hastie *et al.*, 2001; Rousseauw *et al.*, 1983). The response variable was the absence of myocardial infarction (group = 0) or the presence of myocardial infarction (group = 1). Predictor variables were age, prevalence of "type a" behavior, tobacco use, and family history ("a" stands for "absent"). The tree comprises a collection of "nodes." At each node, we ask a question and if the answer is "yes" we move to the left, otherwise we move to the right. At the top node, we ask whether the man's age is less than 50.5. If it is, the man goes to the left; if not, he goes to the right. A similar procedure is followed for subsequent nodes until a stopping criterion is met, at which point the node is "terminal." Numbers at the bottom of the tree represent the class assigned to men who end up in the terminal nodes.

Each node involves a single predictor variable, and we say we "split" on that variable. To construct the tree, we need to split each node, which means we need to decide which variable to split on and at what value to split. To split a node, we look at every possible split on every available predictor. We choose the split that gives the best value of some criterion such as the gini index (Breiman *et al.*, 1984). Usually, the trees are grown to be quite large and are then "pruned" back to prevent overfitting (Breiman *et al.*, 1984). Classification trees are popular for a wide range of problems, in part because the tree diagrams are easily understood. More information on classification trees is given in Breiman *et al.* (1984). For microarray data, we often have thousands of genes, and finding the best possible split at each node can be computationally expensive. Moreover, it has been suggested (Breiman, 2001; Dietterich, 2000) that we can get more accurate results by combining a variety of suitably chosen classification trees.

Trees in a Random Forest

A Random Forest combines a collection of classification trees that differ from each other in two key ways. First, each tree is fit to an independent bootstrap sample from the original data set. To get the bootstrap sample, we randomly sample microarrays *with replacement* from original data until our sample is as large as the original. Some microarrays appear once in the bootstrap sample, some twice, some more often, and some not at all. The microarrays that do not make it into the bootstrap sample are called "out-of-bag" data and form a natural test set for the tree that is fit to the bootstrap sample. The trees also differ because we do not choose the best possible split on all genes. Instead, we take a random sample of just a few genes, independently for each node, and find the best split on the selected genes. The number of genes to be selected at each node is usually chosen to be the square root of the total number of genes. More information about how to choose this value is available in Breiman and Cutler (2005) by looking at the parameter "mtry." The trees are grown until each node contains microarrays from only one class (we say they are "pure"), and the trees are not pruned.

Combining Trees

If we have a new microarray that has been suitably preprocessed to be on the same scale as original data, we pass it down each tree in the forest and each tree provides its best guess at the class. The most popular class, over all the trees in the forest, is the one we use as the Random Forest prediction. This procedure is called "plurality voting." The votes themselves give an idea about which other classes are contenders. For example, suppose the Random Forest has 1000 trees of which 547 say "class 1," 398 say "class 2," and the rest say "class 3." Then the Random Forest prediction is "class 1," with "class 2" a possible contender and "class 3" out of the running. For a microarray that is part of the original data set, the procedure is modified by only voting the trees for which this particular microarray is out of bag.

Error Rate Estimates

Out-of-bag data are used to give an internal estimate of what the misclassification rate will be if the Random Forest is used to predict the classes for a new data set from the same population as the original (Breiman and

Cutler, 2005). To get this out-of-bag error rate, we use each tree to classify the corresponding out-of-bag data (those that did not make it into the bootstrap sample used to get that particular tree). For each tree, we compute error rates for each class and an overall error rate, and we average over all the trees in the forest.

Gene Importance

One interesting aspect of gene expression data is that genes with large expression values or those with highly variable expression values are not always the genes that are important for distinguishing the classes. Random Forests uses an unusual but intuitive measure of the importance of each gene in distinguishing the classes. Consider a single tree and think about the microarrays that are out of bag for this tree. When we pass the out-of-bag microarrays down the tree, we get the out-of-bag error rate for the tree. Now think about randomly permuting the expression values of a particular gene so that each out-of-bag microarray gets a random expression value for this particular gene and all the other genes are kept at their original values. Now we pass the *modified* out-of-bag data down the tree and compute its error rate. If the new error rate is about the same as before, the gene does not appear to be contributing to accurate classification. If, however, the new error rate is higher than before, the gene expression values were useful for accurate classification. The gene importance measure is obtained by averaging the increase in the error rate over all the trees in the forest and this average is used to rank the genes.

Unbalanced Data

Unbalanced data sets, where the class of interest is much smaller than the other classes, are becoming more frequent. A naive classifier will work on getting the large classes right while getting a high error rate on the small class. Random Forests has an effective way of weighting the classes to give balanced results in highly unbalanced data (Breiman and Cutler, 2005). One reason to do this is that the important genes may be different when we force the method to pay greater attention to the small class. Even in the balanced case, the weights can be adjusted to give lower error rates to decisions that have a high misclassification cost. For example, it is often more serious to conclude incorrectly that someone is healthy than it would be to conclude incorrectly that someone is sick.

Proximities

One of the difficult aspects of microarray data analysis is that with thousands of genes, it is not obvious how to get a good "feel" for data or a good impression of what is going on. Are there interesting patterns or structures, such as subgroups within the known classes? Are there outliers? In a multiclass situation, are some of the groups separated while others overlap? Such questions are overwhelming if we try to examine them in obvious ways. Random Forests provides a way to look at data to give some insight into these questions and to show fascinating and unsuspected aspects of data. We do it by computing a measure of *proximity* or *similarity* between each pair of microarrays. We define the proximity between two microarrays as the proportion of the time that they end up in the same terminal node, where the proportion is taken over the trees in the forest. If two microarrays are always in the same terminal node, their proximity will be 1. If they are never in the same terminal node, their proximity will be 0. From these proximities, we derive a distance matrix and use a technique called "classical multidimensional scaling" (see, e.g., Cox and Cox, 2001). Multidimensional scaling takes any set of distances between microarrays and creates a set of points that can be plotted in two or three dimensions. Each point represents one of the microarrays and the distances between the points represent, as closely as possible, the distances between the corresponding microarrays. The resulting picture allows us to "look" at data in a new way.

A natural question at this point is whether it would be just as good to use multidimensional scaling on a conventional distance, such as Euclidean distance or one of the other distances used commonly in cluster analysis (Gollub and Sherlock, 2006). This can certainly be done, but one of the difficulties is that a conventional distance can be dominated by noisy and uninformative genes that can drown out the effects of the genes that are useful. In any case, it may be useful to have an additional view that may illuminate different features of the data.

Proximities are also used to detect outliers and provide a very effective method for filling in missing data (Breiman and Cutler, 2005).

Unsupervised Learning and Clustering

This section describes how Random Forests can be used for unsupervised learning. The presentation is much more brief because unsupervised learning is much more exploratory than classification and not as well understood.

Gollub and Sherlock (2006) describe standard statistical methods for cluster analysis in the microarray context. In the microarray context, we might cluster either the microarrays or the genes. *Clustering the microarrays* involves separating the microarrays into groups or "clusters" so that microarrays in the same cluster have similar gene expression patterns, whereas those from different clusters have quite different expression patterns. For example, we might cluster the microarrays in a medical example where we think there may be distinct types of people in the population. *Clustering the genes* involves finding groups of genes that have similar expression patterns across the microarrays, which in this context might represent different experimental conditions. In this case the goal might be to organize data to facilitate understanding or to identify coregulated genes.

Unsupervised learning is sometimes equated with clustering, but it can be viewed in the more general light of discovering multivariate structure. Cluster structure is one form of multivariate structure, but not the only one. One of the basic assumptions of all clustering methods is that there really *are* clusters. If there are not clusters, cluster analysis might not make sense but it might still make sense to ask whether there is an important multivariate structure.

Random Forests can be used for unsupervised learning without assuming a cluster structure (Breiman and Cutler, 2005). For simplicity, we describe the procedure for exploring structure in the genes and note that structure in the microarrays can be explored in an analogous way. To use Random Forests for unsupervised learning, we label real data "class 1" and generate synthetic data, which are labeled "class 2." Then we use Random Forests to see if we can separate the two classes. Synthetic data are generated from real data by randomly permuting the expression values for each gene independently. In this way, we form a new data set that maintains the distributions of the individual expression values while destroying their multivariate structure. If the original expression values have no multivariate structure, synthetic data will look similar to original data and Random Forests will misclassify about half of the time. If, however, the misclassification rate is much lower than 50%, there is evidence of some interesting structure and we can use all the Random Forests tools (variable importance, proximities, and multidimensional scaling plots) to investigate the structure. In fact, the proximities can be used with a sensible clustering method to determine clusters if a cluster structure turns out to be present.

Case Study: Prostate Cancer Data Set

We illustrate Random Forests using a data set on prostate cancer (Singh *et al.*, 2002). These data have 6033 gene expression values for 102 arrays (50 normal samples and 52 tumor samples). We used the normalization described by Dettling (2004). Random Forests was run with 500 trees and 100 randomly chosen genes at each node. Code is available upon request. The out-of-bag error rate was 7%, which is consistent with Dettling (2004), who cited a 9% cross-validation error rate. The four most important genes

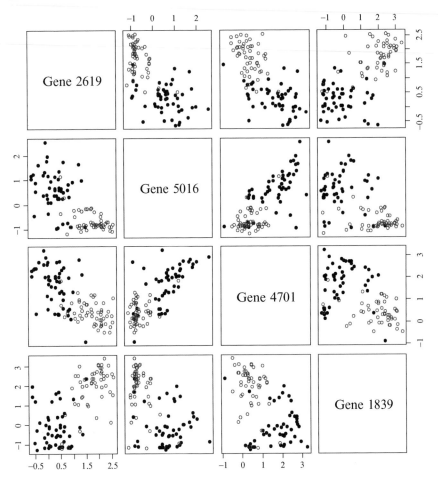

FIG. 2. Prostate cancer data: the four most important genes selected by Random Forests and their relationship to the groups. Solid black circles represent controls and open red circles represent tumors. (See color insert.)

identified by Random Forests are plotted in Fig. 2. Solid black circles represent controls and open red circles represent tumors. It is clear, from both the error rates and the pictures, that Random Forests is able to classify data very well and also to identify genes useful in this process. We compare to performing a principal components analysis on the same gene expression data. The first two principal components are shown in Fig. 3. It is apparent that the dimensions of greatest variability in these data have very little to do with the two groups.

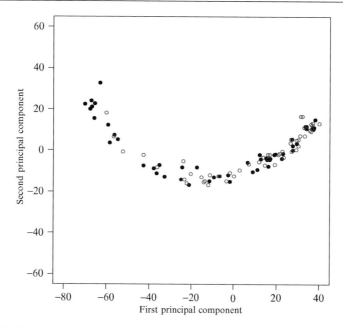

FIG. 3. Prostate cancer data: the first two principal components of data and their relationship to the groups. Solid black circles represent controls and open red circles represent tumors. (See color insert.)

Conclusion

Random Forests provides a new and powerful approach to understanding gene expression data. According to Díaz-Uriarte and Alvarez de Andrés (2006): "Because of its performance and features, random forest and gene selection using random forest should probably become part of the "standard tool-box" of methods for class prediction and gene selection with microarray data."

Open source FORTRAN software for Random Forests is available from www.math.usu.edu/~adele/forests. A commercial version, with an easy-to-use interface, is available from www.salford-systems.com. An R package is also available, written by Liaw and Wiener (2001).

References

Ayroles, J. F., and Gibson, G. (2006). Analysis of variance of microarray data. *Methods Enzymol.* **411,** 214–233.
Breiman, L. (2001). Random forests. *Machine Learn.* **45**(1), 5–32.

Breiman, L., and Cutler, A. (2005). www.math.usu.edu/~adele/forests.

Breiman, L., Friedman, J. H., Olshen, R. A., and Stone, C. J. (1984). "Classification and Regression Trees." Chapman and Hall, New York.

Cox, T. F., and Cox, M. A. A. (2001). "Multidimensional Scaling," 2nd Ed. Chapman and Hall/CRC.

Dettling, M. (2004). BagBoosting for tumor classification with gene expression data. *Bioinformatics* **20**(18), 3583–3593.

Díaz-Uriarte, R., and Alvarez de Andrés, S. (2006). Gene selection and classification of microarray data using random forest. *BMC Bioinform.* **7**, 3.

Dietterich, T. G. (2000). An experimental comparison of three methods for constructing ensembles of decision trees: Bagging, boosting and randomization. *Machine Learn.* **40**, 139–157.

Gollub, J., and Sherlock, G. (2006). Clustering microarray data. *Methods Enzymol.* **411**, 194–213.

Hastie, T., Tibshirani, R., and Friedman, J. H. (2001). The Elements of Statistical Learning: Data Mining, Inference, and Prediction." Springer, New York.

Irizarry, R. A., Hobbs, B., Collin, F., Beazer-Barclay, Y. D., Antonellis, K. J., Scherf, U., and Speed, T. P. (2003). Exploration, normalization, and summaries of high density oligonucleotide array probe level data. *Biostatistics* **4**, 249–264.

Liaw, A., and Wiener, M. (2002). Classification and regression by Random Forest. R News: The Newsletter of the R Project (http://cran.r-project.org/doc/Rnews/) **2**(3), 18–22.

Rousseauw, J., du Plessis, J., Benade, A., Jordann, P., Kotze, J., Jooste, P., and Ferreira, J. (1983). Coronary risk factor screening in three rural communities. *South Afr. Med. J.* **64**, 430–436.

Singh, D., Febbo, P. G., Ross, K., Jackson, D. G., Manola, J., Ladd, C., Tamayo, P., Renshaw, A. A., D'Amico, A. V., Richie, J. P., Lander, E. S., Loda, M., Kantoff, P. W., Golub, T. R., and Sellers, W. R. (2002). Gene expression correlates of clinical prostate cancer behavior. *Cancer Cell* **1**(2), 203–209.

Tusher, V. G., Tibshirani, R., and Chu, G. (2001). Significance analysis of microarrays applied to the ionizing radiation response. *Proc. Natl. Acad. Sci. USA* **98**, 5116–5121.

Further Reading

Dudoit, S., Fridlyand, J., and Speed, T. (2002). Comparison of discrimination methods for the classification of tumors using gene expression data. *J. Am. Stat. Assoc.* **97**(457), 77–87.

R Development Core Team (2004). R: A language and environment for statistical computing. R Foundation for Statistical Computing Vienna, Austria. ISBN 3-900051-07-0, URL http://www.r-project.org/.

Author Index

A

Aach, J., 87, 109, 117, 124, 139, 326, 354, 359
Aas, T., 199
Abernathy, K., 38, 80
Abeygunawardena, N., 117, 143, 371
Afshari, C., 215, 222, 224, 228
Agilent Technologies, 82
Aharonov, R., 19, 21, 22, 24
Ahlquist, P., 61
Aidinis, V., 22
Albala, J. S., 410, 414
Albert, T., 290
Aldape, K. D., 130, 199, 200
Alder, H., 15, 16, 19, 21, 22
Alexander, H. R., 2
Alexander, W. S., 367
Aliferis, C., 319
Alizadeh, A., 175
Allan, D. L., 218
Allen, L., 357, 358
Allison, D. B., 22
Alon, U., 405
Al-Shahrour, F., 341
Alsobrook, J., 64
Altman, R., 304
Altman, R. B., 178, 327
Altschul, S. F., 365, 366
Altshuler, D., 343
Alvarez, F., 234, 242
Alvarez de Andrés, S., 422, 431
Alvarez-Garcia, I., 15, 16
Alvarez-Saavedra, E., 16, 18, 19, 21, 23
Amarzguioui, M., 100
Ambros, V., 15
Ambs, S., 16, 21, 22
Amin, N., 413
Amtmann, A., 387
Anderson, M. L. M., 35
Anderson, N. L., 173
Anderson, R. L., 100
Andersson, A., 100
Andrada, R., 412

Andreeff, M., 21, 22
Andrews, G., 409
Andrews, J., 34, 144, 233, 316
Andruss, B. F., 1, 14
Angenendt, P., 93
Ansorge, W., 109, 117, 124, 139, 326, 354, 359
Antonellis, K. J., 60, 231, 257, 290, 422
Aparicio, O. M., 368
Aprelikova, O., 68
Apweiler, R., 327, 342, 376
Aqeilan, R. I., 15, 16, 21, 22
Aragon, A. D., 80, 91, 94, 95
Archacki, S., 1
Archer, K., 39, 40, 41
Ares, M., Jr., 178
Argyropoulos, C., 22
Armengaud, P., 387
Armour, C. D., 274
Aronow, B. J., 117, 326, 360, 383
Arribas-Prat, R., 341
Arya, S., 252
Ashburner, M., 179, 327, 328, 342, 376, 401, 412, 417
Astola, J., 39
Astrand, M., 20, 23, 60, 298
Attie, A. D., 256
Attwood, T. K., 376
Auer, H., 39, 40
Augustus, M., 64
Avniel, A., 19, 21, 22, 24
Ayivi-Guedehoussou, N., 410, 414
Ayroles, J. F., 144, 160, 167, 177, 214, 340, 423

B

Babak, T., 19, 21, 23
Bach, M. K., 6
Bader, J., 314
Badiee, A., 58, 74, 87, 100
Baelde, H. J., 11
Baid, J., 75
Bailey, D. K., 272, 283, 291
Bairoch, A., 327, 376

433

Bajorek, E., 73
Baker, H. E., 8
Baker, S. C., 69
Balakrishnan, R., 376, 412
Balazsi, G., 80
Baldi, P., 304
Baliga, N. S., 413
Ball, C. A., 71, 109, 117, 124, 139, 179, 326, 327, 342, 353, 354, 359, 360, 371, 373, 383, 384, 401, 412, 417
Ballas, C., 14
Ballin, J., 290
Bammler, T., 61, 63
Bandyopadhyay, S., 409
Bangham, R., 292
Bankaitis-Davis, D., 14
Barabasi, A.-L., 80
Barad, O., 18, 19, 21, 22, 24
Barchas, J. D., 2
Barczak, A., 63
Bard, P. Y., 274
Barkai, N., 380, 405
Barker, D., 218
Barker, J. L., 199
Barna, M., 388
Baron, R., 178
Barrell, D., 342, 376
Barrera, L. O., 272, 275, 277
Barrett, J., 64
Barrett, T., 96, 117, 137, 143, 257, 352, 353, 364, 373
Barrette, T. R., 1
Bartel, B., 15
Bartel, D. P., 15, 19, 410
Barton, S. C., 19
Barzilai, A., 19, 21, 22, 24
Basehore, L. S., 68
Baskerville, S., 15, 19
Bassett, D., 117, 326, 360, 383
Basson, M., 15
Bateman, A., 15, 24, 376
Bates, D. M., 74, 120, 212, 230, 377
Bates, S. E., 173
Bauer, S. R., 69
Baumhueter, S., 233
Beare, 223
Beazer-Barclay, Y. D., 60, 231, 257, 290, 422
Bechhofer, S., 327, 331
Becker, J. W., 233, 251
Becker, S., 3

Bednarski, D., 319
Behbahani, B., 53, 56, 58
Behrendtz, M., 100
Beier, D., 233
Beier, M., 38
Beissbarth, T., 340, 341, 347
Bekiranov, S., 272, 283, 291
Bell, G. I., 272
Bell, G. W., 272, 408
Bell, I., 283, 291
Bell, J., 314
Bell, R., 410
Bellis, M., 343, 373, 387, 389, 393
Benade, A., 425
Ben-Dor, A., 175, 205
Bengtsson, H., 92
Benjamini, Y., 27, 132, 229, 264, 344, 347
Bennett, H. A., 274
Bennett, L., 215, 222, 224, 228
Benson, D. A., 364
Bento, M., 93
Bentwich, I., 19, 21, 22, 24
Bentwich, Z., 18, 19, 21, 22, 24
Berardini, T., 376
Berezikov, E., 15
Berg, J., 290
Berger, M. S., 199, 200
Berges, C. J., 243
Bergh, J., 11
Bergmann, S., 380
Berk, L. S., 3
Bern, E. M., 234, 242
Bernat, J. A., 273, 280
Berndt, M., 64
Bernhart, D., 117, 326, 360, 383
Bernstein, B. E., 272
Berriman, M., 376
Berriz, G. F., 410, 414
Bertone, P., 227, 290, 291, 292, 300
Best, A., 39, 40, 41
Bettinger, J. C., 15
Beule, D., 80
Bex, C., 410, 414
Beyer, R. P., 61, 63, 69
Bhagabati, N. K., 134, 206, 207, 211, 223, 289, 290, 341, 384
Bhatt, D., 21, 22
Bhattacharya, S., 61, 63
Bhavsar, P., 342
Bibikova, M., 218

Bichi, R., 15
Bidlingmaier, S., 292
Bienenstock, E., 321
Bilgin, M., 292
Billard, L., 322
Binkley, G., 384, 412
Binns, D., 342, 376
Bischof, H., 2
Bishop, M. J., 252
Biswal, S., 63, 96
Biswas, M., 376
Bit, S. L., 3
Bittner, M. L., 164
Bjohle, J., 11
Björkbacka, H., 100
Black, S., 73
Blades, N. J., 228
Blake, J. A., 179, 327, 342, 376, 401, 412, 417
Blakesley, R., 273
Blalock, E. M., 1
Blanchard, A. P., 52, 53, 63
Blanchette, C., 1
Blatter, M. C., 327
Blencowe, B. J., 19, 21, 23
Blidy, A. D., 3
Bloomfield, C. D., 21, 22, 387
Boeckmann, B., 327
Boer, J. M., 341
Bogart, K., 34, 144
Boguski, M., 347
Boito, C., 367
Boldrick, J. C., 178, 232, 344
Bollen, A. W., 199, 200
Bolstad, B. M., 20, 23, 60, 63, 74, 120, 212, 230, 231, 298, 377, 380
Bonferroni, C. E., 288, 344
Boon, W. M., 341
Boone, C., 409
Boorman, G. A., 61, 63
Borg, Å., 100
Borg, A. L., 11
Borisevich, I., 341
Borisovsky, I., 206, 207, 211, 289, 290, 384
Bork, P., 376
Borkhardt, A., 15
Borovecki, F., 1
Borresen-Dale, A. L., 199
Bosak, S., 410, 414
Bostian, K. A., 233

Botstein, D., 68, 173, 175, 179, 199, 200, 206, 207, 274, 291, 304, 327, 341, 342, 364, 376, 384, 401, 412, 417
Boucher, L., 410
Bouffard, G., 233, 273, 316
Boutros, M., 60, 304
Bouzou, B., 1
Bowman, E., 21, 22
Bowtell, D., 38, 39, 41, 42
Boxem, M., 410, 414
Boyer, L. A., 410
Boyer, N., 234
Boyles, A., 61, 63
Bozso, P., 64
Bracht, J., 15
Bradford, B. U., 61, 63
Bradley, P., 376
Brady, L., 233
Braisted, J. C., 134, 206, 207, 211, 223, 289, 290, 341, 384
Braman, J., 68
Branham, W. S., 78
Brashler, J. R., 6
Bray, D., 409
Brazma, A., 71, 96, 109, 117, 120, 124, 137, 139, 143, 326, 353, 354, 359, 360, 370, 371, 373, 377, 381, 383
Breckenridge, N., 233
Breiman, L., 422, 424, 425, 426, 427, 428, 429
Breitkreutz, 223
Breitkreutz, A., 410
Breitkreutz, B. J., 410
Breitling, R., 387
Brennecke, J., 15
Brenton, J. D., 69
Brockington, M., 367
Bromley, B., 69
Brooksbank, C., 71, 371
Brors, B., 179, 341
Brown, B., 222
Brown, C. S., 80
Brown, D., 15, 16, 18, 19
Brown, E. L., 51, 56, 340
Brown, M. P., 178
Brown, P., 34, 175, 283, 291, 304
Brown, P. O., 50, 173, 199, 200, 206, 207, 274, 340, 341, 364, 370, 384, 387
Brown, S. C., 367
Browne, L. J., 233
Brubaker, S., 283, 291

Bruijn, J. A., 11
Brune, K., 3
Bryan, J., 208
Bryant, S. H., 364
Bucher, P., 376
Buck, K., 314
Buck, M. J., 275, 277
Buckley, M. J., 80, 147, 150, 151
Buetow, K. H., 335
Bugelski, P. J., 80
Bulak Arpat, A., 85
Bullrich, F., 15, 19
Bult, C. J., 342, 376
Bumgarner, R. E., 61, 63
Bunow, B., 173
Buolamwini, J. K., 173
Burchard, J., 52, 53, 63
Burge, C. B., 15, 410
Burgess, S. M., 367
Burrill, J., 69
Busch, K., 15
Bushel, P., 215, 222, 224, 228
Bushel, P. R., 61, 63
Bushnell, S., 178
Bussemaker, H. J., 387
Bussey, H., 409
Bussow, K., 291
Butcher, L., 290
Butcher, P. D., 223
Butler, H., 179, 327, 342, 401, 412, 417
Butte, A. J., 63, 193
Byrne, D., 64
Byrne, M. C., 51, 340
Byrom, M. W., 15, 16, 18

C

Cai, T., 342
Caldeira, S., 199
Caligiuri, M. A., 387
Calin, G. A., 15, 16, 19, 21, 22
Call, K., 178
Callow, M. J., 177, 232
Calvin, J. T., 233
Cam, M., 64
Camon, E., 342, 376
Campbell, M. J., 203, 283
Campbell, R. E., 95
Canese, K., 364
Caoile, C., 73

Caplen, N., 21, 22
Carazo, J. M., 177
Cardon, L., 314
Cardoso, W. V., 15, 21, 23
Carey, V. J., 74, 119, 120, 160, 167, 206, 212,
    230, 231, 349, 377
Carlsson, E., 343
Carmack, C. E., 52, 54, 58
Carmel, J. B., 100
Carr, D. B., 199
Carrington, J. C., 15
Carroll, R. J., 85
Carter, M. G., 52, 54, 58
Casamayor, A., 292
Casciano, D., 78
Cash, A., 34, 144
Castelnau, C., 234
Cattoretti, G., 388
Causton, H. C., 69, 71, 109, 117, 124, 139, 326,
    353, 354, 359, 371, 373
Cavalieri, D., 71, 371
Cavalli, G., 373, 393
Cavet, G., 52, 53, 63
Cawley, S., 291
Cerrina, F., 290
Cevik, S., 410, 414
Chakraburtty, K., 274
Chambers, J. M., 132
Chan, C., 147, 233
Chan, J. A., 15
Chan, W. C., 175
Chan, Y. M., 73
Chandramouli, G., 64
Chang, J., 290, 300
Chang, S., 15
Chang, T. K., 56
Chapman, K. M., 388
Chaturvedi, K., 61, 63
Chatziioannou, A. A., 22
Chee, M. S., 51, 340
Chen, E., 151, 153, 162, 167, 298
Chen, F., 15, 21, 23
Chen, J. J., 78, 218
Chen, K. C., 1
Chen, M.-H., 324
Chen, T.-S., 151
Chen, X., 15
Chen, Y., 164, 273, 280
Cheng, A., 15, 16, 18
Cheng, J., 283, 291

Cherayil, B. J., 100
Cherry, J. M., 164, 179, 327, 342, 376, 384, 401, 412, 417
Chervitz, S., 71, 117, 326, 353, 360, 373, 383
Chessel, D., 383
Chettier, R., 410
Chinnaiyan, A. M., 1
Chisholm, R., 376
Cho, J., 349
Cho, R. J., 203, 283
Choe, S. E., 60, 304
Choi, D., 61, 63
Christie, K. R., 376, 412
Chu, G., 131, 167, 178, 194, 214, 303, 319, 423
Chu, T.-M., 215, 220, 227
Chudin, E., 56
Chumakov, K., 69
Chung, P.-C., 151
Church, D. M., 364
Church, G. M., 60, 87, 203, 304
Churchill, G. A., 80, 214, 218, 219, 225, 228, 229, 231, 232, 233, 308
Ciafre, S. A., 15
Ciaravino, G., 199
Cimmino, A., 15, 16, 21, 22
Clare, A., 341
Clark, J., 376
Cleveland, W., 242
Cleveland, W. S., 116, 299
Coffey, E., 52, 53, 63, 274
Coffey, E. M., 82
Cohen, P., 173
Cohen, S. M., 15
Colantuoni, C., 300
Cole, M. F., 410
Coller, H., 387
Collin, F., 20, 60, 231, 257, 290, 380, 422
Collins, F. S., 273, 280
Collins, P. J., 218
Comanor, L., 234, 235, 242
Conjeevaram, H. S., 234, 242
Conklin, B., 341
Conley, M. P., 69
Connolly, T., 178
Conrad, R., 18, 19
Considine, E. C., 179, 381
Consortium, ERC, 69
Constantine-Paton, M., 19
Contreras, B., 343
Contrino, S., 117

Conway, A., 283
Cooper, M., 203
Cope, L. M., 20, 60, 75, 380
Copley, R. R., 376
Corbeil, J., 1, 352
Corn, R. M., 93
Costanzo, M. C., 376, 412
Costello, C., 64
Costello, J., 34, 144
Costouros, N. G., 2
Costoya, J. A., 388
Cotter, T. G., 179, 381
Coulson, R., 117
Courcelle, E., 376
Cover, T., 319
Covitz, P. A., 335
Cox, M. A. A., 428
Cox, T. F., 428
Craig, J., 314
Crawford, G. E., 218, 270, 273, 280, 373
Cristianini, N., 178, 319
Croce, C. M., 15, 16, 19, 21, 22
Crompton, T., 45, 47
Cronin, M., 69
Cronin, M. T., 387
Cronkhite, J. T., 388
Crubezy, M., 327
Cui, X., 22, 214, 218, 228, 229
Culhane, A. C., 179, 377, 381
Cummins, J. M., 18
Cunningham, M. L., 61, 63
Cuppen, E., 15
Currier, T., 206, 207, 211, 289, 290, 384
Cusick, M. E., 410, 414
Cutler, A., 422, 426, 427, 428, 429
Cutler, P., 93

D

Dahlquist, K., 341
Dai, H., 52, 53, 63, 82, 274
Dalgaard, P., 343
Dalma-Weiszhausz, D. D., 129, 136, 218, 227, 257, 347
Daly, M. J., 273, 280, 343
D'Amico, A. V., 318, 429
Das, U., 376
Davidson, G. S., 179
Davies, P. F., 2
d'Avignon, C., 319

Davis, A. P., 179, 270, 327, 342, 401, 412, 417
Davis, C., 52, 53, 63
Davis, D., 373
Davis, R., 34, 283
Davis, R. W., 340, 387
Davis, S., 218, 273, 280
Davison, A., 322
Davison, T. S., 14
Day, G., 233
Dean, N. M., 19
Dean, R., 292
De Coronado, S., 335
Dee, S., 75
de Heer, E., 11
de Hoon, M. J., 204, 206, 207, 210
de Kort, F., 92
de la Cruz, N., 376
de La Roche Saint-Andre, V., 100
Dell'Acuila, M. L., 15
Dement, W. C., 2
Demougin, P., 389
den Dunnen, J. T., 92
Deng, H. W., 61
Deng, Q., 73
Deng, S., 61, 63
Dennis, G., 341
Dennis, G., Jr., 179, 190
Denton, D. A., 341
Denys, M., 73
de Plessis, J., 425
DeRisi, J. L., 274, 370
Dettling, M., 74, 120, 212, 230, 377, 426, 429
Deutsch, E., 117, 326, 360, 383
Devlin, S. J., 116
de Vos, W. M., 100
De Witte, A., 218
Dhanasekaran, S. M., 1
Dharap, S., 38, 80
Di, X., 75
Diaz, L. A., Jr., 18
Díaz-Uriarte, R., 341, 422, 431
Dickinson, D. P., 367
Dickson, M., 73
DiCuccio, M., 364
Diehl, F., 38
Diehn, M., 199, 200
Diekhans, M., 73
Dietrich, F. S., 15
Dietterich, T. G., 426
Dike, S., 283, 291

Di Leva, G., 16, 21, 22
Dimitrov, D., 64
Dimmer, E., 342, 376
Dmitrovsky, E., 176, 202
Dobbs, L., 357, 358
Doctolero, M., 147, 233, 251
Dolan, M., 376
Dolatshad, N. F., 367
Dolédec, S., 383
Dolinski, K., 179, 327, 342, 376, 401, 412, 417
Domrachev, M., 143, 353, 373
Donahue, C. P., 15
Dong, H., 51, 340
Dong, Q., 412
Doniger, S., 341
Dono, M., 15, 16, 19
Doose, S., 94
Dopazo, J., 177, 341
Dorschner, M. O., 273
Dostie, J., 15
Doucette-Stamm, L., 410, 414
Dougherty, E. R., 164
Douglas, G., 64
Doursat, R., 321
Downey, T., 64, 167, 178, 256, 341
Downing, J. R., 16, 18, 21, 23, 387
Drabkin, H., 376
Draghici, S., 342
Drenkow, J., 283, 291
Dressman, H., 1, 61, 63
Dreyfuss, G., 15
Dreze, M., 410, 414
Dricot, A., 410, 414
Drucker, H., 243
Dudekula, D. B., 52, 54, 58
Dudley, A. M., 87
Dudoit, S., 74, 80, 100, 120, 147, 151, 162, 164,
    177, 178, 212, 230, 232, 255, 298, 319, 344,
    377, 432
Duffy, N., 319
Dugas, M., 295
Duke, K., 179
Dumitru, C. D., 15, 19
Dumur, C., 39, 40, 41
Dunlea, S., 233
Duran, G. E, 199
Durbin, R., 376
Durinck, S., 377
Dwight, S. S., 179, 327, 342, 376, 384, 401, 412, 417
Dzobnov, E. J., 376

E

Eads, B., 34, 144
Eastman, P. S., 233, 251
Eastman, S., 147, 233, 316
Easton, K. L., 327
Ebersole, B., 64
Ebert, B. L., 16, 18, 21, 23
Eberwine, J. H., 2
Echols, N., 290, 300
Eckhardt, B. L., 100
Eddy, S. R., 15
Edén, P., 100
Edgar, R., 71, 96, 117, 137, 143, 257, 352, 353, 364, 373
Edwards, P., 147, 233
Efron, B., 322, 323
Egyhazi, S., 11
Eickhoff, H., 64, 80, 291
Eierman, D., 3
Eiken, H. G., 58
Eikmans, M., 11
Eilbeck, K., 376
Einat, P., 19, 21, 22, 24
Einav, U., 19, 21, 22, 24
Eisen, M. B., 173, 175, 199, 206, 207, 223, 341, 364, 384
Eizinger, A., 179
Elbashir, S. M., 18
Elcock, F. J., 80
Elespuru, R., 69
Elgin, S. C., 273
Ellis, B., 74, 120, 212, 230, 377
Ellison, S. L. R., 66
Emanuelsson, O., 167, 273
Endoh, H., 15
Engel, S., 376
Engel, S. R., 412
Enright, A. J., 15, 24
Eppig, J. T., 179, 327, 342, 376, 401, 412, 417
Epstein, J. A., 364
Eric, L., 82
Erickson, J. N., 372
Eriksson, K. F., 343
Erle, D., 63
Ernst, M., 316
Esau, C., 19
Escobar, J., 73
Eser, A., 233
Esquela-Kerscher, A., 15, 16

Estrada, E., 409
Estreicher, A., 327
Eurchem/CITAC, 66
Everall, I. P., 1
Everhart, J. E., 234
Everitt, B., 203
Ewen, M. E., 343
Ewing, R., 164
Eynon, B. P., 233
Eystein Lonning, P., 199

F

Fabbri, M., 15, 16, 21, 22
Falquet, L., 376
Fan, H. P., 3, 4
Fan, J.-B., 218
Fang, H., 78
Fang, J., 85, 92
Fannin, R. D., 61, 63
Farace, M. G., 15
Fare, T. L., 82
Farh, K. K., 410
Farin, F. M., 61, 63
Farne, A., 117
Febbo, P. G., 318, 429
Federhen, S., 364
Feierbach, B., 376
Feigal, D. W., Jr., 69
Feinbaum, R. L., 15
Feldman, A. L., 2
Fellenberg, K., 179, 341
Felli, N., 15
Felten, D. L., 3
Fergerson, R. W., 327
Ferng, J., 233
Fero, M., 68, 69
Fero, M. J., 199
Ferracin, M., 15, 16, 19, 21, 22
Ferrando, A. A., 16, 18, 21, 23
Ferreira, J., 425
Ferreira-Gonzalez, A., 39, 40, 41
Fielden, M. R., 233
Fields, S., 410
Finch, C. E., 368
Fink, G. R., 411
Fink, J. L., 352
Finkelstein, D., 8
Finkelstein, D. B., 164
Fioretos, T., 100

Fisher, R. A., 261
Fishman, G., 315
Fisk, D. G., 376, 412
Fleischmann, W., 376
Flores-Morales, A., 367
Flowers, D., 73
Foat, B. C., 387
Fodor, S., 283, 290, 291
Fodor, S. P., 387
Fodstad, O., 100
Fojo, T., 173
Follettie, M. T., 51, 340
Fontes, M., 100
Ford, H. L., 343
Ford, L. P., 15
Fornace, A. J., Jr., 173
Fotopulos, D., 73
Foulger, R., 376
Foy, C., 69
Fraenkel, E., 272
Fragoso, G., 335
Francesco, N. M., 2
Francois, P., 93
Frank, B. C., 53, 56, 58, 63, 64, 96, 151, 153, 162, 167, 298
Franklin, G., 410, 414
Fraser, I., 411
Fraughton, C., 410, 414
Fred, A. L., 388
Freedman, D., 343
Freedman, J. H., 61, 63
Freeman, M. W., 100
Free Software Foundation, 100
Freidman, J. H., 424, 425
Frey, B. J., 23
Fridlyand, J., 319, 432
Friedman, J., 321
Friedman, J. H., 424, 425
Friedman, J. M., 173
Friend, S. H., 52, 53, 63, 173, 274
Friendly, M., 222
Frierson, H. F., Jr., 367
Frokjaer-Jensen, C., 15
Frostvik Stolt, M., 11
Frueh, F. W., 78
Fry, R. C., 61, 63
Fu, D., 64
Fuchs, E., 15
Fuess, S., 367
Fuhrman, S., 199

Fujibuchi, W., 117, 353
Fujii, T., 100
Fujimoto, S. Y., 233
Furey, T. S., 73, 178, 319
Furness, L. M., 233
Fuscoe, J. C., 69, 78
Futschick, M., 45, 47

G

Gaasenbeek, M., 387
Gaasterland, T., 71, 109, 117, 124, 139, 326, 354, 359, 371
Gabrielian, A., 283
Gabrielson, E., 63, 96
Gachotte, D., 274
Gaidukevitch, Y. C., 147
Galardi, S., 15
Gallo, M. V., 51, 340
Gamliel, N., 19
Ganesh, M., 283, 291
Gangi, L., 64
Ganter, B., 233
Gao, F., 387
Gao, J., 367
Gao, W., 341
Gao, X., 69
Garber, M., 304
Garbers, D. L., 388
Garcia, C., 73
Garcia, E. W., 218
Garcia, J. G. N., 63, 96
Garcia, T., 178
Garcia Lara, G., 117
Garner, H. R., 94
Garrard, W. T., 273
Garrett, C., 39, 40, 41
Garrison, B., 367
Garzon, R., 16, 21, 22
Gaspard, R., 38, 80, 153
Gassmann, M., 11
Gasteiger, E., 327
Gaudet, P., 376
Gautier, L., 74, 120, 212, 230, 377
Gavras, H., 64, 96
Gay, C., 38, 80
Ge, H., 410
Ge, Y., 74, 120, 212, 230, 377
Geddes, J. W., 1
Geisert, E. J., 64

Geisler, S., 199
Geli, V., 100
Geman, D., 319
Geman, S., 321
Gene Ontology Consortium, 376
Gentleman, R. C., 74, 100, 120, 212, 230, 377, 380
Gentry, J., 74, 120, 212, 230, 377
Geoghegan, J., 63, 64
George, R. A., 150
Gerhard, D., 283, 291
Gerhold, D. L., 69
Germino, G., 63, 96
Gerstein, M., 290, 291, 292, 293, 300, 303
Gerstein, M. B., 167, 227, 273
Gerton, J. L., 274
Getz, G., 16, 18, 21, 23
Ghosh, D., 1
Ghosh, K., 69
Ghosh, S., 283, 291
Gibbons, F. D., 205, 410, 414
Gibson, G., 144, 160, 167, 177, 214, 215, 219, 222, 224, 227, 228, 230, 340, 423
Giempmans, B. N. G., 95
Gifford, D. K., 272, 410
Gilad, S., 19, 21, 22, 24
Gilad, Y., 227
Gilles, P., 69
Gingeras, T., 283, 290, 291
Gingeras, T. R., 272
Ginsburg, D., 273, 280
Gish, K., 405
Gish, W., 365, 366
Giuily, N., 234
Glasner, M. E., 15
Glatt, S. J., 1
Glenisson, P., 109, 117, 124, 139, 326, 354, 359
Gloekler, J., 93
Glynn, E. F., 274
Goble, C. A., 327, 328, 331
Goda, S., 40
Gojobori, T., 373
Goldberg, D. S., 409, 410, 414
Golenbock, D. T., 100
Gollub, J., 160, 164, 193, 194, 341, 343, 364, 380, 423, 428
Golub, T. R., 16, 18, 21, 23, 50, 123, 176, 193, 202, 318, 343, 387, 429
Gomez, M., 73
Goncalves, J., 117, 326, 360, 383

Gong, L., 233
Gonzales, E., 73
Gonzalez, R., 357, 358
Goodman, R. H., 15
Goodman, R. M., 93
Goodsaid, F., 69, 78
Goodwin, P. C., 80
Goralski, T., 147
Gordon, D. B., 272
Gorski, T., 290
Goto, S., 342
Goyens, P., 234, 242
Grafe, A., 15
Grahlmann, S., 38
Grant, G. R., 2, 335
Granzow, M., 11
Graves, T. A., 73
Green, E. D., 273
Green, J. M., 64, 66
Green, R. D., 272, 275, 277, 290
Greenberg, A. H., 6
Greene, A. S., 53, 56, 58
Greenhalgh, C. J., 367
Gregory, R. I., 16
Grellhesl, D. M., 388
Greshock, J. D., 100
Gribskov, M., 352
Griffin, C., 63, 96, 357, 358
Griffiths-Jones, S., 15, 24, 376
Grigorenko, E., 49
Grimson, A., 410
Grimsrud, T. E., 93
Grimwood, J., 73
Grisafi, P., 411
Griswold, M. D., 388
Grocock, R. J., 15, 24
Grompe, M., 367
Groop, L. C., 343
Gross, D. S., 273
Grosshans, H., 15, 16, 18
Gruber, T. R., 327
Grundy, W. N., 178
Grunstein, M., 34
Gruvberger, S., 100
Gryczynski, I., 85, 92
Guenther, M. G., 410
Gulari, E., 74
Gunderson, K. L., 218
Gunnersen, J., 341
Guo, L., 78

Guo, M., 15
Guo, X., 69
Gurbuz, Y., 64
Guryev, V., 15
Gustafson, S., 335
Gutman, S. I., 69
Gwinn, M., 376

H

Ha, I., 15
Haaland, D. M., 80, 91, 94, 95
Haarslev, V., 327
Haas, S., 347
Hackett, J., 69
Hackett, J. L., 69
Haft, D., 376
Hagerstrom, T., 11
Hahn, M., 38
Hajkova, P., 19
Haley, B., 15
Halfon, M. S., 60, 304
Hall, C., 290
Hall, D., 292
Hall, M. N., 389
Hall, P., 322
Halmen, K., 100
Hamadeh, H., 215, 222, 224, 228
Hammer, R. E., 388
Hammond, S. M., 16, 19
Hamra, F. K., 388
Han, J., 64, 78
Han, M., 1
Han, T., 78
Hannick, L., 376
Hansen, A., 368
Hanspers, K., 63
Hao, T., 410, 414
Harano, T., 15
Hardiman, G., 38
Hardin, D., 319
Harfe, B. D., 15
Harper, A., 61, 63
Harrington, C. A., 75
Harris, C. C., 16, 21, 22
Harris, M. A., 179, 327, 376, 401, 412, 417
Harris, S. C., 78
Hart, P., 319
Hart, R. P., 100
Harte, N., 342, 376

Hartel, F., 335
Haskill, S., 3
Hasseman, J., 153, 298
Hasseman, J. P., 151, 162, 167
Hastie, T., 175, 199, 319, 424, 425
Hauser, N. C., 179, 341
Haussler, D., 73, 178, 319
Hawrylycz, M., 273
Hay, B. A., 15
Haydu, L., 73
Haynor, D. R., 176, 205
He, H., 335
He, Y., 18
He, Y. D., 52, 53, 63, 82, 274
Hegde, P., 38, 80
Heldrup, J., 100
Helmberg, W., 364
Helt, G., 283, 291
Hendricks, D. A., 234
Hennetin, J., 343, 373, 387, 393
Henry, G., 300
Herdman, C., 14
Hergenhahn, M., 341
Hernandez-Boussard, T., 199, 384
Herrero, J., 177
Hersch, S. M., 1
Herzel, H., 80
Herzyk, P., 387
Hesselberth, J. R., 410
Hewett, M., 327
Heyer, L. J., 175
Heymans, K., 410, 419
Higgins, D. G., 179, 381
Hilburn, J., 64
Hill, A. A., 56
Hill, D. E., 410, 414
Hill, D. P., 179, 327, 376, 401, 412, 417
Hillan, K., 8
Hillier, L. W., 73
Hilmer, S. C., 63, 96
Hilton, D. J., 367
Hinds, J., 223
Hingamp, P., 71, 109, 117, 124, 139, 326, 353, 354, 359, 371, 373
Hinkly, D., 322
Hipfner, D. R., 15
Hirozane-Kishikawa, T., 410, 414
Hirschhorn, J. N., 343
Hirschman, J. E., 376, 412
Ho, M. H., 75

Hobbs, B., 20, 60, 231, 257, 290, 380, 422
Hobbs, R. M., 388
Hobert, O., 15
Hochberg, Y., 27, 132, 229, 264, 344, 347
Hockett, R. D., 69
Hoerlein, A., 341
Hoffman, E., 63, 96
Hoffmann, R., 295
Hogarth, P., 1
Höglund, M., 100
Hogness, D. S., 34
Hoheisel, J., 38
Hoheisel, J. D., 179, 341
Holcombe, D., 66, 73
Holloway, A., 38, 41
Holloway, E., 117, 143, 371
Holm, S., 347
Holmes, C. P., 387
Holstege, F. C., 51, 71, 109, 117, 124, 139, 326, 354, 359
Holt, I. E., 273, 280
Hong, E. L., 376, 412
Hong, H., 78
Hoofnagle, H., 234
Horak, C., 290, 291, 300
Horn, M., 291
Hornik, K., 74, 120, 212, 230, 377
Hornstein, E., 15
Horrocks, I., 327, 331
Horton, H., 51, 340
Horvitz, H. R., 15, 16, 18, 19, 21, 23
Hosack, D. A., 179, 190, 341
Hotelling, H., 257
Hothorn, T., 74, 120, 212, 230, 377
Houfek, T., 292
Houstis, N., 343
Hovig, E., 74, 87, 100
Howe, E. A., 134, 206, 207, 211, 223, 341, 384
Hsieh, W.-P., 227
Hu, C., 233
Hu, J. K., 15
Huang, E., 1
Huang, F., 11
Huang, H., 368
Huang, J., 100
Huang, W., 290
Huang, Y., 367
Huard, C., 387
Hubbell, E., 75, 389
Huber, W., 23, 74, 120, 212, 230, 377, 380

Hubley, R., 117, 326, 360, 383
Huebert, D. J., 272
Hughes, J., 38
Hughes, J. D., 203
Hughes, J. E., 80
Hughes, R. E., 410
Hughes, T. R., 19, 21, 23, 52, 53, 63, 274
Hulo, N., 376
Humbert, R., 273
Humble, M. C., 61, 63
Hummerich, L., 38
Hunter, C. P., 56
Hunter, L., 406
Hunter, S., 15
Hurban, P., 19, 21, 22, 24, 61, 63
Hwang, J. T., 228
Hyde, L., 341

I

Iacobuzio-Donahue, C. A., 92
Iacus, S., 74, 120, 212, 230, 377
Ibrahim, J. G., 324
Icahn, C., 71
Ideker, T., 409, 410, 413
Idury, R., 233
Ihaka, R., 120
Ihmels, J., 377, 380
Ikeo, K., 373
Ikonomi, P., 69
Imoto, S., 204, 206, 207, 210
Impey, S., 15
Iordan, D., 117, 326, 360, 383
Iorio, M. V., 15, 16, 21, 22
Ireland, A., 376
Irizarry, R. A., 20, 23, 60, 63, 69, 74, 75, 96, 120, 212, 224, 230, 231, 246, 257, 290, 298, 377, 380, 422
Ishida, S., 1
Ishi-i, J., 373
Israel, M. A., 199, 200
Issel-Tarver, L., 179, 327, 401, 412, 417
Iuliano, R., 16, 21, 22
Iyer, V. R., 291, 370

J

Jacks, T., 16, 18, 21, 23
Jackson, D. G., 318, 429
Jackson, S., 178

Jaenisch, R., 410
Jaffee, H. A., 75
Jain, A. K., 388
Jaiswal, P., 376
James, R. J., 15
Jan, C., 410
Jansen, R., 292
Jarnagin, K., 233
Jarvis, R., 15, 16, 18
Jatkoe, T. A., 55, 58
Jedicka, A. E., 96
Jedlicka, A. E., 63
Jeffrey, S. S., 199
Jenner, R. G., 410
Jensen, R. V., 1
Jenssen, T., 63
Jeong, H., 1
Jiang, M., 179
Jin, H., 384
Jin, W., 219
Jing, L., 61, 63
Job, C., 341
Johansson, B., 100
Johnsen, H., 199
Johnson, C., 3
Johnson, C. D., 14
Johnson, G. S., 173
Johnson, S. M., 15, 16, 18
Johnson, S. W., 199
Johnston, R., 147, 233, 251
Johnston, R. J., Jr., 15
Johnston, W. K., 410
Johnstone, S. E., 410
Jones, A. R., 52, 53, 63, 274
Jönsson, G., 92, 100
Jooste, P., 425
Jordan, I. K., 368
Jordann, P., 425
Jovin, T. M., 95
Judo, M. S. B., 233
Juhl, H., 18
Jurgensen, S., 14

K

Kadin, J. A., 342
Kahn, D., 376
Kakinohana, O., 100
Kaloper, M., 384
Kamal, M., 272

Kamberova, G., 42, 44
Kampa, D., 291
Kanapin, A., 376
Kane, M. D., 55, 58
Kanehisa, M., 342
Kans, J. A., 364
Kantoff, P. W., 318, 429
Kao, M. C., 368
Kapanidis, A. N., 94
Kapranov, P., 283, 291
Kapushesky, M., 96, 117, 120, 137, 143, 353, 370, 371, 377, 381
Karaca, M., 68
Karlsson, E. K., 272
Karov, Y., 19, 21, 22, 24
Karube, Y., 15
Kasarskis, A., 179, 327, 384, 401, 412, 417
Kaufman, L., 243
Kaufmann, W. K., 61, 63
Kavanagh, T. J., 61, 63
Kawamoto, S., 9
Kawasaki, E. S., 63, 64, 69, 96
Kay, K. A., 80
Kay, M. A., 367
Kaysser-Kranich, T., 64, 69
Keating, M., 15
Keenan, M. R., 93, 94
Keiger, K., 18, 19
Kelley, B. P., 410
Kemmeren, P., 143, 371, 377
Kendall, S. L., 223
Kenton, D. L., 364
Kenzelmann, M., 341
Keppel, G., 177
Kerr, K., 69
Kerr, K. F., 56, 59, 60, 61, 63
Kerr, M. K., 80, 214, 219, 225, 229, 232, 233, 308
Kessler, D. A., 82
Khanlou, N., 1
Khovayko, O., 364
Khrapko, K., 15
Kibbe, W., 376
Kidd, M. J., 274
Kilian, K. A., 82
Kim, C., 64
Kim, I., 63
Kim, I. F., 96, 109, 117, 124, 139, 326, 354, 359
Kim, S. K., 179, 414
Kim, S. Y., 368
Kim, T. H., 272, 275, 277

King, A. M., 274
King, K. S., 15
King, O. D., 409
King, R. D., 341
Kinzler, K. W., 18
Kipps, T., 15
Kipps, T. J., 15, 16, 21, 22
Kiraly, M., 179
Kiser, G., 69
Kishore, R., 376
Kitareewan, S., 176, 202
Kittrell, F. S., 343
Klacansky, I., 352
Klaeren, R., 341
Klapa, M., 206, 207, 211, 289, 290, 384
Kleerebezem, M., 100
Klein, M. E., 15
Klein, T. E., 327
Klewe-Nebenius, A., 341
Klisovic, M., 39, 40
Klitgord, N., 410, 414
Kloppel, G., 64
Kluger, Y., 293
Klus, G., 64
Knublauch, H., 327
Ko, M. S., 52, 54, 58
Kobayashi, M., 51, 340
Kobayashi, S., 52, 53, 63
Koch, J. E., 82
Koch, W. H., 69
Koehn, S., 100
Kohane, I. S., 63, 64, 123, 193
Kohanen, T., 176
Kohavi, R., 321
Kohn, K. W., 173
Kohonen, T., 202
Kolaja, K. L., 233
Koller, D., 414
Kollias, G., 22
Konishi, H., 15
Koo, S., 19
Koonin, E. V., 368
Korn, B., 341
Korn, E. L., 177
Kornacker, K., 39, 40
Korner, C., 377
Kosaka, A., 56
Koshland, D. E., 274
Kosik, K. S., 15
Kostukovich, E., 206, 207, 211, 289, 290, 384

Koth, L., 63
Kothapalli, R., 63
Kotula, P. G., 93, 94
Kotze, J., 425
Krainc, D., 1
Kramer, H., 38
Krause, A., 347
Krawetz, S., 342
Kreder, D. E., 56
Kreiner, T., 314
Kremen, W. S., 1
Krestyaninova, M., 376
Krichevsky, A. M., 15
Kriel, D., 51
Kristensen, G. B., 199
Kruglyak, S., 175
Kuhbacher, T., 64
Kuiper, M., 410, 419
Kulbokas, E. J. III, 272
Kull, M., 377
Kulldorff, M., 316
Kumamoto, K., 21, 22
Kumar, R. M., 410
Kunjathoor, V. V., 100
Kunugi, R., 100
Kuo, F. C., 221
Kuo, W., 63
Kurachi, K., 1
Kurschner, C., 410
Kusba, J., 85, 92
Kwitek, A. E., 53, 56, 58
Kwok, P. Y., 73
Kwong, K. Y., 153
Kzallasi, Z., 64

L

Labourier, E., 15, 16, 18, 19
Lachaux, A., 234, 242
LaCount, D. J., 410
Ladd, A., 39, 40, 41
Ladd, C., 318, 429
Laing, W., 341
Lakowicz, J. R., 85, 92
Lamb, J., 16, 18, 21, 23, 343
Lamborn, K. R., 199, 200
Lamesch, P., 410, 414
Lan, H., 256
Lan, N., 292
Lander, E. S., 176, 202, 272, 318, 343, 387, 429

Landfield, P. W., 1
Landsman, D., 283
Lane, C., 412
Lane, H. C., 179, 190, 341
Lanza, G., 16, 21, 22
Lao, K., 19
Laosinchai-Wolf, W., 18, 19
Lapage, M., 117
Lapidus, J. A., 61, 63
Lara, G. G., 143, 371
Larkin, J. E., 64, 96
Lasarev, M. R., 61, 63
Lash, A. E., 117, 143, 147, 353, 373
Lassen, C., 100
Latham, G. J., 1, 11
Lau, D. T. Y., 234
Laurence, T., 94
Laurent, M., 218
Laurenzi, I., 303
Laurie, G. W., 367
Laurila, E., 343
Lavrov, S., 373, 393
Lawlor, S., 341
Leary, R. J., 18
Lebruska, L. L., 218
Ledoux, P., 117, 353
Lee, H., 63, 96
Lee, K. Y., 69
Lee, M. A., 100
Lee, M. D., 233
Lee, M. L., 221, 233
Lee, N., 38, 80, 298
Lee, N. H., 53, 56, 58, 151, 162, 167
Lee, R. C., 15
Lee, T. I., 410
Lee, V., 342, 376
Lefkowitz, S. M., 52, 53, 63
Lehar, J., 343
Lehrach, H., 64, 80, 291
Leiber, M., 11
Leisch, F., 74, 120, 212, 230
Lekso, L. J., 69
Lempicki, R. A., 179, 190, 341
Lendeckel, W., 18
Lenth, R., 222
Lepage, M., 326, 360, 383
LePage, R., 322
Lesage, G., 409
Letunic, I., 376
Levine, A. J., 405

Levine, S. S., 410
Levy, R., 175
Levy, S., 319
Lewis, B. P., 410
Lewis, F., 8
Lewis, S., 179, 327, 376, 401, 412, 417
Lewitter, F., 408, 410
Li, A., 218
Li, C., 15, 21, 23, 74, 120, 212, 223, 230, 246
Li, F. L. C., 377
Li, J., 61, 63, 134, 151, 162, 167, 206, 207, 211, 223, 289, 290, 298, 341, 384
Li, N., 410, 414
Li, S., 410, 414
Li, W., 275
Li, Y. J., 61, 63
Liang, W., 134, 151, 162, 167, 206, 207, 211, 223, 289, 290, 298, 384
Liang, Y., 199, 200
Liao, J. C., 80
Liao, P.-S., 151
Liaw, A., 431
Libutti, S. K., 2
Lichter, P., 38
Lieb, J. D., 274, 275, 277
Liew, C. C., 1
Lightfoot, S., 11
Liles, M. R., 93
Lilja, P., 117
Lim, J., 410, 414
Lim, L. P., 15, 410
Limpicki, R., 64
Lin, D., 178, 298
Lin, D. M., 162, 164
Lin, W., 100
Lindblad-Toh, K., 272
Lindgren, D., 100
Lindren, C. M., 343
Linsley, P. S., 52, 53
Lipman, D. J., 364, 365, 366
Lipshutz, R., 134, 290
Littell, R. C., 218
Liu, C., 8, 69
Liu, C.-G., 15, 16, 19, 21, 22
Liu, F., 100
Liu, J., 335
Liu, P. Y., 61
Liu, W. M., 75, 389
Liu, X., 274
Liu, X. S., 275

Liu, Z. L., 69, 206, 207, 211, 289, 290, 384
Livstone, M. S., 412
Llamosas, E., 410, 414
Lobenhofer, E. K., 19, 21, 22, 24, 61, 63
Lockery, S., 15
Lockhart, D. J., 51, 283, 290, 340
Lockner, R., 64
Loda, M., 318, 429
Loh, M. L., 387
Lomax, J., 376
Long, A., 304
Long, J., 283, 291
Longacre, T. A., 199
Longman, C., 367
Longnecker, M., 215
Lonsdale, D., 376
Lopez, F., 73
Lopez, R., 342, 376
Lord, P., 327
Loughran, T. J., 63
Lovlie, R., 58
Lovrecic, L., 1
Lu, A., 283
Lu, J., 15, 16, 18, 21, 23, 55, 58, 147, 233
Lu, T., 64
Lu, X., 61, 63
Lu, Y., 61, 73
Lu, Z., 90
Lucas, A., 69
Lucas, A. B., 218
Lueking, A., 291
Lum, P. Y., 274
Lund, J., 179
Luo, S., 273, 280
Luscombe, N., 167, 273, 290, 300
Luster, A. D., 100
Luts, L., 100
Luu, P., 162, 164, 255, 298
Luu, T. V., 53, 56, 58
Lyianarachchi, S., 39, 40
Lyman, 215
Lyng, H., 74, 87, 100

M

Mack, D., 405
Madabusi, L. V., 1
Madden, T. L., 364
Madore, S. J., 55, 58
Maechler, M., 74, 120, 212, 230, 377

Maere, S., 410, 419
Maglott, D. R., 73, 364
Magrane, M., 342
Mah, N., 64
Mailman, M. D., 335
Maira, G., 15
Mak, R. H., 16, 18, 21, 23
Malek, R. L., 53, 56, 58, 61, 63
Malicka, J., 85, 92
Malik, A., 80
Malley, J., 233, 316
Manduchi, E., 2, 335
Mane, S., 63
Mangiola, A., 15
Manly, B. F. J., 177
Manohar, C. F., 69
Manola, J., 318, 429
Manova, K., 388
Mansfield, J. H., 15
Mao, M., 52, 53, 63
Mao, X., 342
Marcellin, P., 234
Marcucci, G., 21, 22, 39, 40
Margulies, E. H., 273, 280
Mariani, T., 64
Marincola, F. M., 2
Marino-Ramirez, L., 368
Markel, S., 117, 326, 360, 383
Markesbery, W. R., 1
Markham, N. R., 74
Markiel, A., 413
Markowitz, V., 109, 117, 124, 139, 326, 354, 359
Marks, J. R., 1
Marks, W. L., 117, 326, 360, 383
Maron-Katz, A., 403
Marsala, M., 100
Marshall, B., 376
Marshall, M., 15
Martin, D. E., 389
Martin, M., 214, 219, 225, 233, 308
Martin, M. J., 327
Martinez, M. J., 80, 91, 94, 95
Martinez, R. V., 343
Martinez-Murillo, F., 63, 96
Martinot-Peignoux, M., 234
Marton, M. J., 52, 53, 63, 82, 274
Mas, V., 39, 40, 41
Masiello, C., 273
Masirov, J., 176
Maslen, J., 342

Masuda, N., 9
Masys, D., 352
Matese, J. C., 71, 109, 117, 124, 139, 179,
 199, 326, 327, 353, 354, 359, 373, 384,
 401, 412, 417
Matzke, M., 15
Maughan, N. J., 8
McArthur, M., 273
McCarrey, J. R., 388
McClure, J., 43
McCormick, M., 290
McLean, D. J., 388
McMahon, S., 272
McManus, M. T., 15
McShane, L. M., 177
McSorley, C., 233
Means, T., 100
Mecham, B., 64
Megee, P. C., 274
Mei, R., 75, 389
Meiri, E., 19, 21, 22, 24
Mellgard, B., 64
Meloon, B., 19
Melton, D. A., 410
Menzel, W., 11
Mesirov, J., 202
Mesirov, J. P., 343, 387
Mestril, R., 100
Metzler, M., 15
Meyer, C. A., 275
Meyer, M. R., 52, 53, 63, 82, 274
Michael, J. R., 93, 94
Michael, M. Z., 15
Michaels, G. S., 199
Michalet, X., 94
Michaud, J., 60
Michelson, A. M., 60, 304
Michoud, K., 327
Miles, M. F., 130
Miller, M., 117, 326, 360, 383
Miller, P., 292
Miller, W., 365, 366
Milligan, G., 203
Milliken, G. A., 218
Milstein, S., 410, 414
Milton, S., 61, 63
Minor, J., 147, 233, 235, 236, 239, 241, 242,
 246, 248, 251
Minor, J. M., 90, 120, 136, 145, 220, 233, 234,
 235, 242

Minton, N., 40
Miska, E. A., 15, 16, 18, 19, 21, 23
Mistrot, C., 274
Mitchell, T., 292
Mitelman, F., 100
Mitsudomi, T., 15
Mittmann, M., 51, 340
Miyada, C. G., 129, 136, 218, 227,
 257, 347
Miyada, G., 69
Miyano, S., 204, 206, 207, 210, 368
Miyasato, S., 412
Miyoshi, S., 15
Modrusan, Z., 69
Molenaar, D., 100
Molla, M., 290
Moller, R., 327
Monden, M., 9
Monks, A. P., 173
Montgomery, R., 233, 251
Montini, E., 367
Moore, K. J., 100
Mootha, V. K., 343
Morgan, T. E., 368
Morris, Q., 19, 21, 23
Morsberger, L., 63, 96
Mörse, H., 100
Moseley, J. M., 100
Moueiry, A., 61
Mount, D. M., 252
Mourelatos, Z., 15
Mouse Genome Database Group, 342
Moustakas, A., 22
Mueller, O., 11
Mukherjee, G., 117
Mulder, N. J., 376
Mullikin, J. C., 273
Mundodi, S., 376
Mungall, C., 376
Munroe, D., 64, 367
Muntoni, F., 367
Murakami, Y., 100
Murillo, F., 380
Murray, H. L., 272, 410
Musen, M. A., 327
Mutter, G. L., 8
Myers, E. W., 365, 366
Myers, R. M., 73
Myers, T. G., 173
Myklebost, O., 74, 87, 100

N

Nadler, S. T., 256
Nagalla, S. R., 61, 63
Nagarwalla, N., 316
Nagino, M., 15
Naiman, D., 233, 315, 316, 319, 324
Naiman, D. Q., 312
Nair, R. V., 233
Nakai, H., 367
Nakayama, J., 100
Narasimhan, B., 167, 178, 319
Nash, R. S., 376, 412
Nasim, S., 39, 40, 41
Natoli, A. L., 100
Natsoulis, G., 233
Naylor, T. L., 100
Neal, S. J., 144, 167
Negre, N., 373, 393
Negrini, M., 15, 16, 19, 21, 22
Nelder, J. A., 234
Nelson, B. P., 93
Nelson, C., 52, 54, 58
Neuberg, D., 8
Neuhauser, M., 387
Neutzner, A., 179
Neville, M. C., 406
Nevins, J. R., 1
Newsom, D., 39, 40
Newton, M. A., 61
Ng, H., 291
Ngai, J., 298
Ngal, J., 162, 164
Ngau, W. C., 117, 353
Nguyen, D. V., 85
Nguyen, P., 233
Ni, L., 376
Ni, X., 90
Nicholas, M. K., 199, 200
Nicholls, N., 14
Nicholson, S. M., 233
Nickerson, D. A., 73
Nicolas, A., 100
Nielsen, T., 94
Nikiforidis, G., 22
Nikolaus, S., 64
Nilsson, B., 100
Nimura, Y., 15
Nitsch, R., 15
Nobel, A. B., 275, 277

Noble, W. S., 178
Noch, E., 15
Nolan, J., 204, 206, 207, 210
Norstedt, G., 367
Norton, J., 290
Notterman, D. A., 405
Novoradovskaya, N., 68
Novoradovsky, A., 68
Noy, N. F., 327
Nunez-Iglesias, J., 368
Nuttall, R., 147, 316
Nuttall, R. L., 233, 251
Nuwaysir, E., 290

O

O'Carroll, D., 15
O'Connell, C., 69
O'Connor, P. M., 173
O'Connor, S. M., 15
Oddy, L., 73
Odom, D. T., 272
O'Donovan, C., 327
O'Donovan, M., 314
O'Dwyer, P. J., 199
Oelmueller, U., 14
Oezcimen, A., 117, 143, 371
Oh, M., 80
Ohkawa, H., 364
Ohnishi, T., 9
Ohno-Machado, L., 63
Okamoto, A., 21, 22
Okubo, K., 9
Oliver, B., 147, 233, 316
Oliver, D. E., 327
Oliver, S., 409
Olivier, J. M., 383
Olofsson, T., 100
Olshen, R. A., 424, 425
Olson, J., 302
Olson, J. A., Jr., 1
Olsson, H., 100
Oltvai, Z. N., 80
Olyarchuk, J. G., 342
O'Malley, J. P., 61, 63
Orchard, S. E., 376
Ordija, C. M., 100
Orr, W., 64
Orwig, K. E., 388
Osada, H., 15

Ostell, J., 364
Ota, I., 410
Otsu, N., 151
Ott, R., 215
Ott, S., 368
Oughtred, R., 412
Owen, M., 314
Ozier, O., 410, 413
Ozyildirim, A. M., 367

P

Paarlberg, J., 64
Page, G. P., 22
Pagliarini, R., 18
Pagni, M., 376
Pahl, A., 3
Palaniappan, C., 64
Palmer, A. E., 95
Palmer, V. S., 61, 63
Palumbo, T., 16, 21, 22
Pan, W., 177
Pandolfi, P. P., 388
Parish, T., 223
Parisi, M., 147, 233, 316
Park, J., 412
Parker, B. S., 100
Parker, J., 19
Parkes, H., 69
Parkinson, H., 71, 96, 109, 117, 120, 124, 137, 139, 143, 179, 325, 326, 343, 353, 354, 359, 370, 371, 373
Parmigiani, G., 92
Pasolli, H. A., 15
Pasquinelli, A. E., 15
Passador-Gurgel, G., 219
Passerini, A. G., 2
Patel, S., 283, 291
Pattee, P., 61, 63
Patterson, N., 343
Patterson, P., 343
Patterson, T., 78
Paules, R. S., 61, 63, 215, 222, 224, 228
Paull, K. D., 173
Pavlidis, P., 178
Pearson, C. I., 233
Pease, A. C., 387
Peck, D., 16, 18, 21, 23
Peltier, H. J., 11
Peng, V., 162, 164, 298

Pepperkok, R., 94
Peralta, E., 19
Perera, R. J., 19
Perkins, R. G., 78
Perou, C. M., 19, 61, 63, 68, 199
Perriere, G., 179, 381
Pesich, R., 68
Petersen, D., 63, 64
Petersen, P., 303
Peterson, C., 100
Peterson, D., 96
Petricoin, E. F. III, 69
Petrocca, F., 16, 21, 22
Pevsner, J., 300
Peyruc, D., 376
Pfeffer, R., 64
Pfughoefft, M., 94
Pham, H., 233
Phan, I., 327
Phang, T. L., 406
Phansalkar, A., 410
Phatak, A. G., 6
Phillips, K., 61, 63
Piccolboni, A., 283, 291
Pichiorri, F., 16, 21, 22
Pienta, K. J., 1
Pilbout, S., 327
Pinaud, F., 94
Pirrung, M., 283
Pisani, R., 343
Pitas, A., 290
Pizarro, A., 117, 326, 360, 383
Pizarro, A. D., 335
Plasterk, R. H., 15
Polacek, D. C., 2
Pollock, T., 290
Ponting, C. P., 376
Pontius, J. U., 364
Popov, D., 206, 207, 211, 289, 290, 384
Porter, D., 290
Porter, N. M., 1
Poustka, A., 23, 341, 380
Powell, S., 2
Powers, P., 18, 19
Prandini, P., 367
Pretlow, T. G., 2, 6
Priebe, C., 315, 316, 324
Primig, M., 368, 398
Prueitt, R. L., 16, 21, 22
Pruitt, K. D., 73, 364

Prydz, H., 100
Puigserver, P., 343
Puri, R. K., 64, 69, 78
Purves, R., 343

Q

Qi, R., 38, 80
Qian, J., 15, 21, 23, 293
Qian, M., 2
Qin, L. X., 56, 59, 60, 61, 63
Qiu, J., 228
Qiu, Y., 61, 63
Qu, C., 272, 275, 277
Quackenbush, J., 22, 38, 53, 56, 58, 63, 64, 69, 71, 80, 96, 100, 109, 117, 124, 134, 139, 151, 153, 162, 163, 165, 167, 193, 206, 207, 211, 223, 224, 289, 290, 298, 326, 341, 353, 354, 359, 371, 373, 384
Quigley, S. D., 61, 63
Quirke, P., 8
Quist, H., 199

R

Raaka, B., 64
Rabaglia, M. E., 256
Rabert, D., 56
Råde, J., 100
Ragg, T., 11
Rahman, N., 100
Rai, K., 15
Rainen, L., 14
Rakic, P., 19
Ramage, D., 413
Ramaswamy, S., 50, 343
Ramirez, L., 73
Ramirez-Yanez, M., 218
Randall-Maher, J., 73
Rassenti, L., 15, 16, 21, 22
Rattan, S., 15
Raychaudhuri, S., 178
Raymond, C. K., 18
Rayner, T., 117
R Development Core Team, 100, 432
Read, J., 283
Rector, A., 327
Reguly, T., 410
Reid, L., 69

Reimers, M., 119, 160, 167, 206, 212, 230, 231, 349, 377
Reinert, K. L., 15, 16, 18
Reinhart, B. J., 15
Reisdorf, W. C., 341
Ren, B., 272, 275, 277
Renshaw, A. A., 318, 429
Restall, C. M., 100
Retterer, J., 73
Rezantsev, A., 206, 207, 211, 289, 290, 384
Rhee, S. J., 100
Rhee, S. Y., 376
Richardson, J. E., 179, 327, 342, 401, 412, 417
Richie, J. P., 318, 429
Richmond, T. A., 272, 275, 277, 290
Richter, J., 376
Rico-Bautista, E., 367
Ridderstraele, M., 343
Rietdorf, J., 94
Rifkin, S. A., 227
Riley, R. M., 219
Rinaldi, N. J., 272
Ring, J. E., 401
Ringborg, U., 11
Ringnér, M., 100
Ringwald, M., 71, 179, 327, 353, 371, 373, 376, 412, 417
Rinn, J., 291, 303
Ripault, M. P., 234
Ripley, B., 243
Ritz, C., 100
Rivals, E., 347
Roberts, B. S., 18
Roberts, C. J., 274
Roberts, E. A., 234, 242
Robinson, A., 109, 117, 124, 139, 326, 354, 359, 360, 383
Rocca-Serra, P., 71, 117, 143, 371
Rodland, M., 61, 63
Rodriguez, A., 73
Rodriguez, A. L., 80, 91, 95
Rodriguez, M., 63
Rogaev, E. I., 15
Rogers, S., 73
Rogojina, A., 64
Rohlf, F. J., 215
Rohlin, L., 80
Roldo, C., 16, 21, 22
Rolfe, P. A., 272

Roman-Roman, S., 178
Roos-van Groningen, M. C., 11
Rosa, G. J. M., 218, 220
Rosas, H. D., 1
Rosenberg, J., 410, 414
Rosenthal, P., 234, 242
Ross, D. T., 199
Ross, K., 429
Rossini, A. J., 74, 120, 212, 230, 377
Rosslein, M., 66
Roter, A. H., 233
Roth, F. P., 205, 409, 410, 414
Rougvie, A. E., 15
Rouillard, J. M., 74
Rousseauw, J., 425
Royce, T. E., 167, 273, 290, 291, 300
Rozowsky, J. S., 167, 273, 291, 303
Rual, J. F., 410, 414
Rubin, D. L., 327
Rubin, G. M., 179, 327, 376, 412, 417
Rubin, M. A., 1
Rubin-Bejerano, I., 411
Rubinstein, L. V., 173
Ruch, R., 94
Rudnev, D., 117, 353
Rudolph, M., 406
Ruff, D. W., 3
Russell, 51
Russell, D., 35, 37, 39
Russell, R. B., 15
Rusyn, I., 61, 63
Ruvkun, G., 15
Ruzzo, W. L., 176, 205
Ryan, K. W., 193
Ryder, T. B., 69, 75
Ryltsov, A., 206, 207, 211, 289, 290, 384

S

Saal, L. H., 99, 100, 132, 135
Sabatino, G., 15
Sabo, P. J., 273
Sabripour, M., 22
Saeed, A. I., 134, 151, 162, 167, 206, 207, 211, 223, 289, 290, 298, 341, 384
Sahasrabudhe, S., 410
Sahni, H., 335
Salazar, A., 73
Saldanha, A.J., 206, 210
Salit, M., 55, 63, 66, 69

Salomonis, N., 341
Salowsky, R., 11
Salzberg, S. L., 73
Samaha, R. R., 69
Samanta, M., 291
Sambrook, J., 35, 37, 38, 39, 42
Samson, L. D., 61, 63
Sansone, S. A., 71, 117, 143, 353, 371, 373
SantaLucia, J., Jr., 74
Sarkans, U., 96, 109, 117, 120, 124, 137, 139, 143, 326, 353, 354, 359, 360, 370, 371, 377, 383
Sarkar, D., 61
Sarles, J., 234, 242
Sasik, R., 1
Sausville, E. A., 173
Sawitzki, G., 74, 120, 212, 230, 377
Scacheri, P. C., 218, 270, 373
Scafe, C., 291
Scarpa, A., 16, 21, 22
Schaefer, C., 335
Schaner, M. E., 199
Scharpf, R. B., 92
Scheideler, M., 341
Schelter, J. M., 52, 53, 63
Schena, M., 34, 80, 81, 85, 89, 134, 283, 340, 387
Scherf, U., 60, 69, 78, 231, 257, 290, 422
Schermer, M. J., 82
Schlingemann, J., 38
Schmid, W., 341
Schmutz, J., 73
Schneider, M., 327
Schoenfeld, L. W., 410
Schram, J., 14
Schreiber, J., 272
Schreiber, S., 64
Schreiber, S. L., 272
Schrenzel, J., 93
Schriml, L. M., 364
Schroeder, A., 11
Schroeder, M., 412
Schuchhardt, J., 80
Schueler, K. L., 256
Schuetz, G., 341
Schuler, G., 347
Schuler, G. D., 364
Schulman, B. R., 15
Schultz, N., 388
Schultz, R. A., 94

Schulze-Kremer, S., 109, 117, 124, 139, 326, 353, 354, 359
Schumacher, S., 15
Schummer, M., 319
Schwartz, D. A., 61, 63
Schwarz, E. M., 376
Schwikowski, B., 410, 413
Scott, A., 63, 96
Scott, H. S., 60, 341
Scudiero, D. A., 173
Sealfon, S., 64
Segal, E., 414
Segal, M. R., 61
Seidl, T., 295
Seigfried, T., 376
Seike, M., 21, 22
Seiler, A., 15
Sekinger, E., 291
Selengut, J. D., 376
Sellers, W. R., 318, 429
Sementchenko, V., 283, 291
Sendera, T. J., 64, 69
Senger, M., 117, 326, 360, 383
Senske, R., 387
Seo, J., 207
Sequeira, E., 364
Sequera, D. E., 193
Sequerra, R., 410, 414
Servant, F., 376
Sestan, N., 19
Sethuraman, A., 376, 412
Setterquist, R. A., 69
Sever, N. I., 16, 21, 22
Sevignani, C., 15, 16, 19, 21, 22
Shaffer, J., 344
Shaffer, J. P., 178, 232
Shah, A., 342
Shah, R., 1
Shah, S., 42, 44
Shai, O., 23
Shalon, D., 34, 283, 340, 387
Sham, P., 314
Shamir, R., 175, 205, 403
Shaner, N. C., 95
Shannon, K. W., 52, 53, 63, 218
Shannon, P., 413
Shao, J., 322
Shao, Q.-M., 324
Sharan, R., 403, 409, 410
Sharma, A., 15, 117

Sharon, E., 19, 21, 22, 24
Sharov, A. A., 52, 54, 58
Sharov, V., 134, 151, 153, 162, 167, 206, 207, 211, 223, 289, 290, 298, 341, 384
Shaw, C., 349
Shelton, J., 15, 18, 19
Shelton, M., 234, 242
Shen, R., 218
Sherlock, G., 71, 109, 117, 124, 139, 160, 179, 193, 194, 326, 327, 341, 343, 353, 354, 359, 360, 364, 371, 373, 380, 383, 384, 412, 417, 423, 428
Sherman, B. T., 179, 190, 341
Sherry, S. T., 364
Shi, C., 2
Shi, L., 64, 69, 70, 74, 78
Shi, Y., 61, 63, 74
Shiboleth, Y. M., 19, 21, 22, 24
Shiekhattar, R., 16
Shima, J. E., 388
Shimizu, M., 15, 16, 19, 21, 22
Shin, J. L., 61, 63
Shingara, J., 15, 16, 18, 19
Shippy, R., 64, 69
Shneiderman, B., 207
Shoemaker, D. D., 52, 53, 63, 274
Shojatalab, M., 96, 117, 120, 137, 143, 326, 353, 360, 370, 371, 383
Shtutman, M., 19, 21, 22, 24
Shumlevich, I., 39
Sibson, R., 204
Sieber, S. O., 61, 63
Siegel, A. F., 410
Sigrist, C. J. A., 376
Sihag, S., 343
Sikic, B. I., 199
Sikic, T. L., 199
Sikorski, R. S., 410, 414
Silventoinen, V., 376
Simon, C., 410, 414
Simon, J., 274
Simon, R., 177
Sinclair, M. B., 80, 91, 94, 95
Singer, M. A., 272, 275, 277
Singh, D., 318, 429
Singh, J., 290
Sirotkin, K., 364
Sistare, F., 69
Sistare, F. D., 69
Sjoblom, T., 18

Skatri, P., 342
Skinner, M. K., 388
Sklar, J., 221
Skomedal, H., 199
Skoog, L., 11
Skrzypek, M., 412
Slack, F. J., 15, 16, 18, 24
Slade, D., 274
Slifer, S., 61, 63
Sloan, E. K., 100
Slonim, D., 176, 193, 202
Slonim, D. K., 56, 123, 387
Small, C. L., 388
Smeekens, S. P., 75
Smirnova, L., 15
Smith, C., 74, 120, 212, 230, 377
Smith, S., 199
Smith, V., 8
Smola, A., 243
Smolyar, A., 410, 414
Smyth, G. K., 60, 74, 120, 212, 230, 231, 341, 377
Snesrud, E., 38, 80
Snuffin, M., 206, 207, 211, 289, 290, 384
Snyder, M., 167, 273, 290, 291, 292, 300, 303
Socci, N. D., 173
Sokal, E. M., 234, 242
Sokal, R. R., 215
Solas, D., 283, 387
Sollier, J., 100
Somerville, S., 164
Somogyi, R., 199
Soneji, S., 223
Song, B., 64
Sorger, P. K., 80
Soriano, J. V., 69
Sorlie, T., 199
Soukas, A., 173
Soustelle, C., 100
Southern, E. M., 35
Spana, E. P., 372
Spang, R., 1
Specer, F., 96
Speed, T., 177, 290, 298, 319, 432
Speed, T. P., 20, 23, 60, 63, 75, 80, 147, 150, 151, 162, 164, 231, 232, 233, 255, 257, 341, 380, 422
Speer, M. C., 61, 63
Spellman, P. T., 71, 109, 117, 124, 139, 173, 199, 206, 207, 326, 341, 353, 354, 359, 360, 364, 371, 373, 383, 384

Spencer, F., 63, 380
Spencer, P. S., 61, 63
Spiegelman, B., 343
Spitznagel, E. J., 64
Sproles, D. I., 61, 63
Staaf, J., 100
Stack, R., 233, 251
Stamatoyannopoulos, J. A., 273
Stanley, K. L., 100
Starchenko, G., 364
Stark, A., 15
Stark, C., 410
Starr, B., 412
Statnikov, A., 319
Staudt, L. M., 50, 64, 175
Steen, V. M., 58
Steffen, M. A., 87
Steibel, J. P., 218, 220
Steinbach, P. A., 95
Steinmetz, L., 283
Stephaniants, S. B., 52, 53, 63, 274
Stephens, R. M., 21, 22
Sterky, F., 164
Stern, D., 283, 291
Sternberg, P., 376
Stevens, J. R., 422
Stevens, R., 327, 328, 331
Stewart, J., 71, 109, 117, 124, 139, 326, 353, 354, 359, 360, 371, 383
Stocker, S., 11
Stockwell, B. R., 410
Stoeckert, C., 71, 109, 117, 124, 139, 353, 354, 359, 373
Stoeckert, C. J., 179
Stoeckert, C. J., Jr., 2, 71, 117, 325, 326, 335, 360, 371, 373, 383
Stoecket, C., 343
Stoehr, J. P., 256
Stoker, N. G., 223
Stokke, T., 74, 87, 100
Stolc, V., 291
Stone, C. J., 424, 425
Storey, J. D., 230
Storm, T. A., 367
Stoughton, R., 52, 53, 63, 274
Stoughton, R. B., 82
Stratton, M., 100
Strausberg, R., 291
Stroup, W. W., 218
Strovel, J., 64

Struhl, K., 291
Stryer, L., 283
Stuart, J. M., 178, 179, 327, 414
Stumpf, C. R., 55, 58
Sturme, M. H., 100
Sturn, A., 206, 207, 211, 289, 290, 384
Su, Z., 78
Subramanian, A., 343
Sugnet, C. W., 178
Suhre, K., 100
Suk, W. A., 61, 63
Sukhwani, M., 388
Sullivan, E. J., 387
Sullivan, R. C., 61, 63
Sultana, R., 64
Sultano, R., 96
Sultmann, H., 23, 380
Sumwalt, T., 290
Sun, D., 233
Sun, Y., 19
Surani, M. A., 19
Sussman, M., 290
Suzek, T. O., 117, 353, 364
Svendsrud, D. H., 74, 87, 100
Sweet-Cordero, A., 16, 18, 21, 23
Swenberg, J. A., 61, 63
Swiatek, M., 117, 326, 360, 383
Szafranska, A. E., 18

T

Tabin, C. J., 15
Tai, D., 273, 280
Tai, Y., 63
Tainsky, M., 342
Takahashi, T., 15
Takamizawa, J., 15
Tamayo, P., 123, 176, 193, 202, 318, 343, 387, 429
Tammana, H., 283, 291
Tamura, T., 373
Tan, P., 64
Tan, S., 233
Tan, S. S., 341
Tanaka, H., 15
Tanaka, T., 21, 22
Tang, F., 19
Tang, X., 15, 21, 23
Tanimoto, E. Y., 129, 136, 218, 227, 257, 347
Tanner, M. A., 3

Tapscott, S., 302
Tarakhovsky, A., 15
Tatematsu, Y., 15
Tateno, Y., 71, 353, 373
Tatusov, R., 364
Tatusova, T., 73
Tatusova, T. A., 364
Tavaria, M. D., 100
Tavazoie, S., 203
Taylor, R., 71, 109, 117, 124, 139, 326, 353, 354, 359, 371, 373
Taylor, S. L., 73
Tempelman, R. J., 218, 220, 222
Teng, N. N., 199
Tennant, R. W., 61, 63
Thanawala, M. K., 412
Theesfeld, C. L., 376, 412
Theilhaber, J., 178
Thelin, A., 64
Then, F., 1
Thiagarajan, M., 134, 206, 207, 211, 223, 289, 290, 341, 384
Thiolouse, J., 383
Thode, S., 233
t Hoen, P. A. C., 92
Thomas, J., 302
Thomas, J. D., 55, 58
Thomson, J. M., 19
Thorsen, T., 199
Tian, R., 61, 63
Tibshirani, R., 131, 167, 175, 178, 194, 199, 203, 214, 230, 303, 319, 322, 423, 424, 425
Tierney, L., 74, 120, 212, 230, 377
Tigue, Z., 357, 358
Timlin, J. A., 79, 80, 91, 94, 95, 136, 145
Todd, S. A., 61, 63
Tokiwa, G. Y., 82
Tollet-Egnell, P., 367
Tolley, A. M., 233
Tomida, S., 15
Tonellato, P., 376
Tong, A. H., 409
Tongprasit, W., 291
Torelli, S., 367
Torgerson, W. S., 259
Torrente, A., 377, 381
Tothill, R., 38, 41
Townsend, M., 19
Troein, C., 99, 100, 132, 135
Troendle, J. F., 177

Trollinger, D., 14
Troup, C., 117, 326, 360, 383
Troup, D. B., 117, 353
Troyanskaya, O., 199, 304
Trush, V., 206, 207, 211, 289, 290, 384
Tryon, V., 14
Tsai, M., 73
Tsamardinos, I., 319
Tseng, G. C., 80
Tsien, R. Y., 95
Tsuang, M. T., 1
Tu, D., 322
Tu, I. P., 199
Tu, S. W., 327
Tucher, V. G., 131
Tucker, C. J., 61, 63
Tucker-Kellogg, G., 56
Tugendreich, S., 233
Turk, G. C. T., 66
Tuschl, T., 15, 18
Tusher, V. G., 167, 178, 194, 214, 303, 423
Tyers, M., 410

U

Uhde-Stone, C., 218
Unal, E., 274
Urban, A., 291
Usary, J., 68
Uzumcu, M., 388

V

Vainer, M., 147, 233, 251
Valencia, A., 177
Vallon-Christersson, J., 92, 99, 100, 132, 135
van Bakel, H., 51
VanBuren, V., 52, 54, 58
Vance, C. P., 218
van de Belt, J., 15
Vandenhaute, J., 410, 414
van de Rijn, M., 199
van Dongen, S., 15, 24
Van Gelder, R. N., 2
van Holst Pellekaan, N. G., 15
Van Houten, B., 61, 63
van Laar, R., 38, 41
van Laar, R. K., 100
van Ommen, G. J. B., 92
van Osdol, W. W., 173

Vapnik, V., 243
Varambally, S., 1
Vasicek, T. J., 273, 280
Vaudaux, P., 93
Vaughan, E. E., 100
Vaughan, R., 376
Vecchione, A., 16, 21, 22
Velculescu, V. E., 18
Vendetti, J., 327
Venkatesan, K., 410, 414
Vernooy, S. Y., 15
Vidal, M., 410, 414
Viehmann, S., 15
Vignali, M., 410
Vilker, V., 69
Vilo, J., 109, 117, 124, 139, 143, 326, 353, 354, 359, 371, 377
Vingron, M., 23, 109, 117, 124, 139, 179, 326, 341, 347, 353, 354, 359, 380
Vinsavich, A., 206, 207, 211, 289, 290, 384
Visone, R., 16, 21, 22
Viswanadhan, V. N., 173
Vladimirova, A., 233
Vogelstein, B., 18
Volinia, S., 15, 16, 21, 22
Volkert, T. L., 272
von Heydebreck, A., 23, 380
von Zastrow, M. E., 2
Vranizan, K., 341

W

Wagar, E. A., 69
Wagner, L., 364
Walhout, A. J., 410
Walker, R., 56
Walker, W. A., 100
Walker, W. L., 52, 53, 63
Wallace, J. C., 273
Wallace, R., 290
Walter, G., 291
Walther, G., 203
Wang, B., 147
Wang, B. B., 233, 251
Wang, C., 51, 340
Wang, E., 2
Wang, H. Y., 53, 56, 58
Wang, J., 367
Wang, J. T., 413
Wang, L., 90

Wang, M., 100
Wang, N., 85
Wang, Q., 1
Wang, S., 151, 162, 167, 298
Wang, X., 100, 147
Wang, Y., 82
Wang, Y. C., 199
Ward, M. R., 100
Warren, D., 63, 96
Warren, K., 3
Warrington, J. A., 8, 69, 129, 136, 218, 227, 257, 347
Watson, N., 199, 200
Watts, G., 64
Webb, B. D., 273, 280
Weber, B. L., 100
Weber, J. M., 80, 91, 95
Webster, T. A., 75
Wedderburn, R. W. M., 234
Wei, L., 342
Weidhaas, J. B., 24
Weinstein, J., 64
Weinstein, J. N., 173
Weir, B., 215, 220, 227
Weis, B. K., 61, 63
Weiss, S., 94
Weissman, S., 291
Welch, B. L., 177
Welsh, J., 352
Wen, X., 199
Weng, S., 384, 412
Werner-Washburne, M., 80, 91, 94, 95
Wernisch, L., 223
West, A., 52, 53, 63
West, M., 1
Westwood, T., 144, 167
Wetmore, D., 64
Wetzle, 125
Wheeler, D. L., 364
Wheeler, J., 73
Wheeler, R., 291
Whetzel, P. L., 179, 325, 335, 343
White, J., 71, 117, 151, 162, 167, 206, 207, 211, 289, 290, 298, 326, 353, 360, 373, 383, 384
White, J. A., 134, 206, 207, 211, 223, 341, 384
White, K. P., 219, 227
White, N., 19
White, R., 376
Whitehouse, N., 349
Whitfield, M. L., 68

Whitley, M. Z., 56
Whitmore, G. A., 221, 233
Whittle, J., 273, 280
Wiener, M., 431
Wienholds, E., 15
Wietzorrek, A., 223
Wightman, B., 15
Wilda, M., 15
Wildsmith, S. E., 80
Wilhite, S. E., 117, 353
Wilkinson, D., 39, 40, 41
Williams, A., 64, 66, 291
Williams, J., 412
Williams, M., 69
Wilmer, F., 69
Wilson, M., 63, 69
Wilson, R. K., 73
Winegarden, N., 71, 353, 373
Winkler, H., 2
Winkley, A. J., 80
Winslow, R., 319
Winzeler, E., 283
Wison, M., 96
Wistow, G. J., 367
Wit, E., 43
Wittes, R. E., 173
Wodicka, L., 283
Wojcik, S. E., 15, 16, 21, 22
Wolber, P. K., 69, 218
Wolf, Y. I., 368
Wolfinger, E. D., 215, 222, 224, 228
Wolfinger, R. D., 215, 218, 219, 220, 222, 224, 227, 228, 230
Wolfsberg, T. G., 273, 280, 283
Wolgemuth, D. J., 388
Wolpert, D., 321
Wolski, E., 80
Wong, A., 368
Wong, B., 291
Wong, S. L., 409, 410, 414
Wong, W. H., 80, 246, 368
Wong, W. K., 68, 78
Wong, Y. C., 273
Wood, V., 376
Woodcock, J., 69
Wortman, J., 376
Wrobel, G., 38, 398
Wroe, C. J., 327, 328
Wu, C., 273
Wu, H., 229, 231

Wu, S. X., 56
Wu, W., 80
Wu, X., 69, 367
Wu, Y., 272, 275, 277
Wu, Z., 75, 246, 380
Wuestehube, L. J., 234
Wulczyn, F. G., 15
Wurmbach, E., 64
Wyatt, R., 94
Wylie, B. N., 179
Wyrich, R., 14

X

Xiao, P., 61
Xiao, Y., 61
Xie, Q., 78
Xu, P., 15, 64
Xu, Z. A., 78
Xuan, S., 61, 63

Y

Yakhini, Z., 175, 205
Yallop, R., 80
Yamanaka, M., 291
Yanagisawa, K., 15
Yanaihara, N., 16, 21, 22
Yandell, B. S., 256
Yang, I., 153, 298
Yang, I. V., 50, 151, 162, 167
Yang, J., 233, 341
Yang, J. Y., 73, 74, 120, 212, 230, 377
Yang, M., 15
Yang, Y., 63, 96, 298
Yang, Y. H., 61, 80, 147, 150, 151, 162, 164,
    177, 231, 232, 233, 255, 357, 358
Yant, S. R., 367
Yaschenko, E., 364
Yatabe, Y., 15
Yauk, C., 64
Ybarra, S., 405
Ye, S. Q., 63, 96
Yeakley, J. M., 218
Yeatman, T., 151, 162, 167, 298
Yekta, S., 15
Yekutieli, D., 344, 347
Yeung, K. Y., 176, 205
Yi, M., 21, 22
Yi, R., 15

Yoder, S., 63
Yokota, J., 21, 22
Yool, A., 2
Yooseph, S., 175
Yorke, J. A., 73
Yoshii, A., 19
Young, A., 273, 349
Young, B. D., 35
Young, G. P., 15
Young, R. A., 272, 410
Young, W., 100
Yu, H. G., 167, 273, 274, 293
Yu, W., 63, 96
Yu, Y., 153
Yuan, B., 410
Yue, H., 147, 233, 251
Yue, L., 341
Yuen, T., 64

Z

Zadro, R., 69
Zaharevitz, D. W., 173
Zahnow, C. A., 343
Zahrieh, D., 8
Zakeri, H., 73
Zamore, P. D., 15
Zar, J. H., 177
Zarbl, H., 61, 63
Zavaleta, J. R., 94
Zedeck, S., 177
Zeger, S., 300
Zhang, J., 74, 120, 212, 230, 377
Zhang, L., 130
Zhang, L. V., 409, 410, 414
Zhang, S., 74
Zhang, W., 19, 21, 23, 39
Zhang, Z., 15
Zhao, L., 302
Zheng, M., 272, 275, 277
Zhong, W., 303
Zhou, D., 273, 280
Zhou, J., 1
Zhou, X. J., 368
Zhou, Z., 233
Zhu, H., 291, 292
Zhu, Q., 176, 202
Zhu, X., 167, 273, 291
Ziman, M., 52, 53, 63
Zimmerman, T., 94

Zinn, K. E., 218
Zizlsperger, N., 272
Zoghbi, H. Y., 410, 414
Zong, Y. W., 74
Zoon, K. C., 69
Zou, F., 256

Zou, K., 303
Zucker, J. P., 410
Zuker, M., 74
Zupo, S., 15, 16, 19
Zuzan, H., 1

Subject Index

A

ACME, *see* Algorithm for Capturing
    Microarray Enrichment
Affymetrix GeneChip, *see* GeneChip
Algorithm for Capturing Microarray
    Enrichment
    chromatin immunoprecipitation and
        DNase hypersensitivity microarray
        data analysis
    assumptions, 275, 277
    data format, 277
    data properties, 273–274
    data quality assessment, 279–280
    p values, 277–278, 280–281
    plotting function, 280–281
    probe resolution optimization, 278–279
    window size and threshold
        optimization, 278
    overview, 271
    R programming, 271
Analysis of variance
    biological replication, 221
    data extraction and
        normalization, 222–226
    false discovery rate, 229–230
    fixed versus random effects, 215–218
    flow diagram in microarray design and
        analysis, 216–217
    gene-specific analysis of variance, 226–228
    history of use with microarrays, 214–215
    MeV data mining, 177–178, 188–189
    Partek Genomics Solution
        false discovery rate, 264–265
        gene list of interest generation, 264
        mixed model, 261–263
        nested relationships, 263
        random versus fixed effects, 263
        results analysis, 264
    power analysis, 221–222
    principles, 215
    quality control, 240, 242–243
    significance scoring, 305–308

significance thresholds, 228–230
single-channel array analysis, 220
software, 230–231
technical replication, 221
two-color array analysis, 218–220
ANOVA, *see* Analysis of variance
ArrayExpress
    components, 373–374
    Expression Profiler
        clustering analysis, 380–381
        comparative group analysis, 381, 383
        data selection, 278–379
        data transformation, 380
        differentially expressed gene
            identification, 380
        exporting graphics and results, 383
        functions, 377–378
        normalization, 380
    goals, 373
    growth and content type, 371–373
    origins, 370–371
    prospects, 384
    query
        data warehouse database, 375–377
        repository query interface, 374–375
    submission of data, 383–384
    users, 373–374

B

BASE, *see* BioArray Software Environment
Between group analysis, Expression
    Profiler, 381, 383
BGA, *see* Between group analysis
BioArray Software Environment
    data analysis
        BioAssays and sets, 114–115
        filtering, 115–116
        MAGE-ML file export, 117–118
        normalization, 116
    demo availability, 100–101
    filtering, 103–104
    hardware requirements, 101

461

BioArray Software Environment *(cont.)*
  installation, 101–102
  menus
    Array LIMS, 108–109
    array menus, 109–110
    File format, 106–108
    Hybridizations, 111–112
    Jobs, 116–117
    Plates, 109
    plug-ins, 117
    Protocols, 106
    Reporters, 108
    sample menus, 110–111
    Scans, 112
    Upload, 106–107, 112
  overview, 100
  ownership and access rights, 104–105
  user administration, 105–106
  user interface, 102–103
Bioconductor
  containers
    metadata, 124–126
    overview, 120–121
    preprocessed microarray data, 121–122
    processed microarray data, 122–124
  distribution, 120–121
  documentation strategies, 127
  multiple comparisons, 131–132
  objectives and prospects, 132–133, 212
  preprocessing of arrays
    GeneChip, 129–131
    quality control, 127–129
  R programming, 212, 230, 349
  workflows, 120, 126–127
Boolean clustering, large data sets, 399, 401

C

CDF, *see* Cumulative distribution function
ChIP, *see* Chromatin immunoprecipitation
Chromatin immunoprecipitation, microarray
  combination
  data analysis
    Algorithm for Capturing Microarray
      Enrichment
      assumptions, 275, 277
      data format, 277
      p values, 277–278, 280–281
      plotting function, 280–281
      probe resolution optimization, 278–279

      window size and threshold
        optimization, 278
    comparison of approaches, 274–275
    data properties, 273–274
    data quality assessment, 279–280
    principles, 272–273
    tiled arrays, 271
Cluster
  data transformation, 208–209
  filtering, 206, 208
  hierarchical clustering, 209
  normalization, 209
  version 3.0, 210–211
Clustering, microarray data analysis
  agglomerative hierarchical clustering
    average linkage, 201
    centroid linkage, 201
    complete linkage, 201
    drawbacks, 201–202
    node comparison rules, 199
    overview, 198–199
    single linkage, 199
  cluster affinity search technique, 175, 205
  computational considerations, 204–205
  data partitioning
    K-means clustering, 203
    number of partitions, 204
    self-organizing maps, 202–203, 205
  distance metrics
    correlation metrics, 195
    Euclidian distance, 197–198
    Kendall's τ, 197
    Manhattan distance, 198
    Pearson correlation, 196
    Spearman rank correlation, 197
  Expression Profiler, 380–381
  K-means/K-medians
    clustering, 173–175, 203
  large data set analysis
    Boolean clustering, 399, 401
    combinatorial clustering, 398–399
    gonad development data set for
      analysis, 388–389
    overview of problem, 387–388
    rank difference analysis of microarray
      method, 389–390
    relationships between experimental
      points, 392–395
    replicate reproducibility, 390–391
    similarity of comparison results, 396–397

total variation evolution across ordered comparisons, 395
transcriptional networks, 401, 403, 405–406
MeV data mining
  agglomerative methods, 173
  biological theme discovery, 179–180
  classification algorithms and supervised clustering approaches, 178
  confidence testing of clustering results, 175–176
  data representation and distance metrics, 169–171
  data visualization and component analysis, 178–179
  divisive clustering, 173–175
  hierarchical clustering, 173
  machine learning-based clustering, 176–177
  overview, 171–173
  significant gene list extraction, 177–178
Partek Genomics Solution, 260
permutation test, 316–317
Random Forests, 428–429
COA, *see* Correspondence analysis
Correspondence analysis, Expression Profiler, 381, 383
Cumulative distribution function, pixel statistics, 237, 239
Cyber T, significance scoring, 303–304
Cytoscape
  Gene Ontology annotation visualization, 417, 419
  interaction data combination with expression data, 416–417
  network visualization, 413–414
  prospects, 419
  viewing and filtering of networks, 414–416

D

Distance metrics, *see* Clustering, microarray data analysis
DNA microarray
  chromatin immunoprecipitation combination, *see* Chromatin immunoprecipitation
  clustering, *see* Clustering, microarray data analysis
  data acquisition, 288–290

databases, *see* ArrayExpress; Gene Expression Ominbus
DNase hypersensitivity, *see* DNase hypersensitivity, microarray combination
Monte-Carlo approach to statistical validation, *see* Random data set
nomenclature, 284
overview of steps, 79–81
quality control, *see* Quality control, microarrays
RNA controls, *see* External RNA controls, DNA microarray
RNA extraction, *see* RNA extraction
scanning, *see* Scanning, microarrays
software, *see* Algorithm for Capturing Microarray Enrichment; BioArray Software Environment; Bioconductor; Cluster; Partek Genomics Solution; Random Forests; TM4 software suite; XCluster
standards
  data exchange standards, 71–72
  gene expression process model
    array content, 73–74
    data analysis, 74–75
    detection, 74
    molecular biology segment, 72–73
    overview, 64–65
  measurement uncertainty, 67–68, 71
  method validation, 66–71
  prospects, 75
  traceability, 66, 68
  variability in experiments, 64–66
troubleshooting
  background fluorescence, 42–44
  hybridization quality assessment, 44–45, 48
  overview, 34–35, 37
  printing, 38–39
  sample preparation and labeling, 39–42
DNase hypersensitivity, microarray combination
data analysis
  Algorithm for Capturing Microarray Enrichment
    assumptions, 275, 277
    data format, 277
    p values, 277–278, 280–281
    probe resolution optimization, 278–279

DNase hypersensitivity, microarray
combination (cont.)
window size and threshold
optimization, 278
comparison of approaches, 274–275
data properties, 273–274
data quality assessment, 279–280
principles, 273
tiled arrays, 271

E

EASE, see Expression analysis systematic
explorer
Euclidian distance, distance metrics, 197–198
Expression analysis systematic explorer,
biological theme discovery, 179–180, 190
Expression Profiler, see ArrayExpress
External RNA controls, DNA microarray
applications, 50–51
array performance assessment, 55–56, 58
commercial availability, 52–54
data analysis methodology evaluation
using spike-in data sets, 59–61
External RNA Controls Consortium, 55
synthesis, 51, 58–59

F

False discovery rate
analysis of variance, 229–230, 264–265
Gene Ontology, 344
large data set analysis, 390, 396, 399
significance analysis of microarrays, 303
Family-wise error rate, Bonferroni
correction, 288
FDR, see False discovery rate
Fold change statistic, significance scoring, 301

G

GeneChip, Bioconductor
preprocessing, 129–131
Gene Expression Ominbus
browsing, 361
DataSet clusters, 364–365
downloading, 361
Entrez
advanced features, 364
DataSets, 362

profiles, 362, 364
qualifier fields, 363
GEO BLAST, 366
group A versus B tool, 367
links, 365–366
Profile Neighbors, 365
profile chart interpretation, 356–357, 359
prospects, 367–368
purpose, 353
query and analysis, 361–362, 364–367
Sequence Neighbors, 365
size, 353–354
structure
DataSets, 355–356
overview, 354–355
submitter-supplied data, 355
submission of data, 359–360
subset effect flags, 366
Gene expression terrain map, MeV, 179
Gene Ontology
Boolean clustering, 399, 401
branches, 328
Cytoscape and annotation
visualization, 417, 419
language, 328
microarray data interpretation
GOstat tool
algorithm, 348
description, 345–346
operation, 347–348
output, 348–349
visualization, 349
overview, 340–343
statistically over-represented terms
within group of genes, 343–345
tool resources, 349–350
ontology definition, 326–327
tools, 327–328
Gene shaving, microarray data
clustering, 175
GEO, see Gene Expression Ominbus
Globin, transcript removal from whole blood
samples, 4–6
GO, see Gene Ontology
GOstat, see Gene Ontology

H

Hierarchical clustering, see Clustering
HSS, see Hyperspectral scanning

Hybridization, microarrays, troubleshooting, 44–45, 48
Hyperspectral scanning, microarray scanning, 93–95

K

Kendall's τ, distance metrics, 197
KMC, *see* K-means/K-medians clustering
K-means clustering, data partitioning, 203
K-means/K-medians clustering, MeV, 173–175
KNNC, *see* K-nearest neighbor classifier
K-nearest neighbor classifier, MeV, 178

L

LeukoLOCK, RNA extraction from blood
    filter processing and cell lysis, 7
    RNA isolation and elution, 8
    sample collection and leukocyte capture, 7

M

Madam, *see* TM4 software suite
MAGE, *see* Microarray and Gene Expression
MAGE-ML, *see* Microarray Gene Expression Markup Language
Manhattan distance, distance metrics, 198
Mean, calculation, 285
Median, definition, 286
Median absolute difference, definition, 286
MeV, *see* TM4 software suite
MIAME, *see* Minimum Information about a Microarray Experiment
Microarray and Gene Expression, data exchange standards, 72
Microarray Gene Expression Data Society Ontology
    availability, 329
    class hierarchy, 330–331
    construction, 328–329
    databases and applications, 333
    Enterprise Vocabulary Service, 335–336
    ontology definition, 326–327
    prospects, 338
    RAD Study Annotator, 335
    releases and management, 336–337
    users, 331, 335
    uses, 328

Microarray Gene Expression Markup Language
    file export
        BioArray Software Environment, 117–118
        Madam, 143
    standards, 326
Microarray printing, troubleshooting, 38–39
Micro-RNA
    databases, 15
    functions, 14–15
    microarray analysis
        challenges, 16–17
        data analysis
            messenger RNA expression profiling comparison, 19–22
            normalization, 22–23
        human lung and placenta tissue analysis
            global normalization, 27
            hierarchical clustering, 28–29
            sample preparation, 23–24
            sample size calculation, 24
            scanning, 24
            statistical differential analysis, 27–28
            threshold calculation, 27
        micro-RNA
            labeling, 19
            purification, 18
        one-color versus two-color arrays, 17
        prospects, 29–30
    oncogenesis role, 15–16
Midas, *see* TM4 software suite
Minimum Information about a Microarray Experiment, standards, 71–72, 326
miRNA, *see* Micro-RNA
MO, *see* Microarray Gene Expression Data Society Ontology
Monte-Carlo simulation, *see* Random data set
Multiple testing, overview, 287–288

N

Network
    clustering of large data sets in transcriptional networks, 401, 403, 405–406
    definition, 408–409
    microarray applications, 409–410

Network *(cont.)*
  visualization tools
    Cytoscape
    Osprey, 410–413
Normalization, microarrays
  analysis of variance data, 222–226
  array location bias correction, 299–300
  background correction, 295–296
  Cluster software, 209
  Expression Profiler, 380
  gene set approach, 297
  micro-RNA microarrays, 22–23
  Partek Genomics Solution, 257
  principles, 283–284
  printing effects, 292–293
  quantile approach, 29
  signal intensity bias correction, 298–299
  spiked control approach, 297–298
  spot concentration approach, 300
  Spotfinder
    flip-dye consistency normalization and
      filtering, 165
    iterative linear regression
      normalization, 164
    iterative log-mean centering
      normalization, 163–164
    lowess normalization, 162–163
    overview, 162, 184–185
    ratio statistics normalization, 164
    standard deviation regularization, 164
    total intensity normalization, 162
  total intensity approach, 296–297
  two-channel experiments and log
    ratio, 293–295

O

Osprey
  availability, 410
  data set reduction, 412–413
  example of use, 411–412
  interaction database, 410–411, 413
  prospects, 413
Outlier, definition, 285
OWL, Web Ontology Language, 327

P

Partek Genomics Solution
  analysis of variance

false discovery rate, 264–265
  gene list of interest generation, 264
  mixed model, 261–263
  nested relationships, 263
  random versus fixed effects, 263
  results analysis, 264
data importing and normalization, 257
exploratory data analysis, 257
gene locus visualization, 267–268
hierarchical clustering, 260
multidimensional scaling, 259–260
obese mouse microarray
  experiment, 256–257
poststatistical analysis, 266–267
principal components analysis, 257–260
resources, 269
single gene analysis, 266
statistical analysis overview, 256
PCA, *see* Principal components analysis
Pearson correlation, distance metrics, 196
Permutation test, *see* Random data set
Photomultiplier tube, settings for microarray
  scanning, 86–88
PMT, *see* Photomultiplier tube
Principal components analysis
  MeV, 178–179
  Partek Genomics Solution, 257–260
  quality control, 249
Printing, *see* Microarray printing
Protein microarray
  antibody microarrays, 284
  data acquisition, 291–292
  normalization, *see* Normalization,
    microarrays
  protein-binding microarrays, 284
  significance scoring, 308–309
p value
  Algorithm for Capturing Microarray
    Enrichment, 277–278, 280–281
  statistical significance, 286–287

Q

QTClust, microarray data clustering, 175
Quality control, microarrays
  analysis of variance, 240, 242–243
  Bioconductor, 127–129
  block effects, 240–241
  matrix model of cross-hybridization noise
    equations, 246–247

overview, 246
multichannel analysis
    bias sources, 247–248
    curvilinear pattern fitting, 250
    metrics from array patterns and
        reference channel, 251–252
    principal components analysis, 249
    reference pattern corrections, 250
    two-channel error propagation, 250–251
overview, 233–235
pixel statistics
    approaches, 239–240
    cumulative distribution function,
        237, 239
    probe resolution, 235–236
    signal patterns, 236–237
probe signals
    conceptual models, 243–244
    population models, 244–245
    processing including control
        probes, 245–246
signal noise, 234–235
smooth patterns, 240–242
Spotfinder
    parameters, 152–153
    visualization, 153–155
structure analysis for high-dimensional
        data
    algorithm, 253
    enhancement, 253–254
    overview, 252
    recurrent strategy, 252–253

R

Random data set
    gene clustering, 316–317
    genetic association tests, 314–316
    Monte-Carlo approach to statistical
        validation, 312, 324
    permutation test, 316
    random permutations, 312–314
    supervised classification
        bootstrap error estimates, 322–323
        class predictor, 318
        classification error estimation, 320
        k-nearest neighbor classification, 319
        learning algorithm, 318
        n-fold cross-validation, 321–322
        permutation tests, 323

training separation from testing, 320–321
Random Forests
    availability, 431
    classification
        error rate estimates, 426–427
        overview, 423–424
        trees
            combining, 426
            example, 424–425
            nodes, 425–426
            random forest trees, 426
    functional overview, 422–423
    gene importance measure, 427
    prostate cancer microarray
        analysis, 429–430
    proximities, 427–428
    unbalanced data, 427
    unsupervised learning and
        clustering, 428–429
Rank difference analysis of microarray
        method, large data set
    clustering, 389–390
RDAM, see Rank difference analysis of
        microarray method
Resonance light scattering, microarray
        scanning, 93
RLS, see Resonance light scattering
RNA, microarrays
    micro-RNA, see Micro-RNA
    RNA extraction, see RNA extraction
RNA extraction
    blood samples
        advantages, 2–3
        collection and preservation, 3–4
        fractionation, 6
        globin transcript removal from whole
            blood, 4–6
        LeukoLOCK
            filter processing and cell lysis, 7
            RNA isolation and elution, 8
            sample collection and leukocyte
                capture, 7
        goals, 1
        micro-RNA, see Micro-RNA
        quality assessment, 2, 11–12
        RNAlater stabilization, 2
        solid tissue samples
            biopsy specimens, 10–11
            formalin-fixed, paraffin-embedded
                sections

RNA extraction *(cont.)*
  deparaffinization, 9–10
  filter cartridge binding, 10
  overview, 8–9
  protease digestion, 10
  troubleshooting, 39–42

S

SAM, *see* Significance analysis of
    microarrays
Scanning, microarrays
  array handling, 82, 85
  commercial scanners, 82–84
  hyperspectral scanning, 93–95
  image display settings, 88–89
  instrumentation errors
    scanner bias, 92
    signal contamination, 89–92
  overview, 80–82
  photomultiplier tube settings, 86–88
  pixel dwell time, 85
  resonance light scattering, 93
  spatial resolution, 85–86
  surface plasmon resonance, 93
  variability among slides, scanners, and
    laboratories, 95–96
Self-organizing maps
  Cluster software, 209, 211
  data partitioning, 202–203, 205
  machine learning-based
    clustering, 176–177
  software, 205–212
Self-organizing tree algorithm, machine
    learning-based clustering, 176–177
Significance analysis of microarrays
  false discovery rate, 303
  fudge factor, 303
  MeV, 167, 178
  supervised clustering, 194
SOTA, *see* Self-organizing tree algorithm
Spearman rank correlation, distance
    metrics, 197
Spotfinder, *see* TM4 software suite
SPR, *see* Surface plasmon resonance
Standard deviation, calculation, 285
Support vector machine, MeV, 178
Surface plasmon resonance, microarray
    scanning, 93
SVM, *see* Support vector machine

T

Template matching, MeV, 178
Tiling microarray
  Algorithm for Capturing Microarray
    Enrichment analysis, *see* Algorithm
    for Capturing Microarray Enrichment
  data acquisition, 290–291
  normalization, *see* Normalization,
    microarrays
  significance scoring, 308
TM4 software suite
  availability, 136
  development, 136–137
  Madam
    administration tools, 145
    data entry and page editing, 139–142
    database, 138–139
    functions, 135, 137
    interface, 139
    MAGE-ML file writing, 143
    related tools, 144–145
    report generation, 142–143
  MeV
    analysis branching and scripting,
      182, 184
    cluster operations, 182
    cluster viewers, 181–182
    data mining
      agglomerative clustering, 173
      biological theme discovery, 179–180
      classification algorithms and
        supervised clustering
        approaches, 178
      confidence testing of clustering
        results, 175–176
      data visualization and component
        analysis, 178–179
      divisive clustering, 173–175
      hierarchical clustering, 173
      machine learning-based
        clustering, 176–177
      overview, 171–173
      significant gene list
        extraction, 177–178
    data representation and distance
      metrics, 169–171
    file formats and data
      transformations, 181
    function, 136, 167, 169, 211

history log, 184
interface orientation, 180–182, 184
Midas
    filtering modules
        background filtering, 165–166
        cross file trim, 166
        flag filtering, 165
        in-slide replicate analysis, 166
        low-intensity filtering, 166
    functions, 136, 160–161
    gene identification modules
        one-class *t*-test and statistical analysis
            of microarrays, 167
        slice analysis, 166–167
    graphs and reports, 167
    normalization modules
        flip-dye consistency normalization and
            filtering, 165
        iterative linear regression
            normalization, 164
        iterative log-mean centering
            normalization, 163–164
        lowess normalization, 162–163
        overview, 162, 184–185
        ratio statistics normalization, 164
        standard deviation regularization, 164
        total intensity normalization, 162
    pipeline building, 161–162
overview of components, 135–136
Spotfinder
    annotation import, 159
    functions, 135–136, 145–146
    grids
        adjustment, 158
        alignment examination, 158–159
        construction, 156–158
        loading from file, 156
        operations, 157–158
        processing, 158
    image analysis goals, 146
    image loading, 156
    parameter reporting, 152
    postprocessing, 159
    program settings, 155–156
    quality control
        parameters, 152–153
        visualization, 153–155
    spot location problem-solving
        approaches, 147

steps in spot finding
    background correction, 152
    digitizing, 151–152
    grid composition, 149
    grid expansion and shrinking, 149–150
    spot detection, 150–151
    version 3.0, 147–149
walk-through of sample analysis
    analysis pipeline, 185–186
    biological theme discovery, 190
    Investigation panel, 186
    normalization, 184–185
    output reports, 186
    parameter modification, 185–186
    statistical analysis and
        clustering, 187–190
    study overview, 184
Traceability, standards in DNA
    microarrays, 66, 68
Transcriptional networks, *see* Network
TRN, *see* Gene expression terrain map
t-test
    Cyber T, 303–304
    MeV data mining, 177
    significance scoring, 301–303

U

Uncertainty, standards in DNA
    microarrays, 67–68, 71

V

Validation, standards in DNA
    microarrays, 66–71
Variance, calculation, 285

W

Wilcoxon rank sum test, significance
    scoring, 304
Wilcoxon signed rank test, significance
    scoring, 304–305

X

XCluster, features, 210

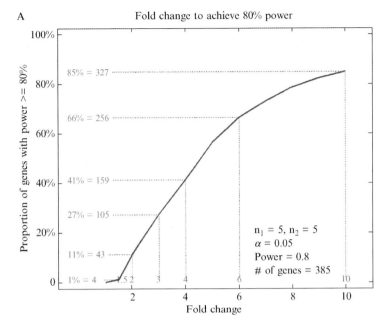

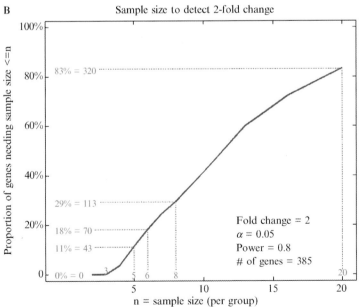

Davison *et al.*, Chapter 2, Fig. 1. Power analysis. Two types of plots are generated for power analysis. (A) The fold difference can be determined statistically at a given power and replication level. (B) The effect of varying the number of replicates and the resulting number of miRNA that can be expected to find dissimilar by twofold and at a power of 0.8.

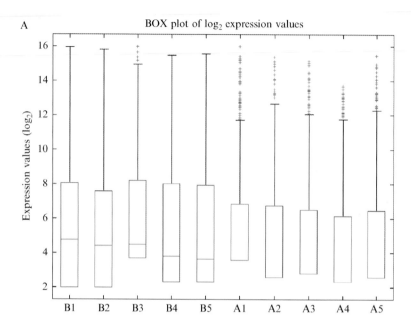

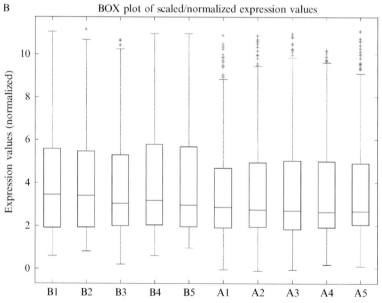

DAVISON *ET AL.*, CHAPTER 2, FIG. 2. Box-whisker plots: The distribution of expression values for lung (replicates A1–A5) and placenta (replicates B1–B5) samples. The ends of the box represent the 25 and 75th percentiles, and the red line bisecting each box is the median signal for each array. (A) After thresholding and log_2 transformation. (B) After VSN transformation.

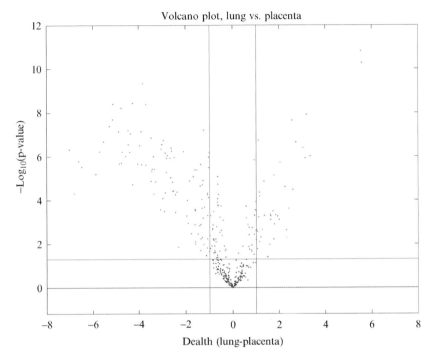

DAVISON *ET AL.*, CHAPTER 2, FIG. 3. Volcano plots. This plot represents both the magnitude of differences between groups and the statistical significant of those differences. To delineate the degree of change between groups there are two vertical red lines at Δh (A-B) $= \pm 1$. There is a single horizontal red line at a negative $\log_{10}$ (p value) $= 1.3$, which corresponds to an unadjusted p value of 0.05. miRNA above this line can be considered statistically significant at a p value of 0.05 in the absence of correction for multiple testing. miRNA found statistically significant after 5% FDR correction are color coded red instead of blue.

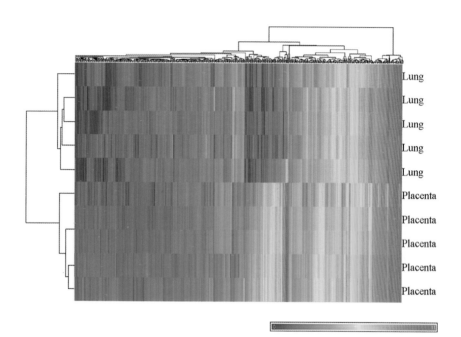

Davison *et al.*, Chapter 2, Fig. 4. Hierarchical clustering. Data are represented as a dendrogram or tree graph with the closest branches of the tree representing genes with similar gene expression patterns. The most common implementation is the agglomerative hierarchical clustering, which starts with a family of clusters with one sample each and merges the clusters iteratively, based on some distance measure, until there is only one cluster left. In this example, Euclidean distance is used as the distance metric. Array and/or sample qualities can be approximated using hierarchical clustering. Ideally, common samples should cluster into similar classes.

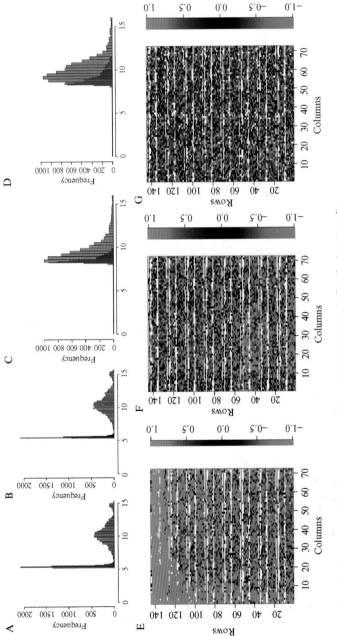

EADS *ET AL.*, CHAPTER 3, FIG. 2. *(continued)*

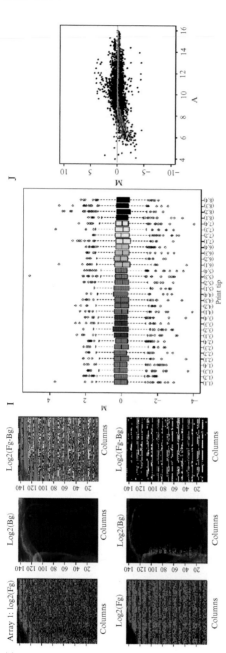

Eads *et al.*, Chapter 3, Fig. 2. Histogram of frequency of log₂ (intensity) values in a hybridization of higher quality (A and B) and of lower quality (C and D). Note the distinct separation of the low-intensity and higher intensity counts for the higher quality array, and the normal distribution of signal. In the lower quality array, signal (lighter color) and background (darker color) are spread across much of the same range, which makes spot finding difficult or impossible. For each channel, noise should be near a log₂ (intensity) of 6 or 7, corresponding to about 130 counts of background in a 2¹⁶-bit scanning system. (E, F, and G) Spatial plots of log₂ ratio values for all pixels on the array. Ratios for each pixel are plotted according to x–y coordinates of pixel location (note scale bar). (E) The array displays high red background to the left and top of the array and is red on the right side as well. This array will require aggressive normalization to remove the spatial artifact. (F) An array with lower overall background and with much less spatial bias. (G) The same array as E after normalization using Olin (Futschik and Crompton, 2004). (H) False-color image of foreground and background in red and green channels plotted across the x–y dimensions of the array. This displays log₂ intensity values in foreground, background, and log₂(foreground−background) and can reveal otherwise invisible trends. (I) Box plot of spot ratios within each print-tip group. Such plots are helpful for revealing artifacts related to printing and how well normalization copes with their removal. (J) M-A (ratio-intensity) plot of all spots on the array, with regression lines colored according to the subarray. In this case, intensity is centered around a mean of zero, which is reasonable, and there is relatively little intensity-dependent bias in average ratio values.

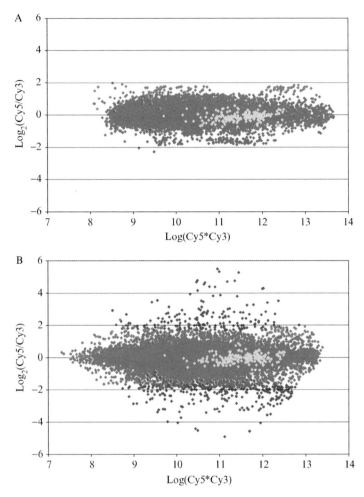

Yang, Chapter 4, Fig. 1. Ratio–intensity (RI) plots of hybridizations of a mouse lung RNA sample to itself (A) or to the Universal Mouse Reference RNA (Stratagene) (B). Ten *A. thaliana* cRNA controls were spiked into mouse lung and reference RNA samples at 1:1 (yellow; 50:50, 100:100, 200:200, and 300:300 in terms of copies per cell), 3:1 (red; 30:10, 120:40, and 450:150), and 1:3 (green; 40:120, 80:240, and 100:300) Cy5:Cy3 ratios. Genes that are identified as differentially expressed in B are shown in blue. RNA was labeled by incorporation of the aminoallyl linker during reverse transcription followed by a coupling of Cy3- or Cy5-NHS esters, and labeled targets were hybridized to a spotted oligonucleotide microarray containing the mouse Operon set (probes for ~17,000 mouse genes) printed once and probes for the 10 *A. thaliana* genes printed multiple times on the array.

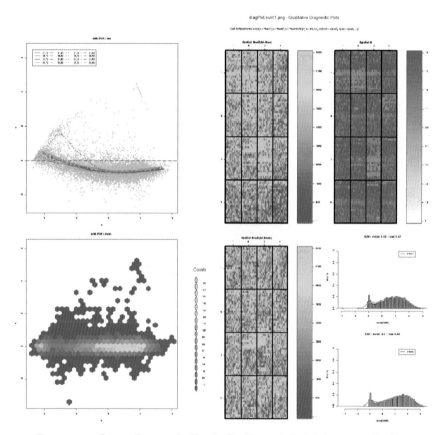

REMIERS AND CAREY, CHAPTER 8, FIG. 1. Quality control plots from *arrayQuality*.

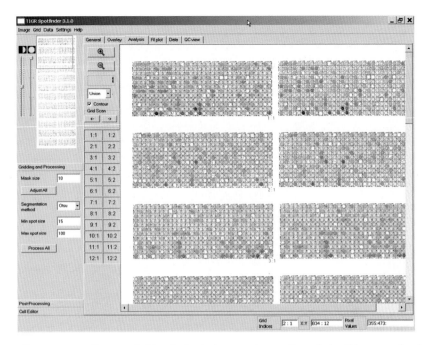

SAEED *ET AL.*, CHAPTER 9, FIG. 4. Analysis page shown after whole slide processing is complete. Contour lines are colored red for good spots and green for bad ones.

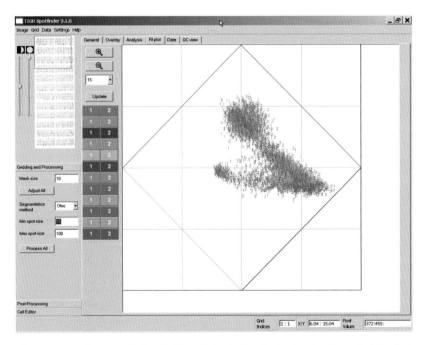

SAEED *ET AL.*, CHAPTER 9, FIG. 5. RI plot view in Spotfinder showing the ratio-intensity log graph for the whole analyzed slide. The four lines forming a diamond are the limits of the log-ratio plot. Red and blue lines are the full saturation limits lines in one channel, whereas the other channel has a valid number. Yellow and green lines are the zero values limit lines at least in one channel.

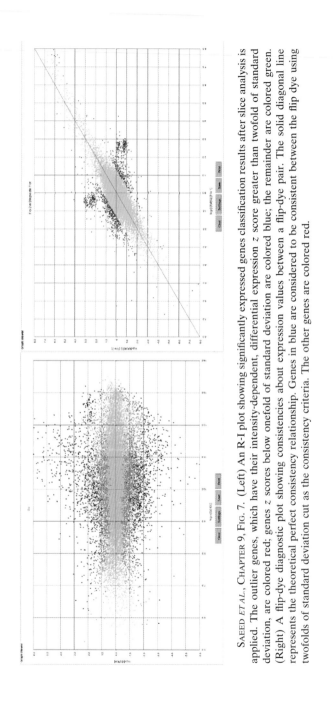

SAEED *ET AL.*, CHAPTER 9, FIG. 7. (Left) An R-I plot showing significantly expressed genes classification results after slice analysis is applied. The outlier genes, which have their intensity-dependent, differential expression z score greater than twofold of standard deviation, are colored red; genes z scores below onefold of standard deviation are colored blue; the remainder are colored green. (Right) A flip-dye diagnostic plot showing consistencies about expression values between a flip-dye pair. The solid diagonal line represents the theoretical perfect consistency relationship. Genes in blue are considered to be consistent between the flip dye using twofolds of standard deviation cut as the consistency criteria. The other genes are colored red.

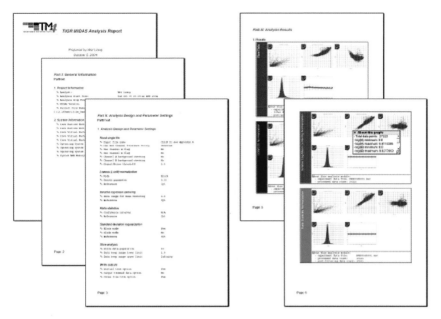

SAEED *ET AL.*, CHAPTER 9, FIG. 8. Midas PDF analysis report.

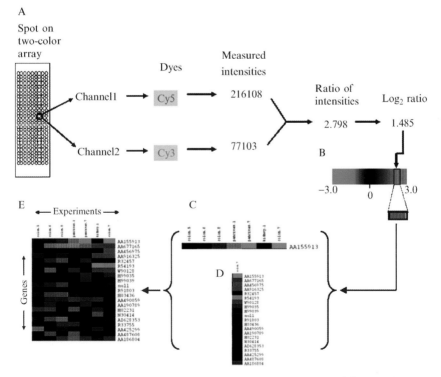

SAEED *ET AL.*, CHAPTER 9, FIG. 9. Data representations in MeV. (A) Numerical and (B) false-color representations of an expression element, (C) a gene expression vector, (D) an experiment expression vector, and (E) an expression matrix.

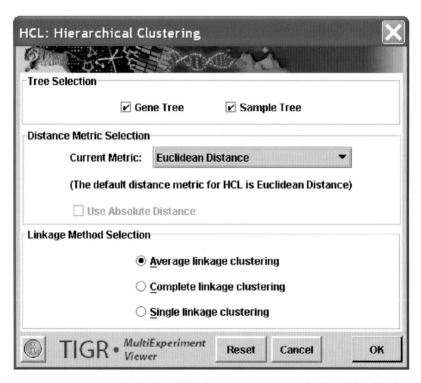

SAEED *ET AL.*, CHAPTER 9, FIG. 10. HCL algorithm parameter selection dialog. The lower left of each algorithm dialog contains the parameter information button.

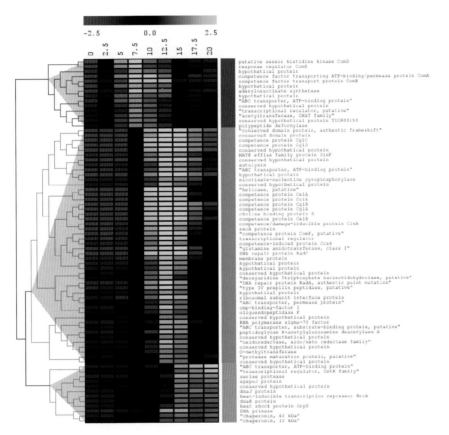

SAEED *ET AL.*, CHAPTER 9, FIG. 11. Hierarchical cluster of time course data using the Pearson correlation distance metric. Prominent patterns of expression have been selected as clusters.

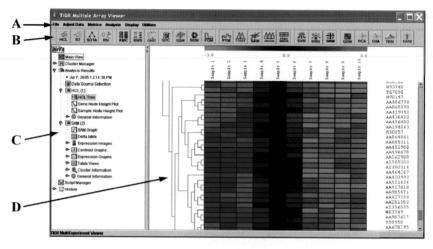

SAEED *ET AL.*, CHAPTER 9, FIG. 12. Graphical interface of MeV: (A) main menu bar, (B) algorithm toolbar, (C) result navigation tree, and (D) viewer panel. A hierarchical tree viewer is shown in the viewer panel.

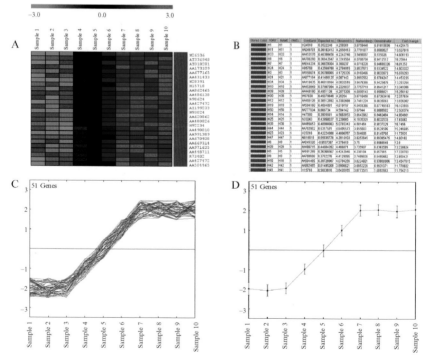

SAEED *ET AL.*, CHAPTER 9, FIG. 13. Cluster viewer examples, (A) Expression image, (B) cluster table viewer, (C) expression graph, and (D) centroid graph.

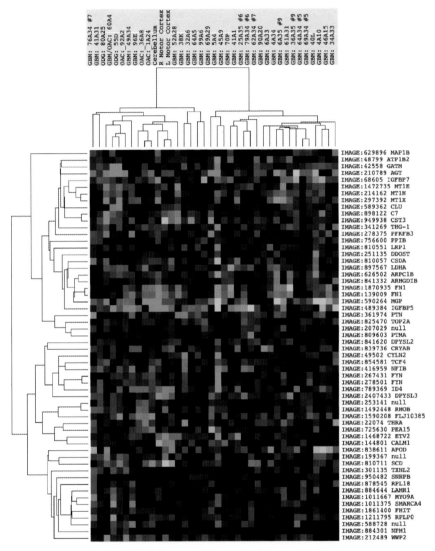

GOLLUB AND SHERLOCK, CHAPTER 10, FIG. 1. Clustered gene expression data (heavily filtered data from Liang *et al.*, 2005). Blue and yellow indicate a negative and a positive log ratio, respectively. Clustering was carried out on both genes and samples (arrays) using a centered Pearson correlation metric and centroid linkage. The dendrograms or "trees" group similar vectors into nodes; the height of each node in the tree indicates the overall similarity of its members. Genes and samples both divide into two, relatively internally consistent groups.

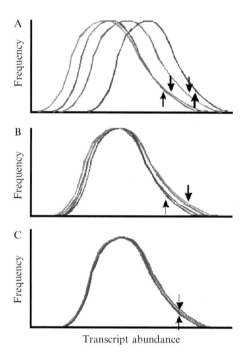

AYROLES AND GIBSON, CHAPTER 11, FIG. 2. Effect of normalization on inference of differential expression. (A) Frequency distributions of two arrays with two channels (dyes) each have different means and variance so that a gene at the 10th percentile (arrows) has a different apparent level of transcript abundance on each array. (B) Centralization by subtracting the mean of each channel reduces these effects, but remaining differences in variance still result in apparent differential expression between the red and yellow samples and the blue and green samples. (C) Further normalization to equalize the variance and remove skew may result in similar relative fluorescence intensity for equally ranked genes on each array.

A

B

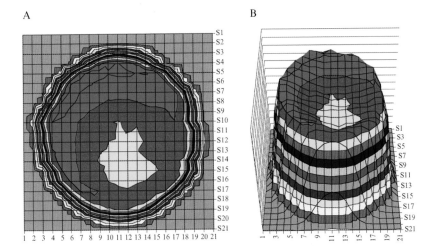

Red = best signals

Green = effect of chemical solutions

Yellow = effect of chemical solutions and pen performance

MINOR, CHAPTER 12, FIG. 1. Pixel signal patterns. (A) Signal contours. (B) Three-dimensional perspective plot.

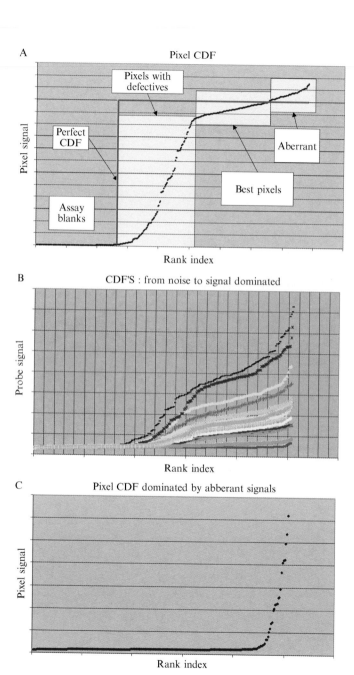

MINOR, CHAPTER 12, FIG. 2. Pixel signal CDF profiles.

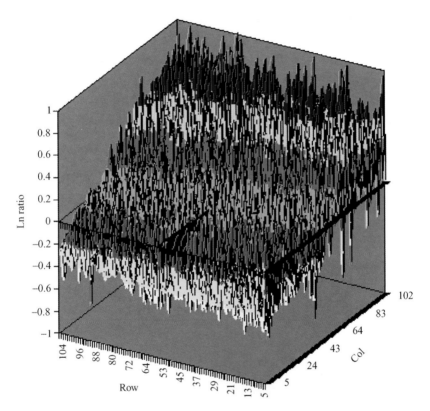

MINOR, CHAPTER 12, FIG. 3. Technical patterns in biological measurements.

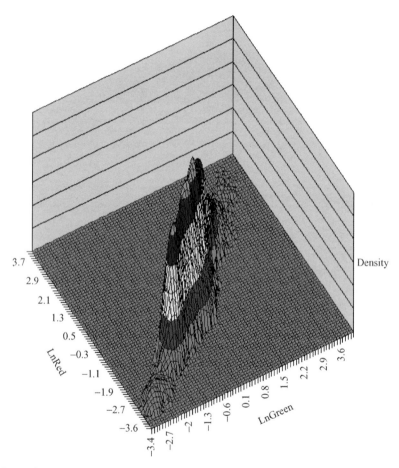

MINOR, CHAPTER 12, FIG. 5. Two-dimensional histogram of two-channel log signal patterns.

RNA prep#	Slide barcode	Hyp date	Quality score	Statistical calibration					Negative controls					Complex sample					Concordance			
				Lower Asymptote	Upper Asymptote	Slope	Ed_{50} calibration curve	Mean square error	Degrees of freedom	Mean signal	Std. dev. signal	CV signal	Efficiency	Degrees of freedom	Mean signal	Std. dev. signal	CV signal	Efficiency	Concordance	Location shift	Scale shift	Precision
R74490	T00126381	1/18/2002	Reject	−6.12	7.01	0.27	2.55	0.74	20	−2.01	0.45	−0.22	0.41	6549	−0.38	1.54	−4.07	9842	0.82	0.17	1.08	0.84
R71318	T00126629	1/18/2002	Reject	−1.46	7.04	0.49	4.72	0.42	20	−0.85	0.46	−0.54	0.41	5413	−0.12	1.07	−9.32	9821	0.71	0.00	1.54	0.78
R72304	T00127044	1/18/2002	Reject	−5.23	7.71	0.23	3.91	0.77	22	1.73	0.39	0.23	0.44	6913	−0.14	1.47	−10.18	9830	0.77	0.03	1.13	0.78
R73501	T00128045	1/18/2002	Reject	−3.28	2.81	0.08	−8.13	1.05	31	0.63	0.92	1.48	0.63	7361	0.20	1.77	8.74	9841	0.25	−0.17	0.94	0.25
R73261	T00128050	1/18/2002	Reject	−4.43	6.13	0.50	2.82	0.21	29	−2.63	0.39	−0.14	0.60	7320	0.08	1.57	20.82	9831	0.55	−0.06	1.06	0.56
R74540	T00128511	1/18/2002	Reject	−7.36	6.73	0.33	0.92	0.72	36	−1.74	0.57	−0.33	0.73	6881	−0.19	1.53	−8.11	9817	0.87	0.06	1.09	0.87
R74506	T00126398	1/10/2002	Review	−7.79	7.55	0.00	1.62	0.50	26	−2.31	0.54	−0.20	0.53	6900	−0.37	1.62	−4.37	9040	0.89	0.17	1.02	0.91
R74545	T00126608	1/18/2002	Review	5.67	2.73	1.65	0.01	1.49	15	−2.50	0.61	0.24	0.31	7374	0.03	1.71	51.75	9849	0.98	0.04	0.97	0.98
R74546	T00126609	1/18/2002	Review	−4.93	1.89	1.23	0.18	1.14	17	−2.53	0.46	−0.18	0.35	7441	−0.09	1.68	−18.82	9830	0.98	0.01	0.98	0.98
R74482	T00128480	1/17/2002	Review	−4.54	−0.90	0.21	−2.47	0.24	23	−2.36	0.43	−0.18	0.47	7009	−0.43	1.62	−3.60	9819	0.89	0.20	1.03	0.91
R74484	T00128482	1/17/2002	Review	−4.54	−0.13	0.07	2.56	0.27	20	−2.60	0.38	−0.14	0.40	7335	−0.09	1.69	−19.76	9755	0.99	0.01	0.98	0.99
R74543	T00128544	1/18/2002	Review	−7.91	6.05	0.34	0.81	0.65	31	−2.30	0.74	−0.32	0.64	7046	−0.24	1.67	−7.05	9841	0.89	0.11	0.99	0.50
R74539	T00128510	1/18/2002	0.63	−7.16	5.91	0.43	1.01	0.33	22	−2.50	0.60	−0.24	0.45	7419	−0.07	1.69	−24.33	9822	0.98	0.00	0.98	0.98
R74500	T00120499	1/10/2002	0.60	−7.50	5.74	0.49	0.45	0.50	36	−2.13	0.77	−0.36	0.73	7041	0.02	1.60	106.34	9034	0.96	−0.04	0.99	0.96
R74529	T00128498	1/18/2002	0.88	7.23	5.91	0.44	0.99	0.35	21	−2.55	0.61	0.24	0.42	7373	0.06	1.69	27.16	9823	0.98	0.00	0.98	0.98
R74507	T00128399	1/18/2002	0.88	−7.43	6.36	0.39	1.12	0.50	31	−2.41	0.70	−0.29	0.64	7269	−0.12	1.69	−13.55	9832	0.97	0.04	0.98	0.97
R73168	T00127047	1/18/2002	0.98	−7.38	6.33	0.41	1.10	0.26	24	−2.48	0.52	−0.21	0.50	7456	−0.07	1.69	−24.99	9832	0.98	0.00	0.98	0.98
R74500	T00120395	1/18/2002	0.99	−7.44	6.28	0.42	1.13	0.22	21	−2.62	0.46	−0.17	0.43	7423	−0.06	1.70	−26.23	9842	0.99	0.00	0.98	0.99

Quality scores assigned to each array

Reject / Review / Pass

MINOR, CHAPTER 12, FIG. 6. Example of AutoQC report with heat map of metrics profile.

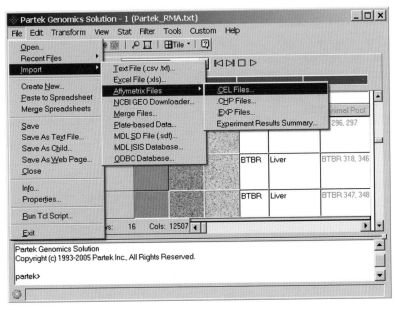

DOWNEY, CHAPTER 13, FIG. 1. Importing Affymetrix data into Partek GS. Note that each sample appears in one row and that each gene appears in a single column. The first three columns show thumbnail images for raw chip data, data after RMA normalization, and residuals of the RMA correction, respectively. Double clicking on the thumbnail images will display the full-size image for inspection.

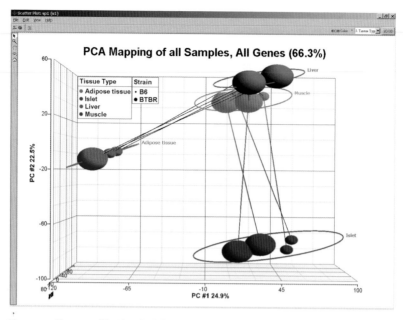

DOWNEY, CHAPTER 13, FIG. 2. PCA mapping of the samples from 12,488-dimensional "gene space" to three dimensions for interactive visualization. Samples are colored by tissue type and sized by strain. Lines connect samples from the same animal pool, and ellipsoids are drawn around each tissue group.

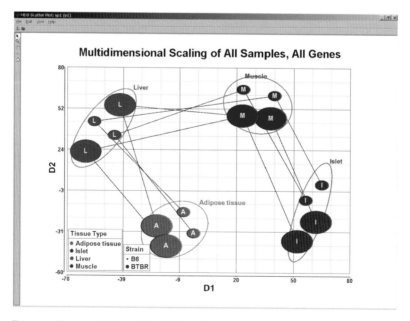

DOWNEY, CHAPTER 13, FIG. 3. Multidimensional scaling shows similar patterns as PCA—samples are grouped by tissue type and within each tissue type (except liver), the samples are differentiated by strain.

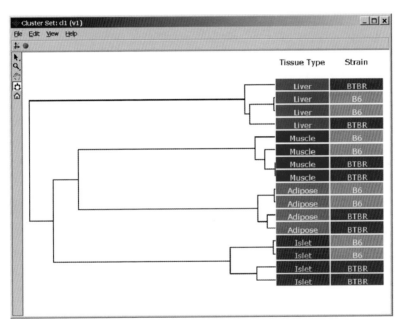

DOWNEY, CHAPTER 13, FIG. 5. Hierarchical clustering of samples using Euclidean distance and average linkage. Each branch is annotated with tissue type and strain. Note that samples cluster primarily by tissue and secondarily by strain (with the exception of liver tissues).

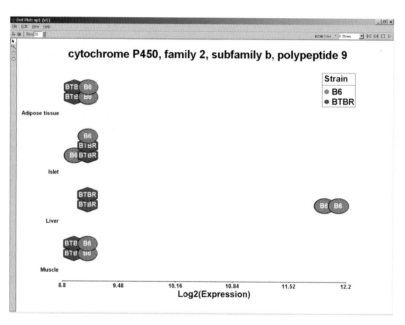

DOWNEY, CHAPTER 13, FIG. 7. A "dot plot" is a very effective visualization tool when the sample size is relatively small, such as in this experiment. There are 16 "dots," one for each sample. The x axis represents the $\log_2$ of the expression for this gene. The points are colored (and shaped) by strain and separated on the y axis by tissue. For clarity, dots are also labeled with the strain. Note that there is very little noise in these data and that the 16 samples provide a very good estimate of the noise variance.

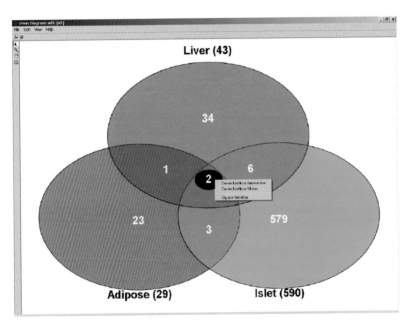

Downey, Chapter 13, Fig. 8. Often researchers are interested in creating gene lists that are a combination of other lists. The list manager of Partek allows the researcher to combine these lists in several ways. This shows the Venn diagram tool of Partek, which allows the scientist to interactively create gene lists from the intersection or union of two or more lists.

A

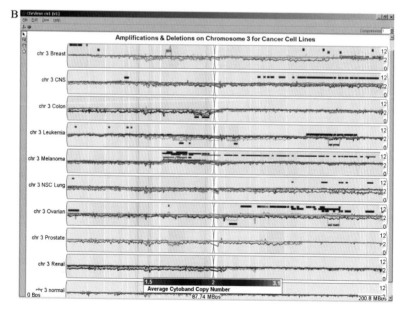

B

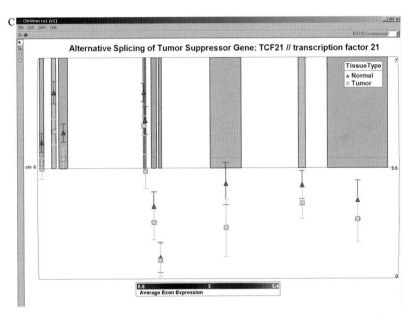

Downey, Chapter 13, Fig. 9. (A) Genes are displayed in the Partek genome browser. Genes are colored by *p* value, and the heights of the lines indicate fold change. In this screen shot, one line is drawn for each chromosome. (B) In this view, chromosomal copy number amplifications and deletions from a tumor sample can be visualized. Regions of statistically significant alterations are indicated with red (amplification) and blue (deletion) rectangles. In this screen shot, only chromosome 8 is displayed. (C) This view examines individual exons on a single gene. This gene exhibits alternative splicing (detected using analysis of variance). Overall, all genes on the exon are upregulated in the tumor relative to the normal; however, there is a single exon near the center of the gene that is expressed higher in the normal group than the tumor group. This is an indication of alternative splicing.

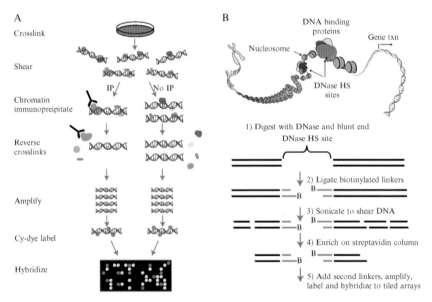

SCACHERI *ET AL.*, CHAPTER 14, FIG. 1. Overview of ChIP-chip (A) and DNase-chip (B) (see text for details).

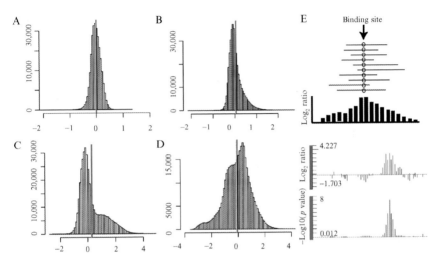

SCACHERI ET AL., CHAPTER 14, FIG. 2. Examples of "one-tailed" data and "neighbor effect." (A–D) Histograms showing various degrees of enrichment from ChIP-chip experiments using no antibody (no enrichment) (A), antibodies to the MLL1 transcription factor (moderate enrichment) (B), and antibodies to a histone modification (H3K4me3) (high enrichment) (C). (D) Background noise caused by streaks on the array can affect the distribution dramatically, making it difficult to estimate the degree of enrichment. Red lines in A–D denote the mean center of the distribution. Note that this line is shifted to the right in instances where obvious enrichment occurred. Frequency is plotted on the vertical axis; $\log_2$ ratio measurements (ChIP DNA/total genomic DNA) are plotted on the horizontal axis. (E) Theoretical (top) and actual (bottom) examples of the "neighbor effect" principle. (Bottom) $\log_2$ ratio measurements (red) and ACME processed (green) data from a ChIP-chip experiment using antibodies to H3K4me3. Data represent the average of three biological replicates. Probes on the array were spaced at a density of 1 probe per 180 bp.

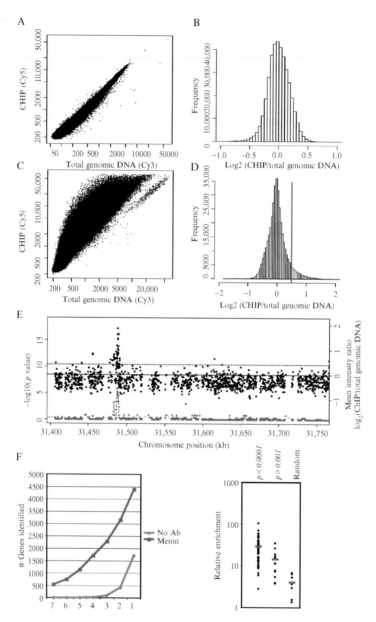

SCACHERI *ET AL.*, CHAPTER 14, FIG. 3. Analysis of data from tiled arrays using ACME. In this example, promoter-specific arrays (NimbleGen) containing 385,0000 probes tiled at a resolution of 1 probe per 100–180 bp were used to identify genomic-binding sites for the tumor suppressor protein menin. (A and C) Scatter plots of single-array intensities of oligonucleotide probes obtained from chromatin immunoprecipitation with antibodies to menin (C) and a control

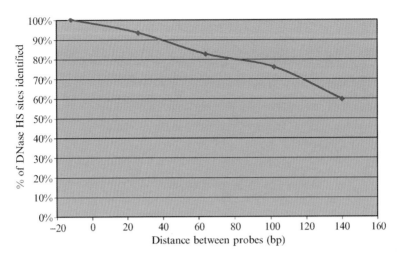

SCACHERI ET AL., CHAPTER 14, FIG. 5. Sensitivity diminishes as probe resolution decreases. The sensitivity of DNase-chip to detect valid DNase HS sites was calculated for different probe-spacing patterns. Valid DNase HS sites were identified from a previous study (Crawford et al., 2006). DNase-chip was performed on tiled arrays that contained 50-mer probes that overlapped by 12 bases. A significant signal ($p < 0.001$) was detected using ACME for data that included data from all probes, every second probe, every third probe, every fourth probe, and every fifth probe. User-defined window sizes were modified so that different spacing patterns had at least 13 probes per window on average.

experiment with no antibody (A). Successful experiments are identified as those that show enrichment of multiple probes in the Cy5 channel (chromatin immunoprecipitated DNA) over the Cy3 channel (total genomic DNA). (B and D) Histograms of normalized intensity ratios from A and C. Compared to the no antibody control (B), the histogram plotted from the menin ChIP (D) shows a distinct tail at the right-hand end, or positive direction. ACME slides a window of user-defined size along tiled regions and tests statistically whether each window contains a higher than expected number of probes above a user-defined threshold (indicated by the red bar). In this example, the window size is set at 1000 bp and the threshold at 90%. (E) Plots showing data before and after processing with ACME. Normalized $\log_2$ ratio measurements from three experimental replicates were averaged and plotted in black, with corresponding chromosome coordinates on the horizontal axis and the mean intensity ratio on the right vertical axis. Points in red indicate corresponding significance values for each data point following processing by ACME (left vertical axis). The dashed line denotes the 90% threshold level. (F) Number of promoters ACME reported to be bound by menin at various p value cutoffs. Compared to the negative control experiment (blue), significant enrichment of multiple promoters was detected for the menin ChIP-chip experiment. Also note that the false-positive rate increases as the p values decrease in significance. (G) Real-time PCR validation of promoters determined by ACME to be bound menin. Mean values are indicated by red bars. Compared to randomly selected regions of the genome, promoter regions identified by ACME to be enriched for menin binding at $p < 0.0001$ were enriched more than sevenfold. These data indicate that sites ACME reported to be enriched at $p < 0.0001$ reliably represent sites of menin occupancy. Users should empirically determine the optimal p value cutoff for their studies using similar methods.

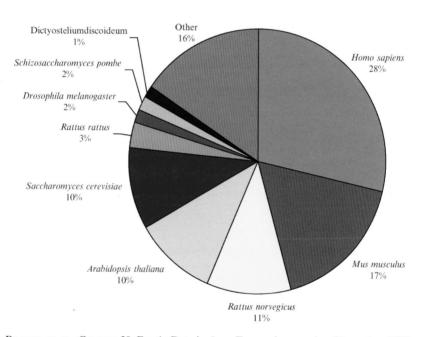

BRAZMA *ET AL.*, CHAPTER 20, FIG. 1. Data in ArrayExpress by organism (December 2005).

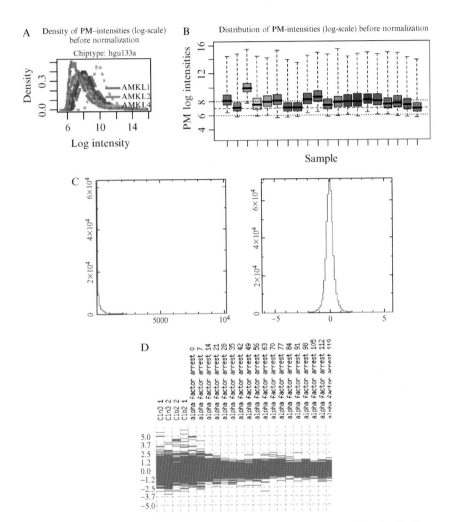

BRAZMA *ET AL.*, CHAPTER 20, FIG. 7. Expression Profiler descriptive statistics visualizations. (A) PM intensity plot. (B) PM intensity box plot. (C) Distribution density histograms (one- and two-channel experiments, absolute and log-ratio data). (D) Multigene line plot.

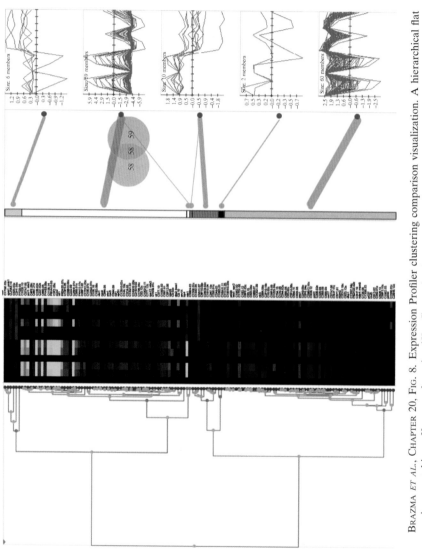

BRAZMA *ET AL.*, CHAPTER 20, FIG. 8. Expression Profiler clustering comparison visualization. A hierarchical flat comparison, matching a K-means clustering ($K = 5$) to a dendrogram. Overlaps between matching clusters are shown with Venn diagrams.

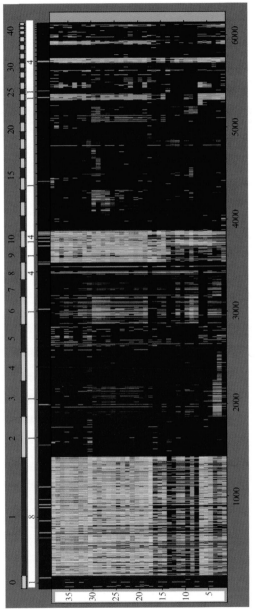

HENNETIN AND BELLIS, CHAPTER 21, FIG. 9. Gene clustering on the transcriptional network with an interrogation list of 38 probe sets. Represented on the *X* axis are 6135 probe sets with a positive or negative correlation score greater than 0% with at least one probe set of the interrogation list plotted on the *Y* axis. Each pair of probe sets (*x,y*) has its CORR value displayed on a heat map (from black for CORR = 0% to white for CORR = 100%). The upper bar delimits the clusters found by CLICK (cluster zero contains the probe sets that could not be clusterized adequately).The second upper bar displays the position and the number of probe sets of the interrogation list found in each cluster.

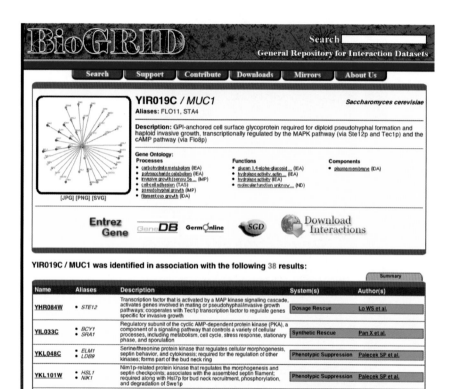

Bell and Lewitter, Chapter 22, Fig. 1. The BioGRID web site (http://www.thebiogrid.org) contains downloadable interaction data. In addition, one can visualize interaction networks for one gene at a time by entering the appropriate identifier.

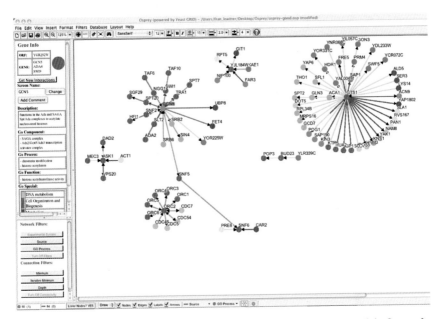

BELL AND LEWITTER, CHAPTER 22, FIG. 2. A display of an interaction network in Osprey for seven genes of interest. The nodes (circles) are color coded by GO category. The edges (lines) are color coded by the publication source of interaction data. The window on the upper left gives annotation about a highlighted node. Filtering of networks is available from the window on the lower left.

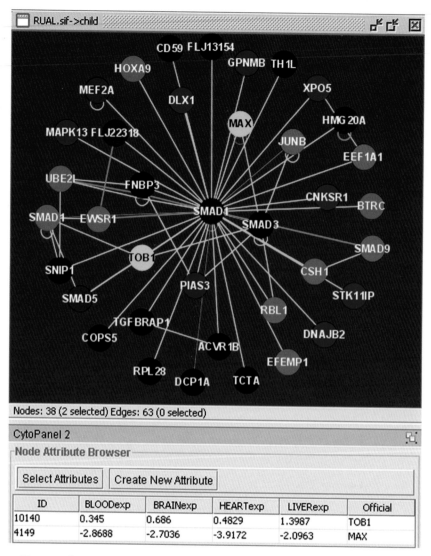

BELL AND LEWITTER, CHAPTER 22, FIG. 7. An interaction network with nodes colored to indicate expression ratios in a selected tissue. The two selected nodes are listed in CytoPanel 2 (below), with corresponding expression ratios and other information shown (for those fields chosen by going to the Select Attributes button).

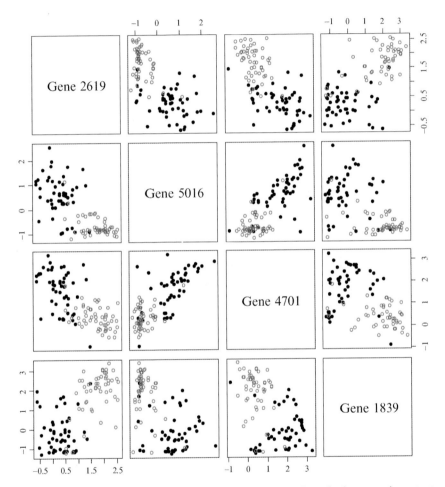

CUTLER AND STEVENS, CHAPTER 23, FIG. 2. Prostate cancer data: the four most important genes selected by Random Forests and their relationship to the groups. Solid black circles represent controls and open red circles represent tumors.

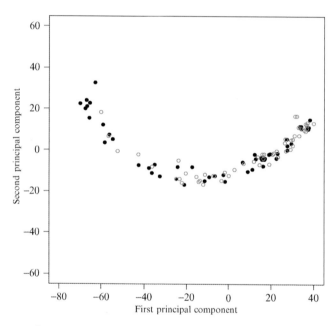

CUTLER AND STEVENS, CHAPTER 23, FIG. 3. Prostate cancer data: the first two principal components of data and their relationship to the groups. Solid black circles represent controls and open red circles represent tumors.